Lecture Notes in Computer Science 11638

Commenced Publication in 1973
Founding and Former Series Editors:
Gerhard Goos, Juris Hartmanis, and Jan van Leeuwen

More information about this series at http://www.springer.com/series/7407

Charles J. Colbourn · Roberto Grossi ·
Nadia Pisanti (Eds.)

Combinatorial Algorithms

30th International Workshop, IWOCA 2019
Pisa, Italy, July 23–25, 2019
Proceedings

 Springer

Editors
Charles J. Colbourn
Arizona State University
Tempe, AZ, USA

Roberto Grossi
University of Pisa
Pisa, Italy

Nadia Pisanti
University of Pisa
Pisa, Italy

ISSN 0302-9743 ISSN 1611-3349 (electronic)
Lecture Notes in Computer Science
ISBN 978-3-030-25004-1 ISBN 978-3-030-25005-8 (eBook)
https://doi.org/10.1007/978-3-030-25005-8

LNCS Sublibrary: SL1 – Theoretical Computer Science and General Issues

This Springer imprint is published by the registered company Springer Nature Switzerland AG
The registered company address is: Gewerbestrasse 11, 6330 Cham, Switzerland

Preface

This volume contains the contributed papers presented at IWOCA 2019, the 30th International Workshop on Combinatorial Algorithms, held at the Dipartimento di Informatica, Università di Pisa, Italy, during July 23–25, 2019. The conference brings together researchers on diverse topics related to combinatorial algorithms, such as algorithms and data structures; complexity theory; graph theory and combinatorics; combinatorial optimization; cryptography and information security; algorithms on strings and graphs; graph drawing and labeling; computational algebra and geometry; computational biology; algorithms for big data and network analytics; probabilistic and randomized algorithms; algorithm engineering; and new paradigms of computation.

IWOCA is an annual conference series that began 29 years ago as a regional conference (AWOCA) in Australasia. In 2007, it became an international conference, so far held in Australia, Canada, Czech Republic, Finland, France, India, Indonesia, Italy, Japan, Singapore, South Korea, UK, and USA.

Highlights of the scientific program for IWOCA 2019 included invited talks by three eminent researchers, Professor Marinella Sciortino (University of Palermo), Professor Stéphane Vialette (CNRS and University Paris-Est), and Professor Ugo Vaccaro (University of Salerno). We appreciate the excellent talks given by our invited speakers.

The IWOCA 2019 Program Committee engaged the talents of 40 computer scientists and mathematicians. Reflecting the international character of IWOCA, members of the Program Committee are eminent researchers from Australia, Austria, Canada, Chile, Czech Republic, Denmark, Finland, France, Germany, Greece, Hong Kong, Israel, Italy, Japan, The Netherlands, Poland, Romania, South Korea, UK, and USA. Needless to say, the success of IWOCA relies on the careful and thorough work by the Program Committee members. We are very grateful for their dedication, expertise, and timely evaluations facilitated by the use of the EasyChair conference management system. We also express sincere thanks to the authors who submitted papers for consideration. The 73 submitted papers underwent rigorous review by the Program Committee followed by in-depth discussion, resulting in the selection of the 36 high-quality papers that you find in this volume. We appreciate the commitment of Springer to IWOCA, again publishing the proceedings of IWOCA 2019 in the LNCS series. We also appreciate Springer's support for a Best Paper Award.

The IWOCA Steering Committee (Maria Chudnovsky, Charles Colbourn, Costas Iliopoulos, and Bill Smyth) provided guidance and support that we acknowledge with thanks. Last, but certainly not least, we express our sincere thanks to the members of the local Organizing Committee (Anna Bernasconi, Alessio Conte, Roberto Grossi, Veronica Guerrini, Andrea Marino, Nadia Pisanti, Nicola Prezza, and Giovanna Rosone) for making the conference a great success.

July 2019

Charles J. Colbourn
Roberto Grossi
Nadia Pisanti

Organization

Program Chairs

Charles Colbourn   Arizona State University, USA
Roberto Grossi   Università di Pisa, Italy
Nadia Pisanti   Università di Pisa, Italy

Steering Committee

Maria Chudnovsky   Princeton University, USA
Charles Colbourn   Arizona State University, USA
Costas Iliopoulos   King's College, UK
Bill Smyth   McMaster University, Canada

Program Committee

Hiroki Arimura   Hokkaido University, Japan
Hideo Bannai   Kyushu University, Japan
Philip Bille   Technical University of Denmark
Paola Bonizzoni   University of Milano Bicocca, Italy
Gerth Stolting Brodal   Aarhus University, Denmark
Maria Chudnovsky   Princeton University, USA
Charles Colbourn   Arizona State University, USA
Dalibor Froncek   University of Minnesota - Duluth, USA
Travis Gagie   Diego Portales University, Chile
Serge Gaspers   UNSW Sydney and Data61, CSIRO, Australia
Dora Giammarresi   University of Rome Tor Vergata, Italy
Roberto Grossi   University of Pisa, Italy
Jan Holub   Czech Technical University in Prague, Czech Republic
Costas Iliopoulos   King's College London, UK
Artur Jeż   University of Wroclaw, Poland
Gregory Kucherov   CNRS and University of Paris-Est, France
Gad Landau   University of Haifa, Israel
Thierry Lecroq   University of Rouen, France
Christos Makris   University of Patras, Greece
Sebastian Maneth   University of Bremen, Germany
Sabrina Mantaci   University of Palermo, Italy
Lucia Moura   University of Ottawa, Canada
Patric R. J. Östergård   Aalto University, Finland
Kunsoo Park   Seoul National University, South Korea
David Pike   Memorial University of Newfoundland, Canada
Nadia Pisanti   University of Pisa, Italy

Solon Pissis	CWI Amsterdam, The Netherlands
Alexandru Popa	University of Bucharest, Romania
Rajeev Raman	University of Leicester, UK
Frank Ruskey	University of Victoria, Canada
Marinella Sciortino	University of Palermo, Italy
Rahul Shah	Louisiana State University, USA
Dimitris Simos	SBA Research, Austria
Blerina Sinaimeri	Inria, France
Douglas Stinson	University of Waterloo, Canada
Alexandru I. Tomescu	University of Helsinki, Finland
Stéphane Vialette	CNRS and University of Paris-Est, France
Lusheng Wang	City University of Hong Kong, SAR China
Ian Wanless	Monash University, Australia

Additional Reviewers

Abedin, Paniz
Alzamel, Mai
Anselmo, Marcella
Arseneva, Elena
Ayad, Lorraine
Barbay, Jérémy
Belazzougui, Djamal
Bereg, Sergey
Bernardini, Giulia
Birmele, Etienne
Boucher, Christina
Bulteau, Laurent
Cazaux, Bastien
Chauve, Cedric
Conte, Alessio
Dahn, Christine
Defrain, Oscar
Della Vedova, Gianluca
Devismes, Stéphane
Di Giacomo, Emilio
Dudek, Bartlomiej
Dósa, György
Favrholdt, Lene
Frisch, Christoph
Garg, Mohit
Garn, Bernhard
Ghazawi, Samah
Golumbic, Martin
Gutin, Gregory

Hermelin, Danny
Holm, Jacob
Husic, Edin
I, Tomohiro
Inenaga, Shunsuke
Irving, Robert
Jeż, Łukasz
Kampel, Ludwig
Kimbrel, Tracy
Klein, Rolf
Knop, Dušan
Kociumaka, Tomasz
Kupavskii, Andrey
Kurita, Kazuhiro
Le Gall, Francois
Lefebvre, Arnaud
Leithner, Manuel
Liao, Chung-Shou
Luo, Kelin
Malucelli, Federico
Mandric, Igor
McCauley, Samuel
Minato, Shin-Ichi
Mincu, Radu-Stefan
Moura, Phablo
Muggli, Martin
Najeebullah, Kamran
Nakashima, Yuto
Nicholson, Patrick K.

Nicoloso, Sara
Nisse, Nicolas
Nussbaum, Yahav
Pajak, Dominik
Paluch, Katarzyna
Patterson, Murray
Peng, Richard
Pivac, Nevena
Pokorski, Karol
Prieur-Gaston, Elise
Pálvölgyi, Dömötör
Renault, Marc
Restivo, Antonio
Rizzi, Raffaella
Rizzi, Romeo

Roy, Sanjukta
Sadakane, Kunihiko
Schulz, Andreas S.
Sciortino, Marinella
Semple, Charles
Shmoys, David
Soto, Mauricio
Spooner, Jakob
Valinezhad Orang, Ayda
Valla, Tomáš
Versari, Luca
Veselý, Pavel
Weller, Mathias
Zhu, Binhai

Abstracts of Invited Talks

BWT Variants: A Combinatorial Investigation

Marinella Sciortino

University of Palermo, Italy
marinella.sciortino@community.unipa.it

Abstract. The Burrows–Wheeler transform (BWT) is a reversible transformation that was introduced in 1994 in the field of data compression and it has also become a fundamental tool for self-indexing data structures. It is a context-dependent transformation that produces a permutation of the input text that is likely to have runs of equal letters (clusters) longer than the ones in the original text. Such a combinatorial property of BWT make it a versatile tool also in several other applications, especially in bioinformatics [4, 6]. In [5], a complexity measure that counts the number of equal-letter runs produced by the BWT is introduced, by exploring how the number of clusters of the BWT output varies depending on the number of clusters of the input.

Over the years other variants of the BWT have been proposed, without, however, obtaining transformations entirely capable of playing the same variegated role as the BWT. Very recently, a whole new class of transformations, called *local ordering trasformation*, has been introduced [3]. They have all the same prerogatives as BWT, i.e., they can be computed and inverted in linear time, they produce provably highly compressible output, and they support linear time pattern search directly on the transformed text. Such a class is a special case of a more general family of transformations based on context-adaptive alphabet orderings. This more general class includes also the alternating BWT (ABWT), another invertible transformation recently introduced in connection with a generalization of Lyndon words. Algorithmic and combinatorial issues on ABWT have been investigated in [1, 2].

In this talk an overview of the aforementioned transformations is presented, by focusing on some distinctive algorithmic and combinatorial features that could represent effective tools for investigating and handling the structure of the input text.

References

1. Giancarlo, R., Manzini, G., Restivo, A., Rosone, G., Sciortino, M.: The Alternating BWT: An Algorithmic Perspective (Submitted)
2. Giancarlo, R., Manzini, G., Restivo, A., Rosone, G., Sciortino, M.: Block sorting-based transformations on words: beyond the magic BWT. In: Hoshi, M., Seki, S. (eds.) DLT 2018. LNCS, vol. 11088, pp. 1–17. Springer, Cham (2018). https://doi.org/10.1007/978-3-319-98654-8_1

3. Giancarlo, R., Manzini, G., Rosone, G., Sciortino, M.: A new class of searchable and provably highly compressible string transformations. In: CPM 2019, Leibniz International Proceedings in Informatics (LIPIcs), vol. 12, pp. 1–12 (2019)
4. Louza, F.A., Telles, G.P., Gog, S., Zhao, L.: Computing burrows-wheeler similarity distributions for string collections. In: Gagie, T., Moffat, A., Navarro, G., Cuadros-Vargas, E. (eds.) SPIRE 2018. LNCS, vol. 11147, pp. 285–296. Springer, Cham (2018). https://doi.org/10.1007/978-3-030-00479-8_23
5. Mantaci, S., Restivo, A., Rosone, G., Sciortino, M., Versari, L.: Measuring the clustering effect of BWT via RLE. Theor. Comput. Sci. **698**, 79–87 (2017)
6. Prezza, N., Pisanti, N., Sciortino, M., Rosone, G.: SNPs detection by eBWT positional clustering. Algorithms Mol. Biol. **14**(1), 3:1–3:13 (2019)

On Square Permutations

Stéphane Vialette

CNRS and University of Paris-Est, France

Abstract. Given permutations π and σ_1 and σ_2, the permutation π is said to be a *shuffle* of σ_1 and σ_2, in symbols $\pi \in \sigma_1 \bullet \sigma_2$, if π (viewed as a string) can be formed by interleaving the letters of two strings p_1 and p_2 that are order-isomorphic to σ_1 and σ_2, respectively. In case $\sigma_1 = \sigma_2$, the permutation π is said to be a *square* and $\sigma_1 = \sigma_2$ is a *square root* of π. For example, $\pi = 24317856$ is a square as it is a shuffle of the patterns 2175 and 4386 that are both order-isomorphic to $\sigma = 2143$ as shown in $\pi = 2_{43}17_85_6$. However, σ is not the unique square root of π since π is also a shuffle of patterns 2156 and 4378 that are both order-isomorphic to 2134 as shown in $\pi = 2_{43}17_856$.

We shall begin by presenting recent results devoted to recognizing square permutations and related concepts with a strong emphasis on constrained oriented matchings in graphs. Then we shall discuss research directions to address square permutation challenges in both combinatorics and algorithmic fields.

Superimposed Codes and Their Applications: Old Results in New Light

Ugo Vaccaro

University of Salerno, Italy

Abstract. Superimposed codes, also known as cover-free families, disjunct matrices, strongly selective families (and, possibly, with different appellatives), were introduced in 1948 by Mooers as a tool for fast and efficient information retrieval in punched card systems. Since then, superimposed codes have found applications in a surprising variety of areas: compressed sensing, cryptography and data security, group testing, computational biology, multi-access communication, database theory, pattern matching, distributed computation, and circuit complexity, among the others.

Owing to the importance of the different scenarios where superimposed codes find applications, and starting from the seminal work of Kautz and Singleton [1964], many researchers have tried to find efficient algorithms to construct good superimposed codes (and still try). In this talk, I will start from classic results in the area, revisit them in a new light, present a few recent results, and discuss novel variations of superimposed codes dictated by modern application scenarios.

Contents

A Note on Handicap Incomplete Tournaments 1
  Appattu Vallapil Prajeesh, Krishnan Paramasivam,
  and Nainarraj Kamatchi

Computing the k-Crossing Visibility Region of a Point in a Polygon 10
  Yeganeh Bahoo, Prosenjit Bose, Stephane Durocher,
  and Thomas Shermer

An Improved Scheme in the Two Query Adaptive Bitprobe Model 22
  Mirza Galib Anwarul Husain Baig, Deepanjan Kesh, and Chirag Sodani

On Erdős–Szekeres-Type Problems for k-convex Point Sets 35
  Martin Balko, Sujoy Bhore, Leonardo Martínez Sandoval,
  and Pavel Valtr

Algorithm and Hardness Results on Liar's Dominating Set and k-tuple
Dominating Set 48
  Sandip Banerjee and Sujoy Bhore

Fixed-Parameter Tractability of $(n − k)$ List Coloring 61
  Aritra Banik, Ashwin Jacob, Vijay Kumar Paliwal,
  and Venkatesh Raman

Finding Periods in Cartesian Tree Matching 70
  Magsarjav Bataa, Sung Gwan Park, Amihood Amir, Gad M. Landau,
  and Kunsoo Park

Parameterized Complexity of Min-Power Asymmetric Connectivity 85
  Matthias Bentert, Roman Haag, Christian Hofer, Tomohiro Koana,
  and André Nichterlein

Solving Group Interval Scheduling Efficiently 97
  Arindam Biswas, Venkatesh Raman, and Saket Saurabh

Call Admission Problems on Trees with Advice (Extended Abstract) 108
  Hans-Joachim Böckenhauer, Nina Corvelo Benz, and Dennis Komm

Power Edge Set and Zero Forcing Set Remain Difficult in Cubic Graphs 122
  Pierre Cazals, Benoit Darties, Annie Chateau, Rodolphe Giroudeau,
  and Mathias Weller

Towards a Complexity Dichotomy for Colourful Components Problems
on k-caterpillars and Small-Degree Planar Graphs   136
   Janka Chlebíková and Clément Dallard

Maximal Irredundant Set Enumeration in Bounded-Degeneracy
and Bounded-Degree Hypergraphs   148
   Alessio Conte, Mamadou Moustapha Kanté, Andrea Marino,
   and Takeaki Uno

Dual Domination   160
   Gennaro Cordasco, Luisa Gargano, and Adele Anna Rescigno

Reaching 3-Connectivity via Edge-Edge Additions   175
   Giordano Da Lozzo and Ignaz Rutter

Cops and Robber on Some Families of Oriented Graphs   188
   Sandip Das, Harmender Gahlawat, Uma Kant Sahoo, and Sagnik Sen

Disjoint Clustering in Combinatorial Circuits   201
   Zola Donovan, K. Subramani, and Vahan Mkrtchyan

The Hull Number in the Convexity of Induced Paths of Order 3   214
   Mitre C. Dourado, Lucia D. Penso, and Dieter Rautenbach

Supermagic Graphs with Many Odd Degrees   229
   Dalibor Froncek and Jiangyi Qiu

Incremental Algorithm for Minimum Cut and Edge Connectivity
in Hypergraph   237
   Rahul Raj Gupta and Sushanta Karmakar

A General Algorithmic Scheme for Modular Decompositions
of Hypergraphs and Applications   251
   Michel Habib, Fabien de Montgolfier, Lalla Mouatadid,
   and Mengchuan Zou

Shortest-Path-Preserving Rounding   265
   Herman Haverkort, David Kübel, and Elmar Langetepe

Complexity and Algorithms for Semipaired Domination in Graphs   278
   Michael A. Henning, Arti Pandey, and Vikash Tripathi

Computing the Rooted Triplet Distance Between Phylogenetic Networks   290
   Jesper Jansson, Konstantinos Mampentzidis, Ramesh Rajaby,
   and Wing-Kin Sung

Parameterized Algorithms for Graph Burning Problem   304
   Anjeneya Swami Kare and I. Vinod Reddy

Extension and Its Price for the Connected Vertex Cover Problem   315
Mehdi Khosravian Ghadikoalei, Nikolaos Melissinos, Jérôme Monnot,
and Aris Pagourtzis

An Improved Fixed-Parameter Algorithm for Max-Cut Parameterized
by Crossing Number   327
Yasuaki Kobayashi, Yusuke Kobayashi, Shuichi Miyazaki,
and Suguru Tamaki

An Efficient Algorithm for Enumerating Chordal Bipartite Induced
Subgraphs in Sparse Graphs   339
Kazuhiro Kurita, Kunihiro Wasa, Takeaki Uno, and Hiroki Arimura

Complexity of Fall Coloring for Restricted Graph Classes   352
Juho Lauri and Christodoulos Mitillos

Succinct Representation of Linear Extensions via MDDs
and Its Application to Scheduling Under Precedence Constraints.   365
Fumito Miyake, Eiji Takimoto, and Kohei Hatano

Maximum Clique Exhaustive Search in Circulant k-Hypergraphs.   378
Lachlan Plant and Lucia Moura

Burrows-Wheeler Transform of Words Defined by Morphisms   393
Srecko Brlek, Andrea Frosini, Ilaria Mancini, Elisa Pergola,
and Simone Rinaldi

Stable Noncrossing Matchings   405
Suthee Ruangwises and Toshiya Itoh

On the Average Case of MergeInsertion.   417
Florian Stober and Armin Weiß

Shortest Unique Palindromic Substring Queries on Run-Length
Encoded Strings   430
Kiichi Watanabe, Yuto Nakashima, Shunsuke Inenaga, Hideo Bannai,
and Masayuki Takeda

A Partition Approach to Lower Bounds for Zero-Visibility Cops
and Robber   442
Yuan Xue, Boting Yang, Farong Zhong, and Sandra Zilles

Author Index   455

A Note on Handicap Incomplete Tournaments

Appattu Vallapil Prajeesh[1], Krishnan Paramasivam[1(✉)],
and Nainarraj Kamatchi[2]

[1] Department of Mathematics, National Institute of Technology Calicut,
Kozhikode 673601, India
prajeesh_p150078ma@nitc.ac.in, sivam@nitc.ac.in
[2] Department of Mathematics, Kamaraj College of Engineering and Technology,
Virudhunagar 625701, India
kamakrishna77@gmail.com

Abstract. An equalized incomplete tournament $EIT(p,r)$ on p teams
which are ranked from 1 to p, is a tournament in which every team plays
against r teams and the total strength of the opponents that every team
plays with is a constant. A handicap incomplete tournament $HIT(p,r)$
on p teams is a tournament in which every team plays against r opponents in such a way that

(i) the total strength of the opponents that the stronger teams play
   with are higher, and
(ii) the total strength of the opponents that the weaker teams play with
   are lower.

Thus, every team has an equal chance of winning in a $HIT(p,r)$. A
d-handicap labeling of a graph $G = (V, E)$ on p vertices is a bijection
$l : V \to \{1, 2, \cdots, p\}$ with the property that $l(v_i) = i$ and the sequence of
weights $w(v_1), w(v_2), \cdots, w(v_p)$ forms an increasing arithmetic progression
with difference d, where $w(v_i) = \sum_{v \in N(v_i)} l(v)$. A graph G is d-handicap
graph if it admits a d-handicap labeling. Thus, existence of an r-regular
d-handicap graph guarantees the existence of a $HIT(p,r)$. In this paper,
we give a method to construct new $(d + k)$-handicap graphs from d-
handicap graphs for all $k \geq 1$ and as an application, we characterize
the d-handicap labeling of Hamming graphs. Further, we give another
method to construct $EIT(p,r)$ from an infinite class of $HIT(p,r)$ by
increasing the number of rounds in $HIT(p,r)$.

Keywords: Tournaments · Distance magic labeling ·
Cartesian product · Hamming graph · d-handicap labeling

2010 AMS Subject Classification: 05C90 · 05C78 · 05C76

1 Motivation

All graphs considered in this paper are simple and finite. We use $V(G)$ for the
vertex set and $E(G)$ for the edge set of a graph G. For more graph-theoretic

© Springer Nature Switzerland AG 2019
C. J. Colbourn et al. (Eds.): IWOCA 2019, LNCS 11638, pp. 1–9, 2019.
https://doi.org/10.1007/978-3-030-25005-8_1

notations and terminologies, we refer to Bondy and Murty [1] and Hammack *et al.* [2].

In Sedláček [3] introduced the concept of magic labeling of graphs. The motivation for magic labeling comes from the concept of magic squares. A magic square is an $n \times n$ array filled with numbers 1 to n^2, each appearing once, such that the sum of each row, column, and the main and main backward diagonal is equal to $n(n^2 + 1)/2$. For more details, one can refer to [4]. Magic labeling has its applications in efficient addressing systems in communication networks, ruler models and radar pulse codes [5,6].

A *complete round-robin tournament* of p teams is a tournament in which each team plays against $p-1$ teams. If the teams are ranked 1 to p according to their strength, depending upon the rankings of the teams, where the *strength* of a team with rank i is defined as $s(i) = p+1-i$. Further, the total strength of all opponents of a team with rank i is defined as $S(i) = \frac{1}{2}p(p+1)-s(i) = \frac{1}{2}(p+1)(p-2)+i$. Thus, the sequence $S(1), S(2), \cdots, S(p)$ or shortly $\{S(i)\}_{i=1}^p$, is an increasing arithmetic progression with difference one. As the complete round-robin tournaments are generally considered to be fair, a tournament, which consists of p teams in which each team plays against $r, (r < p-1)$ opponents and $\{S(i)\}_{i=1}^p$ is again an increasing arithmetic progression with difference one is a *fair incomplete tournament, FIT(p,r)*. For more details, one can refer [7,8]. A disadvantage of such a tournament is that the stronger teams play against, the weaker opponents, while the weaker teams play against, the stronger opponents. This disadvantage is eliminated in *equalized incomplete round-robin tournaments, EIT(p,r)* in which every team plays with exactly r other teams and the total strength of the opponents that team i plays is a constant for every i. It is observed that a $FIT(p,r)$ exists if and only if an $EIT(p, p-r-1)$ exists.

Moreover, if one likes to give each team approximately the same chance of winning, then the weaker teams should play with weaker opponents, whereas the stronger teams should play with stronger opponents. In such case, $\{S(i)\}_{i=1}^p$ is a decreasing arithmetic progression. A tournament in which the above condition is satisfied and every team plays exactly r games is called a *handicap incomplete tournament HIT(p,r)*.

2   Definitions

The concept of distance magic labeling was motivated by the notion of magic squares and equalized incomplete tournaments. The theory was developed by different authors independently. Refer [9,10] for further details.

Definition 1. [9,10] *A distance magic labeling of a graph G on p vertices is a bijection $l : V(G) \to \{1, 2, \cdots, p\}$ such that for any vertex u of G, the weight of u, $w(u) = \sum_{v \in N(u)} l(v)$ is a constant μ. A graph that admits such a labeling is called a distance magic graph.*

Note that the existence of an r-regular distance magic graph on p vertices, guarantees to schedule an $EIT(p,r)$. For further results, refer [11]. Motivated by the real world problem of handicap incomplete tournaments and as a generalization of Definition 1, Froncek introduced the concept of handicap distance d-antimagic labeling of a graph.

Definition 2. [12] *A handicap distance d-antimagic labeling of a graph G with p vertices is a bijection $l : V(G) \to \{1, 2, \cdots, p\}$ with the property that $l(v_i) = i$ and the sequence of the weights $w(v_1), w(v_2), \cdots, w(v_p)$ or shortly $\{w(v_i)\}_{i=1}^{p}$, forms an increasing arithmetic progression with difference d, where $w(v_i) = \sum_{v_j \in N(v_i)} l(v_j)$. A graph G is a handicap distance d-antimagic graph if it allows a handicap distance d-antimagic labeling and handicap distance antimagic graph when $d = 1$.*

In [13], Freyberg refers the same as d-handicap labeling of graphs. It is clear that the existence of a d-handicap, r-regular graph on p vertices guarantees a $HIT(p, r)$.

Recall that the Cartesian product of graphs $G_1, G_2, \cdots, G_k$ is defined as the graph $G_1 \square G_2 \square \cdots \square G_k \cong \square_{i=1}^{k} G_i$ with vertex set $\{(x_1, x_2, \cdots, x_k) : x_i \in V(G_i)\}$, and any two vertices $(x_1, x_2, \cdots, x_k)$ and $(x_1', x_2', \cdots, x_k')$ are adjacent whenever $x_i x_i' \in E(G_i)$ for exactly one i and $x_j = x_j'$ for all $j \neq i$. With respect to the Cartesian product, the kth power of a graph G is denoted by $G^{\square, k} = \square_{i=1}^{k} G$.

A Hamming graph $H_{n,q}$ is isomorphic to $K_q^{\square, n}$ and has the vertices as n-tuples $(b_1, b_2, \cdots, b_n)$, where each $b_i \in \{0, 1, \cdots, q - 1\}$. Note that $H_{n,q}$ is an $n(q-1)$-regular graph on q^n vertices. Also, $H_{n,2}$ is isomorphic to the hypercube, Q_n on 2^n vertices.

3 Known Results

Let $H(p, r, d)$ denote an r-regular, d-handicap graph on p vertices. The following results give an up to date survey of results related to d-handicap graphs.

Theorem 1. [13] *If an $H(p, r, d)$ exists, then all of the following are true.*

(i) $w(x_i) = di + \frac{(r-d)(p+1)}{2}$, *for all $i \in \{1, 2, \cdots, p\}$.*
(ii) *If p is even, then $r \equiv d \mod 2$.*
(iii) *If p is odd, then $r \equiv 0 \mod 2$.*
(iv) $p \geq 4d + 4$.
(v) $d + 2 \leq r \leq p - d - 4$.

The following result completely characterizes the existence of a 1-handicap graph on even number of vertices for all possible r's.

Theorem 2. [14] *An $H(p, r, 1)$ exists when $p \geq 8$ and*

(i) $p \equiv 0 \mod 4$ *if and only if $3 \leq r \leq p - 5$ and r is odd*
(ii) $p \equiv 2 \mod 4$ *if and only if $3 \leq r \leq p - 7$ and $r \equiv 3 \mod 4$,*

except when $r = 3$ and $p \in \{10, 12, 14, 18, 22, 26\}$.

The next theorem gives the existence of 1-handicap graphs on odd number of vertices for some values of r.

Theorem 3. [15] *Let p be an odd positive integer. Then an $H(p, r, 1)$ exists for at least one value of r if and only if $p = 9$ or $p \geq 13$.*

The problem remains still open for the remaining values of r. In 2013, Froncek posted the following problem.

Problem 1. [16] For which triples (p, r, d) there exists an r-regular d-handicap graph on p vertices.

Recently, Froncek proved the following results on 2-handicap graphs.

Theorem 4. [17] *If $p \equiv 0 \mod 16$, then an $H(p, r, 2)$ exists if and only if r is even and $4 \leq r \leq p - 6$.*

Theorem 5. [18] *If $p \equiv 8 \mod 16$ and $p \geq 56$, then an $H(p, r, 2)$ exists if r is even and $6 \leq r \leq p - 50$.*

4 New Results

In this section, we provide a partial solution to Problem 1 by exhibiting infinite number of d-handicap graphs of high regularity r. First, we give a method to build $(d+k)$-handicap graphs from the existing d-handicap graphs, where $k \geq 1$. For simplicity, we denote each vertex by its label. Note that for any x of H, $w_H(x)$ represents the weight of a vertex x in the graph H.

Theorem 6. *Let $q \geq 2$. If an $H(p, d+q, d)$ exists, then a $H(qp, d+2q-1, d+q-1)$ exists.*

Proof. Let $H = H(p, d+q, d)$ be a graph with d-handicap labeling l. Clearly the weight of i, $w_H(i) = di + \frac{q(p+1)}{2}$, where $i \in \{1, 2, \cdots, p\}$. Now, construct q vertex disjoint copies of H, say $H_1, H_2, \cdots, H_q$ and let $G = H_1 \cup H_2 \cup \cdots \cup H_q$. For $k \in \{1, 2, \cdots, q\}$, apply $l + (k-1)p$ to the kth copy of H, that is H_k. For each $i \in \{1, 2, \cdots, p\}$, define $(q-1)$-factors F_i with q vertices $i, i+p, i+2p, \cdots, i+(q-1)p$. Let $F = F_1 \cup F_2 \cup \cdots \cup F_p$. Clearly, $G \cup F$ is $(d + 2q - 1)$-regular on qp vertices (Fig. 1).

Now, it suffices to show that, $\{w(i)\}_{i=1}^{qp}$ is an increasing sequence with difference $d + q - 1$.

Further, for $i \in \{1, 2, \cdots, p\}$ and $k \in \{1, 2, \cdots, q\}$, the weight,

$$w_G(i + (k-1)p) = di + \frac{q(p+1)}{2} + (d+q)(k-1)p,$$

and

$$w_F(i + (k-1)p) = \sum_{\substack{j=1, \\ j \neq k}}^{q} (i + (j-1)p).$$

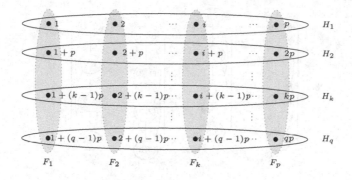

Fig. 1. Graph $G \cup F$.

Therefore, for every $i \in \{1, 2, \cdots, p\}$ and $k \in \{1, 2, \cdots, q\}$,

$$w(i + (k-1)p) = w_G(i + (k-1)p) + w_F(i + (k-1)p)$$
$$= (d + q - 1)(i + (k-1)p) + \frac{q^2 p + q}{2}.$$

Hence, $\{w(i)\}_{i=1}^{qp}$ is an increasing sequence with difference $d + q - 1$. □

Now, for even p, repeatedly applying Theorems 2 and 6, we obtain different classes of $(2k)$-handicap graphs, for $k \geq 1$ as given in the next theorem. Note that, $k \in \{1, 2, \cdots, \frac{p-6}{2}\}$.

Theorem 7. *Let $k \geq 1$. $H(p, 4k, 2k)$ exists,*

(i) *if $p \equiv 0 \mod 16k$ and $p \geq 16k$ or*
(ii) *if $p \equiv 4k \mod 16k$ and $p \geq 68k$ or*
(iii) *if $p \equiv 8k \mod 16k$ and $p \geq 40k$ or*
(iv) *if $p \equiv 12k \mod 16k$ and $p \geq 60k$.*

□

Now, consider the Hamming graph $H_{n,q}$ and represent the vertex $(b_1, b_2, \cdots, b_n)$ of $H_{n,q}$ by the integer i, where i is equal to $b_1 \times q^{n-1} + b_2 \times q^{n-2} + \cdots + b_n \times q^0 + 1$.

For instance, the vertex $(1, 2, 2, 1)$ of $H_{4,3}$ is uniquely represented by the integer $i = 1 \times 3^3 + 2 \times 3^2 + 2 \times 3^1 + 1 \times 3^0 + 1 = 53$. On the other hand, the integer $i = 79$ uniquely represents the vertex $(2, 2, 2, 0)$ of $H_{4,3}$, where $(2, 2, 2, 0)$ is obtained by finding the base three representation of the integer $i - 1 = 78$ (Fig. 2).

Now, for a fixed q, if we consider the graph $H_{2,q}$, for $i \in \{1, 2, \cdots, q\}$, and assign $l_{H_{2,q}}(i) = i$, then the respective weights are,

$$w_{H_{2,q}}(i) = \frac{q(q+1)}{2} - i + (i + q) + (i + 2q) + \cdots + (i + (q-1)q).$$

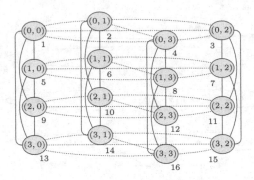

Fig. 2. 2-handicap labeling of $H_{2,4}$.

Further, by varying k from 2 to q, the weights of the remaining vertices of $H_{2,q}$ can be obtained as,

$$w_{H_{2,q}}(i + (k-1)q) = w_{2,q}(i) + (q-1)(k-1)q - (i + (k-1)q) + i,$$

where $i \in \{1, 2, \cdots, q\}$. Thus,

$$w_{H_{2,q}}(i + (k-1)q) = \frac{1}{2}q(q+1) - i + \sum_{j=1}^{q-1}(i + qj) + (q-1)(k-1)q$$

$$-(i + (k-1)q) + i$$

$$= (q-2)(i + (k-1)q) + \frac{1}{2}q(q^2 + 1),$$

where $k \in \{1, 2, \cdots, q\}$.

Hence, for every vertex i of $H_{2,q}$, the weight is,

$$w_{H_{2,q}}(i) = (q-2)i + \frac{1}{2}q(q^2 + 1), \text{ where } i \in \{1, 2, \cdots, q^2\}.$$

Now, if we assume that $H_{r,q}$ is d-handicap with $d = rq - q - r$. That is,

$$w_{r,q}(i) = (rq - q - r)i + \frac{1}{2}q(q^r + 1), \text{ where } i \in \{1, 2, \cdots, q^r\}.$$

Then, one can see that $H_{r,q}$ is a $H(q^r, r(q-1), rq - q - r)$. Hence, we obtain a $H(q^{r+1}, (r+1)(q-1), (r+1)(q-1) - q)$, which is isomorphic to $H_{r+1,q}$ (by the construction in Theorem 6).

Note that, $l_{H_{r+1,q}}(i + (k-1)q^r) = l_{H_{r,q}}(i) + (k-1)q^r$ where $i \in \{1, 2, \cdots, q^r\}, k \in \{1, 2, \cdots, q\}$. Now we get,

$$w_{r+1,q}(i) = ((r+1)q - q - (r+1))i + \frac{1}{2}q(q^{r+1} + 1),$$

where $i \in \{1, 2, \cdots, q^{r+1}\}$.

Hence by the principle of mathematical induction, we obtain,

Theorem 8. *The Hamming graph $H_{n,q}$ is d-handicap with $d = nq - q - n$, where $n, q \in \mathbb{N} - \{1\}$.*　□

Corollary 1. *The hypercube Q_n is d-handicap with $d = n - 2$.*　□

Now, for every p^2, p is odd and $p > 1$, we construct distance magic graphs on p^2 vertices from d-handicap graphs obtained from Theorem 8.

The following technique helps to find a sequence of $H(q^2, k(q-1), q-k)$ for all $k \in \{2, 3, \cdots, q\}$. When $k = q$, we obtain a $(q^2 - q)$-regular distance magic graph on q^2 vertices.

Theorem 9. *Let $q > 1$ be odd. For every k from 2 to q, there exists a $H(q^2, k(q-1), q-k)$.*

Proof. From Theorem 8, we obtain a $H \cong H(q^2, 2q - 2, q - 2)$ for all $q > 1$ and q-odd. Now, construct certain $(q \times q)$-matrices, $M^{q-3}, M^{q-4}, \cdots, M^0$ in the following manner, which precise to define the new set of $(q-1)$-factors, which can be attached to the existing $H(q^2, k(q-1), q-k)$ to obtain $H(q^2, (k+1)(q-1), q-(k+1))$. For any $s \in \{1, 2, \cdots, q-2\}$ and for all i, we define the elements of $M^{q-(s+2)}$ as,

$$m_{i,j}^{q-(s+2)} = (j-1)q + (i + (j-1)s \mod q), \text{ for } j \in \{1, 2, \cdots, q\}.$$

While determining the elements $m_{i,j}^{q-(s+2)}$, if $(i + (j-1)s \mod q)$ is 0, then replace $(i + (j-1)s \mod q)$ by q. Now one can see that each M^{q-k} contains all the elements from 1 to q^2. Further, the weight of each element $m_{i,j}^{q-(s+2)}$ is given as,

$$w(m_{i,t}^{q-(s+2)}) = \sum_{j, j \neq t} m_{i,j}^{q-(s+2)}, \text{ where } j \in \{1, 2, \cdots, q\}.$$

Now, one can see that the weight of each $m_{i,j}^{q-(s+2)}$ is,

$$w(m_{i,j}^{q-(s+2)}) = \frac{(q-1)q}{2}q + \frac{q(q+1)}{2} - m_{i,j}^{q-(s+2)} = \frac{q^3+q}{2} - m_{i,j}^{q-(s+2)}.$$

Hence, if we consider the weight of all the elements $1, 2, \cdots, q^2$, then they form an decreasing arithmetic progression $\frac{q^3+q}{2} - 1, \frac{q^3+q}{2} - 2, \cdots, \frac{q^3+q}{2} - q^2$ respectively, with difference one.

Consider the matrix M^{q-3} and define $(q-1)$-factors F_r's, whose vertices are the elements $m_{r,1}^{q-3}, m_{r,2}^{q-3}, \cdots, m_{r,q}^{q-3}$, $r \in \{1, 2, \cdots, q\}$. Let F be the union of all such $F_1, F_2, \cdots, F_q$. Construct a new graph $G = H \cup F$, with the vertices $1, 2, \cdots, q^2$ and regularity $3(q-1)$. Now, it suffices to show that weights of the vertices form an increasing arithmetic progression with new $d = q - 3$. Further,

$$w_H(i) = (q-2)i + \frac{1}{2}q(q^2 + 1), \text{ where } i \in \{1, 2, \cdots, q^2\}, \text{ and}$$

$$w_F(i) = \frac{q^3 + q}{2} - i, \text{ where } i \in \{1, 2, \cdots, q^2\}.$$

Hence, the weight of each vertex i in G is obtained as,

$$w_G(i) = w_H(i) + w_F(i) = (q-3)i + \frac{q^3 + q}{2}.$$

Thus, we obtain an $H(q^2, 3(q-1), q-3)$. Now to construct $H(q^2, 4(q-1), q-4)$, we follow the same technique given above by replacing M^{q-3} with the matrix M^{q-4}. By repeating these steps, in total $(q-2)$-times we obtain an $H(q^2, q(q-1), 0)$. □

Corollary 2. *There exists a $q(q-1)$-regular distance magic graph on q^2 vertices where $q > 1$ is odd.* □

Example 1. Consider the 3-handicap graph, $H(25, 8, 3)$ obtained from Theorem 8. The following matrices M^2, M^1 and M^0, respectively are involved in the construction of $H(25, 12, 2)$, $H(25, 16, 1)$ and $H(25, 20, 0)$ from $H(25, 8, 3)$.

$$M^2 = \begin{pmatrix} 1 & 7 & 13 & 19 & 25 \\ 2 & 8 & 14 & 20 & 21 \\ 3 & 9 & 15 & 16 & 22 \\ 4 & 10 & 11 & 17 & 23 \\ 5 & 6 & 12 & 18 & 24 \end{pmatrix} \quad M^1 = \begin{pmatrix} 1 & 8 & 15 & 17 & 24 \\ 2 & 9 & 11 & 18 & 25 \\ 3 & 10 & 12 & 19 & 21 \\ 4 & 6 & 13 & 20 & 22 \\ 5 & 7 & 14 & 16 & 23 \end{pmatrix} \quad M^0 = \begin{pmatrix} 1 & 9 & 12 & 20 & 23 \\ 2 & 10 & 13 & 16 & 24 \\ 3 & 6 & 14 & 17 & 25 \\ 4 & 7 & 15 & 18 & 21 \\ 5 & 8 & 11 & 19 & 22 \end{pmatrix}$$

For instance, $G \cong H(25, 12, 2)$ is obtained as $G \cong H \cup F$, where $H \cong H(25, 8, 3)$ and $F = F_1 \cup F_2 \cup \cdots \cup F_5$, where F_i is the 4-factor defined on vertices obtained from the ith row of M^2 (Fig. 3).

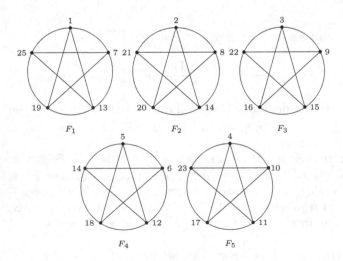

Fig. 3. 4-factors involved in the process of constructing $H(25, 12, 2)$ from $H(25, 8, 3)$.

5 Conclusion and Scope

In this paper, we give a technique to construct new handicap distance $(d + k)$-antimagic graphs from d-handicap graphs for all $k \geq 1$, and hence prove the existence of d-handicap labeling for Hamming graphs. Further, we construct new even-regular distance magic graphs on an odd number of vertices from $H_{2,q}$ for all $q > 1$, odd.

References

1. Bondy, J.A., Murty, U.S.R.: Graph Theory. Springer, New York (2008)
2. Hammack, R., Imrich, W., Klavžar, S.: Handbook of Product Graphs. CRC Press, Boca Raton (2011)
3. Sedláček, J.: Problem 27. Theory of graphs and its applications. (Smolenice 1963), pp. 163–164. Publication House of the Czechoslovak Academy of Sciences, Prague (1964)
4. Colbourn, C., Dinitz, J.: The CRC Handbook of Combinatorial Designs. CRC Press, Boca Raton (2007)
5. Bloom, G.S., Golomb, S.W.: Applications of numbered undirected graphs. Proc. IEEE **65**, 562–570 (1977)
6. Bloom, G.S., Golomb, S.W.: Numbered complete graphs, unusual rulers, and assorted applications. In: Alavi, Y., Lick, D.R. (eds.) Theory and Applications of Graphs. Lecture Notes in Mathematics, vol. 642, pp. 53–65. Springer, Berlin (1978). https://doi.org/10.1007/BFb0070364
7. Froncek, D., Kovár, P., Kovárová, T.: Fair incomplete tournaments. Bull. Inst. Comb. Appl. **48**, 31–33 (2006)
8. Froncek, D.: Fair incomplete tournaments with odd number of teams and large number of games. In: Proceedings of the Thirty-Eighth Southeastern International Conference on Combinatorics, Graph Theory and Computing. Congr. Numer. **187**, 83–89 (2007)
9. Vilfred, V.: $\sum$ –labelled graphs and circulant graphs. Ph.D. thesis, University of Kerala, Trivandrum, India (1994)
10. Miller, M., Rodger, C., Simanjuntak, R.: Distance magic labelings of graphs. Australas. J. Comb. **28**, 305–315 (2003)
11. Arumugam, S., Froncek, D., Kamatchi, N.: Distance magic graphs-a survey. J. Indones. Math. Soc. Special Edition 11–26 (2011)
12. Froncek, D.: Handicap distance antimagic graphs and incomplete tournaments. AKCE J. Graphs Comb. **10**, 119–127 (2013)
13. Freyberg, B.: Distance magic-type and distance antimagic-type labelings of graphs. Ph.D. thesis, Michigan Technological University, Houghton, MI, USA (2017)
14. Kovár, P., Kovárová, T., Krajc, B., Kravčenko, M., Shepanik, A., Silber, A.: On regular handicap graphs of even order. In: 9th International Workshop on Graph Labelings (IWOGL 2016). Electron. Notes Discrete Math. **60**, 69–76 (2017)
15. Froncek, D.: Regular handicap graphs of odd order. J. Comb. Math. Comb. Comput. **102**, 253–266 (2017)
16. Froncek, D.: Incomplete tournaments and handicap distance antimagic graphs. Congr. Numer. **217**, 93–99 (2013)
17. Froncek, D.: Full spectrum of regular incomplete 2-handicap tournaments of order $n \equiv 0(mod16)$. J. Comb. Math. Comb. Comput. **106**, 175–184 (2018)
18. Froncek, D.: A note on incomplete regular tournaments with handicap two of order $n \equiv 8(mod16)$. Opusc. Math. **37**, 557–566 (2017)

Computing the k-Crossing Visibility Region of a Point in a Polygon

Yeganeh Bahoo[1]($\boxtimes$), Prosenjit Bose[2]($\boxtimes$), Stephane Durocher[1]($\boxtimes$), and Thomas Shermer[3]($\boxtimes$)

[1] University of Manitoba, Winnipeg, Canada
{bahoo,durocher}@cs.umanitoba.ca
[2] Carleton University, Ottawa, Canada
jit@scs.carleton.ca
[3] Simon Fraser University, Burnaby, Canada
shermer@sfu.ca

Abstract. Two points p and q in a simple polygon P are k-crossing visible when the line segment pq crosses the boundary of P at most k times. Given a query point q, an integer k, and a polygon P, we propose an algorithm that computes the region of P that is k-crossing visible from q in $O(nk)$ time, where n denotes the number of vertices of P. This is the first such algorithm parameterized in terms of k, resulting in asymptotically faster worst-case running time relative to previous algorithms when k is $o(\log n)$, and bridging the gap between the $O(n)$-time algorithm for computing the 0-visibility region of q in P and the $O(n \log n)$-time algorithm for computing the k-crossing visibility region of q in P.

Keywords: Computational geometry · Visibility · Radial decomposition

1 Introduction

Given a simple n-vertex polygon P, two points p and q inside P are said to be mutually visible when the line segment pq does not intersect the exterior of P. Problems related to visibility are motivated by many applications that require covering a given region using a minimum number of resources, some of which refer to visual coverage (e.g., guarding with cameras [16,21]) or to providing wireless connectivity coverage [19,23]. Unlike the visible-light model, in which a viewer's line of sight typically terminates upon encountering a wall, radio transmissions can pass through some walls, suggesting a more general notion of visibility. Mouad and Shermer [20] introduced a generalized model of visibility in polygons; this model was subsequently extended by Dean et al. [11] and Bajuelos et al. [4] to define *k-crossing visibility*. When p and q are in general position relative to the vertices of P (i.e., no vertex of P is collinear with p and q), p and q are mutually *k-crossing visible* when the line segment pq intersects the boundary of P in at most k points. Various applications require computing the

© Springer Nature Switzerland AG 2019
C. J. Colbourn et al. (Eds.): IWOCA 2019, LNCS 11638, pp. 10–21, 2019.
https://doi.org/10.1007/978-3-030-25005-8_2

region of P that is visible or k-crossing visible from a given query point q in P [1]. This region is called the k-*crossing visibility polygon* of q in P. See Fig. 1.

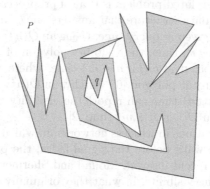

Fig. 1. A polygon P, a point q, and the k-crossing visibility polygon of q in P when $k = 2$

Our goal is to design an algorithm that reduces the time required for computing the k-crossing visibility polygon for a given point q in a given simple polygon P. $O(n)$-time algorithms exist for finding the visibility polygon of q in P (i.e., when $k = 0$) [13,17,18], whereas the best known algorithms for finding the k-crossing visibility polygon of q in P require $\Theta(n \log n)$ time in the worst case for any given k [3]. A natural question that remained open is whether the k-crossing visibility polygon of q in P can be found in $o(n \log n)$ time. In particular, can the problem be solved faster for small values of k? This paper presents the first algorithm parameterized in terms of k to compute the k-crossing visibility polygon of q in P. The proposed algorithm takes $O(nk)$ time, where n denotes the number of vertices of P, resulting in asymptotically faster worst-case running time relative to previous algorithms when k is $o(\log n)$, and bridging the gap between the $O(n)$-time algorithm for computing the 0-visibility polygon of q in P and the $O(n \log n)$-time algorithm for computing the k-crossing visibility polygon of q in P.

The paper begins with an overview of related work, followed by definitions, the presentation of the algorithm, and an analysis of its running time.

2   Related Work

Given a polygon P with n vertices and a query point q inside P, a fundamental problem in visibility is to compute the visibility polygon for q: the portion of P visible from q. This problem was first introduced by Davis and Benedikt [10], who gave an $O(n^2)$-time algorithm. The number of vertices of the visibility polygon of q in P is proportional to the number of vertices of P in the worst case, i.e., $\Theta(n)$ [13,18]. Algorithms for computing the visibility polygon for any given q

and P in $O(n)$ time were given by Gindy and Avis [13], Lee [18], and Joe and Simpson [17].

This paper focuses on finding the k-crossing visibility polygon of q in P without preprocessing P. A related problem is that of preprocessing a given polygon P to construct a query data structure that answers one or more subsequent visibility queries for points given at query time. Using an $O(n^3)$-space data structure precomputed in $O(n^3 \log n)$ time, the visibility polygon of any point q given at query time can be reported in $O(\log n + m)$ time, where m denotes the number of vertices in the output polygon [6]. Finally, an $O(n^2)$-space data structure precomputed in $O(n^2 \log n)$ time can report the visibility polygon of any point q given at query time in $O(\log^2 n + m)$ time [2].

Motivated by applications in wireless networks, in which a radio transmission can pass through some walls before the signal fades, the problem of k-crossing visibility has attracted recent interest. Mouad and Shermer [20] first introduced the concept of k-crossing visibility, in what they originally called the *Superman problem*: given a simple polygon P, a sub-polygon $Q \subseteq P$, and a point q outside P, determine the minimum number of edges of P that must be made opaque such that no point of Q is visible to q. Dean et al. [11] studied pseudo-star-shaped polygons, in which the line of visibility can cross one edge, corresponding to k-crossing visibility where $k = 1$. Bajuelos et al. [4] subsequently explored the concept of k-crossing visibility for an arbitrary given k, and presented an $O(n^2)$-time algorithm to construct the k-crossing visible region of q in P for an arbitrary given point q. Recently, Bahoo et al. [3] examined the problem under the limited-workspace mode, and gave an algorithm that uses $O(s)$ words of memory and reports the k-visibility polygon of q in P in $O(n^2/s + n \log s)$ time. When memory is not constrained (i.e., $\Omega(n)$ words of memory are available) their algorithm computes the k-visibility polygon in $O(n \log n)$ time.

Additional results related to k-crossing visibility include generalizations of the well-known Art Gallery problem to the setting of k-crossing visibility. A set of points W in a polygon P is said to guard P if every point in P is k-crossing visible from some point in W. Each point (guard) in W is called a k-modem. The Art Gallery problem seeks to identify a set of point of minimum cardinality that guards a given polygon P. Aichholzer et al. [1] showed that $\lfloor n/2k \rfloor$ k-modems are sometimes necessary and $\lfloor n/(2k+2) \rfloor$ are always sufficient for guarding monotone polygons. They also proved that a monotone orthogonal polygon can be guarded by $\lfloor n/(2k+4) \rfloor$ k-modems. Duque et al. [12] showed that at most $O(n/k)$ k-modems are needed to guard a simple polygon P; however, given a polygon P, determining the minimum number of modems to guard P is NP-hard [7]. k-crossing visibility can be considered in the plane with obstacles, where the goal is to guard the plane or the boundary of a given region. Ballinger et al. [5] developed upper and lower bounds for the number of k-modems needed to guard a set of orthogonal line segments and other restricted families of geometric objects. Finally, given a set of line segments and a point q, Fabila et al. [14] examined the problem of determining the minimum k such that the entire plane is k-crossing visible from q.

3   Preliminaries and Definitions

3.1   Crossings and k-Crossing Visibility

Two paths P and Q are *disjoint* if $P \cap Q = \varnothing$. To provide a general definition of visibility requires a comprehensive definition for a crossing between a line segment and a polygon boundary, in particular, for the case when points are not in general position.

Definition 1 (Weakly disjoint paths [8]). *Two paths P and Q are weakly disjoint if, for all sufficiently small $\epsilon > 0$, there are disjoint paths $\tilde{P}$ and $\tilde{Q}$ such that $d_{\mathcal{F}}(P, \tilde{P}) < \epsilon$ and $d_{\mathcal{F}}(Q, \tilde{Q}) < \epsilon$.*

$d_{\mathcal{F}}(A, B)$ denotes the Fréchet distance between A and B.

Definition 2 (Crossing paths [8]). *Two paths cross if they are not weakly disjoint.*

Definitions 1 and 2 apply when P and Q are Jordan arcs. We use Definition 2 to help define k-crossing visibility.

Definition 3 (k-crossing visibility). *Two Jordan arcs (or polygonal chains) P and Q cross k times, if there exist partitions $P_1, \ldots, P_k$ of P and $Q_1, \ldots, Q_k$ of Q such that P_i and Q_i cross, for all $i \in \{1, \ldots, k\}$. Points p and q in a simple polygon P are k-crossing visible if the line segment pq and the boundary of P do not cross k times.*

Given a simple polygon P, we refer to the set of points that are k-crossing visible from a point q as the *k-crossing visibility region of q with respect to P*, denoted $\mathcal{V}_k(P, q)$. When the polygon P is clear from the context, we simply refer to set as the k-crossing visibility region of q and denote it as $\mathcal{V}_k(q)$. Our goal is to design an efficient algorithm to compute the k-crossing visibility region of a point q with respect to a simple polygon P.

To simplify the description of our algorithms, we assume that the query point q and the vertices of the input polygon P are in general position, i.e., q, p_i and p_j are not collinear for any vertices p_i and p_j in P. Under the assumption of general position, two points p and q are k-crossing visible if and only if the line segment pq intersects the boundary of P in fewer than k points. That is, Definition 3 is not necessary under general position. All results presented in this paper can be extended to input that is not in general position.

3.2   Trapezoidal and Radial Decompositions

A *polygonal decomposition* of a simple polygon P is a partition of P into a set of simpler regions, such as triangles, trapezoids, or quadrilaterals. Our algorithm uses trapezoidal decomposition and radial decomposition. A *trapezoidal decomposition* (synonymously, *trapezoidation*) of P partitions P into trapezoids and triangles by extending, wherever possible, a vertical line segment from each vertex p of P above and/or below p into the interior of P, until its first intersection with the boundary of P. A *radial decomposition* of P is defined relative to a point q in P. For each vertex p of P, a line segment is extended, wherever possible, toward/away from p into the interior of P on the line determined by p and q, until its first intersection with the boundary of P. A radial decomposition partitions P into quadrilateral and triangular regions. The number of vertices and edges in both decompositions is proportional to the number of vertices in P (i.e., $\Theta(n)$). Note that a trapezoidal decomposition corresponds to a radial decomposition when the point q has its y-coordinate at $+\infty$ or $-\infty$ (outside P). Chazelle [9] gives an efficient algorithm for computing a trapezoidal decomposition of a simple n-vertex polygon in $O(n)$ time.

4 k-Crossing Visibility Algorithm

4.1   Overview

Given as input an integer k, an array storing the coordinates of vertices whose sequence defines a clockwise ordering of the boundary of a simple polygon P, and a point q in the interior of P, our algorithm for constructing the k-crossing visibility polygon of q in P executes the following steps, each of which is described in detail in this section:

1. Partition P into two sets of disjoint polylines, corresponding to the boundary of P above the horizontal line ℓ through q, and the boundary of P below ℓ.
2. Nesting properties of Jordan sequences are used to close each set by connecting disjoint components to form two simple polygons, P_a above ℓ and P_b below ℓ.
3. The two-dimensional coordinates of the vertices of P_a and P_b are mapped to homogeneous coordinates, to which a projective transformation, f_q, is applied, with q as the center of projection.
4. Compute the trapezoidal decompositions of $f_q(P_a)$ and $f_q(P_b)$ using Chazelle's algorithm [9].
5. Apply the inverse tranformation f_q^{-1} on the trapezoidal decompositions to obtain radial decompositions of P_a and P_b.
6. Merge the radial decompositions of P_a and P_b to obtain a radial decomposition of P with respect to q.
7. Traverse the radial decomposition of P to identify the visibility of cells in increasing order from visibility 0 through visibility k, moving away from q and extending edges on rays from q to refine cells of the decomposition as necessary.
8. Traverse the refined radial decomposition to reconstruct and output the boundary of the k-crossing visibility region of q in P.

Steps 1–6 can be completed in $O(n)$ time and Steps 7–8 can be completed in $O(nk)$ time.

4.2   Partitioning P into Upper and Lower Polygons

We begin by describing how to partition the polygon P in two across the line ℓ, where ℓ denotes the horizontal line through q. By our general position assumption, no vertices of P lie on ℓ. Let ϵ denote the minimum distance between any vertex of P and ℓ. Let the *upper polygon*, denoted as P_a (respectively, the *lower polygon*, denoted P_b) refer to the closure of the region of the boundary of P that lies above (respectively, below) ℓ; see Fig. 2. Let $\{x_1, \ldots, x_m\}$ denote the sequence of intersection points of ℓ with the boundary of P, labelled in clockwise order along the boundary of P, such that x_1 is the leftmost point in $P \cap \ell$. This sequence is a *Jordan sequence* [15]. We now describe how to construct P_a and P_b.

Between consecutive pairs (x_{2i-1}, x_{2i}) of the Jordan sequence, for $i \in \{1, \ldots, m/2\}$, the polygon boundary of P lies above ℓ. Similarly, between pairs (x_{2j}, x_{2j+1}), for $j \in \{1, \ldots, m/2 - 1\}$, and between (x_m, x_0), the boundary of P lies below ℓ. We call the former *upper pairs* of the Jordan sequence, and the latter *lower pairs*. These pairs possess the *nested parenthesis* property [22]: every two pairs (x_{2i-1}, x_{2i}) and (x_{2j-1}, x_{2j}) must either *nest* or be *disjoint*. That is, x_{2j-1} lies between x_{2i-1} and x_{2i} in the sequence if and only if x_{2j} lies between x_{2i-1} and x_{2i}.

As shown by Hoffmann et al. [15], the nested parenthesis property for the upper pairs determines a rooted tree, called the *upper tree*, whose nodes correspond to pairs of the sequence. The nodes in the subtree rooted at the pair (x_{2i-1}, x_{2i}) consist of all nodes corresponding to pairs that are nested between x_{2i-1} and x_{2i} in the Jordan sequence order. The leaves of the tree correspond to pairs that are consecutive in the sorted order. If a node (x_{2j-1}, x_{2j}) is a descendant of a node (x_{2i-1}, x_{2i}) in the tree, then the points x_{2j-1} and x_{2j} are nested between x_{2i-1} and x_{2i}. The *lower tree* is defined analogously.

If the boundary of P intersects ℓ in more than two points, the resulting disconnected components must be joined appropriately to form the simple polygons P_a and P_b. To build the lower polygon P_b, we replace each portion of the boundary of P above ℓ from x_{2i-1} to x_{2i} with the following 3-edge path: x_{2i-1}, u, v, x_{2i}. The first edge (x_{2i-1}, u) is a vertical line segment of length $\epsilon/2d_i$, where d_i denotes the depth of the node (x_{2i-1}, x_{2i}) in the tree. The next edge (u, v) is a horizontal line segment whose length is $||x_{2i-1} - x_{2i}||$. The third edge (v, x_{2i}) is a vertical line segment of length $\epsilon/2d_i$. See Fig. 2.

The nesting property of the Jordan sequence ensures that all of the 3-edge paths cross are similarly nested and that none of them intersect. Consider two pairs (x_{2i-1}, x_{2i}) and (x_{2j-1}, x_{2j}). Either they are disjoint or nested. If they are disjoint, then without loss of generality, assume that $x_{2i-1} < x_{2i} < x_{2j-1} < x_{2j}$. Their corresponding 3-edge paths cannot cross since the intervals they cover are disjoint. If they are nested, then without loss of generality, assume that $x_{2i-1} < x_{2j-1} < x_{2j} < x_{2i}$. The only way that the two paths can cross is if the horizontal edge for the pair (x_{2j-1}, x_{2j}) is higher than for the pair (x_{2i-1}, x_{2i}). However, since (x_{2j-1}, x_{2j}) is deeper in the tree than (x_{2i-1}, x_{2i}), the two paths do not cross. Thus, we form the simple polygon P_b by replacing the portions of the boundary above ℓ with these three edge paths. Sorting the Jordan sequence, building the upper tree, computing the depths of all the pairs and adding the 3-edge paths can all be achieved in $O(n)$ time using the Jordan sorting algorithm outlined by Hoffmann et al. [15]. The upper polygon P_a is constructed analogously. We conclude with the following lemma.

Lemma 1. *Given a simple n-vertex polygon P and a horizontal line ℓ that intersects the interior of P such that no vertices of P lie on ℓ, the upper and lower polygons of P with respect to ℓ can be computed in $O(n)$ time.*

4.3   Computing the Radial Decomposition

The two-dimensional coordinates of the vertices of each polygon P_a and P_b are mapped to homogeneous coordinates, to which a projective transformation, f_q, is applied with q as the center of projection. These transformations take constant time per vertex, or $\Theta(n)$ total time. Chazelle's algorithm [9] constructs trapezoidal decompositions of $f_q(P_a)$ and $f_q(P_b)$ in $\Theta(n)$ time, on which the inverse transformation, f_q^{-1} is applied to obtain radial decompositions of P_a and P_b. Merging the radial decompositions of P_a and P_b gives a radial decomposition of the original polygon P without requiring any additional edges. All vertices $x_1, \ldots, x_m$ of the Jordan sequence, all vertices of the three-edge paths, and their adjacent edges are removed. The remaining edges are either on the boundary of P, between two points on the boundary on a ray through q, or between the boundary and q. The entire process for constructing the radial trapezoidation takes $\Theta(n)$ time. This gives the following lemma.

Lemma 2. *The radial decomposition of a simple n-vertex polygon P around a query point q can be computed in $\Theta(n)$ time.*

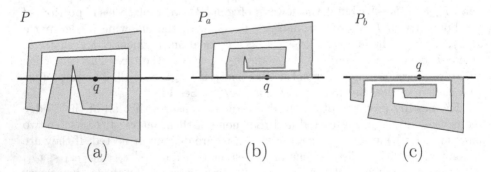

Fig. 2. (a) a polygon P, a point q, and the horizontal line ℓ through q; (b)–(c) the upper polygon P_a and lower polygon P_b of P with the additional 3-edge paths highlighted.

4.4   Reporting the k-Crossing Visible Region

The 0-visibility region of q in P, denoted $\mathcal{V}_0(q)$, is a star-shaped polygon with q in its kernel. A vertex of $\mathcal{V}_0(q)$ is either a vertex v of P or a point x on the boundary of P that is the intersection of an edge of P with a ray emanating from q through a reflex vertex r of P. In the latter case, (r, x) is an edge of $\mathcal{V}_0(q)$ that is collinear with q, called a *window* or *lid*, because it separates a region in the interior of P that is 0-visible from q and an interior region that is not 0-visible. The reflex vertex r is the *base* of the lid and x is its *tip*. There are two types of base reflex vertices. The reflex vertex r is called a *left base* (respectively, *right base*) if the polygon edges incident on r are to the left (respectively, right) of the ray emanating from q through r.

We now describe the algorithm to compute the k-crossing visible region of q in P, denoted $\mathcal{V}_k(q)$. The algorithm proceeds incrementally by computing $\mathcal{V}_{i+1}(q)$ after computing $\mathcal{V}_i(q)$. We begin by computing $\mathcal{V}_0(q)$ in $O(n)$ time using one of the existing linear-time algorithms, e.g. [13,17,18]. Label the vertices of $\mathcal{V}_0(q)$ in clockwise order around the boundary as $x_0, x_1, \ldots, x_m$. Triangulate the visibility polygon by adding the edge (q, x_i) for $i \in \{0, \ldots, m\}$; this corresponds to a radial decomposition of $\mathcal{V}_0(q)$ around q.

If x_i is a left base vertex, then notice that the triangle $\triangle(qx_ix_{i+1})^1$ degenerates to a segment. Similarly, if x_i is a right base vertex, then $\triangle(qx_ix_{i-1})$ is degenerate. If we ignore all degenerate triangles, then every triangle has the form $\triangle(qx_ix_{i+1})$, where (x_i, x_{i+1}) is on the boundary of P. The union of these triangles is $\mathcal{V}_0(q)$. To compute $\mathcal{V}_1(q)$, we show how to compute a superset of triangles whose union is $\mathcal{V}_1(q)$.

We start with an arbitrary triangle $\triangle(qx_ix_{i+1})$ of $\mathcal{V}_0(q)$, where (x_i, x_{i+1}) is on the boundary of P. Note that (x_i, x_{i+1}) is either an edge of P or a segment within the interior of an edge of P. It is this segment (x_i, x_{i+1}) of the boundary that blocks visibility. We show how to compute the intersection of $\mathcal{V}_1(q)$ with the cone that has apex q and bounding rays $\boldsymbol{qx_i}$ and $\boldsymbol{qx_{i+1}}$, denoted $\mathcal{C}(q, x_i, x_{i+1})$. We call this process *extending the visibility* of a triangle. We have two cases to consider. Either at least one of x_i or x_{i+1} is a base vertex or neither is a base vertex. We start with the latter case where neither is a base vertex.

Let Y be the set of vertices of the radial decomposition that lie on the edge (x_i, x_{i+1}). If Y is empty, then (x_i, x_{i+1}) lies on one face of the decomposition in addition to $\triangle(qx_ix_{i+1})$ since neither x_i nor x_{i+1} is a base vertex. We show how to proceed in the case when Y is empty, then we show what to do when Y is not empty. Let f be the face of the decomposition on the boundary of which (x_i, x_{i+1}) lies. By construction, this face is either a quadrilateral or a triangle. In constant time, we find the intersection of the boundary of f excluding the edge containing (x_i, x_{i+1}) with $\boldsymbol{qx_i}$ and $\boldsymbol{qx_{i+1}}$. Label these two intersection points as x_i' and x_{i+1}'. Extending the visibility of $\triangle(qx_ix_{i+1})$ results in $\triangle(qx_i'x_{i+1}')$. Note that $\triangle(qx_i'x_{i+1}')$ is the 1-visible region of q in $\mathcal{C}(q, x_i, x_{i+1})$ and (x_i', x_{i+1}') is on the boundary of P.

---

[1] All indices are computed modulo the size of the corresponding vertex set: $m + 1$ in this case.

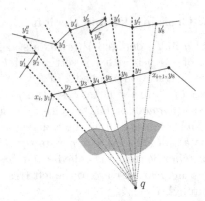

Fig. 3. Edges of the radial decomposition are extended where critical vertices cast a shadow. Portions of the polygon in the blue region that were processed in previous iterations are omitted from the figure. (Color figure online)

We now show how to extend the visibility of $\triangle(qx_ix_{i+1})$ when Y is not empty. Label the points of Y as y_j for $j \geq 1$ in the order that they appear on the edge (x_i, x_{i+1}) from x_i to x_{i+1}; see Fig. 3. Each y_j is an endpoint of an edge of the radial decomposition. Since y_j is a point on the boundary of P, there are 2 faces of the radial decomposition with y_j on the boundary. Let y'_j be the other endpoint of the face on the left of y_j and y''_j be the endpoint for the face on the right. Either $y'_j = y''_j$ or $y'_j \neq y''_j$. In the former case, we simply ignore y''_j. In the latter case, we note that either y'_j is a left base of $\mathcal{V}_0(y_j)$ or y''_j is a right base. See Fig. 3 where y'_2 is a left base and y''_5 is a right base.

Thus, the edges of the radial composition that intersect segment (x_i, x_{i+1}) are of the form (y_j, y'_j) or (y_j, y''_j). Note that y_1 is either x_i or the point closest to x_i on the edge. For notational convenience, if $y_1 \neq x_i$, relabel x_i as y_0. Let f be the face of the radial decomposition on the boundary of which (y_0, y_1) lies. Let y'_0 be the intersection of qy_0 with the boundary of f excluding the edge of f containing (y_0, y_1). We call this operation *extending* x_i. Similarly, if $y_j \neq x_{i+1}$, relabel x_{i+1} as y_{j+1} and compute the edge (y_{j+1}, y'_{j+1}), i.e. *extend* x_{i+1}.

We are now in a position to describe the extension of the visibility of triangle $\triangle(qx_ix_{i+1})$ when neither x_i nor x_{i+1} is a base vertex. The set of triangles are $\triangle(qy'_ky'_{k+1})$ and $\triangle(qy''_ky'_{k+1})$ (when y''_k exists). The union of these triangles is the 1-visible region of q in $\mathcal{C}(q, x_i, x_{i+1})$. Furthermore, notice that each triangle $\triangle(qy'_ky'_{k+1})$ (respectively, $\triangle(qy''_ky'_{k+1})$) has the property that (y'_k, y'_{k+1}) (respectively, (y''_k, y'_{k+1})) is on the boundary of P. This is what allows us to continue incrementally since at each stage we extend the visibility of a triangle $\triangle(qab)$ where (a, b) is on the boundary of P.

Now, if x_i is a base vertex, then it must be a right base. Of the two edges of P incident on x_i, let e be the one further from q. The procedure to extend $\triangle(qx_ix_{i+1})$ is identical except that we only extend x_i when $x_{i+1} \in e$. Similarly, if x_{i+1} is a base vertex, then it must be a left base. Of the two edges of P

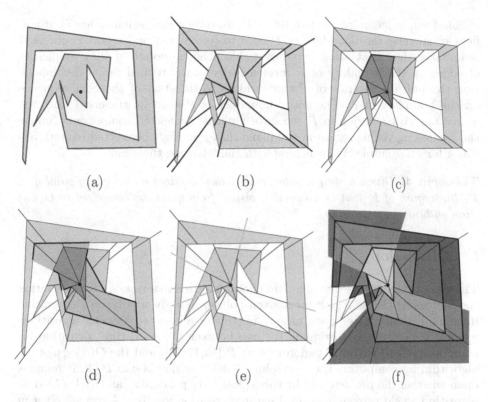

(a)   (b)   (c)

(d)   (e)   (f)

Fig. 4. (a) a simple polygon P and a query point q; (b) the radial decomposition of P; (c) the 0-visibility polygon, $\mathcal{V}_0(q)$, of q in P computed in the first iteration; (d) the 1-visibility polygon, $\mathcal{V}_1(q)$, of q in P computed in the second iteration, with extended edges highlighted in light blue; (e) the refined radial decomposition, with extended edges highlighted in light blue; (f) the 4-visibility polygon, $\mathcal{V}_4(q)$, of q in P computed in the fourth iteration, with the algorithm's output highlighted in black (two components of the boundary of $\mathcal{V}_4(q) \cap P$), and cells of the decomposition with depth ≤ 4 coloured by depth, as computed by the algorithm. (Color figure online)

incident on x_{i+1}, let e be the one further from q. Again, the procedure to extend $\triangle(qx_ix_{i+1})$ is identical except that we only extend x_{i+1} when $x_i \in e$.

The general algorithm proceeds as follows. At iteration i, the visibility region $\mathcal{V}_i(q)$ is represented as a collection of triangles around q with the property that the edge of the triangle opposite q is on the boundary of P and it is the edge blocking visibility. We wish to extend past this edge to compute $\mathcal{V}_{i+1}(q)$ from $\mathcal{V}_i(q)$. To do this, we extend each triangle in $\mathcal{V}_i(q)$. There are at most $O(n)$ triangles at each level. Therefore, the total time to extend all the triangles in $\mathcal{V}_i(q)$ is linear. Thus, we can compute $\mathcal{V}_{i+1}(q)$ from $\mathcal{V}_i(q)$ in $O(n)$ time and computing $V_k(q)$ takes $O(nk)$ time since we repeat this process k times.

The algorithm can report either only the subregion of P that is k-crossing visible from q, i.e., $\mathcal{V}_k(q) \cap P$, or the entire region of the plane that is k-crossing visible from q, including parts outside P. To obtain the region inside P, it suffices to traverse the boundary of P once to reconstruct and report portions of boundary edges that are k-crossing visible. The endpoints of these sequences of edges on the boundary of P meet an edge of the refined radial decomposition through the interior of P that bridges to the start of the next sequence on the boundary of P. The entire boundary of P must be traversed since the k-crossing visible region in P can have multiple connected components (unlike the k-crossing visible region in the plane that is a single connected region). See Fig. 4 for an example. We conclude with the following theorem.

Theorem 4. *Given a simple polygon P with n vertices and a query point q in P, the region of P that is k-crossing visible from q can be computed in $O(kn)$ time without preprocessing.*

5 Discussion

This paper presents the first algorithm parameterized in terms of k for computing the k-crossing visible region for a given point q in a given polygon P, resulting in asymptotically faster worst-case running time relative to previous algorithms when k is $o(\log n)$, and bridging the gap between the $O(n)$-time algorithm for computing the 0-visibility region of q in P [13,17,18], and the $O(n \log n)$-time algorithm for computing the k-crossing visibility region of q in P [3]. It remains open whether the problem can be solved faster. In particular, an $O(n \log k)$-time algorithm would provide a natural parameterization for all k. Alternatively, can a lower bound of $\Omega(n \log n)$ be shown on the worst-case time when k is $\omega(\log n)$?

References

1. Aichholzer, O., Fabila-Monroy, R., Flores-Peñaloza, D., Hackl, T., Huemer, C., Urrutia, J., Vogtenhuber, B.: Modem illumination of monotone polygons. Comput. Geometry **68**, 101–118 (2018)
2. Aronov, B., Guibas, L.J., Teichmann, M., Zhang, L.: Visibility queries and maintenance in simple polygons. Discrete Comput. Geom. **27**(4), 461–483 (2002)
3. Bahoo, Y., Banyassady, B., Bose, P., Durocher, S., Mulzer, W.: A time-space trade-off for computing the k-visibility region of a point in a polygon. Theor. Comput. Sci. 9 p. (2018, to appear)
4. Bajuelos, A.L., Canales, S., Hernández, G., Martins, M.: A hybrid metaheuristic strategy for covering with wireless devices. J. Univ. Comput. Sci. **18**(14), 1906–1932 (2012)
5. Ballinger, B., et al.: Coverage with k-transmitters in the presence of obstacles. J. Comb. Optim. **25**(2), 208–233 (2013)
6. Bose, P., Lubiw, A., Munro, J.I.: Efficient visibility queries in simple polygons. Comput. Geom. **23**(3), 313–335 (2002)
7. Cannon, S., Fai, T., Iwerks, J., Leopold, U., Schmidt, C.: Combinatorics and complexity of guarding polygons with edge and point 2-transmitters. arXiv preprint arXiv:1503.05681 (2015)

8. Chang, H.C., Erickson, J., Xu, C.: Detecting weakly simple polygons. In: Proceedings of the 26th ACM-SIAM Symposium on Discrete Algorithms (SODA 2014), pp. 1655–1670 (2014)
9. Chazelle, B.: Triangulating a simple polygon in linear time. Discrete Comput. Geom. **6**(3), 485–524 (1991)
10. Davis, L.S., Benedikt, M.L.: Computational models of space: isovists and isovist fields. Comput. Graph. Image Process. **11**(1), 49–72 (1979)
11. Dean, J.A., Lingas, A., Sack, J.R.: Recognizing polygons, or how to spy. Vis. Comput. **3**(6), 344–355 (1988)
12. Duque, F., Hidalgo-Toscano, C.: An upper bound on the k-modem illumination problem. arXiv preprint arXiv:1410.4099 (2014)
13. El Gindy, H., Avis, D.: A linear algorithm for computing the visibility polygon from a point. J. Algorithms **2**(2), 186–197 (1981)
14. Fabila-Monroy, R., Vargas, A., Urrutia, J.: On modem illumination problems. In: Proc. XIII Encuentros de Geometría Computacional (EGC 2009) (2009)
15. Hoffmann, K., Mehlhorn, K., Rosenstiehl, P., Tarjan, R.E.: Sorting Jordan sequences in linear time using level-linked search trees. Inf. Control **68**(1–3), 170–184 (1986)
16. Huang, H., Ni, C.C., Ban, X., Gao, J., Schneider, A.T., Lin, S.: Connected wireless camera network deployment with visibility coverage. In: Proceedings of the IEEE International Conference on Computer Communications (INFOCOM 2014), pp. 1204–1212 (2014)
17. Joe, B., Simpson, R.B.: Corrections to Lee's visibility polygon algorithm. BIT Numer. Math. **27**(4), 458–473 (1987)
18. Lee, D.T.: Visibility of a simple polygon. Comput. Vis. Graph. Image Process. **22**(2), 207–221 (1983)
19. Meguerdichian, S., Koushanfar, F., Qu, G., Potkonjak, M.: Exposure in wireless ad-hoc sensor networks. In: Proceedings of the 7th ACM International Conference on Mobile Computing and Networking (MOBICOM 2001), pp. 139–150. ACM (2001)
20. Mouawad, N., Shermer, T.C.: The Superman problem. Vis. Comput. **10**(8), 459–473 (1994)
21. Murray, A.T., Kim, K., Davis, J.W., Machiraju, R., Parent, R.: Coverage optimization to support security monitoring. Comput. Environ. Urban Syst. **31**(2), 133–147 (2007)
22. Rosenstiehl, P.: Planar permutations defined by two intersecting Jordan curves. Graph Theory and Combinatorics, pp. 259–271 (1984)
23. Wang, Y.C., Hu, C.C., Tseng, Y.C.: Efficient deployment algorithms for ensuring coverage and connectivity of wireless sensor networks. In: Proceedings of the 1st IEEE Conference on Wireless Internet (WICON 2005), pp. 114–121 (2005)

An Improved Scheme in the Two Query Adaptive Bitprobe Model

Mirza Galib Anwarul Husain Baig, Deepanjan Kesh[✉], and Chirag Sodani

Indian Institute of Technology Guwahati, Guwahati 781039, Assam, India
{mirza.baig,deepkesh,chirag.sodani}@iitg.ac.in

Abstract. In this paper, we look into the adaptive bitprobe model that stores subsets of size at most four from a universe of size m, and answers membership queries using two bitprobes. We propose a scheme that stores arbitrary subsets of size four using $\mathcal{O}(m^{5/6})$ amount of space. This improves upon the non-explicit scheme proposed by Garg and Radhakrishnan [5] which uses $\mathcal{O}(m^{16/17})$ amount of space, and the explicit scheme proposed by Garg [4] which uses $\mathcal{O}(m^{14/15})$ amount of space. The proposed scheme also answers an open problem posed by Nicholson [8] in the affirmative. Furthermore, we look into a counterexample that shows that our proposed scheme cannot be used to store five or more elements.

Keywords: Data structure · Set membership problem ·
Bitprobe model · Adaptive scheme

1 Introduction

Consider the following static membership problem – given a universe $\mathcal{U}$ containing m elements, we want to store an arbitrary subset $\mathcal{S}$ of $\mathcal{U}$ whose size is at most n, such that we can answer membership queries of the form "Is x in $\mathcal{S}$?" Solutions to problems of this nature are called *schemes* in the literature. The resources that are considered to evaluate the schemes are the size of the data structure devised to store the subset $\mathcal{S}$, and the number of bits read of the data structure to answer the membership queries, called *bitprobes*. The notations for the space used and the number of bitprobes required are s and t, respectively. This model of the static membership problem is called the *bitprobe model*.

Schemes in the bitprobe model are classified as *adaptive* and *non-adaptive*. If the location where the current bitprobe is going to be depends on the answers obtained from the previous bitprobes, then such schemes are called *adaptive schemes*. On the other hand, if the location of the current bitprobe is independent of the answers obtained in the previous bitprobes, then such schemes are called *non-adaptive schemes*. Radhakrishnan *et al.* [9] introduced the notation $(n, m, s, t)_A$ and $(n, m, s, t)_N$ to denote the adaptive and non-adaptive schemes, respectively. Sometimes the space requirement of the two classes of schemes will also be denoted as $s_A(n, m, t)$ and $s_N(n, m, t)$, respectively.

© Springer Nature Switzerland AG 2019
C. J. Colbourn et al. (Eds.): IWOCA 2019, LNCS 11638, pp. 22–34, 2019.
https://doi.org/10.1007/978-3-030-25005-8_3

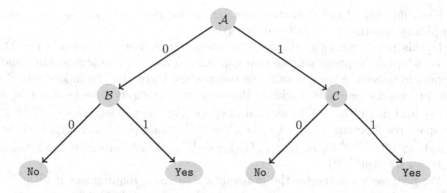

Fig. 1. The decision tree of an element.

1.1   The Bitprobe Model

The scheme presented in this paper is an adaptive scheme that uses two bitprobes to answer membership queries. We now discuss in detail the bitprobe model in the context of two adaptive bitprobes.

The data structure in this model consists of three tables – $\mathcal{A}, \mathcal{B}$, and $\mathcal{C}$ – arranged as shown in Fig. 1. Any element e in the universe $\mathcal{U}$ has a location in each of these three tables, which are denoted by $\mathcal{A}(e), \mathcal{B}(e)$, and $\mathcal{C}(e)$. By a little abuse of notation, we will use the same symbols to denote the bits stored in those locations.

Any bitprobe scheme has two components – the *storage* scheme, and the *query* scheme. Given a subset $\mathcal{S}$, the storage scheme sets the bits in the three tables such that the membership queries can be answered correctly. The flow of the query scheme is traditionally captured in a tree structure, called the *decision tree* of the scheme (Fig. 1). It works as follows. Given a query "Is x in $\mathcal{S}$?", the first bitprobe is made in table $\mathcal{A}$ at location $\mathcal{A}(x)$. If the bit stored is 0, the second query is made in table $\mathcal{B}$, else it is made in table $\mathcal{C}$. If the answer received in the second query is 1, then we declare that the element x is a member of $\mathcal{S}$, otherwise we declare that it is not.

1.2   The Problem Statement

As alluded to earlier, we look into adaptive schemes with two bitprobes ($t = 2$). When the subset size is one ($n = 1$), the problem is well understood – the space required by the data structure is $\Omega(m^{1/2})$, and we have a scheme that matches this bound [1,7].

For subsets of size two ($n = 2$), Radhakrishnan *et al.* [9] proposed a scheme that takes $\mathcal{O}(m^{2/3})$ amount of space, and further conjectured that it is the minimum amount of space required for any scheme. Though progress has been made to prove the conjecture [9,10], it as yet remains unproven.

For subsets of size three ($n = 3$), Baig and Kesh [2] have recently proposed a scheme that takes $\mathcal{O}(m^{2/3})$ amount of space. It has been subsequently proven by Kesh [6] that $\Omega(m^{2/3})$ is the lower bound for this problem. So, the space complexity question for $n = 3$ stands settled.

In this paper, we look into problem where the subset size is four ($n = 4$), i.e. an adaptive bitprobe scheme that can store subsets of size atmost four, and answers membership queries using two bitprobes. Garg and Radhakrishnan [5] have proposed a generalised scheme that can store arbitrary subsets of size $n(<\log m)$, and uses $\mathcal{O}(m^{1-\frac{1}{4n+1}})$ amount of space. For the particular case of $n = 4$, the space requirement turns out to be $\mathcal{O}(m^{16/17})$. Garg [4] further improved the bounds to $\mathcal{O}(m^{1-\frac{1}{4n-1}})$(for $n < (1/4)(\log m)^{1/3}$)., which improved the scheme for $n = 4$ to $\mathcal{O}(m^{14/15})$.

We propose a scheme for the problem whose space requirement is $\mathcal{O}(m^{5/6})$ (Theorem 2), thus improving upon the existing schemes in the literature. Our claim is the following:

$$s_A(4, m, 2) = \mathcal{O}(m^{5/6}).(\text{Theorem 2})$$

The existence of such a scheme also answers in the affirmative an open problem posed by Nicholson [8] which asked if a scheme using the idea of blocks due to Radhakrishnan et $al.$ [9] exists that stores four elements and answers membership queries using two bitprobes. As the description of our data structure in the following section would show that our scheme extends the ideas of blocks and superblocks using a geometric approach to solve the problem.

Finally, in Sect. 5 we provide an instance of a five-element subset of the universe $\mathcal{U}$ which cannot be stored correctly in our data structure, illustrating that a different construction is required to accommodate subsets of larger size.

2   Our Data Structure

In this section, we provide a detailed description of our data structure. To achieve a space bound of $o(m)$, more than one element must necessarily share the same location in each of the three tables. We discuss how we arrange the elements of the universe $\mathcal{U}$, and which of the elements of the universe share the same location in any given table.

Along with the arrangement of elements, we will also talk about the size of our data structure. The next few sections prove the following theorem.

Theorem 1. *The size of our data structure is* $\mathcal{O}(m^{5/6})$.

2.1   Table $\mathcal{A}$

Suppose we are given the following universe of elements –

$$\mathcal{U} = \{\ 1, 2, 3, \ldots, m\ \}.$$

We partition the m elements of the universe into sets of size $m^{1/6}$. Borrowing the terminology from Radhakrishnan et $al.$ [9], we will refer to these sets as $blocks$. It follows that the total number of blocks in our universe is $m^{5/6}$.

The elements within a block are numbered as $1, 2, 3, \ldots, m^{1/6}$. We refer to these numbers as the *index* of an element within a block. So, an element of $\mathcal{U}$ can be addressed by the number of the block to which it belongs, and its index within that block.

In table $\mathcal{A}$ of our data structure, we will have one bit for every block in our universe. As there are $m^{5/6}$ blocks, the size of table $\mathcal{A}$ is $m^{5/6}$.

2.2 Superblocks

The blocks in our universe are partitioned into sets of size $m^{4/6}$. Radhakrishnan *et al.* [9] used the term *superblocks* to refer to these sets of blocks, and we will do the same in our discussion. As there are $m^{5/6}$ blocks, the number of superblocks thus formed is $m^{1/6}$. These superblocks are numbered as $1, 2, 3, \ldots, m^{1/6}$.

For a given superblock, we arrange the $m^{4/6}$ blocks that it contains into a square grid, whose sides are of size $m^{2/6}$. The blocks of the superblock are placed on the integral points of the grid. The grid is placed at the origin of a two-dimensional coordinate space with its sides parallel to the coordinate axes. This gives a unique coordinate to each of the integral points of the grid, and thus to the blocks placed on those points. It follows that if (x, y) is the coordinate of a point on the grid, then $0 \leq x, y < m^{2/6}$.

We can now have a natural way of addressing the blocks of a given superblock – we will use the x-coordinate and the y-coordinate of the point on which the block lies. So, a given block can be uniquely identified by the number of the superblock to which it belongs, and the x and y coordinates of the point on which it lies. Henceforth, we will address any block by a three-tuple of the form (s, x, y), where the s is its superblock number, and (x, y) are the coordinates of the point on which it lies.

To address a particular element of the universe, apart from specifying the block to which it belongs, we need to further state its index within that block. So, an element will be addressed by a four-tuple such as (s, x, y, i), where the first three components specify the block to which it belongs, and the fourth component specifies its index.

2.3 Table $\mathcal{C}$

Table $\mathcal{C}$ of our data structure has the space to store one block for every possible point of the grid (described in the previous section). So, for the coordinate (x, y) of the grid, table $\mathcal{C}$ has space to store one block; similarly for all other coordinates. As every superblock has one block with coordinate (x, y), all of these blocks share the same location in table $\mathcal{C}$. So, we can imagine table $\mathcal{C}$ as a square grid containing $m^{4/6}$ points, where each point can store one block.

There are a total of $m^{4/6}$ points in the grid, and the size of a block is $m^{1/6}$, so the space required by table $\mathcal{C}$ is $m^{5/6}$.

2.4   Lines for Superblocks

Given a superblock whose number is i, we associate a certain number of lines with this superblock each of whose slopes are $1/i$. In the grid arrangement of the superblock (Sect. 2.2), we draw enough of these lines of slope $1/i$ so that every grid point falls on one of these lines. Figure 2 shows the grid and the lines.

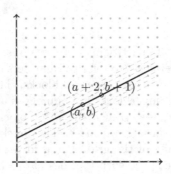

Fig. 2. The figure shows the grid for superblock 2, and some of the lines with slope $1/2$. Note that the line passing through (a, b) intersects the y-axis at a non-integral point.

So, all lines of a given superblock has the same slope, and lines from different superblocks have different slopes. As there are $m^{1/6}$ superblocks, and they are numbered $1, 2, \ldots, m^{1/6}$, so, we have the slopes of the lines vary as

$$0 < i \leq m^{1/6}. \tag{1}$$

There are two issues to consider – the number of lines needed to cover every point of the grid, and the purpose of these lines. We address the issue of the count of the lines in this section, and that of the purpose of the lines in the next.

We introduce the notation $l_i(a, b)$ to denote the line that has slope $1/i$, and passes through the point (a, b). We now define the collection of all lines of slope $1/i$ that we are going to draw for the superblock i.

$$L_i = \left\{ \, l_i(a, 0) \mid a \in \mathbb{Z}, \; -i(m^{2/6} - 1) \leq a < m^{2/6} \, \right\}. \tag{2}$$

In the following three lemmas, we show the properties of this set of lines – they follow from elementary coordinate geometry.

Lemma 1. *Every line of L_i contains at least one point of the grid.*

Lemma 2. *Every point of the grid belongs to some line of L_i.*

Lemma 3. $|\, L_i \,| = (i + 1)(m^{2/6} - 1) + 1$.

Proof. The equality is a direct consequence of the definition of L_i (Eq. 2).

2.5   Table $\mathcal{B}$

In table $\mathcal{B}$, we have space to store one block for every line of every superblock. That means that for a superblock, say i, all of its blocks that fall on the line $l_i(a, b)$ share the same block in table $\mathcal{B}$; and the same is true for all lines of every superblock.

The ith superblock contains $\mid L_i \mid = (i + 1)(m^{2/6} - 1) + 1$ lines (Lemma 3), so the total number of lines from all of the superblocks is

$$\mid L_1 \mid + \mid L_2 \mid + \cdots + \mid L_{m^{1/6}} \mid$$
$$= \sum_{i=1}^{m^{1/6}} \left((i + 1)(m^{2/6} - 1) + 1 \right)$$
$$= \left(\frac{(m^{1/6})(m^{1/6}+1)}{2} + m^{1/6} \right)(m^{2/6} - 1) + m^{1/6}$$
$$= \mathcal{O}(m^{4/6}).$$

As mentioned earlier, we reserve space for one block for each of these lines. Combined with the fact that the size of a block is $m^{1/6}$, we have

$$|\mathcal{C}| = \mathcal{O}(m^{5/6}).$$

2.6   Notations

As described in Sect. 2.2, any element of the universe $\mathcal{U}$ can be addressed by a four-tuple, such as (s, x, y, i), where s is the superblock to which it belongs, (x, y) are the coordinates of its block within that superblock, and i is its index within the block.

Table $\mathcal{A}$ has one bit for each block, so all elements of a block will query the same location. As the block number of the element (s, x, y, i) is (s, x, y), so the bit corresponding to the element is $\mathcal{A}(s, x, y)$; or in other words, the element (s, x, y, i) will query the location $\mathcal{A}(s, x, y)$ in table $\mathcal{A}$.

In table $\mathcal{C}$, there is space for one block for every possible coordinates of the grid. The coordinates of the element (s, x, y, i) is (x, y), and $\mathcal{C}$ has space to store an entire block for this coordinate. So, there is one bit for every element of a block, or, in other words, every index of a block. So, the bit corresponding to the element (s, x, y, i) is $\mathcal{C}(x, y, i)$.

Table $\mathcal{B}$ has a block reserved for every line of every superblock. The element (s, x, y, i) belongs to the line $l_s(x, y)$, and thus table $\mathcal{B}$ has space to store one block corresponding to this line. As the index of the element is i, so the bit corresponding to the element in table $\mathcal{B}$ is $\mathcal{B}(l_s(x, y), i)$.

3   Query Scheme

The query scheme is easy enough to describe once the data structure has been finalised; it follows the decision tree as discussed earlier (Fig. 1). Suppose we want to answer the following membership query – "Is (s, x, y, i) in $\mathcal{S}$?" We would make the first query in table $\mathcal{A}$ at location $\mathcal{A}(s,x,y)$. If the bit stored at that location is 0, we query in table $\mathcal{B}$ at $\mathcal{B}(l_s(x,y),i)$, otherwise we query table $\mathcal{C}$ at $\mathcal{C}(x,y,i)$. If the answer from the second query is 1, then we declare the element to be a member of $\mathcal{S}$, else we declare that it is not a member of $\mathcal{S}$.

4   The Storage Scheme

The essence of any bitprobe scheme is the storage scheme, i.e. given a subset $\mathcal{S}$ of the universe $\mathcal{U}$, how the bits of the data structure are set such that the query scheme answers membership questions correctly. We start the description of the storage scheme by giving an intuition for its construction.

4.1   Intuition

The basic unit of storage in the tables $\mathcal{B}$ and $\mathcal{C}$ of our data structure, in some sense, is a block – table $\mathcal{B}$ can store one block of any line of any superblock, and table $\mathcal{C}$ can store one block of a given coordinate from any superblock. We show next that our storage scheme must ensure that an empty and a non-empty block cannot be stored together in a table.

Suppose, the block (s,x,y) of table $\mathcal{A}$ is non-empty, and it contains the member (s,x,y,i) of subset $\mathcal{S}$. If we decide to store this member in table $\mathcal{B}$, then we have to store the block (s,x,y) in table $\mathcal{B}$. So, we have to set in table $\mathcal{A}$ the following – $\mathcal{A}(s,x,y) = 0$. Thus, (s,x,y,i) upon first query will get a 0 and go to table $\mathcal{B}$. In table $\mathcal{B}$, we store the block (s,x,y) at the storage reserved for the line $l_s(x,y)$. Particularly, we have to set $\mathcal{B}(l_s(x,y),i) = 1$.

If (s,x',y') is a block that is empty, i.e. it does not contain any member of $\mathcal{S}$, and it falls on the aforementioned line, i.e. $l_s(x',y') = l_s(x,y)$, then we cannot store this block in table $\mathcal{B}$, and hence $\mathcal{A}(s,x',y')$ must be set to 1. If this is not the case, and $\mathcal{A}(s,x',y') = 0$, then the first query for the element (s,x',y',i) will get a 0, go to table $\mathcal{B}$ and query the location $\mathcal{B}(l_s(x',y'),i)$ which is same as $\mathcal{B}(l_s(x,y),i)$. We have set this bit to 1, and we would incorrectly deduce that (s,x',y',i) is a member of $\mathcal{S}$.

The same discussion holds true for table $\mathcal{C}$. If we decide to store the block (s,x,y) in table $\mathcal{C}$, we have to set $\mathcal{A}(s,x,y)$ to 1. In table $\mathcal{C}$, we have space reserved for every possible coordinate for a block, and we would store the block at the coordinate (x,y); particularly, we would set $\mathcal{C}(x,y,i)$ to 1. This implies that all empty blocks from other superblocks having the same coordinate cannot be stored in table $\mathcal{C}$, and hence must necessarily be stored in table $\mathcal{B}$. To take an example, if (s',x,y) is empty, then it must stored it table $\mathcal{B}$, and hence $\mathcal{A}(s',x,y) = 0$.

To summarise, for any configuration of the members of subset $\mathcal{S}$, as long as we are able to keep the empty and the non-empty blocks separate, our scheme will work correctly. For the reasons discussed above, we note the following.

1. We have to keep the non-empty blocks and empty blocks separate.

2. We have to keep the non-empty blocks separate from each other; and

3. The empty blocks can be stored together.

Our entire description of the storage scheme would emphasize on how to achieve the aforementioned objective.

4.2 Description

Let the four members of subset S be

$$S = \Big\{ (s_1, x_1, y_1, i_1),\ (s_2, x_2, y_2, i_2),\ (s_3, x_3, y_3, i_3),\ (s_4, x_4, y_4, i_4) \Big\}.$$

So, the relevant blocks are

$$\Big\{ (s_1, x_1, y_1),\ (s_2, x_2, y_2),\ (s_3, x_3, y_3),\ (s_4, x_4, y_4) \Big\},$$

and the relevant lines are

$$\Big\{ l_{s_1}(x_1, y_1),\ l_{s_2}(x_2, y_2),\ l_{s_3}(x_3, y_3),\ l_{s_4}(x_4, y_4) \Big\}.$$

In the discussion below, we assume that no two members of S belong to the same block. This implies that there are exactly four non-empty blocks. The scenario where a block contains multiple members of S is handled in Sect. 4.3.

The lines for the members of S need not be distinct, say when two elements belong to the same superblock and fall on the same line. We divide the description of our storage scheme into several cases based on the number of distinct lines we have due to the members of S, and for each of those cases, we provide the proof of correctness alongside it.

We provide the detailed description of the cases when there are four distinct lines or when there is one line. The extended version of this paper (Baig *et. al* [3]) contains the cases of three lines and two lines. The cases described here would illustrate how to arrange the elements and how to argue its correctness.

Case I. Suppose we have four distinct lines for the four members of S. The slopes of some of these lines could be same, or they could all be different. We know that all lines of a given superblock have the same slope, and lines from different superblocks have different slopes (Sect. 2.4). We also know that if two of these lines, say $l_{s_1}(x_1, y_1)$ and $l_{s_2}(x_2, y_2)$, have the same slope, then the corresponding members of S belong to the same superblock, i.e. $s_1 = s_2$. On the other hand, if their slopes are distinct, then they belong to different superblocks, and consequently, $s_1 \neq s_2$.

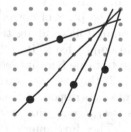

Table $\mathcal{B}$ has space to store one block for every line in every superblock. As the lines for the four members of S are distinct, the space reserved for the lines are also distinct. So we can store the four non-empty blocks in table $\mathcal{B}$, and all of the empty blocks in table $\mathcal{C}$.

To achieve the objective, we set $A(s_j, x_j, y_j) = 0$ for $1 \le j \le 4$, and set the bits in table A for every other block to 1. In table B, we set the bits $B(l_{s_j}(x_j, y_j), i_j) = 1$, for $1 \le j \le 4$, and all the rest of the bits to 0. In table C, all the bits are set to 0.

So, if e is an element that belongs to an empty block, it would, according to the assignment above, get a 1 upon its first query in table A. Its second query will be in table C, and as all the bits of table C are set to 0, we would conclude that the element e is not a member of S.

Suppose, (s, x, y, i) be an element that belongs to one of the non-empty blocks. Then, its coordinates must correspond to one of the four members of S. Without loss of generality let us assume that $s = s_1, x = x_1$, and $y = y_1$.

It follows that $A(s, x, y)$, which is same as $A(s_1, x_1, y_1)$, is 0, and hence the second query for this element will be in table B. The line corresponding to the element is $l_s(x, y)$, which is same as $l_{s_1}(x_1, y_1)$, and hence the second query will be at the location $B(l_s(x, y), i) = B(l_{s_1}(x_1, y_1), i)$. As the four lines for the four members of S are distinct, so $B(l_{s_1}(x_1, y_1), i)$ will be 1 if and only if $i = i_1$. So, we will get a Yes answer for your query if and only if the element (s, x, y, i) is actually the element (s_1, x_1, y_1, i_1), a member of S.

Case II. Let us consider the case when there is just one line for the four members of S. As all of their lines are identical, and consequently, the slopes of the lines are the same, all the elements must belong to the same superblock. So, we have $s_1 = s_2 = s_3 = s_4$.

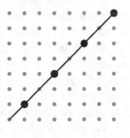

As all the non-empty blocks belong to the same superblock, all of their coordinates must be distinct. Table C can store one block for each distinct coordinate of the grid, and hence we can store the four non-empty blocks there. All the empty blocks will be stored in table B.

To this end, we set $A(s_j, x_j, y_j) = 1$ for $1 \le j \le 4$, and the rest of the bits of table A, which correspond to the empty blocks, to 0. In table B, all bits are set to 0. In table C, the bits corresponding to the four elements are set to 1, i.e. $C(x_j, y_j, i_j) = 1$ for $1 \le j \le 4$. The rest of the bits of table C are set to 0.

The proof of correctness follows directly from the assignment, and the reasoning follows along the lines of the previous case. If the element e belongs to an empty block, it will get a 0 from table A upon its first query, consequently go to table B for its second query, and get a 0, implying e is not a member of S.

If the element (s, x, y, i) belongs to a non-empty block, then its coordinates must correspond to one of the members of S. Without loss of generality, let $s = s_1, x = x_1$, and $y = y_1$.

The first query of the element will be at the location $A(s, x, y) = A(s_1, x_1, y_1)$, and hence it will get a 1 from table A, and go to table C for its second query. In this table, it will query the location $C(x, y, i)$, which is same as $C(x_1, y_1, i)$. As the coordinates of the four members of S are distinct, $C(x_1, y_1, i)$

will be 1 if and only if $i = i_1$. So, we get a 1 in the second query if and only if we have $(s, x, y, i) = (s_1, x_1, y_1, i_1)$, a member of $\mathcal{S}$.

4.3   Blocks with Multiple Members

In the discussion above, we had assumed that each block can contain at most one member of the subset $\mathcal{S}$, and we have shown for every configuration of the members of $\mathcal{S}$, the bits of the data structure can be so arranged that the membership queries are answered correctly.

In general, a single block can contain upto four members of $\mathcal{S}$, and we need to propose a assignment for such a scenario. As has been noted in the previous section, our basic unit of storage is a block and we differentiate between empty and non-empty blocks. At a given location in table $\mathcal{B}$ or $\mathcal{C}$, a block is stored in its entirety, or it isn't stored at all. This implies that the number of members of $\mathcal{S}$ a non-empty block contains is of no consequence, as we always store an entire block. The scheme from the previous section would thus hold true for blocks containing multiple members.

We now summarise the result in the theorem below.

Theorem 2. *There is an explicit adaptive scheme that stores subsets of size at most four and answers membership queries using two bitprobes such that*

$$s_A(4, m, 2) = \mathcal{O}(m^{5/6}).$$

5   Counterexample

We now provide an instance of a five-member subset of the universe $\mathcal{U}$ which cannot be stored correctly using our scheme; that is to say, if the storage scheme does indeed store the five elements in our data structure, queries for certain elements will be answered incorrectly.

5.1   The Arrangment

Consider four lines from four different superblocks which are arranged as shown in Fig. 3. Let us suppose that the four superblocks are s_1, s_2, s_3, and s_4, and the labels of the lines are L_1, L_2, L_3, and L_4, respectively. We will put in $\mathcal{S}$ one element each from the first three superblocks, and two elements from the fourth superblock.

Our subset $\mathcal{S}$ will contain the elements e_1 and e_2 from the superblocks s_1 and s_2, respectively. These elements have the property that the blocks they belong to share the same coordinates, and hence lie on the intersection of the lines L_1 and L_2. The fact that they have the same coordinates also implies thay they share the same location in table $\mathcal{C}$. Let the elements be $e_1 = (s_1, x, y, i_1)$ and $e_2 = (s_2, x, y, i_2)$. We would also have $i_1 \neq i_2$. This would imply that the two non-empty blocks (s_1, x, y) and (s_2, x, y) cannot both be stored in table $\mathcal{C}$.

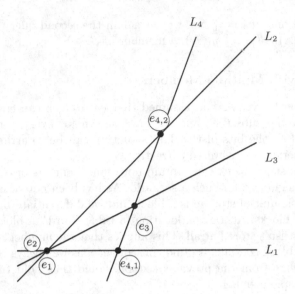

Fig. 3. Counterexample

Consider that block of superblock s_3 that lies on the intersection of the lines L_3 and L_4. We will put one element from that block in our subset $\mathcal{S}$. Let that element be $e_3 = (s_3, x_3, y_3, i_3)$.

Finally we will put two elements of the superblock s_4 in $\mathcal{S}$ – one element from that block of s_4 which lies on the intersection of the lines L_4 and L_1, namely $e_{4,1}$, and another from the block of s_4 which lies on the intersection of the lines L_4 and L_2, namely $e_{4,2}$. These two elements are described as $e_{4,1} = (s_4, x_{4,1}, y_{4,1}, i_{4,1})$ and $e_{4,2} = (s_4, x_{4,2}, y_{4,2}, i_{4,2})$.

5.2   The Contradiction

We can store the element e_1 of superblock s_1 in one of two tables $\mathcal{B}$ and $\mathcal{C}$. Let us assume that we store e_1 in table $\mathcal{B}$. As the block containing e_1 lies on the line L_1, we cannot store any of the other empty blocks on the line L_1 in table $\mathcal{B}$, and hence they must be stored in table $\mathcal{C}$.

The non-empty block of s_4 containing element $e_{4,1}$ which falls on the line L_1, then, cannot be stored in table $\mathcal{C}$, and hence must be stored in table $\mathcal{B}$. So, the other blocks of L_4 must be stored in table $\mathcal{C}$, including the block containing the element $e_{4,2}$.

The non-empty block of s_3 containing the element e_3 falls on the line L_4, and hence must be stored in table $\mathcal{B}$. So, all blocks on the line L_3 must now be store in table $\mathcal{C}$.

The element e_2 of the superblock s_2 falls on the line L_3 and hence must be stored in table $\mathcal{B}$. So, all blocks of line L_2 must be stored in table $\mathcal{C}$.

The block of s_4 containing the element $e_{4,2}$ must be stored in table $\mathcal{B}$ by the same argument as above. But we have already argued that $e_{4,2}$ has to be stored in table $\mathcal{C}$, and hence we arrive at a contradiction.

The preceding argument tells us that we cannot store the element e_1 in table $\mathcal{B}$. So, we must store it in table $\mathcal{C}$. If such is the case, and arguing as above, we can show that this results e_2 being stored in table $\mathcal{B}$, which results in $e_{4,2}$ being stored in table $\mathcal{B}$. This, in turn, results in e_3 being stored in table $\mathcal{B}$, which would force e_1 to be stored in table $\mathcal{B}$.

But we have started with the premise that e_1 is being stored in table $\mathcal{C}$, and again we reach a contradiction. So, we conclude that this arrangement of elements cannot be stored correctly in our data structure, and hence our data structure is not suitable for storing sets of size five or higher.

6   Conclusion

In this paper, we have proposed an adaptive scheme for storing subsets of size four and answering membership queries with two bitprobes that improves upon the existing schemes in the literature. This scheme also resolves an open problem due to Patrick K. Nicholson [8] about the existence of such a scheme that uses the ideas of blocks and superblocks due to Radhakrishnan et al. [9]. The technique used is that of arranging the blocks of a superblock in a two-dimensional grid, and grouping them along lines. We hope that this technique can be extended to store larger subsets by extending the idea of an arrangement in a two-dimensional grid to arrangements in three and higher dimensional grids.

References

1. Alon, N., Feige, U.: On the power of two, three and four probes. In: Proceedings of the Twentieth Annual ACM-SIAM Symposium on Discrete Algorithms, SODA 2009, New York, NY, USA, 4–6 January 2009, pp. 346–354 (2009)
2. Baig, M.G.A.H., Kesh, D.: Two new schemes in the bitprobe model. In: Rahman, M.S., Sung, W.-K., Uehara, R. (eds.) WALCOM 2018. LNCS, vol. 10755, pp. 68–79. Springer, Cham (2018). https://doi.org/10.1007/978-3-319-75172-6_7
3. Baig, M.G.A.H., Kesh, D., Sodani, C.: An improved scheme in the two query adaptive bitprobe model. CoRR abs/1812.04802 (2018). http://arxiv.org/abs/1812.04802
4. Garg, M.: The bit-probe complexity of set membership. Ph.D. thesis, School of Technology and Computer Science, Tata Institute of Fundamental Research, Homi Bhabha Road, Navy Nagar, Colaba, Mumbai 400005, India (2016)
5. Garg, M., Radhakrishnan, J.: Set membership with a few bit probes. In: Proceedings of the Twenty-Sixth Annual ACM-SIAM Symposium on Discrete Algorithms, SODA 2015, San Diego, CA, USA, 4–6 January 2015, pp. 776–784 (2015)
6. Kesh, D.: Space complexity of two adaptive bitprobe schemes storing three elements. In: 38th IARCS Annual Conference on Foundations of Software Technology and Theoretical Computer Science, FSTTCS 2018, Ahmedabad, India, 11–13 December 2018, pp. 12:1–12:12 (2018)

7. Lewenstein, M., Munro, J.I., Nicholson, P.K., Raman, V.: Improved explicit data structures in the bitprobe model. In: Schulz, A.S., Wagner, D. (eds.) ESA 2014. LNCS, vol. 8737, pp. 630–641. Springer, Heidelberg (2014). https://doi.org/10. 1007/978-3-662-44777-2_52

8. Nicholson, P.K.: Revisiting explicit adaptive two-probe schemes. Inf. Process. Lett. **143**, 1–3 (2019)

9. Radhakrishnan, J., Raman, V., Srinivasa Rao, S.: Explicit deterministic constructions for membership in the bitprobe model. In: auf der Heide, F.M. (ed.) ESA 2001. LNCS, vol. 2161, pp. 290–299. Springer, Heidelberg (2001). https://doi.org/ 10.1007/3-540-44676-1_24

10. Radhakrishnan, J., Shah, S., Shannigrahi, S.: Data structures for storing small sets in the bitprobe model. In: de Berg, M., Meyer, U. (eds.) ESA 2010. LNCS, vol. 6347, pp. 159–170. Springer, Heidelberg (2010). https://doi.org/10.1007/978-3-642-15781-3_14

On Erdős–Szekeres-Type Problems
for k-convex Point Sets

Martin Balko[1][✉][iD], Sujoy Bhore[2][iD], Leonardo Martínez Sandoval[3][iD],
and Pavel Valtr[1][iD]

[1] Department of Applied Mathematics, Charles University, Prague, Czech Republic
balko@kam.mff.cuni.cz
[2] Algorithms and Complexity Group, TU Wien, Vienna, Austria
sujoy.bhore@gmail.com
[3] Institut de Mathématiques de Jussieu – Paris Rive Gauche (UMR 7586),
Sorbonne Université, Paris, France
leomtz@im.unam.mx

Abstract. We study Erdős Szekeres-type problems for k-convex point sets, a recently introduced notion that naturally extends the concept of convex position. A finite set S of n points is k-convex if there exists a spanning simple polygonization of S such that the intersection of any straight line with its interior consists of at most k connected components. We address several open problems about k-convex point sets. In particular, we extend the well-known Erdős–Szekeres Theorem by showing that, for every fixed $k \in \mathbb{N}$, every set of n points in the plane in *general position* (with no three collinear points) contains a k-convex subset of size at least $\Omega(\log^k n)$. We also show that there are arbitrarily large 3-convex sets of n points in the plane in general position whose largest 1-convex subset has size $O(\log n)$. This gives a solution to a problem posed by Aichholzer et al. [2].

We prove that there is a constant $c > 0$ such that, for every $n \in \mathbb{N}$, there is a set S of n points in the plane in general position such that every 2-convex polygon spanned by at least $c \cdot \log n$ points from S contains a point of S in its interior. This matches an earlier upper bound by Aichholzer et al. [2] up to a multiplicative constant and answers another of their open problems.

The project leading to this application has received funding from European Research Council (ERC) under the European Union's Horizon 2020 research and innovation programme under grant agreement No. 678765. M. Balko and P. Valtr were supported by the grant no. 18-19158S of the Czech Science Foundation (GAČR). M. Balko and L. Martínez-Sandoval were supported by the grant 1452/15 from Israel Science Foundation. M. Balko was supported by Center for Foundations of Modern Computer Science (Charles University project UNCE/SCI/004). S. Bhore was supported by the Austrian Science Fund (FWF) under project number P31119. L. Martínez Sandoval was supported by the grant ANR-17-CE40-0018 of the French National Research Agency ANR (project CAPPS). This research was supported by the PRIMUS/17/SCI/3 project of Charles University.

© Springer Nature Switzerland AG 2019
C. J. Colbourn et al. (Eds.): IWOCA 2019, LNCS 11638, pp. 35–47, 2019.
https://doi.org/10.1007/978-3-030-25005-8_4

Keywords: Convex position · Point set · k-convex point set · k-convex polygon

1   Introduction

A set of points in the plane is in *convex position* if its points are vertices of a convex polygon. We say that a planar point set is in *general position* if it does not contain a collinear triple of points. A classical result by Erdős and Szekeres [6], called the *Erdős–Szekeres Theorem*, states that every set of n points in the plane in general position contains a set of $\Omega(\log n)$ points in convex position. Moreover, this result is asymptotically tight, with the strongest bounds given in the papers [7,10,15]. The Erdős–Szekeres Theorem, published in 1935, was one of the starting points of both discrete geometry and Ramsey theory. Since then, numerous variants of this result have been studied.

For example, in 1978, Erdős [5] asked about the growth rate of the smallest integers $h(m)$, $m \geq 3$, such that every set P of at least $h(m)$ points in the plane in general position contains an *m-hole* in P, that is, m points in convex position with no point of P in the interior of their convex hull. It is easy to show that $h(3) = 3$ and $h(4) = 5$ and Harborth [9] proved $h(5) = 10$. After this, the question about the existence of the numbers $h(m)$ was settled in two phases. First, in 1983, Horton [11] showed that there are arbitrarily large sets of points with no 7-holes, proving that $h(m)$ does not exist for $m \geq 7$. Around 25 years later, Gerken [8] and Nicolás [13] independently proved that every sufficiently large set of points in the plane in general position contains a 6-hole. In particular, $h(m)$ exists if and only if $m \leq 6$.

In this paper, we study variants of these classical problems for so-called k-convex point sets, a notion that was recently introduced by Aichholzer et al. [2] and that naturally extends the concept of convex position. We also address further open problems about k-convex point sets posed in [2].

Throughout the paper, we consider only finite sets of points in the plane in general position. We use ∂S to denote the boundary of a simple polygon S. For a line segment s, we use $\overline{s}$ to denote the supporting line of s. A line ℓ *crosses* ∂S at a point v if ℓ passes through v from the interior of S to the outside of S. All logarithms in the paper are base two.

2   Preliminaries

In 2012, Aichholzer et al. [1] introduced the following natural extension of convex polygons. For a positive integer k, a simple polygon S with vertices in general position is *k-convex* if no straight line intersects S in more than k connected components. This notion has been later transcribed to finite point sets [2]. A finite set P of points in the plane in general position is *k-convex* if P is a vertex set of a k-convex polygon. In other words, P is k-convex if there exists a spanning simple polygonization of P such that the intersection of any straight line with its interior consists of at most k connected components. It can be shown that a simple polygon S with vertices in general position is k-convex if and only if every line not containing a vertex of S intersects the boundary of S in at most $2k$ points (see Lemma 2).

The notion of k-convexity for point sets satisfies several natural properties. A point set is in convex position if and only if it is 1-convex. Clearly, for each $k \in \mathbb{N}$, every k-convex point set is $(k+1)$-convex. Aichholzer et al. [2, Lemma 2] showed that every subset of a k-convex point set is also k-convex. It is known that every set of n points is k-convex for some $k = O(\sqrt{n})$ and this bound is tight up to a multiplicative constant in the worst case [2, Theorem 2]. Some further results about k-convex polygons and k-convex point sets can be found in [1–3].

Erdős–Szekeres-type questions were among the first problems about k-convex point sets considered in the literature. Aichholzer et al. [2] showed that every set of n points in general position contains a 2-convex subset of size at least $\Omega(\log^2 n)$ [2, Theorem 5] and that this bound is tight up to a multiplicative constant. This result led the authors to pose the following problem.

Problem 1 ([2, Open problem 4]). Let k and n be positive integers. Find the maximum integer $g(k, n)$ such that every set of n points contains a k-convex set of size $g(k, n)$.

Using this notation, their result gives $g(2, n) = \Theta(\log^2 n)$ and the Erdős–Szekeres Theorem gives $g(1, n) = \Theta(\log n)$. No nontrivial bounds were known for $g(k, n)$ with $k \geq 3$.

In a slightly different direction, it was shown that every 2-convex polygon with n vertices contains a 1-convex subset of at least $\lceil \sqrt{n}/2 \rceil$ vertices and that this bound is tight [1, Theorem 14]. In [2], the authors considered related variants of this result and posed the following problem.

Problem 2 ([2, Open problem 3]). Let j, k and n be positive integers. Find the maximum integer $f(k, n)$ such that every k-convex set of n points contains a 1-convex subset of size $f(k, n)$. More generally, find the maximum integer $f(k, j, n)$ such that every k-convex set of size n contains a j-convex subset of size $f(k, j, n)$.

By definition, $f(k, n) = f(k, 1, n)$ for all k and n. With this notation, the result by Aichholzer et al. [1, Theorem 14] gives $f(2, n) = f(2, 1, n) = \Theta(\sqrt{n})$. We trivially have $f(1, n) = f(1, 1, n) = n$ for every n. Since every set of n points is $(c\sqrt{n})$-convex for some constant $c > 0$ [2, Theorem 2], the Erdős–Szekeres Theorem gives $g(1, n) = f(k, n) = f(k, 1, n) = \Theta(\log n)$ for each $k \geq c\sqrt{n}$. By the previous results, we also know that, for $k \geq c\sqrt{n}$, we have $g(2, n) = f(k, 2, n) = \Theta(\log^2 n)$ and $g(j, n) = f(k, j, n)$ for each $j \in \mathbb{N}$.

For a point set P, a 2-convex polygon with vertices from P is *empty* in P if it contains no point of P in the interior. Concerning the question of Erdős about m-holes in point sets, Aichholzer et al. [2, Theorem 3] showed that every set P of n points in general position contains a 2-convex polygon that is empty in P and has size at least $\Omega(\log n)$. Using the tightness of the Erdős–Szekeres Theorem, they also proved that there are arbitrarily large point sets P of n points with no empty 2-convex polygon in P of size at least $c \cdot \log^2 n$ for some constant c. There is a gap between these two bounds and thus the authors posed the following problem.

Problem 3 ([2]). Close the gap between the $\Omega(\log n)$ and $O(\log^2 n)$ bounds for the size of empty 2-convex polygons in point sets of size n.

Let us also remark that it was shown by Aichholzer et al. [3] that every 2-convex point set of size n contains an m-hole for $m = \Omega(\log n)$ and that this bound is tight up to a multiplicative constant in the worst case.

The list of problems about k-convex point sets posed by Aichholzer et al. [2] contains several other interesting open questions.

3 Our Results

First, we prove the following extension of the Erdős–Szekeres Theorem for k-convex point sets.

Theorem 1. *Let k be a fixed positive integer. Then*

$$g(k, n) = \Omega(\log^k n).$$

That is, for every $n \in \mathbb{N}$, every set of n points in the plane in general position contains a k-convex subset of size at least $\Omega(\log^k n)$.

Note that Theorem 1 extends the result of Aichholzer et al. [2, Theorem 5] about the existence of large 2-convex point sets in general sets of n points. Unfortunately, we do not have matching upper bounds on the function $g(k, n)$. It follows from the proof of Theorem 13 in [3] that $g(k, n) = O(k\sqrt{n})$ for every $k \geq 3$.

We also address Problem 2. Using a variant of the sets defined by Erdős and Szekeres, we provide asymptotically tight estimates on the function $f(k, n)$ in the case $k \geq 3$.

Theorem 2. *There is a constant c such that, for every positive integer n, there are 3-convex sets of n points in the plane in general position with no 1-convex subset of size larger than $c \cdot \log n$.*

More precisely, for every $t \geq 3$, there is a 3-convex set of 2^{t-2} points in the plane in general position with no 1-convex subset of size t.

Thus $f(k, n) = O(\log n)$ for every integer k with $k \geq 3$. It follows from the Erdős–Szekeres Theorem that this bound is asymptotically tight. That is, $f(k, n) = \Theta(\log n)$ for $k \geq 3$. Therefore Theorem 2 asymptotically settles the first part of Problem 2. The statement in the second sentence of Theorem 2 implies the more precise bound $f(3, n) \leq \lceil \log(n) + 1 \rceil$. It also shows that the corresponding best known bound in the Erdős–Szekeres Theorem can be achieved by 3-convex sets. A famous conjecture of Erdős and Szekeres [7] states that this bound is tight for general sets. If true, the conjecture of Erdős and Szekeres together with our result would give the precise values $f(k, n) = \lceil \log(n) + 1 \rceil$ for any $n, k \geq 3$.

In the proof of Theorem 2 we define planar point sets which might be of independent interest. We call them *(combinatorial) Devil's staircases*.

Aichholzer et al. [1, Theorem 14] showed that every 2-convex point set of size n contains a 1-convex subset of size at least $\Omega(\sqrt{n})$. Their result together with Theorem 2 gives the following estimate on the function $f(k, 2, n)$ for $k \geq 3$.

Corollary 1. *There is a constant c such that, for every positive integer n, there are 3-convex sets of n points in the plane in general position with no 2-convex subset of size larger than $c \cdot \log^2 n$.*

In particular, $f(k, 2, n) = O(\log^2 n)$ for every integer k with $k \geq 3$. Aichholzer et al. [2, Theorem 5] also showed that every set of n points contains a 2-convex subset of size at least $\Omega(\log^2 n)$. Thus the bound from Corollary 1 is tight up to a multiplicative constant, settling the second part of Problem 2 in the case $j = 2$. The second part of Problem 2 remains open for $j \geq 3$.

Concerning empty 2-convex polygons in general sets of n points, we show that so-called *Horton sets* do not contain large empty 2-convex polygons. More specifically, we derive the following bound.

Theorem 3. *There is a constant $c > 0$ such that, for every positive integer n, there are sets of n points in the plane in general position that contain no empty 2-convex polygon on at least $c \cdot \log n$ vertices.*

The upper bound from Theorem 3 matches the earlier lower bound [2, Theorem 3] up to a multiplicative constant. In other words, Theorem 3 yields a solution to Problem 3.

Aichholzer et al. [2] proved that, for all positive integers k and l, the union of a k-convex point set T and an l-convex point set S is $(k + l + 1)$-convex. Moreover, they showed that if a k-convex polygonization of T and an l-convex polygonization of S intersect, then $T \cup S$ is $(k + l)$-convex. Aichholzer et al. [2] found a set of 10 points that is not 2-convex and is a union of two 1-convex sets, showing that the first bound is tight for $k = l = 1$. Besides the case $k = l = 1$, no matching bound is known and Aichholzer et al. asked [2, Open problem 2] whether there are examples for general integers k and l such that the union of a k-convex point set and an l-convex point set is not $(k + l)$-convex. We prove the following almost matching bound.

Proposition 1. *For all positive integers k and l, there are point sets T_k and S_l such that T_k is k-convex, S_l is l-convex, and $T_k \cup S_l$ is not $(k + l - 1)$-convex.*

It follows from the proof of Proposition 1 that the bound $k + l$ by Aichholzer et al. [2] on the convexity of a union of a k-convex point set with an l-convex point set with intersecting polygonizations is tight in the worst case. The proof also gives an explicit construction, for every $k \in \mathbb{N}$, of a k-convex set that is not $(k-1)$-convex. Such an example seemed to be missing in the literature.

The proofs of the second part of Theorems 2, 3, and Proposition 1 are in the full version of the paper.

4   Proof of Theorem 1

For a fixed positive integer k, we show that every set of n points in the plane in general position contains a k-convex subset of size at least $\Omega(\log^k n)$. We first state two auxiliary statements. The first one, the *Erdős–Szekeres Lemma*, is a classical result proved by Erdős and Szekeres [6].

Lemma 1 ([6]). *For every $n \in \mathbb{N}$, every sequence of $(n-1)^2 + 1$ real numbers contains a non-increasing or a non-decreasing subsequence of length at least n.*

The main idea of the proof of Theorem 1 is inspired by the approach of Aichholzer et al., who proved the lower bound $\Omega(\log^2 n)$ for the case $k = 2$ [2, Theorem 5]. The key ingredient of the proof is the so-called *Positive Fraction Erdős–Szekeres Theorem* proved by Bárány and Valtr [4]. We use a version of the theorem that was used by Suk [15] and that is based on a bound proved by Pór and Valtr [14] (Theorem 4 below). Before stating it, we first introduce some notation.

A set of points in the plane with distinct x-coordinates is a *cup* if the points lie on the graph of a strictly convex function. Similarly, it is a *cap* if the points lie on the graph of a strictly concave function. Given a cap or a cup $C = \{c_1, \ldots, c_l\}$ with points of C ordered according to the increasing x-coordinates, the *support* of C is the collection of open regions $T_1, \ldots, T_l$, where each T_i is the region outside of the convex hull of C bounded by the line segment $c_i c_{i+1}$ and by the lines $\overline{c_{i-1} c_i}$ and $\overline{c_{i+1} c_{i+2}}$, where $c_0 = c_l$, $c_{l+1} = c_1$, and $c_{l+2} = c_2$; see part (a) of Fig. 1. The *base* of each region T_i is the line segment $c_i c_{i+1}$ and we call the line $\overline{c_i c_{i+1}}$ the *base line* of T_i.

Theorem 4 ([4,14,15]). *Let $l \geq 3$ be an integer and P be a finite set of points in the plane in general position such that $|P| \geq 2^{32l}$. Then there is a set C of l points of P such that C is a cap or a cup and the regions $T_1, \ldots, T_{l-1}$ from the support of C satisfy $|T_i \cap P| \geq \frac{|P|}{2^{32l}}$ for every $i \in \{1, \ldots, l-1\}$.*

Let k be a fixed positive integer and let P be a set of n points in the plane in general position with n sufficiently large with respect to k. A curve C in the plane is *x-monotone* or *y-monotone* if every vertical or horizontal line, respectively, intersects C in at most one point. We proceed by induction on k and show that there is a k-convex subset Q of P of size at least $\Omega(\log^k n)$ such that the polygonization of Q has the boundary formed by a union of an x-monotone curve and one edge. We make no serious effort to optimize the constants.

First, Theorem 1 for $k = 1$ follows from the Erdős–Szekeres Theorem [6], as stated at the beginning of the introduction, because each set of points in convex

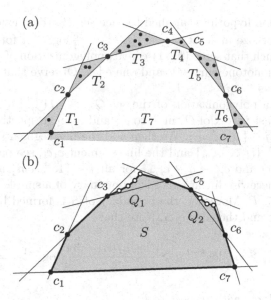

Fig. 1. (a) An illustration of the statement of the Positive Fraction Erdős–Szekeres Theorem (Theorem 4). (b) A construction of the polygon S for $k = 2$. In this example, we have $l = 7$. The described procedure gives $m = 2$, $i_1 = 3$, and $i_2 = 5$, because σ_3, σ_5 are non-decreasing.

position is a union of a cap and a cup that intersect only in two points. This finishes the base case.

Now, for the induction step, assume $k \geq 2$. Without loss of generality we assume that no two points of P have the same x-coordinate. By Theorem 4 applied with $l = \lfloor \log n/64 \rfloor$, there is a set $C = \{c_1, \ldots, c_l\}$ of l points from P such that C is a cap or a cup and the regions $T_1, \ldots, T_{l-1}$ from the support of C satisfy $|T_i \cap P| \geq \frac{|P|}{2^{32l}} \geq \sqrt{n}$. Let $\prec$ be the ordering of the points from P according to their increasing x-coordinates. Note that $c_1 \prec c_2 \prec c_3 \prec \cdots \prec c_l$. By symmetry, we assume that C is a cap.

For every odd i with $3 \leq i < l - 1$, we apply Lemma 1 to the sequence of distances of points from $T_i \cap P$ to the base line of T_i, ordered by $\prec$. For each such sequence, we obtain a non-increasing or a non-decreasing subsequence σ_i of length at least $\sqrt{|T \cap P_i|} \geq n^{1/4}$. By the pigeonhole principle, there are $m \geq (l - 3)/4$ subsequences $\sigma_{i_1}, \ldots, \sigma_{i_m}$ with odd indices $1 < i_1 < \cdots < i_m < l - 1$ such that all these sequences are non-increasing or all non-decreasing. By symmetry, we may assume that $\sigma_{i_1}, \ldots, \sigma_{i_m}$ are all non-decreasing. In the other case we would proceed analogously, considering the ordering of $\prec^{-1}$. For every $j \in \{1, \ldots, m\}$, let P_j be the set of points from $T_{i_j} \cap P$ that determine the distances in σ_{i_j}. In particular, the distances of the points of P_j to the base line of T_{i_j} are non-decreasing in $\prec$ and $|P_j| \geq n^{1/4}$.

By the induction hypothesis applied to each set P_j, there is a $(k-1)$-convex subset Q_j of P_j of size at least $c\log^{k-1}(n^{1/4}) = \frac{c}{4}\log^{k-1} n$ for some constant $c = c(k-1) > 0$ such that some $(k-1)$-convex polygonization of Q_j is formed by a union of an x-monotone curve O_j and one edge. Observe that $c_{i_j} \prec q \prec c_{i_j+1}$ for every $q \in Q_j$, as $1 < i_j < l - 1$.

We construct a polygonization of the set $Q = C \cup \bigcup_{j=1}^{m} Q_j$ by connecting the first and the last vertex of O_j in $\prec$ to c_{i_j} and c_{i_j+1}, respectively, with a line segment for each $j \in \{1, \ldots, m\}$. We then add the line segments $c_i c_{i+1}$ for each $i \in \{1, \ldots, l-1\} \setminus \{i_1, \ldots, i_m\}$ and the line segment $c_1 c_l$; see part (b) of Fig. 1. Since $c_1 \prec \cdots \prec c_l$ and $c_{i_j} \prec q \prec c_{i_j+1}$ for all $j \in \{1, \ldots, m\}$ and $q \in Q_j$, the resulting closed piecewise linear curve is a boundary of a simple polygon S with the vertex set $Q \subseteq P$. Moreover, the boundary of S is formed by a union of an x-monotone curve and the edge $c_1 c_l$. Note that

$$|Q| > \sum_{j=1}^{m} |Q_j| \geq m\frac{c}{4}\log^{k-1} n \geq \frac{\lfloor \log n/64 \rfloor - 3}{4}\frac{c}{4}\log^{k-1} n = \Omega(\log^k n)$$

for n sufficiently large with respect to k.

It remains to prove that the polygon S is k-convex. We start with the following simple observation that restricts the set of lines we have to check.

Lemma 2. *For every $k \in \mathbb{N}$, a simple polygon S with vertices in general position is k-convex if and only if every line not containing a vertex of S intersects ∂S in at most $2k$ points.*

Proof. First, if S is k-convex, then each line ℓ intersects S in at most k connected components. If l contains no vertex of S then each such a component is a line segment with endpoints in ∂S and with interior contained in the interior of S. Thus ℓ intersects ∂S in at most $2k$ points.

On the other hand, if S is not k-convex, then there is a line ℓ that intersects S in more than k connected components. We say that a component of $S \cap \ell$ is *regular* if it contains no vertex of S. Suppose for simplicity that ℓ is horizontal. Since the vertices of S are in general position, at most two components are not regular. Every regular component intersects ∂S in exactly two points. It follows that if all components are regular then ℓ intersects ∂S in at least $2k + 2$ points.

Suppose now that there is a unique component A containing one or two vertices of S. Then either moving ℓ a little bit up or moving it a little bit down turns the component A into one or more regular components and the other components remain regular. Consequently, the perturbed line ℓ contains no vertex of S and intersects ∂S in at least $2k + 2$ points.

Finally, suppose that there are two non-regular components A and B, each containing exactly one vertex of S. Then A can be turned into a regular component either by slightly perturbing ℓ arbitrarily in such a way that it passes above the point $A \cap \text{vert}(S)$, where $\text{vert}(S)$ denotes the vertex set of S, or by slightly perturbing it arbitrarily in such a way that it passes below the point $A \cap \text{vert}(S)$. A similar statement holds for the component B. We claim that a suitable slight perturbation of ℓ turns each of the components A and B into a regular component. Indeed, it is sufficient to move ℓ a little bit up or down or to rotate it slightly clockwise or counterclockwise around the middle point of the segment connecting the points $A \cap \mathrm{vert}(S)$ and $B \cap \mathrm{vert}(S)$. Thus there is a perturbation of ℓ such that the resulting line does not contain a vertex of S and intersects ∂S in at least $2k + 2$ points. This finishes the proof of Lemma 2.

By Lemma 2, it suffices to show that every line ℓ not containing a vertex of S intersects ∂S in at most $2k$ points. Since such a line ℓ intersects ∂S in an even number of points, it actually suffices to show that it intersects ∂S in at most $2k + 1$ points. Every edge of ∂S is contained in the closure $\mathrm{cl}(T_i)$ of some T_i. Since ℓ intersects at most two regions $\mathrm{cl}(T_i)$, it suffices to prove the following claim.

Lemma 3. *The following two conditions are satisfied.*

(i) For every i, $|\ell \cap \partial S \cap \mathrm{cl}(T_i)| \leq 2k$.
(ii) If ℓ intersects two different regions T_α and T_β then $|\ell \cap \partial S \cap \mathrm{cl}(T_\alpha)| \leq 1$ or $|\ell \cap \partial S \cap \mathrm{cl}(T_\beta)| \leq 1$.

Proof. We first prove part (i) of Lemma 3. If $i \notin \{i_1, \ldots, i_m\}$ then $\mathrm{cl}(T_i)$ contains at most one edge of ∂S. Thus, we have $|\ell \cap \partial S \cap \mathrm{cl}(T_i)| \leq 1 < 2k$ in this case. Otherwise $i = i_j$ for some $j \in \{1, \ldots, m\}$, and then $|\ell \cap \partial S \cap \mathrm{cl}(T_i)| \leq 2k$, since ℓ intersects O_j in at most $2k - 2$ points and it intersects each of the two remaining edges of ∂S contained in $\mathrm{cl}(T_i)$ at most once. Part (i) of Lemma 3 follows.

To show part (ii) of Lemma 3, assume that, say, $1 \leq \alpha < \beta \leq l$. If $\beta = \alpha + 1$ then α or β is even and thus $|\ell \cap \partial S \cap \mathrm{cl}(T_\alpha)| \leq 1$ or $|\ell \cap \partial S \cap \mathrm{cl}(T_\beta)| \leq 1$, as required. Similarly, we have $|\ell \cap \partial S \cap \mathrm{cl}(T_\beta)| \leq 1$ if $\beta = l$ and thus we assume $\beta < l$.

Assume now that $\beta \geq \alpha + 2$. Then ℓ intersects the bases of T_α and of T_β. We claim that $|\ell \cap \partial S \cap \mathrm{cl}(T_\alpha)| \leq 1$. This is obvious if $\alpha \notin \{i_1, \ldots, i_m\}$.

Assume now that $\alpha = i_j$ for some $j \in \{1, \ldots, m\}$. Let $c_{i_j} = q_1 \prec q_2 \prec \cdots \prec q_{s-1} \prec q_s = c_{i_j+1}$ be the points from $Q_j \cup \{c_{i_j}, c_{i_j+1}\}$. We use x to denote the intersection point of ℓ and the base of T_{i_j}. Let ℓ^+ be the open half-plane determined by ℓ containing q_s; see Fig. 2.

Let q_t be a point from $Q_j \cap \ell^+$. The distance of the point q_t to the base line of T_{i_j} is at most as large as such a distance for q_{t+1} by the choice of Q_j. Thus the point q_{t+1} does not lie in the strip between the base line of T_{i_j} and the line ℓ' parallel to this line containing q_t. Since $q_t \in \ell^+$, the line ℓ intersects ℓ' to the left of q_t. It then follows from $i_j = \alpha < \beta < l$ that the intersection of ℓ with ℓ' is to the left of x and thus ℓ intersects the vertical line containing q_t below q_t. Since $q_t \prec q_{t+1}$, the point q_{t+1} is thus separated from ℓ by ℓ' and the vertical line that contains q_t; see Fig. 2. In particular, $q_{t+1} \in \ell^+$ and ℓ does not intersect the edge $q_t q_{t+1}$.

Since the vertices along O_j are ordered according to $\prec$, it follows that at most one edge of S in $\mathrm{cl}(T_{i_j})$ intersects ℓ and we have $|\ell \cap \partial S \cap \mathrm{cl}(T_{i_j})| \leq 1$, which completes the proof of part (ii) of Lemma 3 and thus also the proof of Theorem 1.

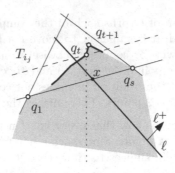

Fig. 2. An illustration of the proof of the fact $|\ell \cap \partial S \cap \operatorname{cl}(T_{i_j})| \leq 1$.

5   Proof of the First Part of Theorem 2

In this section, we construct a 3-convex set of n points with the largest 1-convex subset of size at most $O(\log n)$. Together with the Erdős–Szekeres Theorem, this gives $f(k,n) = f(k,1,n) = \Theta(\log n)$ for all positive integers $k \geq 3$ and n and asymptotically settles the first part of Problem 2. Our example, the so-called Devil's staircase, has a very simple structure and may be of independent interest for reasons discussed in the introduction. In full version of the paper, we also give another example, in which we get a more precise bound described in the second part of Theorem 2. Our examples are modifications of the construction used by Erdős and Szekeres [6] to show the asymptotic tightness of the Erdős–Szekeres Theorem.

A point set D is *deep below* a point set U if the following two conditions are satisfied.

(i) Every point of D lies strictly below each line determined by two points of U, and

(ii) every point of U lies strictly above each line determined by two points of D.

We say that a set S of 2^t points in the plane in general position is a *(combinatorial) Devil's staircase*[1] if S satisfies one of the following two conditions.

(ES1) Either $t = 1$ and the set S consists of two points (x_1, y_1) and (x_2, y_2) with $x_1 < x_2$ and $y_1 < y_2$, or

(ES2) $t \geq 2$ and the set S admits a partition $S = X \cup Y$, where X and Y are both Devil's staircases with 2^{t-1} points. Moreover, X is deep below Y and every point of Y has larger x-coordinate than any point of X.

Let $\{p_1, \ldots, p_n\}$ be the points of a Devil's staircase X_t of size $n = 2^t$ for some $t \in \mathbb{N}$, sorted by increasing x-coordinates. We let $p_1 = (x_1, y_1)$ and $p_n = (x_n, y_n)$ and we define the set $Z_t = X_t \cup \{q\}$ with $q = (x_n, y_1)$.

---

[1] We chose this name, since the set resembles Cantor function, which is also known under the name Devil's staircase [16].

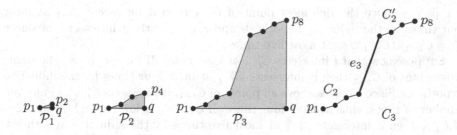

Fig. 3. The polygonizations $\mathcal{P}_1$, $\mathcal{P}_2$, and $\mathcal{P}_3$, and the curve C_3.

Now, we show that the set Z_t is 3-convex. To do so, we consider the following polygonization $\mathcal{P}_t$ of Z_t. Let C_t be an x-monotone piecewise-linear curve formed by the line segments $p_i p_{i+1}$ for each $i \in \{1, \ldots, n-1\}$. Note that, by Properties ES1 and ES2, the chain C_t is also y-monotone. The polygonization $\mathcal{P}_t$ of Z_t is then the polygon whose boundary consists of C_t and the two line segments $p_1 q$ and $p_n q$; see Fig. 3. The polygon $\mathcal{P}_t$ is simple, since C_t has both coordinates increasing if we traverse it from p_1 to p_n. We now prove that $\mathcal{P}_t$ is a 3-convex polygon.

Lemma 4. *Any line ℓ intersects C_t at most five times. Furthermore, if ℓ is non-vertical and passes above the rightmost point of C_t, then it intersects C_t at most four times.*

Proof. We proceed by induction on t. The case $t \leq 2$ is trivial, thus we assume $t \geq 3$. Since X_t is a Devil's staircase, there is a partition $X_t = X_{t-1} \cup X'_{t-1}$ such that X_{t-1} and X'_{t-1} are Devil's staircases of size 2^{t-1}, X_{t-1} lies deep below X'_{t-1}, and X_{t-1} is to the left of X'_{t-1}. Let C_{t-1} and C'_{t-1} be the x- and y-monotone piecewise-linear curves formed by X_{t-1} and X'_{t-1}, respectively.

Let ℓ be a line. Since the points of X_t are in general position, we can, due to Lemma 2, assume that ℓ does not contain a vertex of X_t in the rest of the proof. First, observe that Property ES2 implies that every line that intersects at least two edges of C_{t-1} lies below X'_{t-1}. Similarly, Property ES2 implies that every line intersecting at least two edges of X'_{t-1} is above X_{t-1}. Thus we can assume that ℓ does not intersect both C_{t-1} and C'_{t-1}. Otherwise ℓ intersects both curves at most once and, since C_t contains only a single edge e_t besides C_{t-1} and C'_{t-1}, the line ℓ intersects C_t at most three times.

Since we assume that ℓ does not intersect C_{t-1} or C'_{t-1}, we may also assume that it intersects the other of these two sets at least twice. This will imply restrictions on ℓ.

We assume first that ℓ intersects C_{t-1} at least twice and show that the statement of the lemma is then satisfied. In this case ℓ passes below the rightmost point of C_t, and we only have to show that it intersects C_t at most five times. This is indeed the case because if ℓ passes below the rightmost point of C_{t-1} then it does not intersect e_t and the statement follows from the inductive hypothesis.

If ℓ passes above the rightmost point of C_{t-1} then it intersects C_{t-1} at most four times by the inductive hypothesis and consequently it intersects the curve $C_t = C_{t-1} \cup C'_{t-1} \cup e_t$ at most five times.

Suppose now that ℓ intersects C'_{t-1} at least twice. If ℓ passes above the rightmost point of C'_{t-1} then it intersects C'_{t-1} at most four times by the inductive hypothesis. Since ℓ passes above all points of C_{t-1}, it intersects $C'_{t-1} \cup e_t$ an even number of times, thus at most four times. If ℓ passes below the rightmost point of C'_{t-1} then it intersects C'_{t-1} at most five times by the inductive hypothesis. Since ℓ passes above all points of C_{t-1}, it intersects $C'_{t-1} \cup e_t$ an odd number of times, thus at most five times. This finishes the proof.

Consider a line ℓ containing no point of Z_t. Since ℓ intersects $\partial \mathcal{P}_t$ an even number of times, the first part of Lemma 4 implies that ℓ intersect $\partial \mathcal{P}_t$ at most six times. Lemma 2 then implies that Z_t is a 3-convex point set.

We now show that the largest 1-convex subset of Z_t contains at most $O(t) = O(\log n)$ points. We use an argument analogous to the one used by Erdős and Szekeres [6] (see also Matoušek [12, Sect. 3.1]). Every 1-convex set C of points with distinct x-coordinates is a union of a cup and a cap meeting exactly in the leftmost and the rightmost points of C. To prove the desired bound it is sufficient to show that a Devil's staircase X_t of size $n = 2^t$ contains no cup or cap having more than $t + 1 = \log(n) + 1$ points. A cup in X_1 contains at most two points. Due to the construction of Devil's staircase, every cup in $X_t = X_{t-1} \cup X'_{t-1}$ is either fully contained in one of the smaller Devil's staircases X_{t-1} or X'_{t-1} or it contains at most one point of X'_{t-1}. It follows by induction on t that a cup in X_t contains at most $t + 1$ points. Analogously, every cap in X_t contains at most $t + 1$ points. Thus, every 1-convex subset of X_t contains at most $2t = O(\log n)$ points.

Since any subset of a 3-convex point set is 3-convex [2, Lemma 2] and removing points from Z_t does not increase the size of the largest 1-convex subset, we obtain the first part of Theorem 2.

Acknowledgements. The authors would like to thank Paz Carmi for interesting discussions during the early stages of the research.

References

1. Aichholzer, O., Aurenhammer, F., Demaine, E.D., Hurtado, F., Ramos, P., Urrutia, J.: On k-convex polygons. Comput. Geom. **45**(3), 73–87 (2012)
2. Aichholzer, O., et al.: On k-convex point sets. Comput. Geom. **47**(8), 809–832 (2014)
3. Aichholzer, O., et al.: Holes in 2-convex point sets. Comput. Geom. **74**, 38–49 (2018)
4. Bárány, I., Valtr, P.: A positive fraction Erdős-Szekeres theorem. Discrete Comput. Geom. **19**(3, Special Issue), 335–342 (1998)
5. Erdős, P.: Some more problems on elementary geometry. Austral. Math. Soc. Gaz. **5**(2), 52–54 (1978)

6. Erdős, P., Szekeres, G.: A combinatorial problem in geometry. Compos. Math. **2**, 463–470 (1935)
7. Erdős, P., Szekeres, G.: On some extremum problems in elementary geometry. Ann. Univ. Sci. Budapest. Eötvös Sect. Math. **3–4**, 53–62 (1960/1961)
8. Gerken, T.: Empty convex hexagons in planar point sets. Discrete Comput. Geom. **39**(1–3), 239–272 (2008)
9. Harborth, H.: Konvexe Fünfecke in ebenen Punktmengen. Elem. Math. **33**(5), 116–118 (1978)
10. Holmsen, A.F., Mojarrad, H.N., Pach, J., Tardos, G.: Two extensions of the Erdős-Szekeres problem. Preliminary version: http://arxiv.org/abs/1710.11415 (2017)
11. Horton, J.D.: Sets with no empty convex 7-gons. Canad. Math. Bull. **26**(4), 482–484 (1983)
12. Matoušek, J.: Lectures on Discrete Geometry. Graduate Texts in Mathematics, vol. 212. Springer, New York (2002). https://doi.org/10.1007/978-1-4613-0039-7
13. Nicolás, C.M.: The empty hexagon theorem. Discrete Comput. Geom. **38**(2), 389–397 (2007)
14. Pór, A., Valtr, P.: The partitioned version of the Erdős-Szekeres theorem. Discrete Comput. Geom. **28**(4), 625–637 (2002)
15. Suk, A.: On the Erdős-Szekeres convex polygon problem. J. Am. Math. Soc. **30**(4), 1047–1053 (2017)
16. Thomson, B.S., Bruckner, J.B., Bruckner, A.M.: Elementary Real Analysis. Prentice-Hall, Upper Saddle River (2001)

Algorithm and Hardness Results on Liar's Dominating Set and k-tuple Dominating Set

Sandip Banerjee[1] and Sujoy Bhore[2(✉)]

[1] Department of Computer Science, Hebrew University of Jerusalem,
Jerusalem, Israel
sandip.ndp@gmail.com
[2] Algorithms and Complexity Group, TU Wien, Vienna, Austria
sujoy@ac.tuwien.ac.at

Abstract. Given a graph $G = (V, E)$, the dominating set problem asks for a minimum subset of vertices $D \subseteq V$ such that every vertex $u \in V \setminus D$ is adjacent to at least one vertex $v \in D$. That is, the set D satisfies the condition that $|N[v] \cap D| \geq 1$ for each $v \in V$, where $N[v]$ is the closed neighborhood of v. In this paper, we study two variants of the classical dominating set problem: k-tuple dominating set (k-DS) problem and Liar's dominating set (LDS) problem, and obtain several algorithmic and hardness results. On the algorithmic side, we present a constant factor ($\frac{11}{2}$)-approximation algorithm for the Liar's dominating set problem on unit disk graphs. Then, we design a polynomial time approximation scheme (PTAS) for the k-tuple dominating set problem on unit disk graphs. On the hardness side, we show a $\Omega(n^2)$ bits lower bound for the space complexity of any (randomized) streaming algorithm for Liar's dominating set problem as well as for the k-tuple dominating set problem. Furthermore, we prove that the Liar's dominating set problem on bipartite graphs is W[2]-hard.

1 Introduction

The dominating set problem is regarded as one of the fundamental problems in theoretical computer science which finds its applications in various fields of science and engineering [3,8]. A *dominating set* of a graph $G = (V, E)$ is a subset D of V such that every vertex in $V \setminus D$ is adjacent to at least one vertex in D. The *domination number*, denoted as $\gamma(G)$, is the minimum cardinality of a dominating set of G. Garey and Johnson [6] showed that deciding whether a given graph has domination number at most some given integer k is NP-complete. For a vertex $v \in V$, the open neighborhood of the vertex v denoted as $N_G(v)$ is defined as $N_G(v) = \{u | (u, v) \in E\}$ and the closed neighborhood of the vertex v denoted as $N_G[v]$ is defined as $N_G[v] = N_G(v) \cup \{v\}$.

S. Bhore—Supported by the Austrian Science Fund (FWF) under project number P31119.

C. J. Colbourn et al. (Eds.): IWOCA 2019, LNCS 11638, pp. 48–60, 2019.
https://doi.org/10.1007/978-3-030-25005-8_5

k-tuple Dominating Set (k-DS): Fink and Jacobson [5] generalized the concept of dominating sets as follows.

---

k-tuple Dominating Set (*k-DS*) Problem
Input: A graph $G = (V, E)$ and a non-negative integer k.
Goal: Choose a minimum cardinality subset of vertices $D \subseteq V$ such that, for every vertex $v \in V$, $|N_G[v] \cap D| \geq k$.

The *k-tuple domination number* $\gamma_k(G)$ is the minimum cardinality of a k-DS of G. A good survey on k-DS can be found in [4,12]. Note that, 1-tuple dominating set is the usual dominating set, and 2-tuple and 3-tuple dominating set are known as *double dominating set* [7] and *triple dominating set* [15], respectively. Further note that, for a graph $G = (V, E)$, $\gamma_k(G) = \infty$, if there exists no k-DS of G. Klasing et al. [10] studied the k-DS problem from hardness and approximation point of view. They gave a $(\log |V| + 1)$-approximation algorithm for the k-tuple domination problem in general graphs, and showed that it cannot be approximated within a ratio of $(1 - \epsilon) \log |V|$, for any $\epsilon > 0$.

Liar's Dominating Set (LDS): Slater [17] introduced a variant of the dominating set problem called Liar's dominating set problem. Given a graph $G = (V, E)$, in this problem the objective is to choose minimum number of vertices $D \subseteq V$ such that each vertex $v \in V$ is double dominated and for every two vertices $u, v \in V$ there are at least three vertices in D from the union of their neighborhood set. The LDS problem is an important theoretical model for the following real-world problem. Consider a large computer network where a virus (generated elsewhere in the Internet) can attack any of the processors in the network. The network can be viewed as an unweighted graph. For each node $v \in V$, an anti-virus can: (1) detect the virus at v as well as in its closed neighborhood $N[v]$, and (2) find and report the vertex $u \in N[v]$ at which the virus is located. Notice that, one can make network G virus free by deploying the anti-virus at the vertices $v \in D$, where D is the minimum size *dominating set*. However, in certain situations the anti-viruses may fail. Hence, to make the system virus free it is likely to double-guard the nodes of the network, which is indeed the 2-tuple DS. However, despite of the double-guarding, the anti-viruses may fail to cure the system properly due to some software error or corrupted circumstances. Therefore, for every pair of nodes, it is be important to introduce a guard that sees both of them. This leads us to the Liar's dominating set problem. We define the problem formally below.

Liar's Dominating Set (*LDS*) Problem
Input: A graph $G = (V, E)$ and a non-negative integer k.
Goal: Choose a subset of vertices $L \subseteq V$ of minimum cardinality such that for every vertex $v \in V$, $|N_G[v] \cap L| \geq 2$, and for every pair of vertices $u, v \in V$ of distinct vertices $|(N_G[u] \cup N_G[v]) \cap L| \geq 3$.

1.1 Our Results

In this paper, we obtain seveal algorithmic and hardness results for LDS and k-DS problems on various graph families. On the algorithmic side in Sect. 2, we present a constant factor ($\frac{11}{2}$)-approximation algorithm for the Liar's dominating set (LDS) problem on unit disk graphs. Then, we design a polynomial time approximation scheme (PTAS) for the k-tuple dominating set (k-DS) problem on unit disk graphs. On the hardness side in Sect. 3, we show a $\Omega(n^2)$ bits lower bound for the space complexity of any (randomized) streaming algorithm for Liar's dominating set problem as well as for the k-tuple dominating set problem. Furthermore, we prove that the Liar's dominating set problem on bipartite graphs is W[2]-hard.

2 Algorithmic Results

2.1 Approximation Algorithm for LDS on Unit Disk Graphs

Unit disk graphs are widely used to model of wireless sensor networks. A unit disk graph (UDG) is an intersection graph of a family of unit radius disks in the plane. Formally, given a collection $\mathcal{C} = \{C_1, C_2, \ldots, C_n\}$ of n unit disks in the plane, a UDG is defined as a graph $G = (V, E)$, where each vertex $u \in V$ corresponds to a disk $C_i \in \mathcal{C}$ and there is an edge $(u, v) \in E$ between two vertices u and v if and only if their corresponding disks C_u and C_v contain v and u, respectively. Here, we study the LDS problem on UDG.

Liar's Dominating Set on UDG (*LDS-UDG*) Problem
Input: A unit disk graph $G = (\mathcal{P}, E)$, where $\mathcal{P}$ is a set of n disk centers.
Output: A minimum size subset $D \subseteq \mathcal{P}$ such that for each point $p_i \in \mathcal{P}$, $|N[p_i] \cap D| \geq 2$, and for each pair of points $p_i, p_j \in \mathcal{P}$, $|(N[p_i] \cup N[p_j]) \cap D| \geq 3$.

Jallu et al. [9] studied the LDS problem on unit disk graphs, and proved that this problem is NP-complete. Furthermore, given an unit disk graph $G = (V, E)$ and an $\epsilon > 0$, they have designed a $(1+\epsilon)$-factor approximation algorithm to find an LDS in G with running time $n^{O(c^2)}$, where $c = O(\frac{1}{\epsilon} \log \frac{1}{\epsilon})$. In this section, we design a $\frac{11}{2}$-factor approximation algorithm that runs in sub-quadratic time.

For a point $p \in \mathcal{P}$, let $C(p)$ denote the disk centered at the point p. For any two points $p, q \in \mathcal{P}$, if $q \in C(p)$, then we say that q is a neighbor of p (sometimes we say q is covered by p) and vice versa. Since for any Liar's dominating set D, $|(N[p_i] \cup N[p_j]) \cap D| \geq 3$ ($\forall p_i, p_j \in \mathcal{P}$) holds, we assume that $|\mathcal{P}| \geq 3$ and for all the points $p \in \mathcal{P}$, $|N[p]| \geq 2$. For a point $p_i \in \mathcal{P}$, let $p_i(x)$ and $p_i(y)$ denote the x and y coordinates of p_i, respectively. Let $\mathrm{Cov}_{\frac{1}{2}}(C(p_i))$, $\mathrm{Cov}_1(C(p_i))$, $\mathrm{Cov}_{\frac{3}{2}}(C(p_i))$ denote the set of points of $\mathcal{P}$ that are inside the circle centered at p_i and of radius $\frac{1}{2}$, 1 and $\frac{3}{2}$ unit, respectively.

The basic idea of our algorithm is as follows. Initially, we sort the points of $\mathcal{P}$ based on their x-coordinates. Now, consider the leftmost point (say p_i).

We compute the sets $\text{Cov}_{\frac{1}{2}}(C(p_i))$, $\text{Cov}_1(C(p_i))$ and $\text{Cov}_{\frac{3}{2}}(C(p_i))$. Next, we compute the set $Q = \text{Cov}_{\frac{3}{2}}(C(p_i)) \setminus \text{Cov}_1(C(p_i))$. Further, for each point $q_i \in Q$, we compute the set $S(q_i) = \text{Cov}_1(C(q_i)) \cap \text{Cov}_{\frac{1}{2}}(C(p_i))$. Finally, we compute the set $S = \bigcup S(q_i)$. Moreover, our algorithm is divided into two cases.

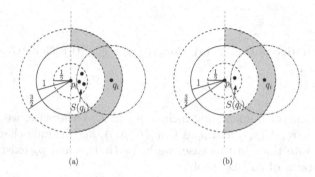

(a)   (b)

Fig. 1. An illustration of Case 1; (a) $|S(q_i)| \geq 3$, (b) $|S(q_i)| \leq 2$.

Case 1 ($S \neq \emptyset$): For each point $q_i \in Q$ such that $S(q_i) \neq \emptyset$, we further distinguish between the following cases.

1. If $|S(q_i)| \geq 3$: we pick two arbitrary points from the set $S(q_i)$, and include them in the output set D (see Fig. 1(a)).
2. If $|S(q_i)| \leq 2$: in this case, we select one point p_a and a possible second point p_b in the output set D (see Fig. 1(b)).

Once these points are selected, we remove the remaining points from $\text{Cov}_1(C(q_i))$ at this step from the set Q. Notice that the points that lie in $\text{Cov}_1(C(q_i))$ are already 1-dominated. Later, we can pick those points if required. However, observe that we may choose a point p_i from $\text{Cov}_{\frac{1}{2}}(C(p_i))$ while in Case 1.2. That would not constitute a LDS. So we maintain a counter t in Case 1. This counter keeps track of how many points we are picking from the set $S(q_i)$ in total, for each point $q_i \in Q$. If t is at least 2, we simply add p_i to the output set and do not enter into Case 2. Otherwise, we proceed to Case 2.

Case 2 ($S = \emptyset$ or $t < 2$): here, we further distinguish between the following cases.

1. If $|\text{Cov}_{\frac{1}{2}}(C(p_i))| \geq 3$: then we choose 2 points arbitrarily in the output set D (see Fig. 2(a)).
2. If $|\text{Cov}_{\frac{1}{2}}(C(p_i))| = 2$: let $p_i, p_x \in \text{Cov}_{\frac{1}{2}}(C(p_i))$ be these points. We include both of them in the output set D. This settles the first condition of LDS for them. However, in order to fulfill the second condition of LDS for p_i and p_x, we must include at least one extra point here. First, we check the cardinality of $X = (\text{Cov}_1(C(p_i)) \cap \text{Cov}_1(C(p_x))) \setminus \{p_i, p_x\}$. If $|X| \neq \emptyset$, then we pick an

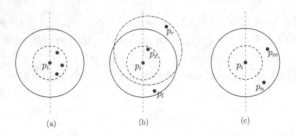

Fig. 2. An illustration of Case 2; (a) $|\mathrm{Cov}_{\frac{1}{2}}(C(p_i))| \geq 3$, (b) $|\mathrm{Cov}_{\frac{1}{2}}(C(p_i))| = 2$, (c) $|\mathrm{Cov}_{\frac{1}{2}}(C(p_i))| = 1$.

arbitrary point p_m from X, and include p_m in D. Otherwise, we include two points $p_l \in \mathrm{Cov}_1(C(p_i))$ and $p_r \in \mathrm{Cov}_1(C(p_x))$, and include them in D (see Fig. 2(b)). Note that, in this case, we know that p_l and p_r exist due to the input constraint of an LDS problem.

3. If $|\mathrm{Cov}_{\frac{1}{2}}(C(p_i))| = 1$: then we check $\mathrm{Cov}_1(C(p_i))$ and include two points p_m and p_n arbitrarily from $\mathrm{Cov}_1(C(p_i)) \setminus \mathrm{Cov}_{\frac{1}{2}}(C(p_i))$ (see Fig. 2(c)).

This fulfills the criteria of LDS of points in $\mathrm{Cov}_{\frac{1}{2}}(C(p_i))$. Then we delete the remaining points of $\mathrm{Cov}_{\frac{1}{2}}(C(p_i))$ from $\mathcal{P}$. Next we select the left-most point from the remaining and repeat the same procedure until $\mathcal{P}$ is empty. The pseudo-code of the algorithm is given in the full version of the paper (see [1]).

Lemma 1. *[⋆]*[1] *The set D obtained from our algorithm, is a LDS of the unit disk graph defined on the points of $\mathcal{P}$.*

Lemma 2. *[⋆] Our algorithm outputs a LDS $D \subseteq \mathcal{P}$ of the unit disk graph defined on the points of $\mathcal{P}$ with approximation ratio $\frac{11}{2}$.*

The algorithm runs in polynomial time (to be precise in sub-quadratic time). Thus, from Lemmas 1 and 2 we conclude the following theorem.

Theorem 1. *The algorithm computes a LDS of the unit disk graph defined on the points of $\mathcal{P}$ in sub-quadratic time with approximation factor $\frac{11}{2}$.*

2.2 PTAS for k-DS on Unit Disk Graphs

In this section we give a PTAS for the k-tuple dominating set on unit disk graphs with a similar approach used by Nieberg and Hurink [13]. It might be possible to design a PTAS by using local search or shifting strategy for the k-tuple dominating set problem on unit disk graphs. However, the time complexity of these algorithms would be high. Thus we use the approach of Nieberg and Hurink [13], that gurantees a better running time.

---

[1] Proof of results labeled with [⋆] have been deferred to the full version [1] due to space constraint.

Let $G = (V, E)$ be an unit disk graph in the plane. For a vertex $v \in V$, let $N^r(v)$ and $N^r[v] = N^r(v) \cup \{v\}$ be the r-th neighborhood and r-th closed neighborhood of v, respectively. For any two vertices $u, v \in V$, let $\delta(u, v)$ be the distance between u and v in G, that is the number of edges of a shortest path between u and v in G. Let $D_k(V)$ be the minimum k-tuple dominating set of G. For a subset $W \subseteq V$, let $D_k(W)$ be the minimum k-tuple dominating set of the induced subgraph on W. We prove the following theorem.

Theorem 2. *There exists a PTAS for the k-tuple dominating set problem on unit disk graphs.*

Proof. The 2-separated collection of subsets is defined as follows: Given a graph $G = (V, E)$, let $S = \{S_1, \ldots, S_m\}$ be a collection of subsets of vertices $S_i \subset V$, for $i = 1, \ldots, m$, such that for any two vertices $u \in S_i$ and $v \in S_j$ with $i \neq j$, $\delta(u, v) > 2$. In the following lemma we prove that the sum of the cardinalities of the minimum k-tuple dominating sets $D_k(S_i)$ for the subsets $S_i \in S$ of a 2-separated collection is a lower bound on the cardinality of $D_k(V)$.

Lemma 3. [⋆] *Given a graph $G = (V, E)$, let $S = \{S_1, \ldots, S_m\}$ be a 2-separated collection of subsets of V then, $|D_k(V)| \geq \sum_{i=1}^{m} |D_k(S_i)|$.*

From Lemma 3, we get the lower bound of the minimum k-tuple dominating set of G. If we can enlarge each of the subset S_i to a subset T_i such that the k-tuple dominating set of S_i (that is $D_k(S_i)$) is locally bounded to the k-tuple dominating set of T_i (that is $D_k(T_i)$), then by taking the union of them we get an approximation of the k-tuple dominating set of G. For each subset S_i, let there is a subset T_i (where $S_i \subset T_i$), and let there exists a bound $(1 + \epsilon)$ $(0 < \epsilon < 1)$ such that $|D_k(T_i)| \leq (1 + \epsilon) \cdot |D_k(S_i)|$. Then, if we take the union of the k-tuple dominating sets of all T_i, this is a $(1 + \epsilon)$-approximation of the k-tuple dominating sets of the union of subsets S_i (for $i = 1, \ldots, m$). Now, we describe the algorithm. Let $V_0 = V$. Consider an arbitrary vertex $v \in V_0$, and begin computing the k-DS of $N^r[v]$, until $D_k(N^{r+2}[v]) > \rho \cdot D_k(N^r[v])$ (for a constant ρ). We iteratively process the remaining graph induced by $V_{i+1} = V_i \setminus N^{\hat{r}_i+2}[v_i]$ (where $\hat{r}_i$ is the first point at ith iteration when the condition is violated).

Lemma 4. *Let $\{N_1, \ldots, N_\ell\}$ be the set of neighborhoods created by the above algorithm (for $\ell < n$). The union $\bigcup_{i=1}^{\ell}(D_k(N_i))$ forms a k-tuple dominating set of G.*

Proof. Consider the set $V_{i+1} = V_i \setminus N_i$, and we know $N_i \subset V_i$. Thus, $V_{i+1} = V_i \cup N_i$. The algorithm stops while $V_{\ell+1} = \emptyset$, which means $V_\ell = N_\ell$. Besides, $\bigcup_{i=1}^{\ell}(N_i) = V$. Thus, if we compute the k-tuple dominating set $D_k(N_i)$ of each N_i, their union clearly is the k-tuple dominating set of the entire graph.   □

These subsets $N^{\hat{r}_i}[v_i]$, for $i = 1, \ldots, \ell$, created by the algorithm form a 2-separated collection $\{N^{\hat{r}_1}[v_1], \ldots, N^{\hat{r}_\ell}[v_\ell]\}$ in G. Consider any two neighborhoods N_i, N_{i+1}. We have computed N_{i+1} on graph induced by $V \setminus V_i$. So, for any two vertices $u \in N_i$ and $v \in N_{i+1}$, the distance is greater than 2. Thus we have the following corollary.

Corollary 1. *The algorithm returns a k-tuple dominating set $\bigcup_{i=1}^{\ell}(D_k(N_i))$ of cardinality no more than ρ the size of a dominating set $D_k(V)$, where $\rho = (1+\epsilon)$.*

It needs to be shown that this algorithm has a polynomial running time. The number of iterations ℓ is clearly bounded by $|V| = n$. It is important to show that for each iteration we can compute the minimum k-tuple dominating set $D_k(N^r[v])$ in polynomial time for r being constant or polynomially bounded. Consider the r-th neighborhood of a vertex v, $N^r[v]$. Let I^r be the maximal independent set of the graph induced by $N^r[v]$. From [13], we have $I^r \leq (2r + 1)^2 = O(r^2)$. The cardinality of a minimum dominating set in $N^r[v]$ is bounded from above by the cardinality of a maximal independent set in $N^r[v]$. Hence, $|D(N^r[v]| \leq (2r + 1)^2 = O(r^2)$. Now, we prove the following lemma.

Lemma 5. $|D_k(N^r[v])| \leq O(k^2 \cdot r^2)$.

Proof. Let I_1^r be the first maximal independent set of $N^r[v]$. We know $|D(N^r[v])| \leq |I_1^r| \leq (2r + 1)^2$. Now, we take the next maximal independent set I_2^r from $N^r[v] \setminus I_1^r$, and take the union of them $(I_1^r \cup I_2^r)$. Notice that every vertex $v \in (N^r[v] \setminus (I_1^r \cup I_2^r))$ has 2 neighbors in $(I_1^r \cup I_2^r)$, so they can be 2-tuple dominated by choosing vertices from $(I_1^r \cup I_2^r)$. Also, every vertex $v \in I_2^r$ can be 2-tuple dominated by choosing vertices from $(I_1^r \cup I_2^r)$, since v itself can be one and the other one can be picked from I_1^r. Additionally, for each vertex $u \in I_1^r$, we take a vertex z from the neighborhood of u in $(N^r[v] \setminus (I_1^r \cup I_2^r))$. Let $W(I_1^r)$ be the union of these vertices. Now, every vertex $v \in N^r[v]$ can be 2-tuple dominated by choosing vertices from $(I_1^r \cup I_2^r \cup W(I_1^r))$. So, $|D_2(N^r[v])| \leq |(I_1^r \cup I_2^r \cup W(I_1^r))|$. $|(I_1^r \cup I_2^r \cup W(I_1^r))| \leq 3 \cdot (2r+1)^2$. Hence, $|D_2(N^r[v])| \leq 3 \cdot (2r+1)^2$. We continue this process k times.

After k steps, we get the union of the maximal independent sets $A = \{I_1^r \cup \ldots \cup I_k^r\}$. Additionally, we get the unions of $B = \{W(I_1^r) \cup W(I_1^r \cup I_2^r) \cup \ldots \cup W(I_1^r \cup \ldots \cup I_{k-1}^r)\}$. Notice that every vertex $v \in N^r[v]$ can be k-tuple dominated by choosing vertices from $(A \cup B)$. The cardinality of $(A \cup B)$ is at most $(2r + 1)^2 \cdot (1 + 3 + \ldots + (2k - 1))$, which is $(2r + 1)^2 \cdot k^2$. We also know $|D_k(N^r[v])|$ is upper bounded by $(A \cup B)$. Thus, $|D_k(N^r[v])| \leq (2r + 1)^2 \cdot k^2 \leq O(k^2 \cdot r^2)$.

Nieberg and Hurink [13] showed that for a unit disk graph, there exists a bound on $\hat{r}_1$ (the first value of r that violates the property $D(N^{r+2}[v]) > \rho \cdot D(N^r[v])$). This bound depends on the approximation ρ not on the size of the of the unit disk graph $G = (V, E)$ given as input. Precisely, they have proved that there exists a constant $c = c(\rho)$ such that $\hat{r}_1 \leq c$, that is, the largest neighborhood to be considered during the iteration of the algorithm is bounded by a constant. Thereby, putting everything together, we conclude the proof. $\square$

3   Hardness Results

3.1   Streaming Lower Bound for LDS

In this section, we consider the streaming model: the edges arrive one-by-one in some order, and at each time-stamp we need to decide if we either *store* the edge or *forget* about it. We now show that any streaming algorithm that solves the LDS problem must essentially store all the edges.

Theorem 3. *Any randomized[2] streaming algorithm for LDS problem on n-vertex graphs requires $\Omega(n^2)$ space.*

Proof. We will reduce from the INDEX problem in communication complexity:

Index Problem
Input: Alice has a string $X \in \{0,1\}^N$ given by $x_1 x_2 \ldots x_N$. Bob has an index $\iota \in [N]$.
Question: Bob wants to find x_ι, i.e., the ιth bit of X.

It is well-known that there is a lower bound of $\Omega(N)$ bits in the one-way randomized communication model for Bob to compute x_i [11]. We assume an instance of the INDEX problem where N is a perfect square, and let $r = \sqrt{N}$. Fix any bijection from $[N] \to [r] \times [r]$. Consequently we can interpret the bit string as an adjacency matrix for a bipartite graph with r vertices on each side. Let the two sides of the bipartition be $V = \{v_1, v_2, \ldots, v_r\}$ and $W = \{w_1, w_2, \ldots, w_r\}$.

From the instance of INDEX, we construct an instance G_X of the LDS. Assume that Alice has an algorithm that solves the k-tuple dominating set problem using $f(r)$ bits. First, we insert the edges corresponding to the edge interpretation of X between nodes v_i and w_j: for each $i, j \in [k]$, Alice adds the edge (v_i, w_j) if the corresponding entry in X is 1. Alice then sends the memory contents of her algorithm to Bob, using $f(r)$ bits.

Bob has the index $\iota \in [N]$, which he interprets as (I, J) under the same bijection $\phi : [N] \to [r] \times [r]$. He receives the memory contents of the algorithm, and proceeds to do the following:

- Add two vertices a and b, and an edge $a - b$
- Add an edge from each vertex of $V \setminus v_I$ to a
- Add an edge from each vertex of $W \setminus w_J$ to a
- Add five vertices $\{u, y, u', y', z\}$ and edges $u - u', y - y', u - z$ and $y - z$.
- Add an edge from each vertex of $V \cup W \cup \{a, b\}$ to each vertex from $\{u, y\}$

Let D be a minimum LDS of G_X. Note that D has to be a double dominating set of G_X. Since u' has only 2 neighbors in G_X, it follows that $\{u, u'\} \subseteq D$. Similarly $\{y, y'\} \subseteq D$. Note that z also has only two neighbors in G_X. Hence, we have that $N[z] \cup N[b] = \{u, y, a, b\}$. Since we must have $|(N[z] \cup N[b]) \cap D| \geq 3$, it follows that at least one of a or b must belong in D. Since $N[b] \subseteq N[a]$, without loss of generality we can assume that $a \in D$. Therefore, so far we have concluded that $\{u, u', y, y', a\} \subseteq D$.

The next two lemmas show that finding the minimum value of a LDS of G_X allows us to solve the corresponding instance X of INDEX.

---

[2] By randomized algorithm we mean that the algorithm should succeed with probability $\geq \frac{2}{3}$.

Lemma 6. $x_\iota = 1$ *implies that the minimum size of a LDS of* G_X *is 6.*

Proof. Suppose that $x_\iota = 1$, i.e., $v_I - w_J$ is an edge in G_X. We now claim that $D := \{u, u', y, y', a\} \cup v_I$ is a LDS of G_X.

First we check that D is indeed a double dominating set of G_X

- For each vertex in $\lambda \in G_X \setminus \{u, u', y, y', z\}$ we have $(N[\lambda] \cap D) \supseteq \{u, y\}$
- $(N[u] \cap D) \supseteq \{u, u'\}$
- $(N[y] \cap D) \supseteq \{y, y'\}$
- $(N[z] \cap D) = \{u, y\}$
- $(N[u'] \cap D) = \{u, u'\}$
- $(N[y'] \cap D) = \{y, y'\}$

We now check the second condition. Let $T = G_X \setminus \{u, u', y, y', z\}$, and $T' = G_X \setminus T$

- For each $\lambda \in T \setminus \{v_I, w_J\}$ and each $\delta \in T'$ we have $(N[\lambda] \cup N[\delta]) \cap D \supseteq \{a, u, y\}$
- For each $\delta \in T'$ we have $(N[v_I] \cup N[\delta]) \cap D \supseteq \{v_I, u, y\}$
- For each $\delta \in T'$ we have $(N[w_J] \cup N[\delta]) \cap D \supseteq \{v_I, u, y\}$
- Now we consider pairs where both vertices are from T'. By symmetry, we only have to consider following choices
  - $(N[u'] \cup N[u]) \cap D = \{u, u', y\}$
  - $(N[u'] \cup N[z]) \cap D = \{u, u', y\}$
  - $(N[u'] \cup N[y]) \cap D = \{u, u', y, y'\}$
  - $(N[u'] \cup N[y']) \cap D = \{u, u', y, y'\}$
- Now we consider pairs where both vertices are from T. By symmetry, we only have to consider following choices
  - For each $\lambda \in V \setminus v_I \cup W \setminus w_J$ we have $(N[\lambda] \cup N[b]) \cap D \supseteq \{u, y, a\}$ and $(N[\lambda] \cup N[a]) \cap D \supseteq \{u, y, a\}$
  - $(N[v_I] \cup N[b]) \cap D \supseteq \{u, y, a\}$
  - $(N[v_I] \cup N[a]) \cap D \supseteq \{u, y, a\}$
  - $(N[w_J] \cup N[b]) \cap D \supseteq \{u, y, a\}$
  - $(N[w_J] \cup N[a]) \cap D \supseteq \{u, y, a\}$
  - $(N[w_J] \cup N[v_I]) \cap D \supseteq \{v_I, y, a\}$
  - For each $\gamma \in V \setminus v_I$ we have $(N[w_J] \cup N[\gamma]) \cap D \supseteq \{v_I, y, a, u\}$
  - For each $\gamma \in W \setminus w_J$ we have $(N[v_I] \cup N[\gamma]) \cap D \supseteq \{v_I, y, a, u\}$
  - For each $\gamma \in V \setminus v_I$ and $\gamma' \in W \setminus w_J$ we have $(N[\gamma] \cup N[\gamma']) \cap D \supseteq \{y, a, u\}$

Hence, it follows that D is indeed a LDS of G_X of size 6.

Lemma 7. $x_\iota = 0$ *implies that the minimum size of a LDS of* G_X *is* ≥ 7.

Proof. Now suppose that $x_\iota = 0$, i.e., v_I and w_J do not have an edge between them in G_X. Let D' be a minimum LDS of G_X. We have already seen above that $\{u, u', y, y', a\} \subseteq D$.

Consider the pair (v_I, z). Currently, we have that $(N[v_I] \cup N[z]) \cap \{u, u', y, y', a\} = \{u, y\}$. Hence, D' must contain a vertex, say $\mu \in N[v_I] \setminus \{u, y\}$. Consider the pair (w_J, z). Currently, we have that $(N[w_J] \cup N[z]) \cap$

$\{u, u', y, y', a\} = \{u, y\}$. Hence, D' must contain a vertex, say $\mu' \in N[w_J] \setminus \{u, y\}$. Since v_I and w_J do not form an edge, we have that $\mu \neq \mu'$. Hence, $|D'| \geq 5+2 = 7$.

Thus, by checking whether the value of a minimum LDS on the instance G_X is 6 or 7, Bob can determine the index x_ι. The total communication between Alice and Bob was $O(f(r))$ bits, and hence we can solve the INDEX problem in $f(r)$ bits. Recall that the lower bound for the INDEX problem is $\Omega(N) = \Omega(r^2)$. Note that $|G_X| = n = 2r + 5 = O(r)$, and hence $\Omega(r^2) = \Omega(n^2)$. $\square$

Corollary 2. *Let $\epsilon > 0$ be a constant. Any (randomized) streaming algorithm that achieves a $(\frac{7}{6} - \epsilon)$-approximation for a LDS requires $\Omega(n^2)$ space.*

Proof. Theorem 3 shows that distinguishing between whether the minimum value of the LDS is 6 or 7 requires $\Omega(n^2)$ bits. The claim follows since $6 \cdot (\frac{7}{6} - \epsilon) < 7$. $\square$

3.2   Streaming Lower Bounds for k-DS

Theorem 4. *For any $k = O(1)$, any randomized (See footnote 2) streaming algorithm for the k-tuple dominating set problem on n-vertex graphs requires $\Omega(n^2)$ space.*

Proof. We reduce from the INDEX problem in communication complexity. We assume an instance of the INDEX problem where N is a perfect square, and let $r = \sqrt{N}$. Fix any bijection from $[N] \rightarrow [r] \times [r]$. Consequently we can interpret the bit string as an adjacency matrix for a bipartite graph with r vertices on each side. Let the two sides of the bipartition be $V = \{v_1, v_2, \ldots, v_r\}$ and $W = \{w_1, w_2, \ldots, w_r\}$. From the instance of INDEX, we construct an instance G_X of the k-tuple dominating set. Assume that Alice has an algorithm that solves the k-tuple dominating set problem using $f(r)$ bits. First, we insert the edges corresponding to the edge interpretation of X between nodes v_i and w_j: for each $i, j \in [k]$, Alice adds the edge (v_i, w_j) if the corresponding entry in X is 1. Alice then sends the memory contents of her algorithm to Bob, using $f(r)$ bits.

Bob has the index $\iota \in [N]$, which he interprets as (I, J) under the same bijection $\phi : [N] \rightarrow [r] \times [r]$. He receives the memory contents of the algorithm, and proceeds to do the following:

- Add (k+1) vertices $A = \{a_1, a_2, \ldots, a_k\}$ and b.
- Add edges $\{a_i - b : 1 \leq i \leq k\}$.
- Add an edge from each vertex of $V \setminus (v_I)$ to each vertex of A.
- Add an edge from each vertex of $W \setminus (w_J)$ to each vertex of A.
- Add an edge v_I to each vertex of $A \setminus a_k$.
- Add an edge from w_J to each vertex of $A \setminus a_k$.

The next lemma shows that finding the minimum value of a k-tuple dominating set of G_X allows us to solve the corresponding instance X of INDEX.

Lemma 8. *[$\star$] The minimum size of a k-tuple dominating set of G_X is $k + 1$ if and only if $x_\iota = 1$.*

Thus, by checking whether the value of minimum k-tuple dominating set on the instance G_X is $k + 1$ or $k + 2$, Bob can determine the index x_ι. The total communication between Alice and Bob was $O(f(r))$ bits, and hence we can solve the INDEX problem in $f(r)$ bits. Recall that the lower bound for the INDEX problem is $\Omega(N) = \Omega(r^2)$. Note that $|G_X| = n = 2r + k + 1 = O(r)$ since $k = O(1)$, and hence $\Omega(r^2) = \Omega(n^2)$.

Corollary 3. *Let $1 > \epsilon > 0$ be any constant. Any (randomized) streaming algorithm that approximates a k-tuple dominating set within a relative error of ϵ requires $\Omega(n^2)$ space.*

Proof. Choose $\epsilon = \frac{1}{k}$. Theorem 8 shows that the relative error is at most $\frac{1}{k+2}$, which is less than ϵ. Hence finding an approximation within ϵ relative error amounts to finding the exact value of the k-tuple dominating set. Hence, the claim follows from the lower bound of $\Omega(n^2)$ of Theorem 8.

3.3 W-Hardness Results for LDS

The LDS problem was is NP-complete on general graphs [17], and NP-complete on bipartite graphs, split graphs (see, e.g., [14,16]). Bishnu et al. [2] showed that LDS problem on planar graphs admits a linear kernel and W[2]-hard on general graphs. We prove that LDS problem is W[2]-hard on bipartite graphs.

Theorem 5. *Liar's dominating set on bipartite graphs is W[2]-hard.*

Proof. We prove this by giving a parameterized reduction from the dominating set problem in general undirected graphs. Let $(G = (V, E), k)$ be an instance of the dominating set, where k denotes the size of the dominating set. We construct a bipartite graph $G' = (V', E')$ from $G = (V, E)$. First, we create two copies of V, namely $V_1 = \{u_1 | u \in V\}$ and $V_2 = \{u_2 | u \in V\}$. Next, we introduce two extra vertices z_1, z_2 in V_1, and two extra vertices z'_1, z'_2 in V_2. Furthermore, we introduce two special vertices $s_{z'_1}, s_{z'_2}$ in V_1 and two special vertices s_{z_1}, s_{z_2} in V_2. The entire vertex set V' is $V_1 \cup V_2$, where $V_1 = \{u_1 | u \in V\} \cup \{z_1, z_2, s_{z'_1}, s_{z'_2}\}$ and $V_2 = \{u_2 | u \in V\} \cup \{z'_1, z'_2, s_{z_1}, s_{z_2}\}$. Now, if there is an edge $(u, v) \in E$, then we draw the edges (u_1, v_2) and (v_1, u_2). We draw the edges of the form (u_1, u_2) in G' for every vertex $u \in V$. Then, we add edges from every vertex in $V_1 \setminus \{z_1, z_2, s_{z'_1}, s_{z'_2}\}$ to z'_1, z'_2, and the edges from every vertex in $V_1 \setminus \{z'_1, z'_2, s_{z_1}, s_{z_2}\}$ to z_1, z_2. Finally we add the edges $(z_1, z'_1), (z_2, z'_2)$ and $(z_1, s_{z_1}), (z_2, s_{z_2}), (z'_1, s_{z_1}), (z'_2, s_{z'_2})$. This completes the construction (see Fig. 3).

We show that G has a dominating set of size k if and only if G' has a LDS of size $k + 8$. Let D denote the dominating set of the given graph G. We claim that $D' = \{u_1 | u \in D\} \cup \{z_1, z_2, s_{z'_1}, s_{z'_2}\} \cup \{z'_1, z'_2, s_{z_1}, s_{z_2}\}$ is a LDS of G'. Note that for any vertex $v \in V'$, $|N_{G'}[v] \cap D'| \geq 2$, since $\{z_1, z_2, z'_1, z'_2\}$ is in D'. This fulfills the first condition of the LDS. Now, for every pair of vertices, $u, v \in V'$, we show that $|(N_{G'}[u] \cup N_{G'}[v]) \cap D'| \geq 3$. If $u, v \in V_1$, we know $|(N_{G'}[u] \cup N_{G'}[v]) \cap D'| \geq 2$ due to z'_1, z'_2. Now, in the dominating set at least one additional vertex dominates

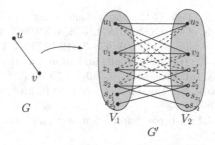

Fig. 3. Construction of G' from G (Illustration of Theorem 5).

them. Thus, $|(N_{G'}[u] \cup N_{G'}[v]) \cap D'| \geq 3$. Similarly, when $u, v \in V_2$ or $u \in V_1$ and $v \in V_2$. This fulfills the second condition of the LDS.

Conversely, let D' be a LDS in G'. Note that, $\{z_1, z_2, z_1', z_2'\}$ are always part of D', since z_1, z_2 are the only neighbors of s_{z_1}, s_{z_2} and z_1', z_2' are the only neighbors of $s_{z_1'}, s_{z_2'}$. These special vertices are taken in the construction to enforce $\{z_1, z_2, z_1', z_2'\}$ to be in D'. Now we know, for any pair of vertices p, q, $|(N_{G'}[p] \cup N_{G'}[q]) \cap D'| \geq 3$. This implies p, q is dominated by at least one vertex or one of them is picked, except $\{z_1, z_2, z_1', z_2'\}$. Otherwise, $|(N_{G'}[p] \cup N_{G'}[q]) \cap D'| < 3$. This violates the second condition of LDS. Now, when p, q are both part of the same edge in G (say $u_2, v_2 \in V_2$; see Fig. 3), we need at least one vertex from $\{u_1, v_1, u_2, v_2\}$ in D'. This means that for every vertex $v \in V$, $|N_G[v] \cap D| \geq 1$. Thus, D is a dominating set of G where the cardinality of D is at most k. $\quad\square$

Acknowledgements. The authors would like to thank Rajesh Chitnis and M. S. Ramanujan for interesting discussions during various stages of the research.

References

1. Banerjee, S., Bhore, S.: Algorithm and hardness results on liar's dominating set and k-tuple dominating set. CoRR, abs/1902.11149 (2019)
2. Bishnu, A., Ghosh, A., Paul, S.: Linear kernels for k-tuple and liar's domination in bounded genus graphs. Discrete Appl. Math. **231**, 67–77 (2017)
3. Chang, G.J.: Algorithmic aspects of domination in graphs. In: Du, D.Z., Pardalos, P.M. (eds.) Handbook of Combinatorial Optimization, pp. 1811–1877. Springer, Boston (1998). https://doi.org/10.1007/978-1-4613-0303-9_28
4. Chellali, M., Favaron, O., Hansberg, A., Volkmann, L.: k-domination and k-independence in graphs: a survey. Graphs Comb. **28**(1), 1–55 (2012)
5. Fink, J.F., Jacobson, M.S.: n-domination in graphs. In: Graph theory with Applications to Algorithms and Computer Science, pp. 283–300. Wiley, Hoboken (1985)
6. Garey, M.R., Johnson, D.S.: Computers and Intractability, vol. 29. W. H. Freeman, New York (2002)
7. Harary, F., Haynes, T.W.: Double domination in graphs. Ars Comb. **55**, 201–214 (2000)
8. Haynes, T.W., Hedetniemi, S., Slater, P.: Fundamentals of Domination in Graphs. CRC Press, Boca Raton (1998)

9. Jallu, R.K., Jena, S.K., Das, G.K.: Liar's dominating set in unit disk graphs. In: Wang, L., Zhu, D. (eds.) COCOON 2018. LNCS, vol. 10976, pp. 516–528. Springer, Cham (2018). https://doi.org/10.1007/978-3-319-94776-1_43

10. Klasing, R., Laforest, C.: Hardness results and approximation algorithms of k-tuple domination in graphs. Inf. Process. Lett. **89**(2), 75–83 (2004)

11. Kushilevitz, E., Nisan, N.: Communication Complexity. Cambridge University Press, Cambridge (1997)

12. Liao, C.S., Chang, G.J.: k-tuple domination in graphs. Inf. Process. Lett. **87**(1), 45–50 (2003)

13. Nieberg, T., Hurink, J.: A PTAS for the minimum dominating set problem in unit disk graphs. In: Erlebach, T., Persinao, G. (eds.) WAOA 2005. LNCS, vol. 3879, pp. 296–306. Springer, Heidelberg (2006). https://doi.org/10.1007/11671411_23

14. Panda, B.S., Paul, S.: Liar's domination in graphs: complexity and algorithm. Discrete Appl. Math. **161**(7–8), 1085–1092 (2013)

15. Rautenbach, D., Volkmann, L.: New bounds on the k-domination number and the k-tuple domination number. Appl. Math. Lett. **20**(1), 98–102 (2007)

16. Roden, M.L., Slater, P.J.: Liar's domination in graphs. Discrete Math. **309**(19), 5884–5890 (2009)

17. Slater, P.J.: Liar's domination. Networks **54**(2), 70–74 (2009)

Fixed-Parameter Tractability of $(n - k)$ List Coloring

Aritra Banik[1], Ashwin Jacob[3(✉)], Vijay Kumar Paliwal[2],
and Venkatesh Raman[3]

[1] National Institute of Science Education and Research, HBNI, Bhubaneswar, India
aritra@niser.ac.in
[2] D. E. Shaw India Private Limited, Hyderabad, India
getvijaypaliwal@gmail.com
[3] The Institute of Mathematical Sciences, HBNI, Chennai, India
{ajacob,vraman}@imsc.res.in

Abstract. We consider the list-coloring problem from the perspective of parameterized complexity. The classical graph coloring problem is given an undirected graph and the goal is to color the vertices of the graph with minimum number of colors so that end points of each edge gets different colors. In list-coloring, each vertex is given a list of allowed colors with which it can be colored.

In parameterized complexity, the goal is to identify natural parameters in the input that are likely to be small and design an algorithm with time $f(k)n^c$ time where c is a constant independent of k, and k is the parameter. Such an algorithm is called a fixed-parameter tractable (fpt) algorithm. It is clear that the solution size as a parameter is not interesting for graph coloring, as the problem is NP-hard even for $k = 3$. An interesting parameterization for graph coloring that has been studied is whether the graph can be colored with $n - k$ colors, where k is the parameter and n is the number of vertices. This is known to be fpt using the notion of crown reduction. Our main result is that this can be generalized for list-coloring as well. More specifically, we show that, given a graph with each vertex having a list of size $n - k$, it can be determined in $f(k)n^{O(1)}$ time, for some function f of k, whether there is a coloring that respects the lists.

1 Introduction

The graph coloring problem is one of the fundamental combinatorial optimization problems with applications in scheduling, register allocation, pattern matching and many other active research areas. Given a graph $G = (V, E)$, the k-coloring problem is asking whether there is a way to assign at most k colors/labels to vertices of a graph such that no two adjacent vertices share the same color. Such a coloring is also known as a *proper k-coloring*. The smallest number of colors needed to color a graph G is called its chromatic number, and is denoted by $\chi(G)$. Determining whether a graph is 3-colorable is NP-hard [9] while the 2-coloring problem has a linear time algorithm. It is even hard to

© Springer Nature Switzerland AG 2019
C. J. Colbourn et al. (Eds.): IWOCA 2019, LNCS 11638, pp. 61–69, 2019.
https://doi.org/10.1007/978-3-030-25005-8_6

approximate the chromatic number in polynomial time. The 3-coloring problem remains NP-complete even on 4-regular planar graphs [6]. There are some generalizations and variations of ordinary graph colorings which are motivated by practical applications such as PRECOLORING EXTENSION, LIST COLORING etc. In this paper, we focus on the LIST COLORING problem defined as follows.

LIST COLORING PROBLEM
Input: A graph $G = (V, E)$ and a LIST L of $|V|$ many set of colors with $L(v)$ being the entry for $v \in V$
Question: Is there an assignment of colors $c : V \to \cup_{v \in V} L(v)$ such that it respects the lists L, i.e. for any vertex v, $c(v) \in L(v)$ and for any two adjacent vertices u and v, $c(v) \neq c(u)$?

A list L is ℓ-REGULAR if each set contains exactly l colors. ℓ-REGULAR LIST COLORING problem is to decide whether $G = (V, E)$ has a coloring that respects L, where L is ℓ-REGULAR.

Note that when all the lists $L(v) = [k]$, the problem becomes the k-COLORING problem.

We begin with the notions of parameterized complexity before we explain our results.

Parameterized Complexity. The goal of parameterized complexity is to find ways of solving NP-hard problems more efficiently than brute force: here the aim is to restrict the combinatorial explosion to a parameter that is hopefully much smaller than the input size. A *parameterization* of a problem is assigning a positive integer parameter k to each input instance. Formally we say that a parameterized problem is *fixed-parameter tractable* (FPT) if there is an algorithm $\mathcal{A}$ (called a *fixed-parameter tractable algorithm*), a computable function f, and a constant c such that, given problem instance (x,k), $\mathcal{A}$ correctly solves the problem in time bounded by $f(k) \cdot |(x, k)|^c$, where x is the input and k is the parameter [5]. There is also an accompanying theory of parameterized intractability using which one can identify parameterized problems that are unlikely to admit FPT algorithms. These are essentially proved by showing that the problem is $W[1]$-hard [5].

A parameterized problem is called *slice-wise polynomial* (XP) if there exists an algorithm $\mathcal{A}$, two computable functions f, g such that, given problem instance (x,k), the algorithm $\mathcal{A}$ correctly solves the problem in time bounded by $f(k) \cdot |(x, k)|^{g(k)}$, where x is the input and k is the parameter [5]. The complexity class containing all slice-wise polynomial problems is called XP. We say that a parameterized problem is PARA-NP-hard if the problem is NP-hard for some fixed constant value of the parameter. PARA-NP-hard problems are not in XP unless $P = NP$.

Literature and Previous Work. As 3-coloring is NP-hard, the k-coloring problem is PARA-NP-hard when parameterized by the number of colors. Hence various other parameterizations have been studied for the COLORING problem. Some include structural parameterizations like the size of the vertex cover [11], treewidth [5], deletion distance to a graph class $\mathcal{G}$ where COLORING is solvable in polynomial time such as bipartite graphs, chordal graphs, complete graphs [3,12].

It is interesting to see if the COLORING problem is FPT for some parameterization, whether LIST COLORING problem which is a generalization of COLORING also has an FPT algorithm. For example, q-COLORING problem parameterized by the vertex cover size k has a $\mathcal{O}(q^k)$ algorithm. The same result can be extended to q-REGULAR LIST COLORING [11]. But there are also parameterizations where an FPT result in COLORING problem does not extend to LIST COLORING. For example, while COLORING is FPT when parameterized by the treewidth of the graph [5], LIST COLORING problem is W-hard for the same parameter [8]. See [12] for a summary of results on parameterizations of COLORING and LIST COLORING.

Using crown reduction (see Definition 1), it can be shown that it is fixed-parameter tractable to determine whether a graph can be colored with at most $n-k$ colors (here k is the parameter) [4]. We ask whether this result can be generalized to LIST COLORING by asking whether $(n-k)$-REGULAR LIST COLORING is FPT parameterized by k.

A previous result by Arora and a subset of authors [1] showed that $(n-k)$-REGULAR LIST COLORING is in XP. In this paper, we improve this result by showing that the problem is in FPT.

2 Preliminaries

Terminology and Notation. We use notations from the book of Diestel [7] for graph-related topics. Here we only define a few frequently used notations. Given a graph G, $V(G)$ and $E(G)$ denote its vertex-set and edge-set, respectively. The complement $\overline{G}$ of a graph G is the graph on $V(G)$ with edge set $[V]^2 \setminus E(G)$. Two vertices u and v of a graph G is neighbor (non-neighbor) if and only if $(u,v) \in E(G)$ $((u,v) \notin E(G))$. For any vertex $v \in V(G)$, we denote the set of neighbors (non-neighbors) of v by $N_G(v)$ $(\overline{N}_G(v))$ or briefly by $N(v)$ $(\overline{N}(v))^1$. For any $W \subseteq V(G)$ we define $N_G(W) = \{\cup_{v \in W} N_G(v)\}$ $(\overline{N}_G(W) = \{\cup_{v \in W}\overline{N}_G(v)\})$. For any subset $W \subseteq V(G)$, let $G[W]$ be the graph induced by the set of vertices W. Let $\mathcal{G}$ be any class of graphs. For any integer k, let $\mathcal{G} + k$ be the set of graphs G such that there exist a set of vertices $W \subseteq V(G)$ where $|W| = k$ and $G[V \setminus W] \in \mathcal{G}$.

Definition 1 *(Crown Decomposition). A crown decomposition of a graph G is a partitioning of $V(G)$ into sets C, H and R such that*

- *C is non-empty.*
- *C induces an independent set in G.*
- *There are no edges from C to R.*
- *G contains a matching of size $|H|$ between C and H.*

---

[1] In the rest of the paper, we drop the index referring to the underlying graph if the reference is clear.

Here C is said to be the crown and H is the head.

Theorem 2.1 *(Hall's theorem, [7]). Let G be an undirected bipartite graph with bipartition V_1 and V_2. The graph G has a matching saturating V_1 if and only if for all $X \subseteq V_1$, we have $|N(X)| \geq |X|$.*

Theorem 2.2 [10]. *Let $G = (V, E)$ be an undirected bipartite graph with bipartitions V_1 and V_2. Then in $O(|E|\sqrt{|V|})$ time we can either find a matching saturating V_1 or an inclusion wise minimal set $X \subseteq V_1$ such that $|N(X)| < |X|$.*

Theorem 2.3 [2]. LIST COLORING *can be solved in $\mathcal{O}(2^n)$ time.*

3 $(n-k)$-REGULAR LIST COLORING Is in FPT

We restate the following reduction rules and results from [1] without proofs.

Reduction Rule 1   [1]. *Delete any vertex with degree less than $(n-k)$.*

Lemma 3.1 [1]. *If there exists a set of k colors using which it is possible to color at least $2k$ vertices of G respecting the lists in L, then there is a feasible coloring for G respecting L.*

Lemma 3.2 [1]. LIST COLORING PROBLEM *is polynomial time solvable on a clique.*

Theorem 3.3 [1]. *$(n-k)$-REGULAR LIST COLORING can be solved in $n^{O(k)}$ time.*

We keep applying Reduction Rule 1 till it is no longer applicable and hence from now onwards we assume that every vertex has degree at least $(n-k)$. We can also assume that $n \geq 3k$ for otherwise, we can apply the algorithm of Theorems 3.3 or 2.3 to obtain a fixed-parameter tractable algorithm.

Let $C = \cup_{v \in V} L(v)$. We create a bipartite graph $G_B(V, C, E)$ with bipartization (V, C). There is an edge between v and a color c if $c \in L(v)$.

We start with the following new reduction rule.

Reduction Rule 2. *Let C' be an inclusion wise minimal subset of C such that $|N(C')| < |C'|$ in the graph G_B. Delete all the vertices in $N(C')$ from G.*

Lemma 3.4. *Reduction Rule 2 is safe and can be implemented in polynomial time.*

Proof. We use Theorem 2.2 to obtain the set C' if present in the graph G_B in polynomial time. Let $D = C' \setminus \{c\}$ for any arbitrary vertex $c \in C'$. Since C' is an inclusion wise minimal set satisfying the condition of the rule, for any subset $D' \subseteq D$, $|N(D')| \geq |D'|$. Hence by Hall's Theorem 2.1, there is a matching saturating D.

Let M be a matching saturating D into $N(D)$. We have a crown decomposition in the graph G_B with D as the crown and $N(D)$ as the head. Since $|N(C')| < |C'|$, we have $|N(D)| = |D|$ and $N(D) = N(C')$. Hence there are no unmatched vertices in $N(C')$ with respect to M.

Let us denote for each vertex $v \in N(C')$, m_v as the matching partner in D with respect to M. We show that there exists a list coloring respecting L in G if and only if there exists a list coloring respecting L in $G \setminus N(C')$. Since $G \setminus N(C')$ is a subgraph of G, the forward direction is true. In the converse, suppose there exists a coloring $\mathcal{C}$ that respecting L in the graph $G \setminus N(C')$. We extend this coloring $\mathcal{C}$ to a coloring $\mathcal{C}'$ in G by assigning color m_v to each vertex $v \in N(C')$. We claim that $\mathcal{C}'$ is a proper coloring respecting L. Suppose not. Then there exists a monochromatic edge $(u, v) \in E$. Since $\mathcal{C}$ is a valid coloring, either $u \in N(C')$ or $v \in N(C')$. Since all the vertices in $N(C')$ have different colors, both u and v cannot be in $N(C')$. Hence without loss of generality, assume $v \notin N(C')$. Since all the vertices of $N(C')$ are colored by using colors in C' and $v \notin N(C')$, color of v cannot be $m_u \in C'$ giving a contradiction. □

Note that when $|V| < |C|$, there is no matching in the bipartite graph G_B saturating C. Hence by Theorem 2.2, there exists a non-empty set C' which is inclusion wise minimal subset of C such that $|N(C')| < |C'|$. Hence Reduction Rule 2 can be applied reducing $|V|$ and $|C|$. When the rule can no longer be applied $|V| \geq |C|$. But note that if $|V| = |C|$ and the reduction rule can no longer be applied, there exists a matching M saturating C. We construct a feasible list coloring function $\mathcal{C}$ respecting L with $\mathcal{C}(v) = m_v$ where $v \in V$ and m_v the matching partner of v in M.

Hence we can assume that in the graph G, $|V| = n > |C|$.

For an edge $e = (u, v) \in \overline{G}$ we define a list $L(e) = L(u) \cap L(v)$. We call a matching M in $\overline{G}$ a MULTICOLOR MATCHING if it is possible to choose a distinct color from each $L(e)$ for every edge $e \in M$. Now, we have the following corollary of Lemma 3.1 as the $2k$ end points of the k matching edges can be colored with k colors.

Corollary 3.4.1. *If there exists a multicolor matching of size k in $\overline{G}$, then there is a feasible coloring for G with respecting L.*

Next we show that we can color each vertex of $G(V, E)$ with a different color, or there exists a MULTICOLOR MATCHING of size k or there exists a large clique in G.

Lemma 3.5. *For any $u, v \in V$, $|N_{G_B}(u) \cap N_{G_B}(v)| > n - 2k$ in G_B.*

Proof. Let $u, v \in V$. We have

$$n = |V| > |C|$$
$$\geq |N_{G_B}(u) \cup N_{G_B}(v)|$$
$$= |N_{G_B}(u)| + |N_{G_B}(v)| - |N_{G_B}(u) \cap N_{G_B}(v)|$$
$$\geq 2n - 2k - |N_{G_B}(u) \cap N_{G_B}(v)|$$

from which it follows that $n > 2n - 2k - |N_{G_B}(u) \cap N_{G_B}(v)|$ from which the claim follows.   □

Next we prove the following.

Lemma 3.6. *Either there is a multicolor matching of size k in $\overline{G}$ or there is a clique of size $n - 2k$ in G.*

Proof. Find any maximal matching M in $\overline{G}$. Suppose $|M| < k$. Let V_M be the set of end points of edges in M. The set of vertices $V \setminus V_M$ is an independent set in $\overline{G}$, therefore they form a clique in G. Since $|V \setminus V_M| > n - 2k$, one part of the lemma follows.

If $|M| \geq k$, choose exactly k edges of the matching and let's call this set of edges as M. From Lemma 3.5 we know that between any pair of vertices in V, there are at least $n - 2k$ shared colors. As we have assumed that $n \geq 3k$, we have $n - 2k \geq k$. Hence we can greedily assign an unassigned color to each edge of M starting from an arbitrary color for the first edge in M, resulting in a multicolor matching of size k in $\overline{G}$.   □

If there is a multicolor matching of size k in $\overline{G}$, then by Corollary 3.4.1, G can be list colored. Now, in what follows, we show that the list coloring can be determined in FPT time when G has a clique of size at least $n - 2k$. Towards this, we define the graph class CLIQUE+$f(k)$ whose members G has the property that there exists a subset of $f(k)$ vertices in G whose deletion results in a clique. Notice that $f(k) = 2k$ in the case that we ended up (Fig. 1).

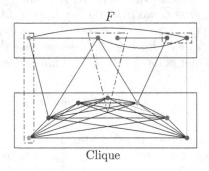

Fig. 1. List coloring in *clique* + $f(k)$

Theorem 3.7. *$(n - k)$-REGULAR LIST COLORING is FPT for the graph class* CLIQUE $+f(k)$.

Proof. Let $G(V, E)$ be a graph in CLIQUE $+ f(k)$ such that $V = D \cup F$ where D induces a clique, and $|F| \leq f(k)$. Any feasible coloring for G partitions V into different color classes where each color class induces an independent set. Now we show that given any partition $\mathcal{V} = \{V_1, V_2 \cdots V_l\}$ of V, in polynomial time, we can determine whether there is a list coloring of the vertices such that all vertices in $V_i \in \mathcal{V}$ are colored with a single color.

First observe that the following properties must be satisfied for $\mathcal{V}$ to be a partition of V into independent sets. We call a partition of V satisfying the following properties a *good partition* of V.

- Each subset V_i must be an independent set in G.
- For each subset $V_i \in \mathcal{V}$, $\cap_{v \in V_i} L(v) \neq \emptyset$
- For each subset $V_i \in \mathcal{V}$, $|D \cap V_i| \leq 1$

For a good partition $\mathcal{V}$, we create the following bipartite graph G_B with bipartization $(\mathcal{V}, C)$, where one partition $\mathcal{V}$ contains a vertex corresponding to each V_i. Recall that $C = \cup_{v \in V} L(v)$ is the set of colors. There is an edge between the vertex corresponding to V_i and c_j if and only if $c_j \in \cap_{v \in V_i} L(v)$. Then the following claim is easy to see.

Claim 1. *There is a feasible coloring of G with exactly the color classes in a good partition $\mathcal{V}$ if and only there is a matching saturating $\mathcal{V}$ in G_B.*

Proof. Let C be the feasible coloring of G. For each color class $V_i \in \mathcal{V}$, we have different colors $\mathcal{C}(V_i)$. Hence by the definition of the edges of G_B, we have a matching saturating $\mathcal{V}$, the matching edges being $(V_i, \mathcal{C}(V_i))$.

In the converse, let M be a matching saturating $\mathcal{V}$. Let c_{V_i} be the matching partner of V_i in M. We construct a coloring function $\mathcal{C} : V \to C$ such that $\mathcal{C}(v) = c_{V_i}$ if $v \in V_i$. Since $\mathcal{V}$ is a good partition, each $V_i \in \mathcal{V}$ is an independent set. Hence we have a feasible coloring of G. □

Let $X = F \cup \overline{N}_{G \backslash F}(F)$ where $\overline{N}_{G \backslash F}(F) = \cup_{v \in F} \overline{N}_{G \backslash F}(v)$, i.e. the union of non neighbors in $V \backslash F$ for each vertex $v \in F$. Define $Y = V \backslash X$. Then every vertex of Y is adjacent to all other vertices of V. Hence each vertex of Y should get a separate color that is separate from the colors of all other vertices of X in any proper coloring. Hence in a good partition $\mathcal{V}$, there will be a color class with the singleton element $\{y\}$ for every $y \in Y$. Let $\mathcal{Y}$ denote the partition of Y of such singleton sets.

Hence to check if there exists a list coloring of G, it suffices to test for every partition $\mathcal{X}$ of X, whether the partition $\mathcal{V}$ formed by the union of $\mathcal{X}$ and $\mathcal{Y}$ forms a feasible coloring. This can be tested using Claim 1 by going over all partitions of X.

The running time is bounded by $B_{|X|} n^{O(1)}$ where $B_{|X|}$ is the number of partitions of X which is $d^{O(d)}$ where $|X| = d$. For any vertex v of F, $|\overline{N}_G(v)| \leq k$ due to reduction Rule 1. Thus $|X| \leq f(k) + k \cdot f(k)$. Hence the overall runtime is $(f(k) + k \cdot f(k))^{O(f(k) + k \cdot f(k))} n^{O(1)}$. □

From Lemma 3.6, we know that either there is a multicolor matching of size k in $\overline{G}$ when we have a list coloring respecting the lists or that there is a clique of size at least $n - 2k$ in G when we can check if there is a list coloring in $(2k^2)^{O(2k^2)}$ time. The running time can be slightly improved (to $(2k^2)^{O(k)}$) by observing that the (non-neighbor) vertices in $\overline{N}(F)$ are part of a clique, and hence in any coloring they need separate colors. So it suffices to consider all partitions of vertices in F and all ways of including vertices of $V \setminus F$ in those partitions such that each color class has at most one vertex from $V \setminus F$. Thus we have the main result of the paper.

Theorem 3.8. $(n - k)$-REGULAR LIST COLORING *is FPT parameterized by k with running time* $(2k^2)^{O(k)} n^{O(1)}$.

4   Conclusion

We have shown that $(n - k)$-REGULAR LIST COLORING is FPT parameterized by k. Another well-studied notion in parameterized complexity is the notion of kernelization, where given an input instance (I, k) of the parameterized problem $\mathcal{Q}$, we use a polynomial time algorithm to convert it to an equivalent instance $(I', k') \in \mathcal{Q}$ where $|I'| \leq g(k)$ for some computable function g. The problem of determining whether there is a coloring using at most $n - k$ colors parameterized by k has a $O(k)$ sized kernel using crown decomposition. While we could use the crown decomposition in a reduction rule to reduce the number of colors and thereby obtain an FPT algorithm, we do not know how to obtain an equivalent instance with polynomial (in k) number of vertices. We leave it as an open problem to see if there exists a polynomial kernel for $(n - k)$-REGULAR LIST COLORING. Also the parameterized complexity of $(n - k)$-REGULAR LIST COLORING parameterized by k where we want to minimize the total number of colors used is left open.

References

1. Arora, P., Banik, A., Paliwal, V.K., Raman, V.: Some (in)tractable parameterizations of coloring and list-coloring. In: Chen, J., Lu, P. (eds.) FAW 2018. LNCS, vol. 10823, pp. 126–139. Springer, Cham (2018). https://doi.org/10.1007/978-3-319-78455-7_10
2. Björklund, A., Husfeldt, T., Koivisto, M.: Set partitioning via inclusion-exclusion. SIAM J. Comput. **39**(2), 546–563 (2009)
3. Cai, L.: Parameterized complexity of vertex colouring. Discrete Appl. Math. **127**(3), 415–429 (2003)
4. Chor, B., Fellows, M., Juedes, D.: Linear kernels in linear time, or how to save k colors in $O(n^2)$ steps. In: Hromkovič, J., Nagl, M., Westfechtel, B. (eds.) WG 2004. LNCS, vol. 3353, pp. 257–269. Springer, Heidelberg (2004). https://doi.org/10.1007/978-3-540-30559-0_22
5. Cygan, M., et al.: Parameterized Algorithms. Springer, Cham (2015). https://doi.org/10.1007/978-3-319-21275-3
6. Dailey, D.P.: Uniqueness of colorability and colorability of planar 4-regular graphs are NP-complete. Discrete Math. **30**(3), 289–293 (1980)
7. Diestel, R.: Graph Theory. Graduate Texts in Mathematics, vol. 173. Springer, Heidelberg (2017). https://doi.org/10.1007/978-3-662-53622-3

8. Fellows, M.R., et al.: On the complexity of some colorful problems parameterized by treewidth. Inf. Comput. **209**(2), 143–153 (2011)
9. Garey, M.R., Johnson, D.S., Stockmeyer, L.J.: Some simplified NP-complete graph problems. Theor. Comput. Sci. **1**(3), 237–267 (1976)
10. Hopcroft, J.E., Karp, R.M.: An $n^{5/2}$ algorithm for maximum matchings in bipartite graphs. SIAM J. Comput. **2**(4), 225–231 (1973)
11. Jansen, B.M.P., Kratsch, S.: Data reduction for graph coloring problems. Inf. Comput. **231**, 70–88 (2013)
12. Paulusma, D.: Open problems on graph coloring for special graph classes. In: Mayr, E.W. (ed.) WG 2015. LNCS, vol. 9224, pp. 16–30. Springer, Heidelberg (2016). https://doi.org/10.1007/978-3-662-53174-7_2

Finding Periods in Cartesian Tree Matching

Magsarjav Bataa[1], Sung Gwan Park[1], Amihood Amir[2], Gad M. Landau[3], and Kunsoo Park[1(✉)]

[1] Seoul National University, Seoul, South Korea
{magsarjav,sgpark,kpark}@theory.snu.ac.kr
[2] Bar-Ilan University, Ramat Gan, Israel
amir@esc.biu.ac.il
[3] University of Haifa, Haifa, Israel
landau@univ.haifa.ac.il

Abstract. In Cartesian tree matching, two strings match if the Cartesian trees of the strings are the same. In this paper we define full, initial, and general periods in Cartesian tree matching, and present an $O(n)$ time algorithm for finding all full periods, an $O(n \log \log n)$ time algorithm for finding all initial periods, and an $O(n \log n)$ time algorithm for finding all general periods of a string of length n.

Keywords: Cartesian tree matching · Parent-distance representation · Period

1 Introduction

Pattern matching is a fundamental problem in many applications, where *matching* is defined in various ways (i.e., various *metrics* for matching). Standard string matching is the basic metric of matching, but there are many generalizations of matching (called *generalized matching*) such as parameterized matching [4,7], jumbled matching [9], order-preserving matching [20,21,24], matching with don't care [15,19], swapped matching [3,18], etc.

The Cartesian tree has been used in many topics such as two-dimensional searching, rank-select data structures, and range minimum queries [16,29]. Recently, Cartesian tree matching was introduced as a new metric of matching, where two strings match if they have the same Cartesian trees [26]. Cartesian tree matching has a weaker matching condition than order-preserving matching [21].

In many metrics of generalized matching, some common problems have been studied. Exact matching is the basic problem in any metric of generalized matching [22]. Another problem is approximate matching, which allows some errors for matching. This problem was studied on jumbled matching [10], order-preserving matching [11], parameterized matching [6], swapped matching [5] and matching with don't care [1]. Yet another problem is to define and build an index data

C. J. Colbourn et al. (Eds.): IWOCA 2019, LNCS 11638, pp. 70–84, 2019.
https://doi.org/10.1007/978-3-030-25005-8_7

structure for a metric of generalized matching. This problem was studied on jumbled matching [2], order-preserving matching [13], matching with don't care [12], and Cartesian tree matching [26].

An interesting problem to study for a metric of generalized matching is to find a *period* from a given string. Even though the period of a string can be computed directly from the border array [22] for standard string matching, it is usually not the case for generalized matching. Furthermore, in contrast to the case of standard string matching, there can be several different definitions of the period for generalized matching. There are several studies on finding the periods in generalized matching. For example, there are many studies about Abelian periods, which is a period in the metric of jumbled matching [14, 23]. Another is to find periods in the metric of order-preserving matching [17].

In this paper we define four different types of periods, called *full period, initial period, general period*, and *sliding period* in Cartesian tree matching, following the definitions in [17]. We also present efficient algorithms to compute the periods in Cartesian tree matching, using the parent-distance representation of a Cartesian tree and the Cartesian suffix tree which were introduced in [26].

In Sect. 2 we give basic notations and definitions for Cartesian tree matching, and formalize the problems of finding periods. In Sect. 3 we present an $O(n)$ algorithm for computing all full periods and an $O(n \log \log n)$ algorithm for computing all initial periods, where n is the length of the given string. In Sect. 4 we present an $O(n \log n)$ algorithm for computing all general periods. Sliding periods are described in Sect. 5.

2 Preliminaries

2.1 Basic Notations

A *string* is defined as a sequence of characters in an alphabet which is a set of integers. We assume that we can compare any two characters in $O(1)$ time. Given a string S, the i-th character of S is denoted by $S[i]$. A substring (i.e., factor) of S which starts from i and ends at j is denoted by $S[i..j]$. Hence, $S[1..i]$ is the prefix of S ending at i, and $S[i..n]$ is the suffix of S starting from i, assuming that n is the length of S. For two integers $a \leq b$, $[\![a, b]\!]$ denotes the set $\{a, a + 1, \ldots, b\}$.

2.2 Cartesian Tree Matching

Given a string S of length n, we define the corresponding Cartesian tree $CT(S)$ [29] as follows:

- If S is empty, $CT(S)$ is an empty tree.
- If S is not empty and $S[i]$ is the minimum value among S, $CT(S)$ is the tree with root $S[i]$, left subtree $CT(S[1..i-1])$, and right subtree $CT(S[i+1..n])$. If there are multiple minimum values, the leftmost one becomes the root.

Cartesian tree matching [26] means that two strings S_1 and S_2 match (which is denoted by $S_1 \approx S_2$) if their Cartesian trees are the same.

Every order-preserving matching [21] is also a Cartesian tree matching, but the converse is not true as follows. If two strings are matched in the metric of order-preserving matching, the indices of minimum values of the two strings are located in the same positions. Furthermore, this property holds recursively in the substrings to the left and to the right of the minimum values. Hence, both strings have the same Cartesian trees. For the other direction, let's consider the text $T = (5, 2, 8, 6, 8, 3, 9, 7, 2, 1)$ and the pattern $P = (2, 1, 4, 3)$. In the metric of order-preserving matching, only $T[1..4]$ matches P. In Cartesian tree matching, both $T[1..4]$ and $T[5..8]$ match P. That is, $T[5..8]$ shows that a Cartesian tree matching may not be an order-preserving matching. Therefore, Cartesian tree matching has a weaker matching condition than order-preserving matching. Both order-preserving matching and Cartesian tree matching satisfy the properties of a *substring consistent equivalence relation (SCER)* [25].

We can represent the Cartesian tree using the *parent-distance representation* [26]. Given a string $S[1..n]$, the corresponding parent-distance representation $PD(S)$ is defined as follows:

$$PD(S)[i] = \begin{cases} i - \max_{1 \leq k < i}\{k : S[k] \leq S[i]\} & \text{if such } k \text{ exists} \\ 0 & \text{otherwise} \end{cases}$$

For example, $S = (5, 2, 8, 6, 8, 3, 9, 7, 2, 1)$ has the parent-distance representation $PD(S) = (0, 0, 1, 2, 1, 4, 1, 2, 7, 0)$. Two strings S_1 and S_2 have the same Cartesian trees if and only if they have the same parent-distance representations, i.e., $CT(S_1) = CT(S_2) \Leftrightarrow PD(S_1) = PD(S_2)$ [26]. Hence we can check whether two strings match by comparing their parent-distance representations. The parent-distance representation of $S[1..n]$ can be built in $O(n)$ time [26]. Furthermore, when $PD(S)$ is given, we can compute the parent-distance representation of any substring of S efficiently by Lemma 1.

Lemma 1 ([26]). Given valid indices i, j, k, we can compute $PD(S[i..j])[k]$ in $O(1)$ time using the following equation:

$$PD(S[i..j])[k] = \begin{cases} 0 & \text{if } PD(S)[i + k - 1] \geq k \\ PD(S)[i + k - 1] & \text{otherwise.} \end{cases}$$

An index data structure called *Cartesian suffix tree* of S is a compact trie built with the parent-distance representations of the suffixes of S. The Cartesian suffix tree of $S[1..n]$ can be built in randomized $O(n)$ time or deterministic $O(n \log n)$ time [26].

2.3 Problem Definition

There are various definitions of periods [14,17,23,25]. In the metric of Cartesian tree matching, we define full, initial, and general periods as follows.

Definition 1 (Full period). If a string S can be decomposed into k factors of length p and all of them have the same Cartesian trees, i.e., $S = V_1 \cdot V_2 \cdots V_k$ and $V_1 \approx V_2 \approx \cdots \approx V_k$, we say that p is a *full period* of S.

Definition 2 (Initial period). If a string S can be decomposed into k factors (i.e., $S = V_1 \cdot V_2 \cdots V_k$) such that $|V_1| = |V_2| = \cdots = |V_{k-1}| = p$, $V_1 \approx V_2 \approx \cdots \approx V_{k-1}$, and the last factor V_k of length $\leq p$ matches a prefix of V_1, we say that p is an *initial period* of S.

Definition 3 (General period). If a non-empty string X exists such that $|X| \leq p$ and p is an initial period of $X \cdot S$, we say that p is a *general period* of S and $p - |X|$ is a *valid shift* of the period p.

Note that a valid shift of a general period p represents the length of the first factor of S if we decompose S into factors such that all factors except the first and the last have the same length p. If we decompose S into such factors, there should exist a string P that matches all the factors except the first and the last, the first factor should match a suffix of P, and the last factor should match a prefix of P.

A general period p can have multiple valid shifts. The set of valid shifts of a general period p is denoted by $\text{Shift}_p \subseteq \{0, 1, \ldots, p - 1\}$. Shift_p is not empty if and only if p is a general period of S. It is easy to see that an initial period is a special case of a general period when $0 \in \text{Shift}_p$.

The problems to solve in this paper are defined as follows: Given a string S of length n, find all full periods and all initial periods; in the case of general periods, find Shift_p for every general period p.

Table 1. Full, initial, and general periods of $S = (5, 2, 8, 6, 8, 3, 9, 7, 2, 1)$

i	1	2	3	4	5	6	7	8	9	10
$S[i]$	5	2	8	6	8	3	9	7	2	1
$p = 2$	0	0	0	0	0	0	0	0	0	0
$p = 4$	0	0	1	2	0	0	1	2	0	0
$p = 6, s = 2$	0	0	0	0	1	0	1	2	0	0
$p = 8, s = 3$	0	0	1	0	1	0	1	2	0	0

Example 1. Table 1 shows the three types of periods for $S = (5, 2, 8, 6, 8, 3, 9, 7, 2, 1)$. Each segment separated by vertical lines shows the parent-distance representation of the corresponding factor.

- The third line shows the only non-trivial full period $p = 2$. Since $PD(S[1..2]) = PD(S[3..4]) = PD(S[5..6]) = PD(S[7..8]) = PD(S[9..10]) = (0,0)$, S has a full period $p = 2$. Note that 1 and 10 are trivial full periods of S.
- The fourth line shows an initial period $p = 4$. Since $PD(S[1..4]) = PD(S[5..8])$ and $PD(S[1..2]) = PD(S[9..10])$, S has an initial period $p = 4$. Note that $1, 2, 8, 9$ and 10 are also initial periods of S. The period $p = 4$ is not an initial period in the metric of order-preserving matching [17], because $S[1..4]$ and $S[5..8]$ do not match in order-preserving matching.
- The fifth line shows a general period $p = 6$ with shift $s = 2$, because $PD(S[3..4]) = PD(S[9..10])$ and $PD(S[1..2]) = PD(S[7..8])$. Note that the first factor has the length of $s = 2$.
- The sixth line shows a general period $p = 8$ with shift $s = 3$. Note that there is no factor with length $p = 8$. By the definition of general periods, however, if we take $X = (7, 8, 6, 10, 9)$ for example, we can check that $X \cdot S$ has an initial period 8. Therefore, we can conclude that S has a general period $p = 8$ with shift $s = 3$. All the general periods of S can be seen in Table 5.

Remark 1. There is one more definition of the period in [17], called *sliding period*. We will discuss it in Sect. 5.

3   Full Periods and Initial Periods

In this section we describe algorithms to compute all the full periods and initial periods of a string S of length n in $O(n)$ and $O(n \log \log n)$ time, respectively.

3.1   Prefix Table and Prefix-Shift Table

Definition 4. For a string S of length n, Pref[i], the i-th element of the *prefix table*, is the length of the longest prefix of $S[i..n]$ that matches a prefix of S in Cartesian tree matching.

This table is a direct analogue of the prefix table in standard string matching. Algorithm 1 shows a linear time algorithm to construct the Pref table (which is derived from the algorithm in [8]), where PARENT-DIST-REP is the procedure to compute the parent-distance representation of S [26].

Algorithm 1. Computing prefix table of S

1: **procedure** PREFIX-TABLE($S[1..n]$)
2:   $PD(S) \leftarrow$ PARENT-DIST-REP($S[1..n]$)
3:   Pref[1] $\leftarrow n$
4:   Pref[2] $\leftarrow 1$
5:   **while** Pref[2] $< n - 1$ **and** $PD(S)[\text{Pref}[2] + 1] = PD(S[2..n])[\text{Pref}[2] + 1]$ **do**
6:       Pref[2] $\leftarrow$ Pref[2] $+ 1$ $\triangleright$ by Lemma 1
7:   $l \leftarrow 2$
8:   **for** $i \leftarrow 3$ **to** n **do**
9:       $w \leftarrow l + \text{Pref}[l] - i$
10:      **if** Pref[$i - l + 1$] $< w$ **then**
11:          Pref[i] $\leftarrow$ Pref[$i - l + 1$]
12:      **else**
13:          $l \leftarrow i$
14:          **while** $w \leq n - i$ **and** $PD(S)[w + 1] = PD(S[i..n])[w + 1]$ **do**
15:              $w \leftarrow w + 1$ $\triangleright$ by Lemma 1
16:          Pref[i] $\leftarrow w$
17:      **return** Pref[1..n]

Algorithm 2. Computing all initial periods of S

1: **procedure** INITIAL-PERIOD($S[1..n]$)
2:   $IP \leftarrow$ an empty list
3:   Pref[1..n] $\leftarrow$ PREFIX-TABLE($S[1..n]$) $\triangleright$ by Algorithm 1
4:   Compute Pref-sh[1..n] from Pref[1..n] $\triangleright$ by Definition 5
5:   $P[1..n] \leftarrow$ Pref-sh[1..n]
6:   PRIME-LIST(n) $\leftarrow$ Compute primes from 2 to n in increasing order
7:   **for** $i \leftarrow n$ **down to** 1 **do**
8:       **for** $j \in$ PRIME-LIST(n) **do**
9:           **if** $i \cdot j > n$ **then**
10:              **break**
11:          $P[i] \leftarrow \min(P[i], P[i \cdot j])$
12:      **for** $p \leftarrow 1$ **to** n **do**
13:          **if** $P[p] \geq p$ **then**
14:              $IP.insert(p)$
15:      **return** IP

Definition 5. A *prefix-shift table*, Pref-sh[1..n], is defined as follows:

$$\text{Pref-sh}[i] = \begin{cases} n & i = n \text{ or Pref}[i + 1] = n - i \\ \text{Pref}[i + 1] & \text{otherwise.} \end{cases}$$

The prefix-shift table can also be computed in $O(n)$ time from the prefix table according to the above definition.

3.2   Computing Initial Periods in $O(n \log \log n)$ Time

Given a string S of length n, we can compute the prefix table and the prefix-shift table in linear time by Algorithm 1 and Definition 5, respectively. If p is an initial period of S, Pref$[qp + 1] \geq p$ or Pref$[qp + 1] = n - qp$ should be satisfied for all $1 \leq q < n/p$. This condition is the same as Pref-sh$[qp] \geq p$ for all $1 \leq q \leq n/p$. We can check it for all $1 \leq p \leq n$ in $O(n(1 + \frac{1}{2} + \cdots + \frac{1}{n})) = O(n \log n)$ time by checking every possible value of q. We can improve the time complexity to $O(n \log \log n)$ by employing Eratosthenes's sieve as in [17].

Algorithm 2 describes the algorithm to compute all the initial periods of a string. Algorithm 2 is different from the algorithm in [17] that computes initial periods in order-preserving matching, though both algorithms use Eratosthenes's sieve, and it is easier to prove the correctness. First, we compute the primes from 2 to n by Eratosthenes's sieve in $O(n \log \log n)$ time [27]. We define $P_m[p]$ as follows:

$$P_m[p] = \min_{1 \leq q \leq n/p} \{\text{Pref-sh}[qp]\}.$$

At the end of the iteration (with i) of the loop in lines 7–11, we maintain the loop invariant that $P[i]$ stores $P_m[i]$. When $i = n$ initially, the invariant holds trivially. At the beginning of the iteration with an arbitrary i, therefore, $P[p]$ for any $p > i$ stores $P_m[p]$. During the loop in lines 8–11, we compute the minimum of Pref-sh$[i]$, $P_m[2i]$, $P_m[3i]$, $P_m[5i]$, etc. (which is $P_m[i]$), and so $P[i]$ becomes $P_m[i]$ at the end of the loop. The loop in lines 7–11 has the same time complexity as Eratosthenes's sieve. Finally, we check whether p is an initial period of S by checking the condition $P[p] \geq p$. Therefore, the time complexity of Algorithm 2 is $O(n \log \log n)$.

Table 2. Process of Algorithm 2 for input $S = (5, 3, 6, 6, 2, 4, 8, 6, 6, 4, 2, 3, 5, 1)$

i	1	2	3	4	5	6	7	8	9	10	11	12	13	14
$S[i]$	5	3	6	6	2	4	8	6	6	4	2	3	5	1
$PD(S)[i]$	0	0	1	1	0	1	1	2	1	4	6	1	1	0
Pref$[i]$	14	1	1	4	1	1	3	1	2	5	1	1	2	1
Pref-sh$[i]$	1	1	4	1	1	3	1	2	14	1	1	14	14	14
$P_m[i]$	1	1	3	1	1	3	1	2	14	1	1	14	14	14

For example, Table 2 shows the process of Algorithm 2 when the input is string $S = (5, 3, 6, 6, 2, 4, 8, 6, 6, 4, 2, 3, 5, 1)$. Since $P[p] \geq p$ for $p = \{1, 3, 9, 12, 13, 14\}$, we can conclude that they are the initial periods of S.

Remark 2. If we need only the smallest initial period which is greater than one, we can compute it in $O(n)$ time using an algorithm similar to the one in [17].

Algorithm 3. Computing all full periods of S

```
 1: procedure FULL-PERIOD(S[1..n])
 2:     FP ← an empty list
 3:     Pref[1..n] ←PREFIX-TABLE(S[1..n])
 4:     Compute Pref-sh[1..n] from Pref[1..n]
 5:     P_gcd[1..n] ← ∞
 6:     for i ← 1 to n do
 7:         p ← gcd(n, i)
 8:         P_gcd[p] ← min(P_gcd[p], Pref-sh[i])
 9:     P[1..n] ← ∞
10:     for p ∈ DIVISORS(n) do
11:         for j ∈ DIVISORS(n) do
12:             if p | j then
13:                 P[p] ←min(P[p], P_gcd[j])
14:     for p ∈ DIVISORS(n) do
15:         if P[p] ≥ p then
16:             FP.insert(p)
17:     return FP
```

3.3 Computing Full Periods in $O(n)$ Time

Given a string S of length n, we first compute the prefix table and the prefix-shift table in linear time. If p is a full period of S, both $p|n$ and $\mathrm{Pref}[qp + 1] \geq p$ for all $1 \leq q < n/p$ should be satisfied. This condition is the same as $p|n$ and $\mathrm{Pref\text{-}sh}[qp] \geq p$ for all $1 \leq q \leq n/p$. We can check it for all $p|n$ in $O(n \log n)$ time by checking every value of q. However, we can improve the time complexity to linear time as follows.

Algorithm 3 describes the algorithm to compute all the full periods of a string. Algorithm 3 is based on the one in [17], but it is easier to prove the correctness. We use the same definition of $P_m[i]$ as in the initial period case, but we also use the fact that we need only compute $P_m[p]$ for $p|n$. Note that the condition of both $p|n$ and $p|i$ is equivalent to $p|\gcd(n, i)$. Therefore, $P_m[p]$ for $p|n$ is the minimum value of $\mathrm{Pref\text{-}sh}[i]$ for every i such that $p|\gcd(n, i)$. We do this computation in two stages. First, we define a new array P_{gcd}:

$$P_{\mathrm{gcd}}[p] = \min_{1 \leq i \leq n} \{\mathrm{Pref\text{-}sh}[i] : p = \gcd(n, i)\}.$$

Then we compute $P_m[p]$ using the fact that $P_m[p] = \min\{P_{\mathrm{gcd}}[j] : p \mid j\}$.

Since we can compute $\gcd(n, i)$ for all $1 \leq i \leq n$ in $O(n)$ time (also the divisors of n) [23], the computation of $P_{\mathrm{gcd}}[p]$ in lines 5–8 can be done in linear time. The computation of $P_m[p] = \min\{P_{\mathrm{gcd}}[j] : p \mid j\}$ in lines 9–13 can be done in $O(n)$ time because there are $O(\sqrt{n})$ divisors of n. Therefore, we can compute the full periods of S in linear time.

For example, Table 3 shows the process of Algorithm 3 when the input is $S = (3, 1, 6, 4, 8, 6, 7, 5, 8, 6, 9, 7)$. Note that we need the values of $P_m[p]$ and $P_{\mathrm{gcd}}[p]$ when $p \mid n$, and so the values of $P_m[p]$ and $P_{\mathrm{gcd}}[p]$ such that $p \nmid n$ are omitted in the table. We can conclude that $p = \{1, 2, 6, 12\}$ are the full periods of S since $P_m[p] \geq p$.

Table 3. Process of Algorithm 3 for input $S = (3, 1, 6, 4, 8, 6, 7, 5, 8, 6, 9, 7)$

i	1	2	3	4	5	6	7	8	9	10	11	12
$S[i]$	3	1	6	4	8	6	7	5	8	6	9	7
$PD(S)[i]$	0	0	1	2	1	2	1	4	1	2	1	2
Pref$[i]$	12	1	5	1	3	1	6	1	4	1	2	1
Pref-sh$[i]$	1	5	1	3	1	12	1	12	1	12	12	12
$P_{\mathrm{gcd}}[i]$	1	5	1	3		12						12
$P_m[i]$	1	3	1	3		12						12

4 General Periods

In this section we describe an algorithm to compute all the general periods of a string S of length n in $O(n \log n)$ time.

4.1 Properties of General Periods

Computing the general periods of a string relies on the following properties of a valid shift, as in the metric of order-preserving matching [17].

Lemma 2 ([17] **Observation 23**). Given a string S of length n and a period p $(1 \leq p \leq n)$, s is a valid shift (i.e., $s \in \text{Shift}_p$) if and only if the following three conditions hold:

(1) p is a full period of $S[s + 1..t]$,
(2) $S[1.. \min(n - p, s)] \approx S[p + 1.. \min(n, p + s)]$, and
(3) $S[1 + \max(0, t - p)..n - p] \approx S[1 + \max(p, t)..n]$,

where $t = n - ((n - s) \bmod p)$.

Definition 6. An *interval representation* $I = [\![i_1, j_1]\!] \cup [\![i_2, j_2]\!] \cup \cdots \cup [\![i_k, j_k]\!]$ is the union of disjoint and ordered intervals (in increasing order); $|I| = k$ is called the *size* of the interval representation.

Definition 7. We define the following shifts from Lemma 2:

- *middle shifts*: Middle$_p$ is a set of shifts that satisfy Condition (1) and is represented by a union of intervals.
- *left shifts*: Left$_p$ is a set of shifts that satisfy Condition (2) and is represented by an interval.

– *right shifts*: Right_p is a set of shifts that satisfy Condition (3) and is represented by an interval or a union of two intervals.

We will compute Middle_p, Left_p, and Right_p separately, and take the intersection of them to compute the valid shifts of a general period p.

Lemma 3 ([17] **Lemma 21**). *Let I be an interval representation.*

(a) Let interval representation G be $[\![i, j]\!] \cap I$. Then $|G| \leq |I|$, and G can be computed in $O(|I|)$ time.
(b) Let $I \bmod p$ denote $\{x \bmod p : x \in I\}$ for a positive integer p. If we have k interval representations $I_1, I_2, \ldots, I_k$ and k positive integers $p_1, p_2, \ldots, p_k \leq n$, we can compute $I_1 \bmod p_1, I_2 \bmod p_2, \ldots, I_k \bmod p_k$ in $O(|I_1| + |I_2| + \cdots + |I_k| + n)$ time.

4.2 Computing Left and Right Shifts

Consider a fixed value of p, which is a candidate for a general period. If $S[1..n-p] \approx S[p+1..n]$ or $p = n$, then any shift value $0 \leq s \leq p-1$ satisfies Condition (2) of Lemma 2, and so we have $\text{Left}_p = [\![0, p-1]\!]$. Note that $S[1..n-p] \approx S[p+1..n]$ is equivalent to $\text{Pref}[p+1] = n-p$. Otherwise, the left shifts of period p can be directly computed from the following equation in $O(1)$ time:

$$\text{Left}_p = \{s \in \{0, 1, \ldots, p-1\} : \text{Pref}[p+1] \geq s\}. \tag{1}$$

The right shifts of p can be computed similarly. But, we need a new definition called the reverse prefix table.

Definition 8. For a string S of length n, $\text{Pref}^R[i]$, the i-th element of the *reverse prefix table*, is the length of the longest suffix of $S[1..i]$ that matches a suffix of S in Cartesian tree matching.

The reverse prefix table can be constructed in $O(n)$ time using Algorithm 1. To compute the table, the input to Algorithm 1 is the reverse of S and the output of the algorithm should be reversed. We also need a minor change in computing the parent-distance representation when S contains equal numbers: the condition $value \leq S[i]$ in line 6 of procedure PARENT-DIST-REP in [26] should be changed to $value < S[i]$. For example, Table 4 shows the prefix table and the reverse prefix table of $S = (5, 2, 8, 6, 8, 3, 9, 7, 2, 1)$.

Table 4. Prefix table and reverse prefix table for $S = (5, 2, 8, 6, 8, 3, 9, 7, 2, 1)$

i	1	2	3	4	5	6	7	8	9	10
$S[i]$	5	2	8	6	8	3	9	7	2	1
$\text{Pref}[i]$	10	1	3	1	4	1	2	2	2	1
$\text{Pref}^R[i]$	1	2	1	2	1	2	1	2	3	10

Similarly to the left shifts, if $S[1..n-p] \approx S[p+1..n]$ (i.e., $\text{Pref}^R[n-p] = n-p$) or $p = n$, then $\text{Right}_p = [\![0, p-1]\!]$. Otherwise, the right shifts of period p can be computed in constant time by the following equation:

$$\text{Right}_p = \{s \in \{0, 1, \ldots, p-1\} : \text{Pref}^R[n-p] \geq (n-s) \bmod p\}$$

Interval representations of the left and right shifts are simple because the left shifts constitute one interval and the right shifts constitute one interval or the union of two intervals.

4.3 Computing Middle Shifts

Since computing the middle shifts is more complicated than the left and right shifts, the following extra definitions are used in the computation.

Definition 9. For a string S of length n and a period p, a substring $S[i..i+2p-1]$ is called a *square* if $S[i..i+p-1] \approx S[i+p..i+2p-1]$. A set of squares is defined as follows:

$$\text{Square}_p = \{i \in [\![1, n-2p+1]\!] : S[i..i+2p-1] \text{ is a square}\}.$$

Definition 10. A complement of Square_p (i.e., a set of *non-squares*) is defined as follows:

$$\text{NSquare}_p = [\![1, n-2p+1]\!] \backslash \text{Square}_p.$$

Definition 11. A set of *non-middle shifts* is defined as follows:

$$\text{NMiddle}_p = \{(x-1) \bmod p : x \in \text{NSquare}_p\}.$$

Once we have the non-middle shifts, the middle shifts are computed by the following equation:

$$\text{Middle}_p = [\![0, p-1]\!] \backslash \text{NMiddle}_p.$$

Note that all the above sets are represented by interval representations.

Theorem 1. Given a string S of length n, Square_p for all $1 \leq p \leq n$ can be computed in $O(n \log n)$ time and the total size of the interval representations is $O(n \log n)$.

Proof. We compute the Cartesian suffix tree of S in $O(n \log n)$ time [26] and use the same approach as the algorithms in [13, 17, 28], since the algorithms are based on the suffix tree. It is proved in [17] that the total size of the interval representations from the algorithms is $O(n \log n)$ (i.e., $\sum_{p=1}^{n} |\text{Square}_p| = O(n \log n)$).

Theorem 2. Given a string S of length n, the non-squares, non-middle shifts and middle shifts can be computed in $O(n \log n)$ time and the total size of the interval representations of the sets are bounded by $O(n \log n)$.

Proof. It is easy to see that $|\mathrm{NSquare}_p| \leq |\mathrm{Square}_p| + 1$ and we can compute it in $O(|\mathrm{Square}_p|)$ time. Thus $\sum_{p=1}^{n} |\mathrm{NSquare}_p| \leq \sum_{p=1}^{n} |\mathrm{Square}_p| + n$. Since $\sum_{p=1}^{n} |\mathrm{Square}_p| = O(n \log n)$ by Theorem 1, $\sum_{p=1}^{n} |\mathrm{NSquare}_p| = O(n \log n)$. Furthermore, we can compute them in $\sum_{p=1}^{n} O(|\mathrm{Square}_p|) = O(n \log n)$ time.

We can see that $|\mathrm{NMiddle}_p| \leq 2|\mathrm{NSquare}_p|$, because each interval is converted into at most two intervals by the modulo operation. Therefore, $\sum_{p=1}^{n} |\mathrm{NMiddle}_p| \leq 2 \cdot \sum_{p=1}^{n} |\mathrm{NSquare}_p| = O(n \log n)$. By using Lemma 3(b), we can compute them simultaneously in $O(n \log n)$ time.

As in $\mathrm{NSquare}_p$, we have $|\mathrm{Middle}_p| \leq |\mathrm{NMiddle}_p| + 1$, which leads to $\sum_{p=1}^{n} |\mathrm{Middle}_p| = O(n \log n)$, and we can compute them in $O(n \log n)$ time. □

4.4 Computing General Periods in $O(n \log n)$ Time

Given a string S of length n, we can compute the prefix table and the reverse prefix table in linear time by Algorithm 1 and Definition 8, respectively. We compute Left_p and Right_p for all $1 \leq p \leq n$ in $O(n)$ time from the prefix table and the reverse prefix table, as described in Sect. 4.2. Then we compute Square_p, $\mathrm{NSquare}_p$, $\mathrm{NMiddle}_p$, and Middle_p successively for all $1 \leq p \leq n$ in $O(n \log n)$ time by Theorems 1 and 2. Finally, we compute all the valid shifts of p as the intersection of Left_p, Right_p, and Middle_p for all $1 \leq p \leq n$ in $O(n \log n)$ time by Lemma 3(a) because $|\mathrm{Left}_p| = 1$ and $|\mathrm{Right}_p| \leq 2$. The valid shifts for all p are represented by interval representations.

For example, Table 5 shows the computation of the general periods for input $S = (5, 2, 8, 6, 8, 3, 9, 7, 2, 1)$. Let's consider the case of $p = 4$. From $\mathrm{Pref}[5] = 4$ in Table 4, we can compute $\mathrm{Left}_4 = [\![0, 3]\!]$ according to Eq. (1). Similarly, we can compute $\mathrm{Right}_4 = [\![0, 2]\!]$ from $\mathrm{Pref}^R[6] = 2$. For computing the squares, we can see that $\mathrm{Square}_4 = [\![1, 1]\!]$ because only $S[1..4] \approx S[5..8]$ holds. From Square_4, we can compute $\mathrm{NSquare}_4 = [\![2, 3]\!]$ using Definition 10, $\mathrm{NMiddle}_4 = [\![1, 2]\!]$ using Lemma 3(b), and $\mathrm{Middle}_4 = [\![0, 0]\!] \cup [\![3, 3]\!]$ from $\mathrm{NMiddle}_4$. By taking

Table 5. Computing the general periods for input $S = (5, 2, 8, 6, 8, 3, 9, 7, 2, 1)$

p	Left_p	Right_p	Square_p	$\mathrm{NSquare}_p$	$\mathrm{NMiddle}_p$	Middle_p	Shift_p
1	$[\![0,0]\!]$	$[\![0,0]\!]$	$[\![1,9]\!]$	$\emptyset$	$\emptyset$	$[\![0,0]\!]$	$[\![0,0]\!]$
2	$[\![0,1]\!]$	$[\![0,1]\!]$	$[\![1,5]\!] \cup [\![7,7]\!]$	$[\![6,6]\!]$	$[\![1,1]\!]$	$[\![0,0]\!]$	$[\![0,0]\!]$
3	$[\![0,1]\!]$	$[\![0,1]\!]$	$\emptyset$	$[\![1,5]\!]$	$[\![0,2]\!]$	$\emptyset$	$\emptyset$
4	$[\![0,3]\!]$	$[\![0,2]\!]$	$[\![1,1]\!]$	$[\![2,3]\!]$	$[\![1,2]\!]$	$[\![0,0]\!] \cup [\![3,3]\!]$	$[\![0,0]\!]$
5	$[\![0,1]\!]$	$[\![0,0]\!] \cup [\![4,4]\!]$	$\emptyset$	$[\![1,1]\!]$	$[\![0,0]\!]$	$[\![1,4]\!]$	$\emptyset$
6	$[\![0,2]\!]$	$[\![2,4]\!]$	$\emptyset$	$\emptyset$	$\emptyset$	$[\![0,5]\!]$	$[\![2,2]\!]$
7	$[\![0,2]\!]$	$[\![2,3]\!]$	$\emptyset$	$\emptyset$	$\emptyset$	$[\![0,6]\!]$	$[\![2,2]\!]$
8	$[\![0,7]\!]$	$[\![0,7]\!]$	$\emptyset$	$\emptyset$	$\emptyset$	$[\![0,7]\!]$	$[\![0,7]\!]$
9	$[\![0,8]\!]$	$[\![0,8]\!]$	$\emptyset$	$\emptyset$	$\emptyset$	$[\![0,8]\!]$	$[\![0,8]\!]$
10	$[\![0,9]\!]$	$[\![0,9]\!]$	$\emptyset$	$\emptyset$	$\emptyset$	$[\![0,9]\!]$	$[\![0,9]\!]$

the intersection of $Left_4$, $Right_4$, and $Middle_4$, we get $Shift_4 = [\![0, 0]\!]$. Therefore, $p = 4$ is a general period of S with shift 0.

5 Concluding Remarks

We have defined full, initial, and general periods in Cartesian tree matching, and presented an $O(n)$ algorithm for computing all full periods, $O(n \log \log n)$ algorithm for computing all initial periods and $O(n \log n)$ algorithm for computing all general periods of a string of length n.

There is another period of a string, called *sliding period*, introduced in [17]. This period is a special case of general period p in which $Shift_p = [\![0, p - 1]\!]$. An $O(n \log \log n)$ expected time or $O(n \log^2 \log n / \log \log \log n)$ worst-case time algorithm based on a suffix tree construction is presented for sliding periods in order-preserving matching [17].

In Cartesian tree matching, the sliding periods can be computed in $O(n \log n)$ time directly from the general periods. If we use the randomized $O(n)$ time algorithm for constructing a Cartesian suffix tree [26], we can compute the sliding periods in randomized $O(n)$ time because all the computations except constructing the suffix tree can be done in linear time using an algorithm similar to the one in [17].

Acknowledgements. M. Bataa, S.G. Park and K. Park were supported by Institute for Information & communications Technology Promotion(IITP) grant funded by the Korea government(MSIT) (No. 2018-0-00551, Framework of Practical Algorithms for NP-hard Graph Problems). A. Amir and G.M. Landau were partially supported by the Israel Science Foundation grant 571/14, and Grant No. 2014028 from the United States-Israel Binational Science Foundation (BSF).

References

1. Akutsu, T.: Approximate string matching with don't care characters. Inf. Process. Lett. **55**(5), 235–239 (1995). https://doi.org/10.1016/0020-0190(95)00111-O
2. Amir, A., Apostolico, A., Hirst, T., Landau, G.M., Lewenstein, N., Rozenberg, L.: Algorithms for jumbled indexing, jumbled border and jumbled square on run-length encoded strings. Theoret. Comput. Sci. **656**, 146–159 (2016). https://doi.org/10.1016/j.tcs.2016.04.030
3. Amir, A., Aumann, Y., Landau, G.M., Lewenstein, M., Lewenstein, N.: Pattern matching with swaps. J. Algorithms **37**(2), 247–266 (2000). https://doi.org/10.1006/jagm.2000.1120
4. Amir, A., Farach, M., Muthukrishnan, S.: Alphabet dependence in parameterized matching. Inf. Process. Lett. **49**(3), 111–115 (1994). https://doi.org/10.1016/0020-0190(94)90086-8
5. Amir, A., Lewenstein, M., Porat, E.: Approximate swapped matching. Inf. Process. Lett. **83**(1), 33–39 (2002). https://doi.org/10.1016/S0020-0190(01)00302-7
6. Apostolico, A., Erdos, P.L., Lewenstein, M.: Parameterized matching with mismatches. J. Discrete Algorithms **5**(1), 135–140 (2007). https://doi.org/10.1016/j.jda.2006.03.014

7. Baker, B.S.: A theory of parameterized pattern matching: algorithms and applications. In: STOC, pp. 71–80 (1993). https://doi.org/10.1145/167088.167115
8. Bland, W., Kucherov, G., Smyth, W.F.: Prefix table construction and conversion. In: Lecroq, T., Mouchard, L. (eds.) IWOCA 2013. LNCS, vol. 8288, pp. 41–53. Springer, Heidelberg (2013). https://doi.org/10.1007/978-3-642-45278-9_5
9. Burcsi, P., Cicalese, F., Fici, G., Liptak, Z.: Algorithms for jumbled pattern matching in strings. Int. J. Found. Comput. Sci. **23**(2), 357–374 (2012). https://doi.org/10.1142/S0129054112400175
10. Burcsi, P., Cicalese, F., Fici, G., Liptak, Z.: On approximate jumbled pattern matching in strings. Theory Comput. Syst. **50**(1), 35–51 (2012). https://doi.org/10.1007/s00224-011-9344-5
11. Chhabra, T., Giaquinta, E., Tarhio, J.: Filtration algorithms for approximate order-preserving matching. In: Iliopoulos, C., Puglisi, S., Yilmaz, E. (eds.) SPIRE 2015. LNCS, vol. 9309, pp. 177–187. Springer, Cham (2015). https://doi.org/10.1007/978-3-319-23826-5_18
12. Cole, R., Gottlieb, L.A., Lewenstein, M.: Dictionary matching and indexing with errors and don't cares. In: STOC, pp. 91–100 (2004). https://doi.org/10.1145/1007352.1007374
13. Crochemore, M., et al.: Order-preserving indexing. Theoret. Comput. Sci. **638**, 122–135 (2016). https://doi.org/10.1016/j.tcs.2015.06.050
14. Crochemore, M., et al.: A note on efficient computation of all Abelian periods in a string. Inf. Process. Lett. **113**(3), 74–77 (2013). https://doi.org/10.1016/j.ipl.2012.11.001
15. Fischer, M.J., Paterson, M.S.: String-matching and other products. Technical report, MIT Cambridge Project MAC (1974)
16. Gabow, H.N., Bentley, J.L., Tarjan, R.E.: Scaling and related techniques for geometry problems. In: STOC, pp. 135–143 (1984). https://doi.org/10.1145/800057.808675
17. Gourdel, G., Kociumaka, T., Radoszewski, J., Rytter, W., Shur, A.M., Walen, T.: String periods in the order-preserving model. In: STACS, pp. 38:1–38:16 (2018). https://doi.org/10.4230/LIPIcs.STACS.2018.38
18. Iliopoulos, C.S., Rahman, M.S.: A new model to solve the swap matching problem and efficient algorithms for short patterns. In: Geffert, V., Karhumäki, J., Bertoni, A., Preneel, B., Návrat, P., Bieliková, M. (eds.) SOFSEM 2008. LNCS, vol. 4910, pp. 316–327. Springer, Heidelberg (2008). https://doi.org/10.1007/978-3-540-77566-9_27
19. Kalai, A.: Efficient pattern-matching with don't cares. In: SODA, pp. 655–656 (2002)
20. Kim, J., Amir, A., Na, J.C., Park, K., Sim, J.S.: On representations of ternary order relations in numeric strings. Math. Comput. Sci. **11**(2), 127–136 (2017). https://doi.org/10.1007/s11786-016-0282-0
21. Kim, J., et al.: Order-preserving matching. Theoret. Comput. Sci. **525**, 68–79 (2014). https://doi.org/10.1016/j.tcs.2013.10.006
22. Knuth, D.E., Morris, J.H., Pratt, V.R.: Fast pattern matching in strings. SIAM J. Comput. **6**(2), 323–350 (1977). https://doi.org/10.1137/0206024
23. Kociumaka, T., Radoszewski, J., Rytter, W.: Fast algorithms for Abelian periods in words and greatest common divisor queries. J. Comput. Syst. Sci. **84**, 205–218 (2017). https://doi.org/10.1016/j.jcss.2016.09.003
24. Kubica, M., Kulczynski, T., Radoszewski, J., Rytter, W., Walen, T.: A linear time algorithm for consecutive permutation pattern matching. Inf. Process. Lett. **113**(12), 430–433 (2013). https://doi.org/10.1016/j.ipl.2013.03.015

25. Matsuoka, Y., Aoki, T., Inenaga, S., Bannai, H., Takeda, M.: Generalized pattern matching and periodicity under substring consistent equivalence relations. Theoret. Comput. Sci. **656**, 225–233 (2016). https://doi.org/10.1016/j.tcs.2016.02.017
26. Park, S.G., Amir, A., Landau, G.M., Park, K.: Cartesian tree matching and indexing. Accepted to CPM (2019). https://arxiv.org/abs/1905.08974
27. Sorenson, J.: An introduction to prime number sieves. Technical report, Department of Computer Sciences, University of Wisconsin-Madison (1990)
28. Stoye, J., Gusfield, D.: Simple and flexible detection of contiguous repeats using a suffix tree. Theoret. Comput. Sci. **270**(1), 843–856 (2002). https://doi.org/10.1007/bfb0030787
29. Vuillemin, J.: A unifying look at data structures. Commun. ACM **23**(4), 229–239 (1980). https://doi.org/10.1145/358841.358852

Parameterized Complexity of Min-Power Asymmetric Connectivity

Matthias Bentert$^{(\boxtimes)}$, Roman Haag, Christian Hofer, Tomohiro Koana, and André Nichterlein

Algorithmics and Computational Complexity, Faculty IV, TU Berlin, Berlin, Germany
{matthias.bentert,andre.nichterlein}@tu-berlin.de,
{roman.haag,hofer,koana}@campus.tu-berlin.de

Abstract. We investigate multivariate algorithms for the NP-hard MIN-POWER ASYMMETRIC CONNECTIVITY (MINPAC) problem that has applications in wireless sensor networks. Given a directed arc-weighted n-vertex graph, MinPAC asks for a strongly connected spanning subgraph minimizing the summed vertex costs. Here, the cost of each vertex is the weight of its heaviest outgoing arc. We present linear-time algorithms for the cases where the number of strongly connected components in a so-called obligatory subgraph or the feedback edge number in the underlying undirected graph is constant. Complementing these results, we prove that the problem is W[2]-hard with respect to the solution cost, even on restricted graphs with one feedback arc and binary arc weights.

1 Introduction

In wireless ad-hoc networks, nodes equipped with limited power supply transmit data using a multi-hop path. We study the problem of minimizing the overall power consumption while maintaining full network connectivity, that is, each node can send messages to each other node using some (multi-hop) route through the network. Formally, we study the following optimization problem.

MIN-POWER ASYMMETRIC CONNECTIVITY (MINPAC)
Input: A strongly connected digraph G and arc weights (costs) $w \colon A(G) \to \mathbb{N}$.
Task: Find a strongly connected spanning subgraph H of G minimizing

$$\sum_{v \in V} \max_{vu \in A(H)} w(vu).$$

Related Work. This problem was initially formalized and shown to be NP-complete by Chen and Huang [8]. Since then, there have been numerous publications on polynomial-time approximation algorithms (an asymptotically optimal $O(\log n)$ approximation [5], a constant approximation factor with symmetric arc weights [3,8], and approximation algorithms for special cases [4,6,7]), and hardness results about special cases [6,9]. To the best of our knowledge, however,

© Springer Nature Switzerland AG 2019
C. J. Colbourn et al. (Eds.): IWOCA 2019, LNCS 11638, pp. 85–96, 2019.
https://doi.org/10.1007/978-3-030-25005-8_8

Table 1. Overview of our results, using the following notation: n—number of vertices, m—number of arcs, c—number of strongly connected components in the obligatory subgraph (see Sect. 2), g—size of a minimum feedback edge set of the underlying undirected graph (see Sect. 3), q—number of different arc weights, x—size of a minimum vertex cover (see Sect. 4), h—size of a minimum feedback arc set (see Sect. 5), PAC is the decision version of MINPAC asking for a solution of cost at most k.

	Result	Reference
Section 2	Dynamic programming solving MINPAC in $O(2^c \cdot c \cdot n + m + 4^c \cdot c^{2c-3/2})$ time	Theorem 10
Section 3	Linear-time data reduction resulting in an equivalent MINPAC instance with at most $20g - 20$ vertices and $42g - 42$ arcs	Theorem 11
Section 4	An $O(xn + m)$-time data reduction resulting in an equivalent MINPAC instance with at most $(q + 1)^{2x} + x$ vertices	Theorem 14
Section 5	PAC is NP-hard for any $h \geq 1$	Theorem 15
	PAC is W[2]-hard parameterized by k, even if the arcs have only cost zero or one and $h = 1$	
	PAC is not solvable in $2^{o(n)}$ time (assuming ETH)	

there has been no work to study MINPAC from a parameterized complexity viewpoint.

In previous work, we investigated the parameterized complexity of the symmetric version of our problem [2], that is, an undirected edge can only be used if both endpoints pay at least the weight of the edge. The asymmetric case turns out to be more involved on a technical level. However, comparable results (as in the symmetric case) are achievable.

Our Contributions. We show algorithmic results for grid-like and tree-like input graphs as well as parameterized hardness for very restricted cases. Table 1 summarizes our results. We discuss the different parameters subsequently.

It is known that the alignment of nodes in some regular grid-like patterns is optimal to fully cover a plane. In such cases, we can assume that the *obligatory arcs*, arcs that are in any optimal solution, induce a small number c of strongly connected components as there are many arcs of minimum weight. In Sect. 2, we present an algorithm that solves MINPAC in linear time when c is a constant.

In Sect. 3, we describe a linear-time algorithm which reduces any input instance to an equivalent instance with at most $20g - 20$ vertices and $42g - 42$ arcs, where g is the feedback edge number of the underlying undirected graph. It follows that the problem can be solved in polynomial time for $g \in O(\log n)$, that is, for very tree-like input graphs. In terms of parameterized complexity, this gives us a partial (weights left unbounded) kernelization of MINPAC with respect to the feedback edge number.

In Sect. 4, we study MINPAC parameterized by the number of different arc weights and vertex cover number. For this combined parameter, we present an exponential-size kernel.

Finally, in Sect. 5 we derive hardness results for PAC, the decision version of MINPAC. We show that even if the input graph has only binary weights and is almost a DAG (a directed acyclic graph with one additional arc), PAC parameterized by the solution cost is W[2]-hard. This is in sharp contrast to the FPT result for the parameter feedback edge number.

Due to the lack of space several details are deferred to a full version.

Preliminaries. For $a \in \mathbb{N}$, we abbreviate $\{1, \ldots, a\}$ by $[a]$. Throughout the paper, we assume that a graph is directed unless stated otherwise. For a graph $G = (V, A)$, we write $V(G)$ to denote V and $A(G)$ to denote A. We denote by $G[V']$ the subgraph induced by $V' \subseteq V(G)$. We use $G + vu$ to denote $(V(G) \cup \{v, u\}, A(G) \cup \{vu\})$ and $G - vu$ to denote $(V(G), A(G) \setminus \{vu\})$. For a vertex $v \in V(G)$, we write $N_G^+(v) = \{u \mid vu \in A\}$ and $N_G^-(v) = \{u \mid uv \in A\}$ to denote the out- and in-neighborhood of v. We define the degree of v as $\deg_G(v) = |N_G^+(v) \cup N_G^-(v)|$. We say that $S \subseteq V(G)$ is strongly connected if there exists a path from each node $u \in S$ to every other node $v \in S$ in $G[S]$. We write S_G to denote the set of strongly connected components. We use U_G to denote the underlying undirected graph of G. We denote the optimal cost of a MINPAC instance I by $\text{OPT}(I)$. The cost of a vertex subset $V' \subseteq V(G)$ in a solution with arcs $A' \subseteq A(G)$ is denoted by $\text{Cost}(V', A', w) = \sum_{v \in V'} \max_{vu \in A'} w(vu)$.

A parameterized problem Π is a set of pairs (I, k), where I denotes the problem instance and k is the parameter. The problem Π is *fixed parameter tractable* (FPT) if there exists an algorithm solving any instance of Π in $f(k) \cdot |I|^c$ time, where f is some computable function and c is some constant. A *reduction to a problem kernel* is a polynomial-time algorithm that, given an instance (I, k) of Π, returns an equivalent instance (I', k'), such that $|I'| + k' \leq g(k)$ for some computable function g. Problem kernels are usually achieved by applying *data reduction rules*. Given an instance (I, k) for MINPAC, our data reduction rules compute in polynomial time a new instance (I', k') of MINPAC and a number d. We call a data reduction rule *correct*, if $\text{OPT}(I) = \text{OPT}(I') + d$.

2   Parameterization by the Number of Strongly Connected Components Induced by the Obligatory Arcs

In this section we present a fixed-parameter algorithm with respect to the number c of strongly connected components (SCCs) induced by *obligatory arcs*—arcs that can be included into any optimal solution with no additional cost. We find the obligatory arcs by means of lower bounds on costs paid by each vertex.

Definition 1. *A* vertex lower bound *is a function* $\ell \colon V(G) \to \mathbb{N}$ *such that for any optimal solution H and any vertex $v \in V(G)$, it holds that*

$$\ell(v) \leq \max_{vu \in A(H)} w(vu).$$

Observe that each vertex $v \in V(G)$ has at least one outgoing arc in any optimal solution. Hence, the cost paid by v in the optimal solution is at least $\min_{vu \in A(G)} w(vu)$. Thus, $\ell(v) \geq \min_{vu \in A(G)} w(vu)$. Moreover, if a vertex v has only one incoming arc uv, then the cost for the vertex u is at least $w(uv)$, and thus $\ell(u) \geq w(uv)$. Clearly, finding more effective but still efficiently computable vertex lower bounds is challenging on its own.

Definition 2. *The* obligatory subgraph G_ℓ *induced by a vertex lower bound ℓ for G is a subgraph $(V(G), A_\ell)$, where $A_\ell = \{vu \mid w(vu) \leq \ell(v)\}$.*

It has been shown that sensor placements are optimal for fully covering an area when sensors are deployed in the triangular lattice pattern [18] or a strip based-pattern [1,15]. In such cases, there are many arcs of minimum weight. Taking these arcs usually suffice to (almost) achieve connectivity. So even the obligatory subgraph induced by the trivial vertex lower bound described above yields a small number of SCCs.

Let ℓ be a vertex lower bound for a graph G. We denote the number of SCCs of the obligatory subgraph G_ℓ by $c = |\mathcal{S}_{G_\ell}|$. The number c of (strongly) connected components in the obligatory subgraph has recently been used as parameter to obtain FPT results [2,17]. In this section, we also provide an FPT result with respect to this parameter. More specifically, we will present an algorithm for MINPAC that runs in $O(2^c \cdot c^2 \cdot n + m + 4^c \cdot c^{2c-3/2})$ time. Our algorithm runs in three phases. In the first phase, it shrinks the graph to a *relevant* subgraph in which each vertex v has at most one arc towards each SCC that does not contain v (Algorithm 1). In the second phase, it uses a dynamic programming procedure to compute the cost to connect each SCC to each subset of other SCCs (Algorithm 2). In the last phase, it exhaustively tries all combinations of connecting SCCs to find an optimal solution (Algorithm 3).

Phase 1. The following lemma specifies the conditions under which we can remove arcs. It plays a central role in this phase. The basic idea herein is to remove, for each vertex $v \in V(G)$ and each SCC S, all but the cheapest arc from v to vertices in S.

Lemma 3. *Let (G, w) be an instance of* MINPAC *and let ℓ be a vertex lower bound. Let $S_v, S_u \in \mathcal{S}_{G_\ell}$ be two distinct SCCs and let $v \in S_v$ and $u, u' \in S_u$ be vertices of G with $w(vu) \leq w(vu')$. Then, it holds that $\mathrm{OPT}((G, w)) = \mathrm{OPT}((G - vu', w))$.*

Algorithm 1 exhaustively removes all arcs vu' which satisfy the preconditions of Lemma 3: The algorithm iterates over each arc in G twice. It finds a minimum-weight arc from each vertex to each SCC in the first iteration. In the second

---

Algorithm 1. A reduction procedure for the first phase

---

1 **Function** Reduction(G, w, ℓ)
2   **foreach** $v \in V(G), S \in \mathcal{S}_{G_\ell}$ **do** $M(v, S) \leftarrow$ **null**
3   **foreach** $vu \in A(G)$ **do**
4   **if** $M(v, S_u) =$ **null** or $w(vu) < w(M(v, S_u))$ **then** $M(v, S_u) \leftarrow vu$
        // $S_u \in \mathcal{S}_{G_\ell}$ denotes the SCC to which u belongs
5   **foreach** $vu \in A(G)$ **do**
6   **if** $S_v = S_u$ **then continue**
7   **if** $vu \neq M(v, S_u)$ **then** Remove vu from G
8   **return** (G, w)

---

iteration, it removes all but one minimum-weight arc that share the initial vertex and the SCC the terminal vertex belongs to.

We show subsequently that the resulting MINPAC instance satisfy the properties listed in the next definition.

Definition 4. *Let (G, w) be an instance of MINPAC and let ℓ be a lower bound. We say that a graph G_ℓ^{rel} and a weight function w_ℓ^{rel} with $V(G) = V(G_\ell^{\mathrm{rel}})$, $A(G_\ell^{\mathrm{rel}}) \subseteq A(G)$, and $w_\ell^{\mathrm{rel}} \colon A(G_\ell^{\mathrm{rel}}) \to \mathbb{N}$ are relevant subgraphs and relevant weight functions induced by ℓ, respectively, if they satisfy the following properties:*

(i) $\mathrm{OPT}((G, w)) = \mathrm{OPT}((G_\ell^{\mathrm{rel}}, w_\ell^{\mathrm{rel}}))$.
(ii) For any SCC $S \in \mathcal{S}_{G_\ell}$, it holds that $G[S] = G_\ell^{\mathrm{rel}}[S]$.
(iii) For any two distinct SCCs $S, S' \in \mathcal{S}_{G_\ell}$ and any vertex $v \in S$, it holds that $|\{vu \in A(G_\ell^{\mathrm{rel}}) \mid u \in S'\}| \leq 1$.

Since it follows from the property (ii) that $\mathcal{S}_{G_\ell} = \mathcal{S}_{G_\ell^{\mathrm{rel}}}$, we will use them interchangeably.

Lemma 5. *Let (G, w) be an instance of MINPAC and let ℓ be a vertex lower bound. Algorithm 1 computes in $O(cn + m)$ time a relevant subgraph G_ℓ^{rel} and a relevant weight function w_ℓ^{rel} induced by ℓ.*

Phase 2. In this phase, we aim to compute an optimal set of arcs to connect each SCC to all other SCCs. We start with some notation.

Definition 6. *Let G_ℓ^{rel} be a relevant subgraph. For any $S \in \mathcal{S}_{G_\ell}$, we define the set of SCCs reachable from S via an arc as*

$$\mathcal{S}_{G,\ell}^{\mathrm{reach}}(S) = \{S' \in \mathcal{S}_{G_\ell} \setminus \{S\} \mid \exists vu \in A(G_\ell^{\mathrm{rel}}), v \in S \wedge u \in S'\}.$$

We say that an arc set B is a *connector* if B connects some SCC S to some set $\mathcal{T} \subseteq \mathcal{S}_{G,\ell}^{\mathrm{reach}}(S)$ of SCCs reachable from S. Then, our goal is to find a connector of minimum cost for each $S \in \mathcal{S}_{G_\ell}$ and each subset $\mathcal{T} \subseteq \mathcal{S}_{G,\ell}^{\mathrm{reach}}(S)$. This allows us to compute an optimal solution with exhaustive search on connections between SCCs in the last phase.

Algorithm 2. A dynamic programming procedure for the second phase

1 **Function** $\mathrm{DP}(G_\ell^{\mathrm{rel}}, w_\ell^{\mathrm{rel}})$
    // $S_v \in \mathcal{S}_{G_\ell}$ denotes the SCC to which $v \in V(G_\ell)$ belongs
    // $\mathcal{T}_B \subseteq \mathcal{S}_{G_\ell}$ denotes $\{S_u \mid \exists vu \in B\}$ for any $B \subseteq A(G)$
2   **foreach** $S = \{v_1, \ldots, v_{n_S}\} \in \mathcal{S}_{G_\ell}$ **do**
        // Initialization phase
3         $B_0 \leftarrow \emptyset,\ D_0(\emptyset) \leftarrow \emptyset$
4   **foreach** $i \in [n_S]$ **do**
5             $B_i \leftarrow B_{i-1} \cup \{v_i u \in A(G_\ell^{\mathrm{rel}}) \mid u \notin S, S_u \notin \mathcal{T}_{B_{i-1}}\}$
6   **foreach** $\mathcal{T} \subseteq \mathcal{T}_{B_i}$ **do** $D_i(\mathcal{T}) \leftarrow \{vu' \in B_i \mid S_{u'} \in \mathcal{T}\}$
        // Update phase
7   **foreach** $i \in [n_S]$ **do**
8   **foreach** $u \in \{u' \notin S \mid v_i u' \in A(G_\ell^{\mathrm{rel}})\}$ **do**
9                 $B_{i,u} \leftarrow \{v_i u' \in A(G_\ell^{\mathrm{rel}}) \mid u' \notin S, w_\ell^{\mathrm{rel}}(v_i u') \leq w_\ell^{\mathrm{rel}}(v_i u)\}$
10   **foreach** $\mathcal{T} \subseteq \mathcal{S}_{G,\ell}^{\mathrm{reach}}(S)$ **do**
11   **if** $\mathcal{T} \not\subseteq \mathcal{T}_{B_{i-1}} \cup \mathcal{T}_{B_{i,u}}$ **then continue**
12   **if** $\mathcal{T} \subseteq \mathcal{T}_{B_{i-1}}$ **then** $D_i(\mathcal{T}) \leftarrow D_{i-1}(\mathcal{T})$
13   **if** $\mathrm{Cost}(S, D_{i-1}(\mathcal{T} \setminus \mathcal{T}_{B_{i,u}}), w_\ell^{\mathrm{rel}}) + w_\ell^{\mathrm{rel}}(v_i u) <$
                        $\mathrm{Cost}(S, D_i(\mathcal{T}), w_\ell^{\mathrm{rel}})$ **then**
14                       $D_i(\mathcal{T}) \leftarrow D_{i-1}(\mathcal{T} \setminus \mathcal{T}_{B_{i,u}}) \cup \{v_i u' \in B_{i,u} \mid S_{u'} \in \mathcal{T}\}$

15     $\mathrm{MCC}(S, \cdot) \leftarrow D_{n_S}$
16   **return** MCC

Definition 7. *Let (G, w) be an instance of MINPAC and let ℓ be a vertex lower bound. A minimum-cost connector is a function $\mathrm{MCC} \colon \mathcal{S}_{G_\ell} \times 2^{\mathcal{S}_{G_\ell}} \to 2^{A(G_\ell^{\mathrm{rel}})}$ such that for any $S \in \mathcal{S}_{G_\ell}$ and any $\mathcal{T} \subseteq \mathcal{S}_{G,\ell}^{\mathrm{reach}}(S)$ the following properties are satisfied:*

(i) *For any $S' \in \mathcal{T}$, there exist vertices $v \in S$ and $u \in S'$ with $vu \in \mathrm{MCC}(S, \mathcal{T})$.*

(ii) *There is no subset $B \subseteq A(G_\ell^{\mathrm{rel}})$ that satisfies the above property and that satisfies $\mathrm{Cost}(S, B, w_\ell^{\mathrm{rel}}) < \mathrm{Cost}(S, \mathrm{MCC}(S, \mathcal{T}), w_\ell^{\mathrm{rel}})$.*

Algorithm 2 computes a minimum-cost connector. For each SCC $S \in \mathcal{S}_{G_\ell}$, we employ dynamic programming over vertices in S and subsets of $\mathcal{S}_{G_\ell}$. This gives us a significant speed-up compared to the naive approach of branching into at worst c different neighbors on each vertex: From $n^{\theta(c)}$ time to $O(2^c \cdot c^2 \cdot n)$ time.

Lemma 8. *Given a relevant subgraph G_ℓ^{rel} and a relevant weight function w_ℓ^{rel}, Algorithm 2 computes a minimum-cost connector MCC in $O(2^c \cdot c^2 \cdot n)$ time.*

Phase 3. We finally present the search tree algorithm for MINPAC in Algorithm 3. The algorithm "guesses" the connections between SCCs of G_ℓ to obtain the optimal solution. To this end, we first try all possible numbers of outgoing arcs from each SCC.

Algorithm 3. An exhaustive search algorithm for MinPAC

1 **Function** Search($G_\ell^{\mathrm{rel}}, w_\ell^{\mathrm{rel}}, \mathrm{MCC}$)
2     $OptCost \leftarrow \infty, \mathcal{C} \leftarrow (\mathcal{S}_{G_\ell}(S_1), \ldots, \mathcal{S}_{G_\ell}(S_c))$
3   **for** $k \leftarrow c$ **to** $2c - 2$ **do**
4   **foreach** $k_1, \ldots, k_c \in \mathbb{N}$ such that $k_1, \ldots, k_c \geq 1$ and $\sum_{i=1}^{c} k_i = k$ **do**
5   **foreach** $\mathcal{T}_{S_1}, \ldots, \mathcal{T}_{S_c}$ such that $\mathcal{T}_{S_i} \subseteq \mathcal{S}_{G,\ell}^{\mathrm{reach}}(S_i)$ and $|\mathcal{T}_{S_i}| = k_i$ for any $i \in [c]$ **do**
6            $H^{\mathrm{aux}} \leftarrow (\{v_1, \ldots, v_c\}, \{v_i v_j \mid S_j \in \mathcal{T}_{S_i}\})$
7   **if** $|\mathcal{S}_{H^{\mathrm{aux}}}| > 1$ **then continue**
8            $Cost \leftarrow 0$
9   **foreach** $S \in \mathcal{S}_{G_\ell}$ **do** $Cost \leftarrow Cost + \mathrm{Cost}(S, \mathrm{MCC}(S, \mathcal{T}_S), w_\ell^{\mathrm{rel}})$
             // We assume that $\mathrm{Cost}(S, \mathrm{MCC}(S, \mathcal{T}_S), w_\ell^{\mathrm{rel}})$ is computed
             for any $S \in \mathcal{S}_{G_\ell}, T \subseteq \mathcal{S}_{G,\ell}^{\mathrm{reach}}(S)$ in Algorithm 2
10   **if** $Cost < OptCost$ **then** $OptCost \leftarrow Cost, \mathcal{C} \leftarrow (\mathcal{T}_{S_1}, \ldots, \mathcal{T}_{S_c})$
11   **return** $(V(G), A(G_\ell) \cup \bigcup_{i=1}^{c} \mathrm{MCC}(S_i, \mathcal{C}[i]))$

Lemma 9. *Given a relevant subgraph G_ℓ^{rel}, a relevant weight function w_ℓ^{rel}, and a minimum-cost connector $\mathrm{MCC}\colon \mathcal{S}_{G_\ell} \times 2^{\mathcal{S}_{G_\ell}} \to 2^{A(G_\ell^{\mathrm{rel}})}$, Algorithm 3 computes an optimal solution of $(G_\ell^{\mathrm{rel}}, w_\ell^{\mathrm{rel}})$ in $O(n + m + 4^c \cdot c^{2c-3/2})$ time.*

Combining Algorithms 1 to 3 we arrive at our main theorem of this section.

Theorem 10. *MinPAC can be solved in $O(c^2 \cdot 2^c \cdot n + m + 4^c \cdot c^{2c-3/2})$ time.*

3   Parameterization by Feedback Edge Number

In this section we describe a kernelization for MinPAC parameterized by the *feedback edge number*. The feedback edge number for an undirected graph is the minimum number of edges that have to be removed in order to make it a tree. We define the feedback edge number for a directed graph G as the feedback edge number of its underlying undirected graph U_G. Note that a minimum feedback edge set can be computed in linear time. In Sect. 5, we will show that the parameter *feedback arc number*, which is the digraph counterpart of the feedback edge number, does not allow the design of an FPT algorithm for MinPAC unless P = NP.

The feedback edge number measures how tree-like the input is. From a theoretical perspective this is interesting to analyze because any instance (G, w) of MinPAC is easy to solve if U_G is a tree. In this case all edges of U_G must correspond to arcs in both directions in G and the optimal solution is G itself. The parameter is also motivated by real world applications in which the feedback edge number is small; for instance, sensor networks along waterways (including canals) are expected to have a small number of feedback edges. In this section we prove the following theorem which states that MinPAC admits a partial kernel with respect to feedback edge number.

Theorem 11. *In linear time, one can transform any instance* $I = (G, w)$ *of* MINPAC *with feedback edge number* g *into an instance* $I' = (G', w')$ *and compute a value* $d \in \mathbb{N}$ *such that* G' *has at most* $20g - 20$ *vertices, and* $42g - 42$ *arcs, and* $\mathrm{OPT}(I) = \mathrm{OPT}(I') + d$.

Corollary 12. MINPAC *can be solved in* $O(2^{O(g)} + n + m)$ *time.*

We will present a set of data reduction rules which shrink any instance of MINPAC to an essentially equivalent instance whose size is bounded as specified in Theorem 11. We simultaneously compute the value d, which we initialize with 0. In this section we assume that $w(vu) = \infty$ if $vu \notin A$.

Our first reduction rule reduces the weights of arcs outgoing from a vertex by the weight of its cheapest outgoing arc. This ensures that each vertex has at least one outgoing arc of weight zero.

Reduction Rule 1. Let v be a vertex with $\delta_v = \min_{vu \in A(G)} w(vu) > 0$. Update the weights and d as follows:

(i) $w(vu) = w(vu) - \delta_v$ for each $vu \in A(G)$.
(ii) $d := d + \delta_v$.

Our next reduction rule discards all degree-one vertices.

Reduction Rule 2. Let v be a vertex with $\deg_G(v) = 1$ and let u be its neighbor. Update (G, w) and d as follows:

(i) $G := G[V(G) \setminus \{v\}]$.
(ii) $w(uv') := \max\{0, w(uv') - w(uv)\}$ for each $uv' \in A(G) \setminus \{uv\}$.
(iii) $d := d + w(vu) + w(uv)$.

Lemma 13. *Reduction Rules 1 and 2 can be exhaustively applied in linear time.*

Proof. For each vertex $v \in V(G)$, set $\ell(v) := \min_{vu \in A(G)} w(vu)$. Let L be a list of degree-1 vertices. We apply the following procedure as long as L is nonempty. Let v be the vertex taken from L and let u be its neighbor. Remove v and its incident arcs from G and set $\ell(u) := \max\{\ell(u), w(uv)\}$ and update $d := d + \max\{w(vu), \ell(v)\}$. If the degree of u becomes 1 after deleting v, then we add u to L. Once L is empty, we update the weight of each remaining arc $w(vu) := \max\{0, w(vu) - \ell(v)\}$. Finally, we update $d := d + \ell(v)$ for each remaining vertex v. It is easy to see that the algorithm runs in linear time.   □

Henceforth, we can assume that Reduction Rules 1 and 2 are exhaustively applied. Thus, the underlying undirected graph U_G will have no degree-one vertices. It remains to bound the number of vertices that have degree two in U_G. Once this is achieved, we can use standard arguments to upper-bound the size of the instance [2].

In order to upper-bound the number of degree-two vertices in U_G, we consider long paths in U_G. A path $P = (v_0, \ldots, v_{h+1})$ in U_G is a *maximal induced path* of G if $\deg_G(v_0) > 2$, $\deg_G(v_{h+1}) > 2$, and $\deg_G(v_i) = 2$ for all $i \in [h]$. We

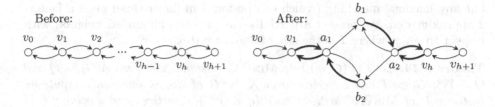

Fig. 1. Illustration of the reduction rule replacing long maximal induced paths. Bold arcs denote arcs of weight 0. For the weights of other arcs in the gadget on the right.

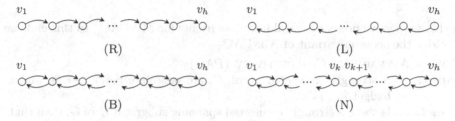

Fig. 2. Sketch of the cases for connectivity inside maximal induced paths.

call the vertices $\{v_i \mid i \in [h]\}$ the inner vertices of P. We will replace the inner vertices of each maximal induced path on at least seven vertices with a fixed gadget (see Fig. 1). The arc-weights in the gadget are chosen such that the four possible ways in which the outermost inner vertices are connected inside the path (see Fig. 2 for a visualization of the four cases) are preserved. We refer to a full version for the details on the reduction rule and the proof of Theorem 11.

4   Parameterization by the Number of Power Levels

It is fair to assume that the nodes cannot transmit signals with arbitrary power levels due to practical limitations [6]. In fact, many researchers have studied approximation algorithms for the MINPAC problem when only two power levels are available [3,4,6,16]. In this section, we consider the case $w\colon A(G) \to Q$, where the set of integers $Q = \{p_1, \ldots, p_q\}$ represents available power levels. The parameter q—"the number of numbers"—has been advocated by Fellows et al. [12]. The problem remains NP-hard even when $q = 2$ [8], as also can be seen in our hardness result (Theorem 15). Thus, fixed-parameter tractability is unlikely with this parameter alone. However, using an additional parameter may alleviate this problem. We consider the *vertex cover number*, as many problems are known to become tractable when this parameter is bounded. Here we define the vertex cover number for a directed graph as the vertex cover number of the underlying undirected graph. Recall that the vertex cover number for an undirected graph is the minimum number of vertices that have to be removed to make it edgeless. Computing a minimum-cardinality vertex cover is NP-hard but any maximal matching (which can be found in linear time) gives a factor-2 approximation. We present a partial kernelization (unbounded weights) with respect to $q + x$, where x is the size of a given vertex cover.

Theorem 14. *Let $I = (G, w)$ be a* MinPAC-*instance where $w: A(G) \to Q$ and $Q \in \mathbb{N}^q$. Given I and a vertex cover X for G of size x, one can compute an instance I' of* MinPAC *with at most $(q + 1)^{2x} + x$ vertices and a value $d \in \mathbb{N}$ such that $\mathrm{OPT}(I) = \mathrm{OPT}(I') + d$ in $O(xn + m)$ time.*

5 Parameterized Hardness

In this section we present several hardness results for MinPAC. To this end, we consider the decision variant of MinPAC.

POWER ASYMMETRIC CONNECTIVITY (PAC)
Input: A strongly connected graph G, arc weights $w: A(V) \to \mathbb{N}$, and a
                budget $k \in \mathbb{N}$.
Question: Is there a strongly connected spanning subgraph H of G, such that
                $\mathrm{Cost}(V(G), A(H), w) \le k$?

We prove that PAC remains NP-hard even if the *feedback arc number* is 1. This complements the result in Sect. 3, where we showed that MinPAC parameterized by the feedback edge number admits an FPT algorithm via a kernelization. Recall that the feedback arc number for a directed graph is the minimum number of arcs that have to be removed to make it a directed acyclic graph. Furthermore, we show that PAC is W[2]-hard with respect to the solution cost k. We also show that PAC cannot be solved in subexponential time in the number of vertices assuming the Exponential Time Hypothesis (ETH) [13], which states that 3-SAT cannot be solved in $2^{o(n+m)}$ time, where n and m are the number of variables and clauses in the input formula. Summarized we show:

Theorem 15. *Even if each arc weight is either one or zero and the feedback arc number is 1,*

(i) PAC *is NP-hard.*
(ii) PAC *is W[2]-hard when parameterized by the solution cost k.*
(iii) *Unless the ETH fails,* PAC *is not solvable in $2^{o(n)}$ time.*

It follows from Theorem 15 (ii) that there (presumably) is no algorithm solving PAC running in $f(k) \cdot n^{O(1)}$ time. Nonetheless, a simple brute-force algorithm solves PAC in $n^{\theta(k)}$ time, certifying that PAC is in the class XP with respect to the parameter solution cost. In order to prove the claims of Theorem 15, we use a reduction from the well-studied SET COVER problem.

SET COVER
Input: A universe $U = \{u_1, \ldots, u_n\}$, a set family $\mathcal{F} = \{S_1, \ldots, S_m\}$ con-
                taining sets $S_i \subseteq U$, and $\ell \in \mathbb{N}$.
Question: Is there a size-ℓ set cover $\mathcal{F}' \subseteq \mathcal{F}$ (that is, $\bigcup_{S \in \mathcal{F}'} S = U$)?

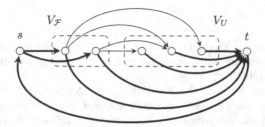

Fig. 3. Illustration of Reduction 1 on a SET COVER instance with universe $U = \{1, 2, 3\}$ and set family $\mathcal{F} = \{\{2, 3\}, \{1, 2\}\}$. Bold arcs denote arcs of weight 0 and other arcs have weight 1.

SET COVER is NP-hard and W[2]-hard with respect to the solution size ℓ [10] and is not solvable in $2^{o(|U|+|\mathcal{F}|)}$ time unless ETH fails [14].

For the reduction, we use one vertex for each element and each subset and one arc to represent the membership of an element in a subset. The construction resembles the one used in MIN-POWER SYMMETRIC CONNECTIVITY [2].

Reduction 1. Given an instance $I = (U, \mathcal{F}, \ell)$ of SET COVER, we construct an instance $I' = (G, w, k = \ell)$ of PAC as follows. We introduce a vertex v_u for every $u \in U$, a vertex v_S for every $S \in \mathcal{F}$, and two additional vertices s and t. We construct a graph such that $V(G) = \{s, t\} \cup V_U \cup V_{\mathcal{F}}$ where $V_U = \{v_u \mid u \in U\}$ and $V_{\mathcal{F}} = \{v_S \mid S \in \mathcal{F}\}$. For the arcs we first add an arc ts of weight 0. We then add arcs sv_S and v_St of weight 0 for every $S \in \mathcal{F}$ and an arc v_ut of weight 0 for every $u \in U$. For every $S \in \mathcal{F}$ and every $u \in S$ we finally add an arc v_Sv_u of weight 1.

Figure 3 illustrates the reduction to PAC. We can assume that arcs of weight zero (bold arcs in the figure) are part of the solution. The idea is that in order to obtain a strongly connected subgraph, one has to select at least one incoming arc for each vertex in V_U such that only k vertices in $V_{\mathcal{F}}$ are affected.

6   Conclusion

We performed a first analysis of the parameterized complexity of MINPAC, leading to first tractability and intractability results, but also to several open questions: Can the running time of the parameterized algorithm with respect to the number c of SCCs in the obligatory subgraph be improved to single-exponential? In both our kernelization results, the weights are left unbounded. To also upper-bound the weights, we are working on adapting the approach of Etscheid et al. [11] to our problem. Resolving the parameterized complexity of MINPAC with respect to the single parameter vertex cover number is another task for future work.

Finally, we remark that our algorithms run in linear time when the respective parameters are bounded. Thus we believe that our results are worthwhile for empirical experiments.

References

1. Bai, X., Kumar, S., Xuan, D., Yun, Z., Lai, T.: Deploying wireless sensors to achieve both coverage and connectivity. In: Proceedings of MobiHoc 2006, pp. 131–142. ACM (2006)
2. Bentert, M., van Bevern, R., Nichterlein, A., Niedermeier, R.: Parameterized algorithms for power-efficient connected symmetric wireless sensor networks. In: Fernández Anta, A., Jurdzinski, T., Mosteiro, M.A., Zhang, Y. (eds.) ALGOSEN-SORS 2017. LNCS, vol. 10718, pp. 26–40. Springer, Cham (2017). https://doi.org/10.1007/978-3-319-72751-6_3
3. Călinescu, G.: Approximate min-power strong connectivity. SIAM J. Discrete Math. **27**(3), 1527–1543 (2013)
4. Călinescu, G.: 1.61-approximation for min-power strong connectivity with two power levels. J. Comb. Optim. **31**(1), 239–259 (2016)
5. Calinescu, G., Kapoor, S., Olshevsky, A., Zelikovsky, A.: Network lifetime and power assignment in ad hoc wireless networks. In: Di Battista, G., Zwick, U. (eds.) ESA 2003. LNCS, vol. 2832, pp. 114–126. Springer, Heidelberg (2003). https://doi.org/10.1007/978-3-540-39658-1_13
6. Carmi, P., Katz, M.J.: Power assignment in radio networks with two power levels. Algorithmica **47**(2), 183–201 (2007)
7. Chen, J., Lu, H., Kuo, T., Yang, C., Pang, A.: Dual power assignment for network connectivity in wireless sensor networks. In: Proceedings of GLOBECOM 2005, p. 5. IEEE (2005)
8. Chen, W., Huang, N.: The strongly connecting problem on multihop packet radio networks. IEEE Trans. Commun. **37**(3), 293–295 (1989)
9. Clementi, A.E.F., Penna, P., Silvestri, R.: On the power assignment problem in radio networks. MONET **9**(2), 125–140 (2004)
10. Downey, R.G., Fellows, M.R.: Fundamentals of Parameterized Complexity. Springer, Heidelberg (2013). https://doi.org/10.1007/978-1-4471-5559-1
11. Etscheid, M., Kratsch, S., Mnich, M., Röglin, H.: Polynomial kernels for weighted problems. J. Comput. Syst. Sci. **84**, 1–10 (2017)
12. Fellows, M.R., Gaspers, S., Rosamond, F.A.: Parameterizing by the number of numbers. Theory Comput. Syst. **50**(4), 675–693 (2012)
13. Impagliazzo, R., Paturi, R.: On the complexity of k- SAT. J. Comput. Syst. Sci. **62**(2), 21 (2001)
14. Impagliazzo, R., Paturi, R., Zane, F.: Which problems have strongly exponential complexity? J. Comput. Syst. Sci. **63**(4), 512–530 (2001)
15. Iyengar, R., Kar, K., Banerjee, S.: Low-coordination topologies for redundancy in sensor networks. In: Proceedings of MobiHoc 2005, pp. 332–342. ACM (2005)
16. Rong, Y., Choi, H., Choi, H.: Dual power management for network connectivity in wireless sensor networks. In: Proceedings of IPDPS 2004. IEEE (2004)
17. Sorge, M., Van Bevern, R., Niedermeier, R., Weller, M.: A new view on rural postman based on Eulerian extension and matching. J. Discrete Algorithms **16**, 12–33 (2012)
18. Zhang, H., Hou, J.C.: Maintaining sensing coverage and connectivity in large sensor networks. Ad Hoc Sensor Wirel. Netw. **1**, 89–124 (2005)

Solving Group Interval Scheduling Efficiently

Arindam Biswas[1]($\boxtimes$), Venkatesh Raman[1], and Saket Saurabh[1,2]

[1] The Institute of Mathematical Sciences, HBNI, Chennai, India
{barindam,vraman,saket}@imsc.res.in
[2] University of Bergen, Bergen, Norway

Abstract. The GROUP INTERVAL SCHEDULING problem models the scenario where there is set $[\gamma] = \{1, \ldots, \gamma\}$ of jobs to be processed on a single machine, and each job i can only be scheduled for processing in exactly one time interval from a group G_i of allowed intervals. The objective is to determine if there is a set of $S \subseteq [\gamma]$ of k ($k \in \mathbb{N}$) jobs which can be scheduled in non-overlapping time intervals.

This work describes a deterministic algorithm for the problem that runs in time $O((5.18)^k n^d)$, where $n = |\bigcup_{i \in [\gamma]} G_i|$ and $d \in \mathbb{N}$ is a constant. For $k \geq d \log n$, this is significantly faster than the best previously-known deterministic algorithm, which runs in time $O((12.8)^k \gamma n)$. We obtain our speedup using efficient constructions of *representative families*, which can be used to solve the problem by a dynamic programming approach.

Keywords: Group · Job · Interval · Scheduling · Graph · Independent · Colourful · Representative · Hash · Fixed · Parameter · FPT · Multivariate

1 Introduction

A ubiquitous problem arising in industrial processes is when there are multiple jobs to be processed on a single machine, and some of the jobs have conflicting time constraints. In this scenario, the next best thing is for the machine to process as many jobs as possible without violating any time constraints. This can be modelled as follows.

GROUP INTERVAL SCHEDULING

Instance: A pair (J, k), where $J = \{G_1, \ldots, G_\gamma\}$ such that G_i ($i \in [\gamma]$) is a set of disjoint intervals of $\mathbb{R}$, and $k \in \mathbb{N}$.
Question: Is there a set of at least k disjoint intervals $S \subseteq \bigcup J$ such that $|S \cap G_i| \leq 1$ ($i \in [\gamma]$)?

The sets G_i represent time constraints: job i ($i \in [\gamma]$) can only be processed during a time interval from the set G_i. In this scheme, picking a set S of disjoint

© Springer Nature Switzerland AG 2019
C. J. Colbourn et al. (Eds.): IWOCA 2019, LNCS 11638, pp. 97–107, 2019.
https://doi.org/10.1007/978-3-030-25005-8_9

intervals such that $|S \cap G_i| \leq 1$ ($i \in [\gamma]$) is equivalent to scheduling the set $\{i \in [\gamma] \mid G_i \in S\}$ of jobs on the machine such that they occupy distinct time intervals.

GROUP INTERVAL SCHEDULING is known to be NP-hard [6] while being polynomial-time solvable (via a reduction to 2-SAT; see [3]) when there are at most 2 intervals per job.

Consider a finite set X and a function $f : X \rightarrow [t]$. A subset $S \subseteq X$ is *colourful* with respect to f if for any $x, y \in S$, $x \neq y \implies f(x) \neq f(y)$. The following problem is an equivalent formulation of JOB INTERVAL SCHEDULING that models constraints among the jobs using a graph and a colouring function on its vertex set.

COLOURFUL INDEPENDENT SET

Instance: A triple (G, ϕ, k), where G is a graph, $\phi : V \rightarrow [\gamma]$ is a colouring and $k \in \mathbb{N}$.
Question: Is there is an independent set $S \subseteq V$ in G of size k which is colourful with respect to ϕ?

Let (J, k) be an instance of GROUP INTERVAL SCHEDULING. Define $V = \bigcup_{i \in [\gamma]} G_i$. Taking V as the vertex set, define $G = (V, E)$, where $E = \{uv \mid u \cap v \neq \emptyset\}$ and define $\phi : V \rightarrow [\gamma]$ by $\phi(v) = i$, where $i \in [\gamma]$ such that $v \in G_i$. This gives an equivalent instance (G, ϕ, k) of COLOURFUL INDEPENDENT SET on interval graphs: k jobs from J can be scheduled on the machine if and only if G has an independent set of size k that is colourful with respect to ϕ.

This alternative formulation of GROUP INTERVAL SCHEDULING is used throughout the remainder of this paper.

Fixed-Parameter Tractability. The results are presented here in the framework of Parameterized Complexity. Consider a computational problem P and let x be a instance of P. Suppose that there is a number $k_x \in \mathbb{N}$ that describes a property of x, e.g. the optimal solution value for x. Such a scheme is called a *parameterization* of P, and k_x is called the *parameter* of x. Attaching the parameter to the problem instance gives us a *parameterized* problem: $\{\langle x, k_x \rangle \mid x \in P\}$.

Given any parameterized problem Q with parameterization k, if there is an algorithm that solves it in time $O(f(k)n^c)$, where $f : \mathbb{N} \rightarrow \mathbb{N}$ is a computable function and $c \in \mathbb{N}$ is a constant, then Q is said to be *fixed-parameter tractable*.

Our Results and Previous Work. We consider two parameterizations of COLOURFUL INDEPENDENT SET: k, the size of the solution sought, and γ, the number of colours used by the colouring ϕ. The question of fixed-parameter tractability was studied by Halldórsson and Karlsson [2] and later by van Bevern et al. [8], which led to the following results.

Proposition 1 (Halldórsson and Karlsson [2]**, Theorem 4).** *Instances* (G, ϕ, k) *of* COLOURFUL INDEPENDENT SET *on interval graphs can be solved deterministically in time* $O(2^\gamma n)$, *where* γ *is the number of colours.*

Proposition 2 (van Bevern et al. [8]**, Theorem 4).** *Instances* (G, ϕ, k) *of* COLOURFUL INDEPENDENT SET *on interval graphs can be solved with error probability* ϵ *in time* $O\left(|\log \epsilon|(5.5)^k n\right)$. *The algorithm can be derandomized to solve the problem deterministically in time* $O\left((12.8)^k \gamma n\right)$, *where* γ *is the number of colours.*

Proposition 1 establishes fixed-parameter tractability with respect to γ while Proposition 2 shows that there is a (randomized) fixed-parameter algorithm with respect to k. This work makes the following improvements.

- We show that the running time of the deterministic algorithm of Proposition 2 can actually be improved to $\left(e^{k+O((\log k)^2)} \log \gamma\right) 2^k n = O\left((5.44)^k (\log \gamma) n\right)$ using smaller families of perfect hash functions (Theorem 1).
- Using efficiently-constructible *representative families*, we obtain an algorithm (Theorem 2) for COLOURFUL INDEPENDENT SET that runs in time $O\left((5.18)^k n^d\right)$ $(d \in \mathbb{N},$ a constant).

2 Preliminaries

In this section, we introduce notation used in the rest of the paper and review some basic concepts concerning matroids, representative families and perfect hash families.

2.1 Basics

- $\mathbb{N}$ denotes the set of natural numbers and $\mathbb{R}$ denotes the set of real numbers.
- For $t \in \{1, 2, \ldots\}$, $[t]$ denotes the set $\{1, \ldots, t\}$ and for $a, b \in \mathbb{R}$ with $a \leq b$, $[a, b]$ denotes the set $\{x \mid x \in \mathbb{R}, a \leq x \leq b\}$.
- Let X be a set and $\mathcal{F}$ be a family of subsets of X. For $x \in X$, define $x + \mathcal{F} = \{\{x\} \cup S \mid S \in \mathcal{F}\}$.
- Let $G = (V, E)$ be a graph and $\phi : V \to [\gamma]$ be a colouring of its vertices.
  - $V(G)$ denotes the vertex set V and $E(G)$ denotes the edge set E.
  - For each $i \in [\gamma]$, define $V_i = \{v \in V \mid \phi(v) = i\}$. A set of vertices $V' \subseteq V$ is called *colourful* if $|V_i \cap V'| \leq 1$ for all $i \in [\gamma]$.
- For a function $g : A \to B$, dom g denotes the set A and rng g denotes the set $B' = \{y \in B \mid \exists x \in B : g(x) = y\}$.

A *matroid* is a pair $(E, \mathcal{I})$ consisting of *ground* set E and a family $\mathcal{I}$ of subsets called *independent* sets that has the following properties.

1. $\emptyset \in \mathcal{I}$.

2. If $X \in \mathcal{I}$ and $Y \subseteq X$, then $Y \in \mathcal{I}$.
3. For any two sets $X, Y \in \mathcal{I}$ with $|X| < |Y|$, there is an element $e \in Y \setminus X$ such that $X \cup \{e\} \in \mathcal{I}$.

Because of Property 3 (also known as the *exchange property*) above, all maximal independent sets in a matroid have the same size. This number is called the *rank* of the matroid.

Given a matroid $\mathcal{M} = (E, \mathcal{I})$ and an integer $k \in \mathbb{N}$, it is easy to see that the pair $\mathcal{M}' = (E, \mathcal{I}')$ with $\mathcal{I}' = \{S \in \mathcal{I} \mid |S| \leq k\}$ is also a matroid. It is called the k-truncation of $\mathcal{M}$. Since the independent sets $S \in \mathcal{I}'$ all satisfy $|S| \leq k$, the rank of $\mathcal{M}'$ is at most k.

Let $A_\mathcal{M}$ be a matrix over some field F whose columns are $A_1, \ldots A_t$. Suppose that there is a injective function $\rho : E \to \{A_1, \ldots, A_t\}$ such that for any $S \subseteq E$, S is independent in $\mathcal{M}$ if and only if the set of columns $\rho(S)$ is linearly independent. In this case, the matrix $A_\mathcal{M}$ is called a representation for $\mathcal{M}$, and $\mathcal{M}$ is said to be representable over the field F.

Definition 1 (Linear Matroid). *A matroid $\mathcal{M}$ is called a* linear *matroid if it has a representation $A_\mathcal{M}$ over some field F.*

2.2   Matroids of Colourful Sets

Let $G = (V, E)$ be a graph, $\phi : V \to [\gamma]$ be a colouring of its vertices and $k \in \mathbb{N}$. Define $\mathcal{I} = \{S \subseteq V \mid S \text{ is colourful and } |S| \leq k\}$ and let $\mathcal{K} = (V, \mathcal{I})$. In the following, we show that $\mathcal{K}$ is a linear matroid with a representation that can be computed efficiently.

Consider the partition $V = V_1 \cup \cdots \cup V_\gamma$, where V_i $(i \in [\gamma])$ comprises vertices of colour i. Define $\mathcal{P} = (V, \mathcal{I}')$ where $\mathcal{I}'$ comprises all sets $S \subseteq V$ such that $|S \cap V_i| \leq 1$ $(i \in [\gamma])$.

Lemma 1. *$\mathcal{P}$ is a linear matroid. A representation $A_\mathcal{P}$ (over $\mathbb{F}_2$) for $\mathcal{P}$ of size $\gamma \times n$ can be computed in time $\mathrm{O}(\gamma n)$.*

Proof (Sketch). It is easy to verify that $\mathcal{P}$ is a *partition* matroid. Consider the $\gamma \times n$ matrix $A_\mathcal{K}$ defined by

$$A_\mathcal{P} = \left(e_1^{|V_1|}, \ldots, e_\gamma^{|V_\gamma|} \right),$$

where e_i $(i \in [t])$ denotes the column vector with a 1 at the ith coordinate and 0's everywhere else. $A_\mathcal{P}$ represents $\mathcal{P}$ and because the column vectors can be computed in time $\mathrm{O}(\gamma)$, the entire matrix can be constructed in time $\mathrm{O}(\gamma n)$. □

Lemma 2. *$\mathcal{K}$ is the k-truncation of $\mathcal{P}$.*

Proof. Let $S \in \mathcal{I}$. Because S is colourful, we have $|S \cap V_i| \leq 1$ $(i \in [\gamma])$, i.e. $S \in \mathcal{I}'$. Conversely, any $S \in \mathcal{I}'$ with $|S| \leq k$ is a colourful set, so $S \in \mathcal{I}$. Thus, $\mathcal{K}$ is the k-truncation of $\mathcal{P}$. □

The next proposition provides a method for computing a truncation of a matroid from its representation. It will be used later to construct a representation for $\mathcal{K}$.

Proposition 3 (Lokshtanov et al. [5], Theorem 1.1). *Let A be an $m \times n$ matrix of rank m over a field F that represents the matroid $\mathcal{M}$. For any natural number $k \leq m$, the k-truncation of $\mathcal{M}$ admits a representation A_k over $F(X)$, the field of fractions of the polynomial ring $F[X]$. This representation can be computed in $\mathrm{O}(mnk)$ operations over F, and given any set of columns of A_k, it can be determined whether they are linearly independent in $\mathrm{O}(m^2k^3)$ operations over F.*

Lemma 3. *$\mathcal{K}$ is a linear matroid of rank k that admits a representation $A_{\mathcal{K}}$ over $\mathbb{F}_2(X)$ which can be computed in time $\mathrm{O}(k\gamma n)$.*

Proof. We show that $\mathcal{K}$ is a linear matroid by computing a representation for it. Note that the ground set of $\mathcal{P}$ has $n = |V|$ elements.

Using the procedure of Lemma 1, obtain representation $A_{\mathcal{P}}$ for $\mathcal{P}$. This takes time $\mathrm{O}(\gamma n)$ and the representation is a 0-1 matrix of size $\gamma \times n$. Now use the procedure of Proposition 3 to obtain the k-truncation $A_{\mathcal{K}}$ of $\mathcal{P}$. This can be done in $\mathrm{O}(k\gamma n)$ operations over $\mathbb{F}_2$, each of which takes time $\mathrm{O}(1)$. Thus, the overall running time of the algorithm is $\mathrm{O}(k\gamma n)$. $\qquad\square$

2.3 Representative Families

Definition 2 (q-Representative Family). *Let $p, q \in \mathbb{N}, M = (E, \mathcal{I})$ be a matroid, and $\mathcal{F} \subseteq \mathcal{I}$ be a family of independent sets of size p. A subfamily $\mathcal{F}' \subseteq \mathcal{F}$ is q-representative of F if the following statement holds. For any $X \subseteq E$ with $|X| \leq q$, if there is a set $Y \in \mathcal{F}$ such that $X \cap Y = \emptyset$ and $X \cup Y \in \mathcal{I}$, then there is a set $Y' \in \mathcal{F}'$ such that $X \cap Y' = \emptyset$ and $X \cup Y' \in \mathcal{I}$.*

Proposition 4 (Fomin et al. [1], Theorem 1.1). *Let $\mathcal{M} = (E, \mathcal{I})$ be a linear matroid of rank $p+q = k$, and $A_{\mathcal{M}}$ be a matrix over some field F that represents it. For any family $\mathcal{R} = \{S_1, \ldots, S_t\}$ of independent sets of size p in $\mathcal{M}$, there is a family $\hat{\mathcal{R}} \subseteq \mathcal{R}$ with at most $\binom{k}{p}$ sets which is q-representative of $\mathcal{R}$. The family $\hat{\mathcal{R}}$ can be found in $\mathrm{O}\left(t\left(p^{\omega}\binom{k}{p} + \binom{k}{p}^{\omega-1}\right)\right)$ operations over F, where $\omega < 2.373$ is the matrix multiplication exponent.*

2.4 Perfect Hash Families

Definition 3 (Perfect Hash Family). *Let $n, k \in \mathbb{N}$ with $n \geq k$. A family of functions $\mathcal{H}_{n,k} \subseteq [k]^{[n]}$ is called an (n, k)-perfect hash family if for any set $S \subseteq [n]$ with $|S| \leq k$, there is a function $f \in \mathcal{H}$ such that f is injective on S.*

Proposition 5 (Naor et al. [7], Theorem 3). *For any $n, k \in \mathbb{N}$ with $n \geq k$, there is an (n, k)-perfect hash family $\mathcal{H}_{n,k}$ of cardinality $e^{k+\mathrm{O}\left((\log k)^2\right)} \log n$ that can be computed in time $e^{k+\mathrm{O}\left((\log k)^2\right)} n \log n$.*

3   Colourful Independent Set on Interval Graphs

Here, we give two algorithms for Colourful Independent Set on interval graphs. The first uses small families of perfect hash functions to make an improvement over the algorithm of Proposition 2. The second algorithm employs a dynamic programming approach using representative families.

Definition 4 (Compact Representation). *Let G be an interval graph and $\mathcal{R} = \{L_v \mid v \in V(G)\}$ be an interval representation for G. $\mathcal{R}$ is called c-compact $(c \in \mathbb{N})$ if the endpoints of every interval in R are in $\{0, \ldots, c\}$. If G has such a representation, it is called c-compact.*

Proposition 6 (van Bevern et al. [8], Observation 2). *Interval graphs of order n are n-compact.*

Proposition 7 (van Bevern et al. [8], Observation 4). *Given an adjacency list representation for an interval graph G with n vertices and m edges, a c-compact representation $\mathcal{R}$ for G that minimizes c can be computed in time $O(n + m)$.*

We begin by observing that the deterministic algorithm of Proposition 2 can be improved on by using slightly more efficient constructions of hash families.

3.1   Using Hash Families

By using the hash families of Theorem 5 with the algorithm of van Bevern et al. [2], we make the following improvement on the derandomization claim of Proposition 2.

Theorem 1. *Instances (G, ϕ, k) of Colourful Independent Set on interval graphs can be solved in time $\left(e^{k+O\left((\log k)^2\right)} \log \gamma\right) 2^k n = O\left((5.44)^k (\log \gamma) n\right)$.*

Lemma 4. *Let (G, ϕ, k) be an instance of Colourful Independent Set on interval graphs with $\gamma = |\mathrm{rng}\ \phi|$ colours. There is a family of colouring functions $\mathcal{C}_{\phi,k} \subseteq [n] \to [k]$ of size $e^{k+O\left((\log k)^2\right)} \log \gamma$ such that (G, ϕ, k) is a YES instance if and only if there is a function $\phi' \in \mathcal{C}_{\phi,k}$ such that (G, ϕ', k) is a YES instance. The family $\mathcal{C}_{\phi,k}$ can be constructed in time $e^{k+O\left((\log k)^2\right)} \gamma \log \gamma$.*

Proof. Let $\mathcal{H}_{\gamma,k}$ be the perfect hash family obtained from Proposition 5 by substituting $n = \gamma$. This family is of size $e^{k+O\left((\log k)^2\right)} \log \gamma$ and can be computed in time $e^{k+O\left((\log k)^2\right)} \gamma \log \gamma$. Define $\mathcal{C}_{\phi,k} = \{\rho \circ \phi \mid \rho \in \mathcal{H}_{\gamma,k}\}$. Clearly, $\mathcal{C}_{\phi,k} \subseteq [n] \to [k]$. Note that $\mathcal{C}_{\phi,k}$ can be obtained by chaining ϕ to each function in $\mathcal{H}_{\gamma,k}$, and this takes time $O(1)$ per function. Thus, $\mathcal{C}_{\phi,k}$ can be computed in time $e^{k+O\left((\log k)^2\right)} \gamma \log \gamma$.

Suppose that (G, ϕ, k) is a YES instance and let S be a colourful independent set in G, i.e. ϕ is injective on S. Consider $R = \phi(S)$. Since $\mathcal{H}_{\gamma,k}$ is (γ, k)-perfect, there is a function $\rho \in \mathcal{H}_{\gamma,k}$ such that ρ is injective on R. Because of this, $\rho \circ \phi \in \mathcal{C}_{\gamma,k}$ is injective on S, i.e. S is a colourful independent set with respect to $\phi' = \rho \circ \phi$. Thus, (G, ϕ', k) is a YES instance.

Conversely, if there is a function $\phi' \in \mathcal{C}_{\gamma,k}$ such that there is a colourful independent S with respect to ϕ' and $|S| \geq k$, then S is also colourful with respect to ϕ. □

The proof of Theorem 1 is now quite straightforward.

Proof (Theorem 1). Let (G, ϕ, k) be an instance of COLOURFUL INDEPENDENT SET on interval graphs. Using the construction of Lemma 4, we obtain a family of colourings $\mathcal{C}_{n,k}$ of size $e^{k+\mathrm{O}((\log k)^2)} \log \gamma$. Consider the following algorithm.

For each colouring $\phi' \in \mathcal{C}_{n,k}$, run the procedure of Proposition 1 on the instance (G, ϕ', k). If the procedure returns YES on any of the instances, then return YES. Otherwise, return NO. The correctness of the algorithm follows from Lemma 4.

Note that for each $\phi' \in \mathcal{C}_{n,k}$, $|\mathrm{rng}\ \phi'| = k$, so the instance (G, ϕ', k) has k colours. Thus, each invocation of the algorithm of Proposition 1 takes time $\mathrm{O}(2^k n)$. The overall running time of the algorithm is therefore

$$\left(e^{k+\mathrm{O}((\log k)^2)} \log \gamma\right) 2^k n + e^{k+\mathrm{O}((\log k)^2)} \gamma \log \gamma = \left(e^{k+\mathrm{O}((\log k)^2)} \log \gamma\right) 2^k n =$$

$$\mathrm{O}\left((5.44)^k (\log \gamma) n\right). \qquad \square$$

3.2   Using Representative Families

In this subsection, we employ a dynamic programming approach using representative families to obtain the following result.

Theorem 2. *Instances (G, ϕ, k) of COLOURFUL INDEPENDENT SET on interval graphs can be solved deterministically in time $\mathrm{O}\left((5.18)^k n^d\right)$, where ω is the matrix multiplication exponent and $d \in \mathbb{N}$ is a constant.*

Consider an instance (G, ϕ, k) of COLOURFUL INDEPENDENT SET. By Proposition 6, G has a c-compact representation with $c \leq n$. Let $\mathcal{D}$ be such a representation. Define L_v ($v \in V$) to be the interval corresponding to v in $\mathcal{D}$ and let $l(v)$ denote the length of L_v. We say that v lies in the interval $[i, j]$ ($0 \leq i < j \leq n$) if $L_v \subseteq [i, j]$. A set $S \subset V$ lies in $[i, j]$ if all its elements lie in $[i, j]$.

Families of Colourful Independent Sets. For $i \in [c]$ and $j \in [k]$, define $\mathcal{R}_j^i$ to be the family of all colourful independent sets in G of size exactly j in the interval $[0, i]$. Consider the matroid $\mathcal{K} = (V, \mathcal{I})$ of sets of colourful vertices defined in Subsect. 2.2. Since the sets in $\mathcal{R}_j^i$ are colourful, they are independent in $\mathcal{K}$. In what follows, we show how to efficiently compute a $(k-j)$-representative family for $\mathcal{R}_j^i$ with respect to $\mathcal{K}$.

For each $(i, j) \in (\{0\} \times [k]) \cup ([c] \times \{0\})$, the family $\mathcal{R}_j^i$ is empty, and is trivially represented by $\hat{\mathcal{R}}_j^i = \emptyset$.

Lemma 5. *Let $i \in [c]$ and $j \in [k]$. For each $r < i$ and $s \le j$, let $\hat{\mathcal{R}}_s^r$ be a $(k-s)$-representative family for $\mathcal{R}_s^r$ with respect to $\mathcal{K}$. Define*

$$\bar{\mathcal{R}}_j^i = \hat{\mathcal{R}}_j^{i-1} \cup \left(\bigcup_{L_v \text{ ends at } i} \left[v + \hat{\mathcal{R}}_{j-1}^{i-(l(v)+1)} \right] \right), \text{ where} \tag{1}$$

$$[\mathcal{R}] = \{S \in \mathcal{R} \mid S \text{ is a colourful independent set in } G\}.$$

The family $\bar{\mathcal{R}}_j^i$ $(k-j)$-represents $\mathcal{R}_j^i$.

Proof. Let $X \subseteq V$ with $|X| \le k - j$ such that there is a set $S \in \mathcal{R}_j^i$ with $S \cap X = \emptyset$ and $S \cup X \in \mathcal{I}$, i.e. S is a colourful independent set in G. We have the following cases.   □

Case 1. S contains a vertex v such that L_v ends at i.

In this case, $S' = S \setminus \{v\}$ is a colourful independent set (in G) appearing in $S_j^{i-(l(v)+1)}$ such that $S' \cap (X \cup \{v\}) = \emptyset$ and $S' \cup (X \cup \{v\}) \in \mathcal{I}$. Let $X' = X \cup \{v\}$. Because $\hat{\mathcal{R}}_{j-1}^{i-(l(v)+1)}$ is a $(k-j+1)$-representative family for $\mathcal{R}_{j-1}^{i-(l(v)+1)}$ and $|X'| \le k - j + 1$, there is a colourful independent set $\tilde{S} \in \mathcal{R}_{j-1}^{i-(l(v)+1)}$ such that $\tilde{S} \cap X' = \emptyset$ and $\tilde{S} \cup X' \in \mathcal{I}$.

Note that $\tilde{S} \cup \{v\} \subseteq \tilde{S} \cup X' \in \mathcal{I}$ is colourful. Since $\tilde{S} \in \mathcal{R}_{j-1}^{i-(l(v)+1)}$ only contains vertices that lie in the interval $[0, i - (l(v) + 1)]$, v has no neighbours in $\tilde{S}$. Because $S^* = \tilde{S} \cup \{v\}$ is a colourful independent set in G, the $[\cdot]$ operator in Eq. 1 preserves it. Thus, there is a set S^* in $\bar{\mathcal{R}}_j^i$ such that $S^* \cap X = \emptyset$ and $S^* \cup X \in \mathcal{I}$.

Case 2. S contains no vertex v such that L_v ends at i.

Observe that S lies entirely in the interval $[0, i-1]$, so it appears in $\mathcal{R}_j^{i-1}$. Since $\hat{\mathcal{R}}_j^{i-1}$ is a $(k-j)$-representative family for $\mathcal{R}_j^{i-1}$, there is a set $\tilde{S} \in \hat{\mathcal{R}}_j^{i-1}$ such that $\tilde{S} \cap X = \emptyset$ and $\tilde{S} \cup X \in \mathcal{I}$.

The procedure ComputeTable constructs a table $S[0..c][0..k]$ using the predefined procedure `ComputeRepresentativesFLPS` that computes representative families according to Proposition 4.

Lemma 6. *The procedure ComputeTable computes $S[0..c][0..k]$ such that each entry $S[i][j]$ is $(k-j)$-representative of $\mathcal{R}_j^i$. The table is computed in time $O(c\chi k^\omega 2^{\omega k} n)$, where χ is the time required to perform field operations over $\mathbb{F}_2(X)$.*

Proof. In Line 1, the initialization step ensures that for each $(i, j) \in (\{0\} \times [k]) \cup ([c] \times \{0\})$, $\mathcal{R}_j^i$ is $(k-j)$-represented by $S[i][j]$. The family constructed in Line 8 is $\bar{\mathcal{R}}_j^i$, which has $t = |S[i-1][j]| + \sum_{L_v \text{ ends at } i} |S[i-(l(v)+1)][j-1]|$ sets.

Each entry $S[i'][j']$ referenced in this step was computed in a previous iteration, and (by Proposition 4) Line 9 ensures that $|S[i'][j']| \le \binom{k}{j'}$. Thus, we

---

Procedure ComputeTable: compute a table of representative families

---

Input: $G, \mathcal{D}, A_\mathcal{K}, c, k$, where G is a graph, $\mathcal{D}$ is a c-compact representation and $k \in \mathbb{N}$

Output: $S[0 .. c][0 .. k]$, where each entry $S[i][j]$ $(k-j)$-represents $\mathcal{R}_j^i$

1   initialize $S[0 .. c][0 .. k]$ with $\emptyset$;

2   **for** $i \in [1 .. c]$ **do**

3   **for** $j \in [1 .. k]$ **do**

4       $S[i][j] \leftarrow S[i-1][j]$;

5   **for** $v \in V(G)$ *such that* L_v *ends at* i **do**

6         $T_v \leftarrow v + S[i - (l(v)+1)][j-1]$;

7         $T_v \leftarrow \{A \in T_v \mid A$ is a colourful independent set$\}$;

8         $S[i][j] \leftarrow S[i][j] \cup T_v$;

9       $S[i][j] \leftarrow$ ComputeRepresentativesFLPS$(A_\mathcal{K}, S[i][j], k-j)$;

10 **return** $S[0 .. c][0 .. k]$;

---

have $t \leq \binom{k}{j} + n\binom{k}{j-1}$. Note that (again because of Proposition 4) $S[i][j]$ is $(k-j)$-representative of $\bar{\mathcal{R}}_j^i$, so it also $(k-j)$-represents $\mathcal{R}_j^i$.

The computation can be carried out using $O\left(t\left(j^\omega\binom{k}{j} + \binom{k}{j}^{\omega-1}\right)\right)$ operations over $\mathbb{F}_2(X)$. This takes time $O\left(t\left(j^\omega\binom{k}{j} + \binom{k}{j}^{\omega-1}\right)\chi\right)$, where χ is the time required to perform field operations over $\mathbb{F}_2(X)$. The expression further simplifies to $O\left(\binom{k}{j}^\omega(j^\omega + 1)\chi n\right) = O\left(\binom{k}{j}^\omega j^\omega \chi n\right)$.

The construction of $S[i][j]$ in Lines 4–8 takes time $O(nt) = O\left(\binom{k}{j}^\omega j^\omega n\right)$, since it only involves copying and adding (single) elements to $O(t)$ sets. Thus, the running time of the double loops is

$$O\left(\chi n \sum_{i=1}^c \sum_{j=1}^k \binom{k}{j}^\omega j^\omega\right) = O\left(\chi k^\omega n \sum_{i=1}^c \sum_{j=1}^k \binom{k}{j}^\omega\right).$$

By straightforward arguments, it can be shown that this expression is $O(c\chi k^\omega 2^{\omega k} n)$. The other steps of the procedure take time $O(n)$, so the overall running time is $O(c\chi k^\omega 2^{\omega k} n)$. $\square$

We are now ready to prove Theorem 2.

Proof (Theorem 2). Using SolveIntervalCIS, we solve (G, ϕ, k). Its correctness follows directly from Lemmas 5 and 6. A c-compact representation $\mathcal{D}$ for G can be computed using the procedure of Proposition 7 in time $O(n^2)$. Then using the procedure of Lemma 3, a representation $A_\mathcal{K}$ for $\mathcal{K}$ can be computed in time $O(k\gamma n)$ (Lemma 3). Finally, ComputeTable$(G, \mathcal{D}, A_\mathcal{K}, c, k)$ takes time $O(c\chi k^\omega 2^{\omega k} n)$ (Lemma 5). All other operations take time $O(n^2)$.

Since the entries of $A_\mathcal{K}$ are computed in $O(k\gamma n)$ field operations over $\mathbb{F}_2$, they at most n^a bits in size for some constant $a \in \mathbb{N}$. A fact we use without proof is that these entries can be interpreted as elements of a field $\mathbb{F}_q$ for some $q \sim 2^{n^a}$ while still maintaining the condition that $A_{\mathcal{K}}$ represents $\mathcal{K}$. Field operations over $\mathbb{F}_q$ take time $O(n^b)$ for some constant $b \in \mathbb{N}$, i.e. $\chi = O(n^b)$.

Thus, ComputeTable$(G, \mathcal{R}, A_{\mathcal{K}}, c, k)$ takes time $O(c\chi k^{\omega} 2^{\omega k} n) = O(ck^{\omega} 2^{\omega k} n^{b+1})$. The matrix multiplication algorithm of Gall [4] has $\omega \leq 2.3728639$, and because of Proposition 6, $c \leq n$. Therefore, the overall running time of SolveIntervalCIS is $O\left((5.18)^k n^d\right)$ for some constant $d \in \mathbb{N}$.   □

Algorithm SolveIntervalCIS. determine if G has a colourful independent set of size k under ϕ

Input: G, ϕ, k, where G is a graph, $\phi : V(G) \to [\gamma]$ is a colouring and $k \in \mathbb{N}$

Output: YES if G has a colourful independent set of size k under ϕ and NO otherwise

1 $\mathcal{D} \leftarrow$ ComputeCompactRepresentation(G);
2 let $c \in \mathbb{N}$ such that $\mathcal{D}$ is c-compact;
3 compute a representation $A_{\mathcal{K}}$ for the matroid of colourful sets of size at most k;
4 $S \leftarrow$ ComputeTable$(G, \mathcal{D}, A_{\mathcal{K}}, c, k)$;
5 **if** $S[c][k]$ *is non-empty* **then**
6     $\lfloor$ **return** YES;

7 **else**
8     $\lfloor$ **return** NO;

---

4 Conclusion

We have designed improved algorithms for COLOURFUL INDEPENDENT SET via two distinct approaches:

- using improved constructions of hash families, and
- using representative families.

The algorithm of Theorem 1 is an improvement over earlier algorithms with regard to the parameter k, i.e. the number of jobs to be scheduled, as well as n, the total number of jobs. On the other hand, SolveIntervalCIS (Theorem 2), which runs in time $O\left((5.18)^k n^d\right)$ and outperforms previous algorithms in the case $k \geq d \log n$.

Using a variant of Proposition 3 (see [5], Theorem 3.15), we were able to obtain the bound $d \leq 4$. An interesting question is to see if the dependence on n in the running time of SolveIntervalCIS could be made quadratic or even linear.

References

1. Fomin, F.V., Lokshtanov, D., Panolan, F., Saurabh, S.: Efficient computation of representative families with applications in parameterized and exact algorithms. J. ACM **63**(4), 29:1–29:60 (2016)
2. Halldórsson, M.M., Karlsson, R.K.: Strip graphs: recognition and scheduling. In: Fomin, F.V. (ed.) WG 2006. LNCS, vol. 4271, pp. 137–146. Springer, Heidelberg (2006). https://doi.org/10.1007/11917496_13
3. Keil, J.M.: On the complexity of scheduling tasks with discrete starting times. Oper. Res. Lett. **12**(5), 293–295 (1992)
4. Le Gall, F.: Powers of tensors and fast matrix multiplication. In: Proceedings of the 39th International Symposium on Symbolic and Algebraic Computation, pp. 296–303. ACM Press, Kobe (2014)
5. Lokshtanov, D., Misra, P., Panolan, F., Saurabh, S.: Deterministic truncation of linear matroids. ACM Trans. Algorithms **14**(2), 14:1–14:20 (2018)
6. Nakajima, K., Hakimi, S.L.: Complexity results for scheduling tasks with discrete starting times. J. Algorithms **3**(4), 344–361 (1982)
7. Naor, M., Schulman, L.J., Srinivasan, A.: Splitters and near-optimal derandomization. In: Proceedings of the 36th Annual Symposium on Foundations of Computer Science, pp. 182–191. IEEE Computer Society Press, Milwaukee, October 1995
8. van Bevern, R., Mnich, M., Niedermeier, R., Weller, M.: Interval scheduling and colorful independent sets. J. Sched. **18**(5), 449–469 (2015)

Call Admission Problems on Trees with Advice
(Extended Abstract)

Hans-Joachim Böckenhauer[(✉)], Nina Corvelo Benz, and Dennis Komm

Department of Computer Science, ETH Zurich, Zurich, Switzerland
{hjb,dennis.komm}@inf.ethz.ch, cnina@student.ethz.ch

Abstract. A well-studied problem in the online setting, where requests have to be answered immediately without knowledge of future requests, is the call admission problem. In this problem, we are given nodes in a communication network that request connections to other nodes in the network. A central authority may accept or reject such a request right away, and once a connection is established its duration is unbounded and its edges are blocked for other connections. This paper examines the admission problem in tree networks. The focus is on the quality of solutions achievable in an advice setting, that is, when the central authority has some information about the incoming requests. We show that $O(m \log d)$ bits of additional information are sufficient for an online algorithm run by the central authority to perform as well as an optimal offline algorithm, where m is the number of edges and d is the largest degree in the tree. In the case of a star tree network, we show that $\Omega(m \log d)$ bits are also necessary (note that $d = m$). Additionally, we present a lower bound on the advice complexity for small constant competitive ratios and an algorithm whose competitive ratio gradually improves with added advice bits to $2\lceil \log_2 n \rceil$, where n is the number of nodes.

1 Introduction

A well-studied problem in the context of regulating the traffic in communication networks is the so-called *call admission problem*, where a central authority decides about which subset of communication requests can be routed. This is a typical example of an *online problem*: every request has to be routed or rejected immediately without the knowledge about whether some forthcoming, possibly more profitable, requests will be blocked by this decision.

We consider the *call admission problem on trees* (short CAT), which is an online maximization problem. An instance $I = (r_1, \ldots, r_k)$ consists of requests $r = (v_i, v_j)$ with $i, j \in \{0, \ldots, n-1\}$ and $v_i < v_j$, representing the unique path in a tree network that connects vertices v_i and v_j. We require that all requests in I are pairwise distinct. The first request contains the network tree $T = (V, E)$, given to the algorithm in form of an adjacency list or matrix. In particular, we study the problem framework in which an accepted connection has an unbounded

© Springer Nature Switzerland AG 2019
C. J. Colbourn et al. (Eds.): IWOCA 2019, LNCS 11638, pp. 108–121, 2019.
https://doi.org/10.1007/978-3-030-25005-8_10

duration and each edge in the network may be used by at most one request, i.e., it has a capacity of 1. Thus, a valid solution $O = (y_1, \ldots, y_k) \in \{0,1\}^k$ for I describes a set $\mathcal{P}(I, O) := \{r_i \mid i \in \{1, \ldots, k\}$ and $y_i = 1\}$ of edge-disjoint paths in T, where $\text{gain}(I, O) := |\mathcal{P}(I, O)|$. Whenever I is clear from the context, we write $\text{gain}(O)$ instead of $\text{gain}(I, O)$.

An *online algorithm* ALG for CAT computes the output sequence (solution) $\text{ALG}(I) = (y_1, \ldots, y_k)$, where y_i is computed from $x_1, \ldots, x_i$. The gain of ALG's solution is given by $\text{gain}(\text{ALG}(I))$. ALG is *c-competitive*, for some $c \geq 1$, if there exists a constant γ such that, for every input sequence I, $\text{gain}(\text{OPT}(I)) \leq c \cdot \text{gain}(\text{ALG}(I)) + \gamma$, where OPT is an optimal offline algorithm for the problem. This constitutes a measure of performance used to compare online algorithms based on the quality of their solutions, which was introduced by Sleator and Tarjan [18].

The downside of *competitive analysis* as a measurement of performance is that it seems rather unrealistic to compare the performance of an all-seeing offline algorithm to that of an online algorithm with no knowledge at all about future requests. This results in this method not really apprehending the hardness of online computation. Moreover, it cannot model information about the input that we may have outside the strictly defined setting of the problem. The advice model was introduced as an approach to investigate the amount of information about the future an online algorithm lacks [6,7,12,13,15]. It investigates how many bits of information are necessary and sufficient to achieve a certain output quality, which has interesting implications for, e.g., randomized online computation [5,9,16]. For lower bounds on this number in particular, we do not make any assumptions on the kind of information the advice consists of.

Let Π be an online maximization problem, and consider an input sequence $I = (x_1, \ldots, x_k)$ of Π. An *online algorithm* ALG *with advice* computes the output sequence $\text{ALG}(I)^\phi = (y_1, \ldots, y_k)$ such that y_i is computed from $\phi, x_1, \ldots, x_i$, where ϕ is the content of the advice tape, i.e., an infinite binary sequence. ALG is *c-competitive with advice complexity* $b(k)$ if there exists a constant γ such that, for every k and for each input sequence I of length at most k, there exists some ϕ such that $\text{gain}(\text{OPT}(I)) \leq c \cdot \text{gain}(\text{ALG}^\phi(I)) + \gamma$ and at most the first $b(k)$ bits of ϕ have been read during the computation of $\text{ALG}^\phi(I)$.

For a better understanding, consider the following example. A straightforward approach to an optimal online algorithm with advice for CAT is to have one bit of advice for each request in the given instance. This bit indicates whether the request should be accepted or not. Thus, ALG reads $|I|$ advice bits and accepts only requests in $\text{OPT}(I)$, i.e., ALG is optimal.

This approach gives us a bound on the advice complexity that is linear in the size of the instance. Opposed to most other online problems however, call admission problems, like CAT, are usually analyzed with respect to the size of the communication network instead of the size of an instance as stated in the general definition. Thus, the advice complexity of this naive optimal algorithm on a tree with n vertices is of order n^2.

Related Work. The call admission problem is a well-studied online problem; for an overview of results regarding classical competitive analysis for this problem on various graph topologies, see Chapter 13 in the textbook by Borodin and El-Yaniv [3]. For the call admission problem on path networks (also called the *disjoint path allocation problem*, short DPA), Barhum et al. [2] showed that $l - 1$ advice bits are both sufficient and necessary for an online algorithm to be optimal, where l is the length of the path. They also generalized the $\log_2 l$-competitive randomized algorithm for DPA presented by Awerbuch et al. [1]. Gebauer et al. [14] proved that, with $l^{1-\varepsilon}$ bits of advice, no online algorithm for DPA is better than $(\delta \log_2 l)/2$-competitive, where $0 < \delta < \varepsilon < 1$. The advice complexity of call admission problems on grids was investigated by Böckenhauer et al. [8].

When considering trees as network structure, we still have the property that the path between two nodes is unique in the network, thus all lower bounds on the advice complexity easily carry over by substituting the length l of the path network by the diameter D of the tree network. These lower bounds can be further improved as shown in Sect. 2. Concerning upper bounds, Borodin and El-Yaniv [3] presented two randomized online algorithms, a $2\lceil \log_2 n \rceil$-competitive algorithm, first introduced by Awerbuch et al. [1], and an $O(\log D)$-competitive algorithm. We will modify the former in Subsect. 3.2 to an online algorithm that reads $\lceil \log_2 \log_2 n - \log_2 p \rceil$ advice bits and is $((2^{p+1} - 2)\lceil \log_2 n/p \rceil)$-competitive, for any integer $1 \leq p \leq \log_2 n$.

Another problem closely related to DPA is the length-weighted disjoint path allocation problem on path networks, where instead of optimizing the number of accepted requests, one is interested in maximizing the combined length of all accepted requests. Burjons et al. [10] extensively study the advice complexity behavior of this problem.

Overview. In Sect. 2, we present the already mentioned lower bound for optimality, which even holds for star trees. We complement this with lower bounds for the trade-off between the competitive ratio and advice, based on reductions from the well-known string guessing problem [4]. Section 3 is devoted to the corresponding upper bounds. In Subsect. 3.1, we present algorithms for computing an optimal solution, both for general trees and for star trees and k-ary trees. As mentioned above, in Subsect. 3.2, we analyze the trade-off between the competitive ratio and advice and estimate how much the competitive ratio degrades by using less and less advice bits. Due to space restrictions, some of the proofs are omitted in this extended abstract.

Notation. Following common conventions, m is the number of edges in a graph and n the number of vertices. The degree of a vertex v is denoted by $d(v)$. Let $v_0, \ldots, v_{n-1}$ be the vertices of a tree T with some order $v_0 < \cdots < v_{n-1}$. This order can be arbitrarily chosen, but is fixed and used as order in the adjacency matrix or adjacency list of the tree. Hence, an algorithm knows the ordering on the vertices when given the network.

For the sake of simplicity, we sometimes do not enforce that $v < v'$, but regard (v, v') and (v', v) as the same request. For a request $r = (v, v')$, the function edges: $V \times V \to \mathcal{P}(E)$ returns, for request r, the set of edges corresponding to the unique path in T that connects v and v'. Let $\text{edges}(r) := \{e_1, \ldots, e_l\}$; we call l the length of request r. Note that all logarithms in this paper are of base 2, unless stated otherwise.

2   Lower Bounds

First we present lower bounds on the number of advice bits for the call admission problem on trees. We first look at optimal algorithms, then we focus on the connection between the competitive ratio and the advice complexity.

2.1   A Lower Bound for Optimality

Barhum et al. [2] proved that solving DPA optimally requires at least $l - 1$ advice bits. As DPA is a subproblem of the call admission problem on trees, this bound also holds for CAT. We can improve on this by considering instances on trees of higher degree. We focus on the simplest tree of high degree, the star tree.

Theorem 1. *There is no optimal online algorithm with advice for CAT that uses less than $\lceil (m/2) \log(m/e) \rceil$ advice bits on trees of m edges.*

Proof Sketch. This proof is based on the partition-tree method as introduced by Barhum et al. [2]. A partition tree of a set of instances $\mathcal{I}$ is defined as a labeled rooted tree such that (i) every of its vertices v is labeled by a set of input sequences $\mathcal{I}_v$ and a number ϱ_v such that all input sequences in $\mathcal{I}_v$ have a common prefix of length at least ϱ_v, (ii) for every inner vertex v of the tree, the sets at its children form a partition of $\mathcal{I}_v$, and (iii) the root r satisfies $\mathcal{I}_r = \mathcal{I}$. If we consider two vertices v_1 and v_2 in a partition tree that are neither an ancestor of each other, with their lowest common ancestor v and any input instances $I_1 \in \mathcal{I}_{v_1}$ and $I_2 \in \mathcal{I}_{v_2}$ such that, for all optimal solutions for I_1 and I_2, their prefixes of length ϱ_v differ, then any optimal online algorithm with advice needs a different advice string for each of the two input sequences I_1 and I_2. This particularly implies that any optimal online algorithm with advice requires at least $\log(w)$ advice bits, where w is the number of leaves of the partition tree. We sketch the construction of the instances that can be used for building such a partition tree.

Consider the star trees S_{2k} for $k \geq 1$ with $2k$ edges. Let $v_0, v_1, \ldots, v_{2k}$ be the vertices in S_{2k}, where v_0 denotes the center vertex. We construct a set $\mathcal{I}$ of input sequences for S_{2k} so that any two input sequences share a common prefix of requests and each input $I \in \mathcal{I}$ has a unique optimal solution $\text{OPT}(I)$, which is only optimal for this particular instance.

Each input instance will be partitioned into k phases. At the end of each of these phases, one vertex will be blocked for all subsequent phases. We can uniquely describe each of our instances $I_{(j_1, \ldots, j_k)}$ by the sequence of these blocked vertices $(v_{j_1}, \ldots, v_{j_k})$. Note that we will not use every possible vertex sequence for our construction. In phase i with $i \in \{1, \ldots, k\}$, we request in ascending order all paths from the non-blocked vertex v_i^* with the smallest index to all other non-blocked vertices, then we block v_i^* and v_{j_i} for all future phases. We note that,

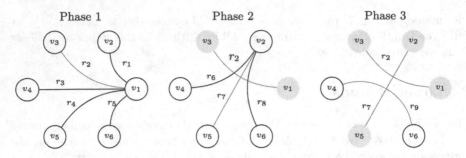

Fig. 1. Input sequence $I_{(3,5,6)} = (r_1, \ldots, r_9)$ on star graph S_6 partitioned into 3 phases. Red vertices are blocked, a red edge indicates the request is in the optimal solution $\text{OPT}(I_{(3,5,6)})$. For the sake of simplicity, the center vertex v_0 is omitted from the drawings; thus all lines represent paths of length 2 (Color figure online).

at the start of phase 1, all vertices are non-blocked. Let $\mathcal{I}_{(j_1,\ldots,j_i)}$ denote the set of all input sequences whose tuples have prefix $(j_1, \ldots, j_i)$ with $i \leq k$. Observe that, by definition, all input sequences in $\mathcal{I}_{(j_1,\ldots,j_i)}$ have the same requests until phase i ends and that tuple $(j_1, \ldots, j_k)$ describes exactly one input sequence in $\mathcal{I}$, i.e., $|\mathcal{I}_{(j_1,\ldots,j_k)}| = 1$. Figure 1 shows an illustration of such an input sequence for the star S_6, i.e., for $k = 3$.

We can prove that, for each input sequence $I_{(j_1,\ldots,j_k)}$, the unique optimal solution is $\text{OPT}(I_{(j_1,\ldots,j_k)}) := \{r_1, \ldots, r_k\}$, where $r_i := (v_i^*, v_{j_i})$ is the request accepted by OPT in each phase i. The next step is to show that, for any two input sequences in $\mathcal{I}$, their unique optimal solution OPT differs. Consider two input sequences $I, I' \in \mathcal{I}$ with $I \neq I'$ and let $(j_1, \ldots, j_k)$ and $(j'_1, \ldots, j'_k)$ be their identifying tuples, respectively. As the two input sequences are non-identical, there must exist some smallest index i so that $j_i \neq j'_i$. In particular, since i marks the first phase at whose end different vertices are blocked in I and I', the requests in phase i must be identical in both input sequences. Let v_i^* be the non-blocked vertex with the smallest index in phase i in both input sequences. By definition of OPT, request $(v_i^*, v_{j_i}) \in \text{OPT}(I)$ and request $(v_i^*, v_{j'_i}) \in \text{OPT}(I')$ with $j_i \neq j'_i$. It thus follows that $\text{OPT}(I) \neq \text{OPT}(I')$ as both requests share an edge.

Since the optimal solution for the common prefix differs between two instances, no online algorithm without advice can be optimal on this set, because, with no additional information on the given instance (i.e., based on the prefix alone) the two instances cannot be distinguished. It follows that the algorithm needs a unique advice string for each instance in the set. Thus, it only remains to bound the number of instances in $\mathcal{I}$. Each instance has a unique label $\mathcal{I}_{(j_1,\ldots,j_k)}$, so that the total number equals the number of tuples $(j_1, \ldots, j_k)$ of pairwise distinct vertex indices, that is,

$$(2k-1) \cdot (2k-3) \cdot (2k-5) \cdots 1 = (2k-1)!! = \frac{(2k)!}{2^k \cdot k!},$$

which we can bound from below using Stirling's inequalities, yielding

$$\frac{(2k)!}{2^k \cdot k!} \geq \frac{(2k)!}{2^k \cdot e\sqrt{k} \cdot (\frac{k}{e})^k} \geq \frac{\sqrt{2\pi 2k} \cdot (\frac{2k}{e})^{2k}}{e\sqrt{k} \cdot (\frac{2k}{e})^k} \geq \frac{\sqrt{4\pi}}{e} \cdot \left(\frac{2k}{e}\right)^k \geq \left(\frac{2k}{e}\right)^k.$$

Using that $m = 2k$, we conclude that at least $\lceil (m/2) \log(m/e) \rceil$ advice bits are necessary for any online algorithm to be optimal on the tree S_m. $\square$

This lower bound is asymptotically larger by a logarithmic factor than the DPA lower bound [2], which suggests that the advice complexity of CAT increases with the degree of the tree network and not only with its size.

2.2   A Lower Bound for Competitiveness

In this section, we present a reduction from an online problem called the string guessing problem to CAT. In the *string guessing problem with unknown history* (q-SGUH), an algorithm has to guess a string of specified length z over a given alphabet of size $q \geq 2$ character by character. After guessing all characters, the algorithm is informed of the correct answer. The cost of a solution $\text{ALG}(I)$ is the Hamming distance between the revealed string and $\text{ALG}(I)$. Böckenhauer et al. [4] presented a lower bound on the number of necessary advice bits depending on the achieved fraction of correct character guesses.

We will use this lower bound for our results, reducing the q-SGUH problem to CAT by assigning each element of the alphabet to an optimal solution for a family of instances on the star tree S_d where $d = q$. The idea is to have a common prefix on all instances and the last requests specifying a unique optimal solution for the instance corresponding to the character in the string. For each character we have to guess, we insert such a star tree $S^{(i)} := S_d$ into our graph such that the graph is connected but the trees do not share edges. We can then look at each subtree independently and join the instances to an instance corresponding to the whole string. Let therefore, for some $z, d \in \mathbb{N}$, $T_{z,d}$ be the set of trees that can be constructed from subtrees $S^{(i)}$ with $i \in \{1, \ldots, z\}$, such that any two subtrees share at most one vertex and the tree is fully connected. For each request of a d-SGUH instance of length z, where the optimal answer would be the string $s_1 \ldots s_z$ we now construct a sequence of requests for $S^{(i)}$ such that choosing the optimal set of requests for $S^{(i)}$ corresponds to correctly guessing s_i. Then, any algorithm that solves a fraction α of all subtrees optimally can be used to achieve a fraction α of correct guesses on the d-SGUH instance.

Theorem 2. *Every online algorithm with advice for CAT which achieves a competitive ratio of $c \leq d/(d-1)$ on any tree $T \in T_{z,d}$, for $d \geq 3$ and $z, d \in \mathbb{N}$, has to read at least*

$$\left(1 - H_d\left(d - \frac{d-1}{c} - 1\right)\right)\frac{m}{d}\log d$$

advice bits, where H_d is the d-ary entropy function. $\square$

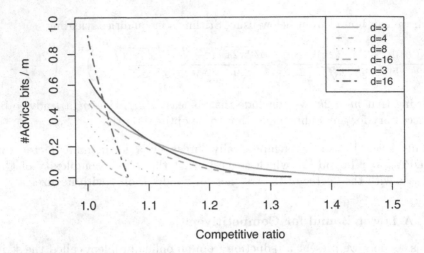

Fig. 2. Lower bound on the number of advice bits stated in Theorem 2 (light green) and Theorem 3 (dark blue) divided by the number m of edges in $T \in \mathcal{T}_{z,d}$. (Color figure online)

We can use the same tree structure to prove a better lower bound for small values of c by changing the reduction instance of CAT. This change increases the alphabet size of q-SGUH that we can reduce to instances on trees in $\mathcal{T}_{z,d}$ for $q = 2^{d-1}$, and allows to prove the following theorem.

Theorem 3. *Every online algorithm with advice for CAT which achieves a competitive ratio of $c \leq d/(d - 1 + 1/2^{d-1})$ on any tree $T \in \mathcal{T}_{z,d}$, for $d \geq 2$ and $z, d \in \mathbb{N}$, has to read at least*

$$\left(1 - H_{2^{d-1}}\left(d - \frac{d}{c}\right)\right) m \cdot \frac{d-1}{d}$$

advice bits, where H_d is the d-ary entropy function. □

Figure 2 depicts the lower bounds of Theorems 2 and 3, respectively.

3 Upper Bounds

In the following, we present online algorithms with advice for the call admission problem on trees. In the first part, the focus will be on optimal algorithms with different advice complexities. In the second part, we discuss an algorithm whose competitiveness gradually improves with added advice bits.

3.1 Optimal Online Algorithms with Advice

The fundamental idea of the following algorithms is to encode the optimal solution using edge labels as advice. A straightforward approach is to give all requests in $\mathrm{OPT}(I)$ an identifying number and label the edges of each request with this identifier. After communicating the labels of all edges, the algorithm will be able to distinguish which request is in $\mathrm{OPT}(I)$ and which is not, by checking whether all the edges of the request have the same label and no other edges have this label. As a result, the algorithm can recognize and only accept requests that an optimal solution accepts.

As for the advice complexity, we need m labels each consisting of a number in $\{0, 1, \ldots, |\mathrm{OPT}(I)|\}$. Since, for any input sequence I, we have $|\mathrm{OPT}(I)| \leq m$, in total $m\lceil \log(m+1) \rceil$ advice bits suffice. If $|\mathrm{OPT}(I)|$ is much smaller than m, we can communicate the size of a label using a self-delimiting encoding [17], using $m\lceil \log(|\mathrm{OPT}(I)| + 1) \rceil + 2\lceil \log\lceil \log(|\mathrm{OPT}(I)| + 1) \rceil \rceil$ advice bits in total.

We will continue to use this idea of an identifier for the following algorithms in a more local manner. Instead of giving each request in $\mathrm{OPT}(I)$ a global identifying number and labeling corresponding edges accordingly, we associate identifiers with an optimal request depending on the vertices incident to the request's edges.

We can picture this labeling scheme with a request as a row of dominoes, where each edge of the request represents one domino and consecutive edges have the same identifying number at their common vertex. Knowing for all edges the incident edges that are part of the same request, we can reconstruct the paths belonging to all requests in $\mathrm{OPT}(I)$ and since we only need local identifiers, this reduces the size of each label.

Theorem 4. *There is an optimal online algorithm with advice for CAT that uses at most $2m\lceil \log(d + 1) \rceil$ advice bits, where d is the maximum degree of a vertex in T.*

Proof Sketch. First, let us define such a local labeling formally. For some input sequence $I = (r_1, \ldots, r_l)$, let $\mathrm{OPT}(I) \subseteq \{r_1, \ldots, r_l\}$ be an arbitrary, but fixed optimal solution for I. Furthermore, for every vertex v, let $\mathrm{OPT}_v(I)$ be the subset of requests in $\mathrm{OPT}(I)$ which occupy an edge incident with v. Observe that $|\mathrm{OPT}_v(I)| \leq d(v)$ as there are $d(v)$ edges incident with v. For all $v \in V$, let $g_v \colon \mathrm{OPT}_v(I) \to \{1, \ldots, d(v)\}$ be an injective function that assigns a number to each request in $\mathrm{OPT}_v(I)$. These numbers serve as local identifiers of each request in $\mathrm{OPT}_v(I)$. We define the label function lb as follows; for $e = \{v, v'\} \in E$, where $v < v'$, let

$$lb(e) := \begin{cases} (lb_v(e),\ lb_{v'}(e)) = (g_v(r),\ g_{v'}(r)) & \text{if } \exists\ r \in \mathrm{OPT}(I) \text{ s.t. } e \in \mathrm{edges}(r), \\ (0, 0) & \text{otherwise.} \end{cases}$$

Thus, if an edge e is used by $r \in \mathrm{OPT}(I)$, the local identifiers of r for the two vertices of e constitute the edge label; if unused, e is labeled $(0, 0)$. Observe that, if $e = \{v, v'\} \in \mathrm{edges}(r)$ and $r \in \mathrm{OPT}(I)$, it follows that $r \in \mathrm{OPT}_v(I)$ and $r \in \mathrm{OPT}_{v'}(I)$, so lb is well-defined. Figure 3 shows an example of a local and a global labeling side by side. Observe that, for vertex v, we have the label 4 for two edges, but no label 2. As g_v may arbitrarily assign an identifier in $\{1, \ldots, d_v\}$,

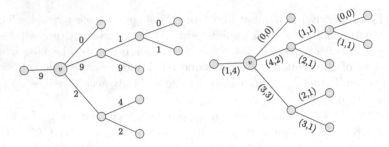

Fig. 3. Examples of a global labeling (left) and a local labeling (right) for the same optimal solution.

the assigned number to a request does not have to be minimal. In the lb-labeled tree T, we call $p = (v'_1, \ldots, v'_l)$ a *labeled path of length* l if p is a path in T with $v'_1 < v'_l$ and $lb_{v'_j}(\{v'_{j-1}, v'_j\}) = lb_{v'_j}(\{v'_j, v'_{j+1}\}) \neq 0$ for all $j \in \{2, \ldots, l-1\}$.

We refer to p as a *complete labeled path* if further no other edges incident to v'_1 or v'_l have label $lb_{v'_1}(\{v'_1, v'_2\})$ or $lb_{v'_l}(\{v'_{k-1}, v'_l\})$, respectively.

Consider an algorithm ALG$'$ that reads the labels of all edges from the advice before starting to receive any request and then computes the set P of all complete labeled paths in T. ALG$'$ then accepts a request $r = (v, v')$ if and only if it coincides with a complete labeled path in P.

We can prove that ALG$'$ accepts all requests in OPT(I), and thus is optimal, by showing that every request in OPT(I) has a coinciding path in P and that all paths in P are pairwise edge-disjoint. It remains to bound the number of advice bits used. For an edge $\{v, v'\}$, we need $\lceil \log(d(v)+1) \rceil + \lceil \log(d(v')+1) \rceil$ advice bits to communicate the label $lb(\{v, v'\})$. Hence, per vertex w, we use $d(w) \cdot \lceil \log(d(w)+1) \rceil$ advice bits. Summing up over all vertices yields the claimed bound. $\qquad\square$

We can further improve this bound by showing that pinpointing an endvertex of a request $r \in$ OPT(I) does not require a unique identifier.

Theorem 5. *There is an optimal online algorithm with advice for CAT that uses at most $(m-1) \lceil \log(\lfloor d/2 \rfloor + 1) \rceil$ advice bits, where d is the maximum degree of a vertex in T.*

Proof Sketch. Consider a function $g_v \colon$ OPT$_v(I) \rightarrow \{0, 1, \ldots, \lfloor d(v)/2 \rfloor\}$ that assigns a non-zero identifier only to requests in OPT$_v$ that occupy two incident edges to v, otherwise it assigns identifier 0. Observe that we halve the number of identifiers needed this way.

Let us now refer to a labeled path $p = (v'_1, \ldots, v'_l)$ as complete if and only if

$$lb_{v'_1}(\{v'_1, v'_2\}) = lb_{v'_l}(\{v'_{l-1}, v'_l\}) = 0. \tag{1}$$

Again, we consider ALG$'$ that reads the advice $lb(e_1), \ldots, lb(e_m)$ for a tree T and computes the set P of all complete labeled paths in T according to Theorem 4.

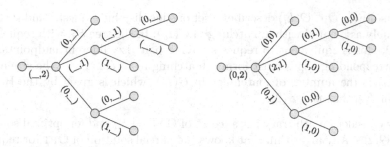

Fig. 4. Example of a labeling as used in Theorem 4 (left) and the inferred full labeling (right) for the same optimal solution.

ALG' then accepts a request $r = (v, v')$ if and only if it coincides with a complete labeled path in P.

We can show, analogously to the proof of Theorem 4, that all paths in P are pairwise edge-disjoint and that each request in OPT(I) has a coinciding path in P (Fig. 4). Thus, as before, ALG' accepts all requests in OPT(I) and is therefore optimal. Finally, we note that not all labels have to be communicated. Consider a vertex v and its incident edges $e'_1, \ldots, e'_{d(v)}$. Assuming that we have all labels $lb_v(e'_1), \ldots, lb_v(e'_{d(v)-1})$, we can infer the last label $lb_v(e'_{d(v)})$ as follows. If there exists only one edge $e \in \{e'_1, \ldots, e'_{d(v)}\}$ with non-zero label $lb_v(e)$, then $lb_v(e'_{d(v)}) = lb_v(e)$, since by definition of $g_v(r)$ and lb there are exactly two edges with the same non-zero label. If there is no such edge, $lb_v(e'_{d(v)}) = 0$ for the same reason. Therefore, for each vertex we only need to communicate the advice for the first $d(v) - 1$ edges and per edge only $\lceil \log(\lfloor d(v)/2 \rfloor + 1) \rceil$ advice bits. In total, ALG' needs at most $\lceil \log(\lfloor d/2 \rfloor + 1) \rceil \cdot (m - 1)$ advice bits, where d is the maximum degree of a vertex in T. □

Note that the central idea behind the algorithms of Theorems 4 and 5 is to identify, for all inner vertices, which incident edges belong to the same request in some fixed optimal solution. We used edge labels as advice to convey this information. In what follows, we will discuss another technique to encode this information for some types of trees. First, let us examine the star tree S_d of degree d; let $I_d = (r_1, \ldots, r_{(d(d-1))/2})$ denote the instance with all possible requests in S_d of length 2.

Lemma 1. *For the instance I_d of CAT on S_d, the size of the set of solutions $\mathcal{O}(I_d)$ is at most*

$$\sum_{j=0}^{\lfloor d/2 \rfloor} \frac{d!}{2^j \cdot j! \cdot (d - 2j)!}. \tag{2}$$

Proof. We construct a graph $G(I_d)$, where the vertex set corresponds to the leaf vertices of S_d. For a request r of I_d, we insert an edge in $G(I_d)$ between the respective vertices. Note that, since I_d consists of all requests between leaf vertices in S_d, we have that $G(I_d)$ is the complete graph K_d on d vertices.

Any solution $O \in \mathcal{O}(I_d)$ describes a set of edge-disjoint requests, and thus can be uniquely associated with a matching in $G(I_d)$: In the tree S_d, with requests of length 2, this is equivalent to requests having pairwise different endpoints, i.e., their corresponding edges must form a matching in $G(I_d)$. Thus, the size of the set $\mathcal{O}(I_d)$ is the number of matchings in $G(I_d)$, which is given by the Hosoya index[1] of K_d, that is, by (2). □

Now consider an arbitrary instance I^* of CAT on S_d and an optimal solution $\text{OPT} \in \mathcal{O}(I^*)$. Any algorithm that knows the partial solution of OPT for requests of length 2 is optimal on I^*, as it can allocate requests of length 2, such that the edges of length-1 requests in OPT are not blocked. Furthermore, note that this partial solution can be described by a solution in the set $\mathcal{O}(I_d)$ of instance I_d. Thus, enumerating the elements in the set $\mathcal{O}(I_d)$ and using the index of the partial solution as advice yields an algorithm that is optimal on I^*.

Corollary 1. *There exists an optimal online algorithm with advice for CAT on S_d that uses at most*

$$\left\lceil \log \left(\sum_{j=0}^{\lfloor d/2 \rfloor} \frac{d!}{2^j \cdot j! \cdot (d-2j)!} \right) \right\rceil \approx \left\lceil \frac{d}{2} \log \left(\frac{d}{e} \right) + \log \left(\frac{e^{\sqrt{d}}}{(4e)^{1/4}} \right) \right\rceil$$

advice bits. □

The asymptotical approximation is given by using Stirling's inequality on the bound of Lemma 1 as shown by Chowla et al. [11]. Thus, the upper bound of Corollary 1 is asymptotically of the same order as the lower bound of Theorem 1 in the previous chapter, which is constructed on a star tree S_d.

We can use the set of solutions $\mathcal{O}(I_d)$ to construct a similar algorithm as in Corollary 1 for k-ary trees of arbitrary height. The idea is to regard each inner vertex of a k-ary tree and its neighbors as a star tree with at most $k+1$ leaves. Since any k-ary tree has at most $l := (k^h - 1)/(k - 1)$ inner vertices, we get subtrees $S_1, \ldots, S_l$ for which we can give advice as described before. Since the advice complexity of Theorem 5 is about twice that of Corollary 1 for a star tree S_{k+1}, this algorithm reduces the amount of advice used for each inner node, improving the upper bound for k-ary trees by a factor of about 2 when compared to Theroem 5.

Theorem 6. *There exists an optimal online algorithm with advice for CAT on k-ary trees of height h that uses at most*

$$\left\lceil \frac{k^h - 1}{k - 1} \cdot \log \left(\sum_{j=0}^{\lfloor (k+1)/2 \rfloor} \frac{(k+1)!}{2^j \cdot j! \cdot (k+1-2j)!} \right) \right\rceil$$

advice bits. □

---

[1] The Hosoya index, or Z-index, describes the total number of matchings in a graph. Note that it counts the empty set as a matching. The above expression for the Hosoya index of K_d is given by Tichy and Wagner [19].

3.2 Competitiveness and Advice

A popular approach to create competitive algorithms for online problems is to divide the requests into classes, and then randomly select a class. Within this class, requests are accepted greedily and requests of other classes are dismissed.

Awerbuch et al. [1] describe a version of the "classify and randomly select" algorithm for the CAT problem based on vertex separators as follows. Consider a tree with n vertices. There has to exist a vertex v_1' whose removal results in disconnected subtrees with at most $n/2$ vertices. Iteratively choose, in each new subtree created after the $(i-1)$-th round, a new vertex to remove and add it to the set V_i; vertices in this set are called level-i vertex separators. This creates disjoint vertex classes $V_1, V_2, \ldots, V_{\lceil \log n \rceil}$. We can now separate incoming requests into levels. A request r is a level-i_V request if $i_V = \min_j(V_j \cap V(r) \neq \emptyset)$ where $V(r)$ is the set of vertices in the path of the request. The algorithm then chooses a level i_V^* uniformly at random and accepts any level-i_V^* request greedily, i.e., an incoming level-i_V^* request is accepted if it does not conflict with previously accepted requests.

This randomized algorithm is $2\lceil \log_2 n \rceil$-competitive in expectation and can be easily adapted to the advice model by choosing the accepted class using advice. When we reduce the number of classes by a factor of $1/p$ for some $p \in \{1, \ldots, \lceil \log n \rceil\}$, the number of advice bits necessary to communicate the level index will decrease, but we can expect the greedy scheduling to perform worse.

Theorem 7. *For any $p \in \{1, \ldots, \lceil \log n \rceil\}$ there is an online algorithm with advice for CAT that uses $\lceil \log \log n - \log p \rceil$ advice bits and is*

$$\left((2^{p+1} - 2) \cdot \left\lceil \frac{\log n}{p} \right\rceil \right)\text{-competitive.}$$

Proof Sketch. We define the set $\mathrm{Join}(i_V, p)$ to include all requests in I of levels $i_V, \ldots, \min\{i_V + (p-1), \log n\}$. We say that request r is in a subtree S if all its edges are in S, and use "block" in the sense of two requests having at least one edge in common. Observe that requests of level i_V or higher have all edges in a subtree created by removing vertices in $V_1, \ldots, V_{i_V - 1}$. We call such a subtree a level-i_V subtree. Let $\mathrm{OPT}(I)$ be an optimal solution to I; it can be proven by induction on p that for all $i_V \in \{1, \ldots, \lceil \log n \rceil\}$, any request r in a subtree of level i_V can block at most $2^{p+1} - 2$ other requests in $\mathrm{OPT}(I) \cap \mathrm{Join}(i_V, p)$. We can conclude that, for a fixed $p \in \{1, \ldots, \lceil \log n \rceil\}$, any greedy scheduler is $(2^{p+1} - 2)$-competitive when requests are restricted to a level $i_{V'}$, for some $i_{V'} \in \{1, \ldots, \lceil (\log n)/p \rceil\}$. This follows directly from the induction hypothesis. The competitive ratio and advice complexity are easily deduced from there on. □

Corollary 2. *There exists a $2\lceil \log n \rceil$-competitive algorithm for CAT that uses $\lceil \log \log n \rceil$ advice bits.* □

References

1. Awerbuch, B., Bartal, Y., Fiat, A., Rosén, A.: Competitive non-preemptive call control. In: Proceedings of SODA 1994, pp. 312–320. SIAM (1994)
2. Barhum, K., et al.: On the power of advice and randomization for the disjoint path allocation problem. In: Geffert, V., Preneel, B., Rovan, B., Štuller, J., Tjoa, A.M. (eds.) SOFSEM 2014. LNCS, vol. 8327, pp. 89–101. Springer, Cham (2014). https://doi.org/10.1007/978-3-319-04298-5_9
3. Borodin, A., El-Yaniv, R.: Online Computation and Competitive Analysis. Cambridge University Press, Cambridge (1998)
4. Böckenhauer, H.-J., Hromkovič, J., Komm, D., Krug, S., Smula, J., Sprock, A.: The string guessing problem as a method to prove lower bounds on the advice complexity. Theoret. Comput. Sci. **554**, 95–108 (2014)
5. Böckenhauer, H.-J., Komm, D., Královič, R., Královič, R.: On the advice complexity of the k-server problem. J. Comput. Syst. Sci. **86**, 159–170 (2017)
6. Böckenhauer, H.-J., Komm, D., Královič, R., Královič, R., Mömke, T.: On the advice complexity of online problems. In: Dong, Y., Du, D.-Z., Ibarra, O. (eds.) ISAAC 2009. LNCS, vol. 5878, pp. 331–340. Springer, Heidelberg (2009). https://doi.org/10.1007/978-3-642-10631-6_35
7. Böckenhauer, H.-J., Komm, D., Královič, R., Královič, R., Mömke, T.: Online algorithms with advice: the tape model. Inf. Comput. **254**, 59–83 (2017)
8. Böckenhauer, H.-J., Komm, D., Wegner, R.: Call admission problems on grids with advice (extended abstract). In: Epstein, L., Erlebach, T. (eds.) WAOA 2018. LNCS, vol. 11312, pp. 118–133. Springer, Cham (2018). https://doi.org/10.1007/978-3-030-04693-4_8
9. Böckenhauer, H.-J., Hromkovič, J., Komm, D., Královič, R., Rossmanith, P.: On the power of randomness versus advice in online computation. In: Bordihn, H., Kutrib, M., Truthe, B. (eds.) Languages Alive. LNCS, vol. 7300, pp. 30–43. Springer, Heidelberg (2012). https://doi.org/10.1007/978-3-642-31644-9_2
10. Burjons, E., Frei, F., Smula, J., Wehner, D.: Length-weighted disjoint path allocation. In: Böckenhauer, H.-J., Komm, D., Unger, W. (eds.) Adventures Between Lower Bounds and Higher Altitudes. LNCS, vol. 11011, pp. 231–256. Springer, Cham (2018). https://doi.org/10.1007/978-3-319-98355-4_14
11. Chowla, S., Herstein, I.N., Moore, W.K.: On recursions connected with symmetric groups I. Can. J. Math. **3**, 328–334 (1951)
12. Dobrev, S., Královič, R., Pardubská, D.: Measuring the problem-relevant information in input. Theoret. Inform. Appl. (RAIRO) **43**(3), 585–613 (2009)
13. Emek, Y., Fraigniaud, P., Korman, A., Rosén, A.: Online computation with advice. Theoret. Comput. Sci. **412**(24), 2642–2656 (2011)
14. Gebauer, H., Komm, D., Královič, R., Královič, R., Smula, J.: Disjoint path allocation with sublinear advice. In: Xu, D., Du, D., Du, D. (eds.) COCOON 2015. LNCS, vol. 9198, pp. 417–429. Springer, Cham (2015). https://doi.org/10.1007/978-3-319-21398-9_33
15. Hromkovič, J., Královič, R., Královič, R.: Information complexity of online problems. In: Hliněný, P., Kučera, A. (eds.) MFCS 2010. LNCS, vol. 6281, pp. 24–36. Springer, Heidelberg (2010). https://doi.org/10.1007/978-3-642-15155-2_3
16. Komm, D.: Advice and Randomization in Online Computation. Ph.D. thesis, ETH Zurich (2012)
17. Komm, D.: An Introduction to Online Computation - Determinism, Randomization Advice. Springer, Heidelberg (2016). https://doi.org/10.1007/978-3-319-42749-2

18. Sleator, D.D., Tarjan, R.E.: Amortized efficiency of list update and paging rules. Commun. ACM **28**(2), 202–208 (1985)
19. Tichy, R.F., Wagner, S.: Extremal problems for topological indices in combinatorial chemistry. J. Comput. Biol. **12**(7), 1004–1013 (2005)

Power Edge Set and Zero Forcing Set Remain Difficult in Cubic Graphs

Pierre Cazals[1]([✉]), Benoit Darties[2], Annie Chateau[2], Rodolphe Giroudeau[2], and Mathias Weller[3]

[1] Université Paris-Dauphine, PSL Research University, CNRS, LAMSADE, 75016 Paris, France
pierre.cazals@dauphine.psl.eu
[2] LIRMM - CNRS UMR 5506 - Univ. de Montpellier, Montpellier, France
{chateau,darties,giroudeau}@lirmm.fr
[3] CNRS - LIGM - Univ. Marne-La-Vallée, Champs-sur-Marne, France
mathias.weller@u-pem.fr

Abstract. This paper presents new complexity and non-approximation results concerning two color propagation problems, namely POWER EDGE SET and ZERO FORCING SET. We focus on cubic graphs, exploiting their structural properties to improve and refine previous results. We also give hardness results for parameterized precolored versions of these problems, and a polynomial-time algorithm for ZERO FORCING SET in proper interval graphs.

Keywords: Synchrophasor · Power Edge Set · Zero Forcing Set · Complexity

1 Introduction

Motivation. In power networks, synchrophasors are time-synchronized numbers that represent both the magnitude and phase angle of the sine waves on network links. A Phasor Measurement Unit (PMU) is an expensive measuring device used to continuously collect the voltage and phase angle of a single station and the electrical lines connected to it. The problem of minimizing the number of PMUs to place on a network for complete network monitoring is an important challenge for operators and has gained a considerable attention over the past decade [4, 7,8,12,13,15,17,19,21,22,25]. The problem is known as POWER DOMINATING SET [25] and we state it below. We model the network as a graph $G = (V, E)$ with $|V| =: n$ and $|E| =: m$. We denote the set of vertices and edges of G by $V(G)$ and $E(G)$, respectively. We let $N_G(v)$ denote the set of neighbors of $v \in V$ in G and $d_G(v) = |N_G(v)|$ its degree in G. Further, we let $N_G[v] = N_G(v) \cup \{v\}$ denote the *closed* neighborhood of v in G, and we let $G[W]$ denote the subgraph of G *induced* by vertices $W \subseteq V(G)$. The problem is described through *monitoring* of nodes of the network, corresponding to monitoring vertices $V(G)$ by PMUs, propagated using the following rules.

C. J. Colbourn et al. (Eds.): IWOCA 2019, LNCS 11638, pp. 122–135, 2019.
https://doi.org/10.1007/978-3-030-25005-8_11

RULE R_1^*: A vertex v of G on which a PMU is placed will be called a *monitored* vertex, and all its neighbors vertices $N_G(v)$ automatically become monitored.
RULE R_2: if all but one neighbor of a monitored vertex are monitored, then this unmonitored vertex will become monitored as well.

Letting $\Gamma_P(G)$ denote the minimum number of PMUs to place on vertices to obtain a full monitoring of the network (using RULE R_2), the decision version of the problem is described as follows:

POWER DOMINATING SET (PDS)
Input: a graph $G = (V, E)$ and some $k \in \mathbb{N}$
Question: Is $\Gamma_P(G) \leq k$?

POWER DOMINATING SET is $\mathcal{NP}$-complete in general graphs [15]. A large amount of literature is devoted to this problem, describing a wide range of approaches, either exact such as integer linear programming [12] or branch-and-cut [21], or heuristic, such as greedy algorithms [17], approximations [4] or genetic algorithms [19]. The problem has also been shown to be polynomial-time solvable on grids [7], but $\mathcal{NP}$-complete in unit-disk graphs [22].

In this paper, we consider two variants of the problem, called POWER EDGE SET (PES) [23,24] and ZERO FORCING SET (ZFS) [3], which respectively consist in placing PMUs on the *links*, and reducing the monitoring range of a PMU placed on a node. This leads us to replace RULE R_1^* in each of these problems as follows (RULE R_2 remains unchanged):

RULE R_1 (PES): two endpoints of an edge bearing a PMU are monitored.
RULE R_1 (ZFS): only the node bearing a PMU is monitored.

We let $pes(G)$ and $zfs(G)$ denote the minimum number of PMUs to place on the edges, resp. nodes, of G to entirely monitor G. Both PES and ZFS can be seen as a problem of color propagation with colors 0 (white) and 1 (black), respectively designating the states *not monitored* and *monitored* of a vertex of G. As input to PES or ZFS, we will consider a connected graph $G = (V, E)$. For each vertex $v \in V$, let $c(v)$ be the color assigned to v (we abbreviate $\bigcup_{v \in X} c(v) =: c(X)$). Before placing the PMUs, we have $c(V) = \{0\}$ and the aim is to obtain $c(V) = \{1\}$ using RULE R_1 and RULE R_2 while minimizing the number of PMUs. See Figs. 1 and 2 for detailed examples illustrating the differences between PES and ZFS.

POWER EDGE SET(PES) | ZERO FORCING SET(ZFS)
Input: a graph G, some $k \in \mathcal{N}$ **Input:** a graph G, some $k \in \mathcal{N}$
Question: Is $pes(G) \leq k$?   **Question:** Is $zfs(G) \leq k$?

Previous work. Assigning a minimum number of PMUs to monitor the whole network is known to be $\mathcal{NP}$-hard in general for both PES and ZFS. For the former, some complexity results and a lower bound on approximation of $1.12 - \epsilon$ with $\epsilon > 0$ have been shown by Toubaline et al. [23], who also present a linear-time algorithm on trees by reduction to PATH COVER. Poirion et al. [20] propose

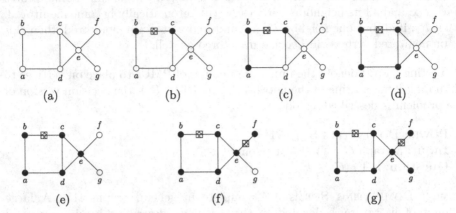

Fig. 1. PMU propagation on PES problem: before any PMU placement, all vertices are white (a). A PMU on $\{b, c\}$ induces $c(b) = c(c) = 1$ (black) by RULE R_1 (b). By applying RULE R_2 on b, we obtain $c(a) = 1$ (c). Then RULE R_2 on a induces $c(d) = 1$ (d), and RULE R_2 on c or d induces $c(e) = 1$ (e). A second PMU is required to obtain a complete coloring. Placing a PMU on $\{e, f\}$ gives us $c(f) = 1$ by RULE R_1 (f). Finally, RULE R_2 on e induces $c(g) = 1$ (g). The set of edges where PMUs have been placed is $S = \{bc, ef\}$, giving (b, c, a, d, e, f, g) as a valid order for G.

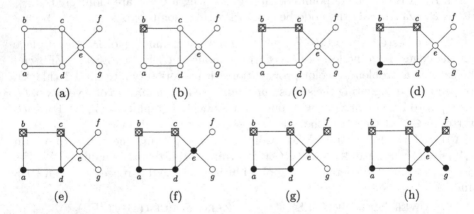

Fig. 2. PMU propagation on ZFS problem: before any PMU placement, all vertices are white (a). Placing one PMU on $\{b\}$ allows to monitor it. (b). Placing a second PMU on $\{c\}$ allows to monitor it (c), and now we can apply RULE R_2 on b, to obtain $c(a) = 1$ (d). Then RULE R_2 on a induces $c(d) = 1$ (e), and RULE R_2 on c or d induces $c(e) = 1$ (f). A third PMU is required to obtain a complete coloring. For example, placing a PMU on f (g) allows to apply RULE R_2 on e to obtain $c(g) = 1$ (h).

a linear program with binary variables indexed by the necessary iterations using propagation rules. Recently, inapproximability results have been proposed on planar or bipartite graphs [5]. In this work, we develop hardness results on complexity and approximation for special cases of POWER EDGE SET and ZERO FORCING SET.

Preliminaries. In the following, we will consider a total order σ of vertices of a graph G as a sequence $(v_1, v_2, \ldots)$ such that v_i occurs before v_j in the sequence if and only if $v_i <_\sigma v_j$.

Definition 1 (valid order). *Let $G = (V, E)$ be a graph, let $S \subseteq E$ (resp. $S \subseteq V$), and let σ be a total order of V, such that for each $v \in V(G)$, there is an edge incident to v in S (resp. $v \in S$) or there is a vertex $u \in N_G(v)$ which verifies $N_G[u] \leq_\sigma v$. Then, $<_\sigma$ is called* valid *for S.*

Given a graph $G = (V, E)$, any set $S \subseteq V$ (or $S \subseteq E$) such that repeated application of RULE R_1 (ZFS) (or RULE R_1 (PES)) and RULE R_2 leads to G being completely monitored is called a *zero forcing set* (or *power edge set*). Using Definition 1, we can formally define the propagation process in G. For instance, in Fig. 1, a valid order for $S = \{bc, ef\}$ is (b, c, a, d, e, f, g).

Observation 1. *Let $G = (V, E)$ be a graph and let $S \subseteq E$ (resp. $S \subseteq V$). Then, S is a power edge set (res. a zero forcing set) if and only if there is a valid order σ on G, with respect to S.*

Note that, for a graph $G = (V, E)$, any set $S \subseteq E$ is a power edge set if and only if $\bigcup_{e \in S} e$ is a zero forcing set for G. It is therefore a natural and unambiguous to also call such an edge set *zero forcing set*.

Finally, we call a vertex v *propagating* to $x \in N_G(v)$ if $c(x) = 0$ and for all $y \in N_G[v] \setminus \{x\}$, we have $c(y) = 1$. Note that each maximal clique of G can contain at most one propagating vertex.

Lemma 1. *Let $G = (V, E)$ be a graph, let S be a zero forcing set of G, and let $\mathcal{C} := \{C_1, \ldots, C_c\}$ be a set of maximal cliques in G covering E. Then $|V \setminus S| \leq c$.*

Proof. Let σ be a valid order for S. We show that each C_i contains at most one edge uv such that $v \notin S$ and $N_G[u] \leq_\sigma v$. Since $\mathcal{C}$ covers E, this implies $|V \setminus S| \leq |\mathcal{C}| = c$. Let $C \in \mathcal{C}$ and let C contain an edge uv such that $N_G[u] \leq_\sigma v$ and $v \notin S$. Then, $C \subseteq N_G[u]$, implying $C \leq_\sigma v$. Thus, v is the last vertex of C with respect to σ and this vertex is *unique*. $\square$

Contribution. The next section is devoted to the NP-completeness for cubic graph for POWER EDGE SET. We show that POWER EDGE SET and ZERO FORCING SET are W[2]-hard parameterized by the size of the solution in Sect. 3. Section 4 is mainly dedicated to inapproximability and we show that there is an $\frac{n}{2}$-approximation for POWER EDGE SET. In the last section, we propose a linear polynomial-time algorithm on proper interval graph for ZERO FORCING SET.

2   Computational Results

Most results presented in this section rely on reductions from graph problems using gadgets for vertices or edges of the original instance. We model the propagation process using the notion of valid order with respect to the solution set, whatever the nature of it: set of edges for PES, of vertices for ZFS.

We present new lower bounds for POWER EDGE SET that hold even in the very restricted case that G is cubic (*i.e.* all vertices in G have degree three). Previous results show that the problem is $\mathcal{NP}$-complete even if G is a subgraph of the grid with bounded degree at most three [5]. In this paper, we show the problem remains $\mathcal{NP}$-complete if G is cubic and planar. The proof is done by reduction from VERTEX COVER (see below) on 3-regular, planar graphs, which is $\mathcal{NP}$-complete [11] but admits a PTAS [1], and a $3/2$-approximation [2].

3-REGULAR PLANAR VERTEX COVER (3RPVC).
Input: a 3-regular planar graph $G = (V, E)$, some $k \in \mathcal{N}$.
Question: Is there a size-k set $S \subseteq V$ covering E, *i.e.* $\forall_{e \in E} \ e \cap S \neq \varnothing$?

Construction 1. *For a given cubic planar graph $G = (V, E)$ with n vertices, we construct a graph G' as follows:*

- *For each $v \in V$, construct H_v (see Fig. 3).*
- *If x is adjacent to y in G, we add exactly one of the edge between x_0, x_1 or x_2 and y_0, y_1 or y_2 to connect H_x and H_y*

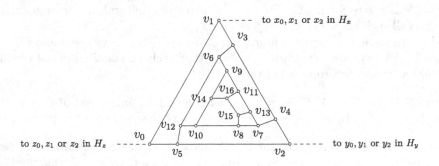

Fig. 3. The gadget H_v for a vertex v.

The graph G' is clearly cubic and planar and Construction 1 is applied in polynomial time. The construction is linear in n and k.

Lemma 2. *The gadget H_v needs at least one PMU to be fully colored: if x_1, y_2 and z_0 are propagating respectively to v_1, v_2 and v_0, then one PMU is sufficient; otherwise two PMUs are needed to fully color H_v.*

Proof. First, if x_1, y_2 and z_0 are propagating respectively to v_1, v_2 and v_0, then, after application of RULE R_2, $c(v_0) = c(v_1) = c(v_2) = 1$. Thus this is the beginning of a valid order: $(v_0, v_1, v_2, v_3, v_5, v_4, v_6, v_7, v_{12}, v_9, v_{10})$. There is no more possible propagation, it is necessary to assign a new PMU. If we place it on the edge $v_{14}v_{16}$, the remainder of a valid order for H_v is: $(v_{14}, v_{16}, v_{11}, v_8, v_{13}, v_{15}, v_{16})$.

Second, we show that H_v may be colored by two PMUs in every case. If PMUs are assigned to the edges $v_{11}v_{13}$ and $v_{15}v_{16}$, we the following order is valid: $(v_{11}, v_{13}, v_{15}, v_{16}, v_7, v_8, v_9, v_{14}, v_4, v_6, v_{10}, v_3, v_{12}, v_3, v_{12}, v_1, v_2, v_5, v_0)$.

Third, we show that even if x_1 and z_0 are propagating to respectively v_1 and v_0, and y_2 is not, we need two PMUs to color H_v. The beginning of the propagation is given by the following order: (v_0, v_1, v_3, v_5). There is no more possible propagation, therefore we have to put one more PMU. As more than two uncolored vertices remain, so we have to initiate propagation with this PMU. So the potential edges are v_6v_{12}, v_4v_2, v_6v_9 or $v_{10}v_{12}$ (other edges won't start a propagation, and we need to color more than two vertices). By exhaustive search, we find that it is impossible to color H_v with only one PMU on any one of these edges. We use the same kind of argument if x_1 and y_2 or y_2 and z_0 propagate. □

Theorem 1. POWER EDGE SET *remains $\mathcal{NP}$-complete on planar cubic graphs.*

Proof. Let G' be the graph obtained by using Construction 1 on $G = (V, E)$, a cubic planar graph. We show that G has a size-k vertex cover iff POWER EDGE SET has a solution of size $n + k$ on G'. Clearly, POWER EDGE SET is in $\mathcal{NP}$.

"$\Rightarrow$": With a size-k vertex cover S for G, we build a power edge set S' for G':

$$S' := \bigcup_{v \in S} \{v_{11}v_{16}, v_{13}v_{15}\} \cup \bigcup_{v \in V \setminus S} \{v_{14}v_{16}\}$$

Then, $|S'| = n + k$ and, by Lemma 2, all vertices of G' are colored by S'.

"$\Leftarrow$": Suppose that G' is colored with $n + k$ PMUs. By Lemma 2, there is at least one PMU on each gadget. Further, if a gadget H_v is colored with a single PMU, then every H_x with $x \in N_G(v)$ is colored with two PMUs inside (by Lemma 2). Then, $\{v \mid H_v$ admits two PMUs$\}$ is a vertex cover for G. □

3 Parameterized Hardness

In what follows, we introduce parameterized versions of our problems and recall the notion of parameterized reduction. Using known results for DOMINATING SET, we deduce hardness results for POWER EDGE SET and ZERO FORCING SET. First, we recall the parameterized DOMINATING SET problem. We obtain hardness results for a restricted version of our problems, when a precoloring exists on a particular set of vertices.

DOMINATING SET (DS)
Input: a graph $G = (V, E)$, some $k \in \mathbb{N}^*$
Question: Does G have a size-k dominating set?
Parameter: k

PRECOLORED ZERO FORCING SET/PRECOLORED POWER EDGE SET
Input: a graph $G = (V, E)$, a set $B \subseteq V$, some $c : V \to \{0, 1\}$ with $c^{-1}(1) = B$, and an integer k
Question: Is there a set $S' \subseteq V$ (resp. $S' \subseteq E$) of size k such that $B \cup S'$ (resp. $B \cup \bigcup_{e \in S'} e$) is a zero forcing set for G?
Parameter: k

We prove the hardness using a parameterized reduction from DOMINATING SET. First, we introduce a gadget which allows to propagate a coloration, but only in one direction. It is called "check-valve".

Definition 2 (Check-valve). *A check-valve $C_{x,y}$ from x to y is a graph $G = (V, E)$, with $V = \{x, y, x_1, x_2, x_3, x_4\}$ and $E = \{xx_1, xx_2, x_1x_3, x_1x_4, x_2x_4, x_3y, x_4y\}$, with a coloring function $c : V \to \mathbb{N}$, such that $c(x) = c(x_1) = 1$ and all other vertices are colored by 0. A check-valve $C_{x,y}$ is illustrated on Fig. 4.*

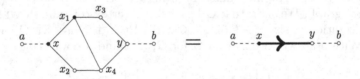

Fig. 4. The check-valve $C_{x,y}$

Observation 2. *Let $C_{x,y}$ be a check-valve inserted between two vertices a and b, depicted by Fig. 4. Then:*

1. *If $c(a) = 1$ then $c(b) = 1$ after exhaustive application of RULE R_2.*
2. *If $c(b) = 1$, and $c(a) = 0$, then $c(a)$ is still 0 after exhaustive application of RULE R_2, and it is necessary to add a PMU in order to have $c(a) = 1$.*

Construction 2. *Let xy be a edge such that $c(x) = 1$ and $c(y) = 0$, we construct the gadget C_{xy}: we add vertices x_1, x_2, x_3 and x_4 and we add edges xx_1, x_1x_2, x_3y, yx_4, x_4x_2 et x_2x. Notice that xy is deleted.*

Construction 3. *For given $G = (V, E)$, construct $G' = (V', E')$ as follows:*

1. *For all $x \in V$, build J_x depicted in Fig. 5, containing a core graph ($\{E_x, R_x, V_x, \ x_1, x_2, x_3, x_4\}$, $\{E_xx_3, E_xx_4, E_xV_x, x_1x_3, x_2x_4, R_xx_1, R_xx_2, R_xV_x\}$) with precolored vertices V_x, x_3 and x_4, and outgoing check-valves $d_G(x)$ many $C_{x_{v_i}^1, x_{v_i}^2}$ connected to E_x, and n many $C_{x_{s_i}^1, x_{s_i}^2}$ connected to R_x.*

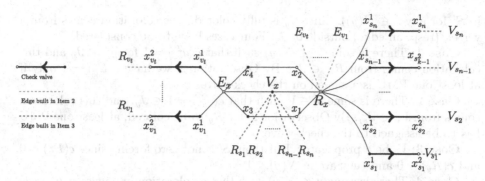

Fig. 5. The gadget J_x for a vertex x. Note that for sake of clarity, some external vertices have been duplicated. Indeed, $\{R_{v_1}, \dots R_{v_t}\}$, where $v_1, \dots, v_t$ are neighbors of x, is included in $\{R_{s_1}, \dots, R_{s_n}\}$.

2. For all $v_i \in N(x)$, add edges $x^2_{v_i} R_{v_i}$ with $x^2_{v_i} \in J_x$ and $R_{v_i} \in J_{v_i}$.
3. For all $s_i \in V$, add edges $x^2_{s_i} V_{s_1}$ with $x^2_{s_i} \in J_x$ and $V_{s_1} \in J_{s_i}$.

Lemma 3. *For all* $x \in V$, *if* $c(E_x) = 1$ *then, after exhaustive application of* RULE R_2, $c(J_x) = 1$ *and* $c(R_v) = 1$ *for all* $v \in N(x)$.

Proof. If $V = \{s_1, \dots, s_n\}$ and $N(x) = \{v_1, \dots, v_t\}$, then the following sequence is a valid order: $(E_x, x_1, x_2, R_x, x^2_{s_1}, \dots, x^2_{s_n}, x^2_{v_1}, \dots, x^2_{v_t}, R_{v_1}, \dots R_{v_t})$.   □

Lemma 4. *Let* $c(R_x) = 1$ *for all* $x \in V$. *Then, after exhaustive application of* RULE R_2, G' *becomes fully colored.*

Proof. Clearly, all vertices in $N(V_x) \backslash \{E_x\}$ are colored by R_x for all $x \in V$. Then, E_x is colored by V_x. By Lemma 3, $c(E_x) = 1$ leads to J_x being fully colored. As $V' = \bigcup_{z \in V} V(J_z)$, G' becomes fully colored.   □

Theorem 2. PRECOLORED ZERO FORCING SET *and* PRECOLORED POWER EDGE SET *are* $W[2]$*-hard wrt. the solution size* k.

Proof. Let $G = (V, E)$ be a graph and let G' the product of Construction 3 on G. We show that G has a size-k dominating set if and only if G' has a size-k zero forcing set (power edge set).

"⟹": Let S be a size-k dominating set for G. A size-k zero forcing set S' for G' is obtained as follows: for all $x \in S$, we place a PMU on E_x, (resp. $E_x x_4$). By Lemmas 3 and 4, G' is fully colored after applying RULE R_2 exhaustively.

"⟸": Let S' be a zero forcing set of size k for G'. Let S be the set of vertices $x \in V(G)$ such that J_x has at least one vertex, resp. one edge, in S' (for each $x, y \in V$, if there is a PMU on the edge $E_x R_y$ or $R_x V_y$ it counts as an edge of J_x). Suppose that S is not a dominating set for G. So, there is some $y \in V$ such that no $u \in V(J_y)$ is in S' and no $v \in V(J_x)$ is in S' for any $x \in N(y)$. (for PES, there is some $y \in V$ such that no $u_1 u_2 \in E(J_y)$ is in S' and no $v_1 v_2 \in E(J_x)$ is in S' for any $x \in N(y)$). Since J_y is fully colored, this coloration comes from a vertex (resp. an edge) outside of J_y. Four cases have to be considered:

Case 1: There is some $v_i \in N(y)$ such that $c(y_{v_i}^2) = 1$ for $y_{v_i}^2 \in J_y$ and this coloration comes from $R_{v_i} \in J_{v_i}$. By Observation 2, we have $c(E_y) = 1$ only if at least one PMU is assigned on the check-valve.

Case 2: There is some $s_i \in V$ such that $c(y_{s_i}^2) = 1, \in J_y$, and this coloration comes from $V_{s_i} \in J_{s_i}$. By Observation 2, for R_y to be colored, at least one PMU has to be assigned to the check-valve.

Case 3: V_y be a propagator. But then, S' is not zero forcing since $c(E_y) = 0$ and $c(R_y) = 0$ and they are in $N(y)$.

Case 4: There is some $v_i \in N(y)$ such that a coloration happens on $R_y \in J_y$ from $E_{v_i} \in J_{v_i}$. Then, either there is some $t \in J_{v_i} \cap S$ and so S is a dominating set, or no PMU is assigned on J_{v_i}, but we already know that E_{v_i} cannot be colored (see Case 1). Consequently, if $c(E_{v_i}) = 1$ then $c(w) = 1$ for some $w \in J_{v_i}$ contradicting S not being a dominating set.

Thus S is a dominating set of G. Further, Construction 3 can be carried out in polynomial time and $|S| = |S'|$, yielding the desired result. $\square$

4   Non-approximation

In this section, we will show that the reductions presented in the proofs of Theorems 1 and 2 are L-reductions.

But above all, it is clear it exists a $\frac{n}{2}$-approximation; it is sufficient to put one PMU incident to each vertex (at most n), and the lower bound for optimal solution is at least two PMUs (in cubic graph) so we obtain a $\frac{n}{2}$-approximation.

Theorem 3. POWER EDGE SET *is $\frac{n}{2}$-approximable*

For the first, by construction, we have $OPT(I') = OPT(I) + n$. Let S be a solution to I, suppose that $n > 3|S|$. By the pigeon hole principle, there is a vertex which cover at least four edges, which is impossible because the degree of each vertex is three, so $n \leq 3|S|$. Thus $OPT(I') \leq 4OPT(I)$.

Moreover, by construction, we have

$$val(g(S') \leq val(S') - n \leq val(S') - OPT(I') + OPT(I)$$

Thus, we construct an L-reduction with $\alpha_1 = 4$ and $\alpha_2 = 1$.

Assuming $\mathcal{P} \neq \mathcal{NP}$, VERTEX COVER is hard to approximate to a factor 1.36 [6] and [9], thus yielding the desired result:

$$|S'| \geq |g(S')| + OPT(I') - OPT(I)$$
$$\geq 1.36OPT(I) + OPT(I') - OPT(I)$$
$$\geq 1.09OPT(I')$$

$\square$

Corollary 1. *Under $\mathcal{P} \neq \mathcal{NP}$, POWER EDGE SET on cubic graph cannot be approximated to within a factor better than 1.09.*

Assuming, VERTEX COVER is hard to approximate to a factor $2 - \epsilon$ [16] and [9], thus yielding the desired result:

$$
\begin{aligned}
|S'| &\geq |g(S')| + OPT(I') - OPT(I) \\
&\geq 2 - \epsilon \, OPT(I) + OPT(I') - OPT(I) \\
&\geq {}^5/_4 - \epsilon \, OPT(I')
\end{aligned}
$$

□

Corollary 2. *Under UGC,* POWER EDGE SET *on cubic graph cannot be approximated to within a factor better than* $5/4$.

Previous results mainly show that POWER EDGE SET do not admit a PTAS algorithm, even on cubic graphs.

For the second, we got $OPT(I) = OPT(I')$, so clearly it is a S-reduction. DOMINATING SET do not admit a polynomial time approximation algorithm with ratio $O(\log n)$ ([18]), so PRECOLORED POWER EDGE SET and PRECOLORED ZERO FORCING SET do not too.

Corollary 3. *Under* $NP \neq DTIME(n^{polylogn})$, PRECOLORED POWER EDGE SET *and* PRECOLORED ZERO FORCING SET *do not admit a polynomial time approximation algorithm with ratio* $O(\log n)$.

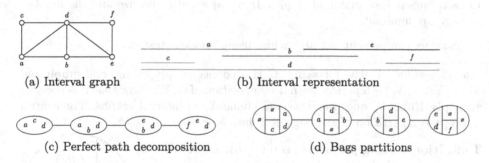

(a) Interval graph   (b) Interval representation

(c) Perfect path decomposition   (d) Bags partitions

Fig. 6. An interval graph (a), with its interval representation (b), a perfect path decomposition of this graph (c) and its bag partition according to Definition 4 (d).

5   ZFS on Proper Interval Graphs

Preliminaries. A graph G is an *interval graph* if it is the intersection graph of a family of intervals on the real line. Each interval is represented by a vertex of G and an intersection between two intervals is represented by an edge between the corresponding vertices (see Fig. 6). G is called *proper interval* if it has an interval representation in which no interval is properly contained in another.

In the following, we use perfect path decompositions to solve POWER EDGE SET on proper interval graphs.

Definition 3. *A* path decomposition *$\mathcal{D}$ of a graph $G = (V, E)$ is a sequence $(X_i)_{i=1...\ell}$ of subsets of V (called* bags*), verifying the following properties:*

(a) for each $xy \in E$, there is some X_i with $x, y \in X_i$ (each edge is in a bag),
(b) for $i \leq j \leq k$, $X_i \cap X_k \subseteq X_j$ (bags containing any $v \in V$ are consecutive).

$\mathcal{D}$ is called perfect *if the number of bags and their sizes are minimal under (a) and (b). The* pathwidth *of $\mathcal{D}$ is the size of the largest X_i minus one.*

Lemma 5. *If G is connected, then $X_i \cap X_{i-1} \neq \varnothing$ for all $i > 1$.*

Proof. Towards a contradiction, assume that $X_i \cap X_{i-1} = \varnothing$. Then, by Definition 3(b), $A := \bigcup_{1 \leq k \leq i-1} X_k$ and $B := \bigcup_{i \leq l} X_l$ are disjoint. Since G is connected, there is an edge xy between A and B, but no bag contains both x and y, contradicting Definition 3(a). □

Lemma 6. *Let G be an interval graph. A perfect path decomposition $\mathcal{D}$ of G can be computed in linear time and*
   each bag of $\mathcal{D}$ is a maximal clique in G.

Proof. Being an interval graph, G admits a linear order of its maximal cliques such that, for each vertex v, all maximal cliques containing v are consecutive [10] and this order can be computed in $O(n + m)$ time [14]. Such a "clique path" naturally corresponds to a perfect path decomposition and we know that vertices of each bags induce maximal cliques. In a clique path, the size and the number of bags are minimal. □

Now we can present our algorithm, using previous results:

The Algorithm. In the following, G is a connected proper interval graph and $\mathcal{D} = (X_1, ..., X_\ell)$ is a perfect path decomposition of G. We show that it is possible to apply RULE R_2 once per maximal clique X_i in interval graphs. The central concept is a partition of the bags of $\mathcal{D}$ into four sets.

Definition 4 (Bag partition, see Fig. 6**).** *Let X_i be a bag in a perfect path decomposition of an interval graph.*

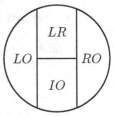

- *IO (Inside Only) is the set $X_i \setminus (X_{i-1} \cup X_{i+1})$.*
- *LO (Left Only) is the set $X_i \cap X_{i-1} \setminus X_{i+1}$.*
- *RO (Right Only) is the set $X_i \cap X_{i+1} \setminus X_{i-1}$.*
- *LR (Left Right), contains all remaining vertices of X_i.*

Note that $RO(X_i)$ and $RO(X_j)$ are disjoint for $i \neq j$. Further, since G is proper interval, $RO(X_i) \neq \varnothing$ for all $i < \ell$. Our algorithm will simply choose any vertex of $RO(X_i) \cup IO(X_i)$ for all i. This can clearly be done in linear time and we show that it is correct and optimal.

Lemma 7. *Let G be a connected interval graph and let $\mathcal{D} = (X_1, \ldots, X_\ell)$ be a perfect path decomposition of G. Let $\overline{S}$ be a set intersecting each $RO(X_i)$ for all $1 \leq i < \ell$ in exactly one vertex and intersecting $IO(X_\ell)$ in exactly one vertex. Then, $S := V \setminus \overline{S}$ is an optimal zero forcing set for G.*

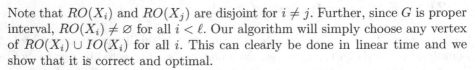

Proof. For each i, let x_i be the ith vertex of $\overline{S}$, that is, $\overline{S} \cap (RO(X_i) \cup IO(X_i)) = \{x_i\}$ for each $X_i \in \mathcal{D}$. We show that the order σ consisting of S in any order followed by $(x_1, \ldots, x_\ell)$ is valid for S. To this end, let $1 \leq j < \ell$. Note that $IO(X_j) \cup LO(X_j) = ((X_j \setminus X_{j+1}) \cap X_{j-1}) \cup ((X_j \setminus X_{j+1}) \setminus X_{j+1}) = X_j \setminus X_{j+1}$. Thus, there is some $u \in IO(X_j) \cup LO(X_j)$ as otherwise, $X_j \subseteq X_{j+1}$ contradicting $\mathcal{D}$ being perfect. Towards a contradiction, assume $N_G[u] \not\leq_\sigma x_j$, that is, there is some $v \in N_G[u]$ with $x_j <_\sigma v$. By construction of σ, there is a $k > j$ such that $v = x_k$. By construction of $\overline{S}$, we have $x_k \in IO(X_k) \cup RO(X_k)$, implying $x_k \notin X_{k-1}$ by definition of RO and IO. Further, since ux_k is an edge of G, there is a bag X_i containing both u and x_k and, since $u \in IO(X_j) \cup LO(X_j)$ we know $i \leq j$. But then, x_k occurs in X_i, not in X_{k-1} but again in X_k, contradicting $\mathcal{D}$ being perfect. It remains to treat x_ℓ, but since x_ℓ is the last vertex of σ, $N_G[u] \leq_\sigma x_\ell$ for all $u \in N_G(x_\ell)$.

Finally, optimality of S is implied by Lemma 1 as $|\overline{S}| = |\mathcal{D}|$. □

Theorem 4. ZERO FORCING SET *is solvable in* $O(n+m)$ *time on proper interval graphs.*

Proof. We know that our algorithm is exact, compute the path decomposition PD can done in linear time (Lemma 6) and partitioning its vertices is easy. So there is a polynomial time algorithm for ZERO FORCING SET in proper interval graph. □

6 Conclusion and Perspectives

In this article, we investigated POWER EDGE SET and ZERO FORCING SET from the point of view of computational complexity. We obtained a series of negative results, refining the previous hardness results and excluding certain exact algorithms. On the positive side, we give a linear-time algorithm in case the input is a proper interval graph and a naive approximation algorithm. There is a big gap between positive and negative result in approximation so further research will be focused on developing efficient polynomial-time approximation algorithms, as well as considering more special cases and structural parameterizations.

References

1. Baker, B.: Approximation algorithms for NP-complete problems on planar graphs. J. ACM **41**(1), 153–180 (1994)
2. Bar-Yehuda, R., Even, S.: On approximating a vertex cover for planar graphs. In: Proceedings of the 14th Annual ACM Symposium on Theory of Computing, 5–7 May 1982, San Francisco, California, USA, pp. 303–309 (1982)
3. Barioli, F., et al.: Zero forcing sets and the minimum rank of graphs. Linear Algebra Appl. **428**, 1628–1648 (2008)
4. Cheng, X., Huang, X., Li, D., Wu, W., Du, D.: A polynomial-time approximation scheme for the minimum-connected dominating set in ad hoc wireless networks. Networks **42**(4), 202–208 (2003)

5. Darties, B., Champseix, N., Chateau, A., Giroudeau, R., Weller, M.: Complexity and lowers bounds for power edge set problem. J. Discrete Algorithms **52–53**, 70–91 (2018)
6. Dinur, I., Safra, S.: On the hardness of approximating minimum vertex cover. Ann. Math. **162**, 439–485 (2005)
7. Dorfling, M., Henning, M.A.: A note on power domination in grid graphs. Discrete Appl. Math. **154**(6), 1023–1027 (2006)
8. Fan, N., Watson, J.-P.: Solving the connected dominating set problem and power dominating set problem by integer programming. In: Lin, G. (ed.) COCOA 2012. LNCS, vol. 7402, pp. 371–383. Springer, Heidelberg (2012). https://doi.org/10.1007/978-3-642-31770-5_33
9. Feige, U.: Vertex cover is hardest to approximate on regular graphs. Technical Report MCS03-15 (2003)
10. Fulkerson, D.R., Gross, O.A.: Incidence matrices and interval graphs. Pac. J. Math. **15**(3), 835–855 (1965). https://projecteuclid.org:443/euclid.pjm/1102995572
11. Garey, M.R., Johnson, D.S.: Computers and Intractability: A Guide to the Theory of NP-Completeness. W. H. Freeman & Co., New York (1979)
12. Gou, B.: Generalized integer linear programming formulation for optimal PMU placement. IEEE Trans. Power Syst. **23**(3), 1099–1104 (2008)
13. Guo, J., Niedermeier, R., Raible, D.: Improved algorithms and complexity results for power domination in graphs. Algorithmica **52**(2), 177–202 (2008)
14. Habib, M., McConnell, R.M., Paul, C., Viennot, L.: Lex-BFS and partition refinement, with applications to transitive orientation, interval graph recognition and consecutive ones testing. Theor. Comput. Sci. **234**(1–2), 59–84 (2000)
15. Haynes, T.W., Hedetniemi, S.M., Hedetniemi, S.T., Henning, M.A.: Domination in graphs applied to electric power networks. SIAM J. Discrete Math. **15**(4), 519–529 (2002)
16. Khot, S., Regev, O.: Vertex cover might be hard to approximate to within 2- ε. J. Comput. Syst. Sci. **74**(3), 335–349 (2008)
17. Li, Y., Thai, M.T., Wang, F., Yi, C., Wan, P., Du, D.: On greedy construction of connected dominating sets in wireless networks. Wirel. Commun. Mobile Comput. **5**(8), 927–932 (2005)
18. Lund, C., Yannakakis, M.: On the hardness of approximating minimization problems. J. ACM (JACM) **41**(5), 960–981 (1994)
19. Milosevic, B., Begovic, M.: Nondominated sorting genetic algorithm for optimal phasor maesurement placement. IEEE Power Eng. Rev. **22**(12), 61–61 (2002)
20. Poirion, P., Toubaline, S., D'Ambrosio, C., Liberti, L.: The power edge set problem. Networks **68**(2), 104–120 (2016)
21. Simonetti, L., Salles da Cunha, A., Lucena, A.: The minimum connected dominating set problem: formulation, valid inequalities and a branch-and-cut algorithm. In: Pahl, J., Reiners, T., Voß, S. (eds.) INOC 2011. LNCS, vol. 6701, pp. 162–169. Springer, Heidelberg (2011). https://doi.org/10.1007/978-3-642-21527-8_21
22. Thai, M.T., Du, D.Z.: Connected dominating sets in disk graphs with bidirectional links. IEEE Commun. Lett. **10**(3), 138–140 (2006)
23. Toubaline, S., D'Ambrosio, C., Liberti, L., Poirion, P., Schieber, B., Shachnai, H.: Complexity and inapproximability results for the power edge set problem. J. Comb. Optim. **35**(3), 895–905 (2018)

24. Toubaline, S., Poirion, P.-L., D'Ambrosio, C., Liberti, L.: Observing the state of a smart grid using bilevel programming. In: Lu, Z., Kim, D., Wu, W., Li, W., Du, D.-Z. (eds.) COCOA 2015. LNCS, vol. 9486, pp. 364–376. Springer, Cham (2015). https://doi.org/10.1007/978-3-319-26626-8_27
25. Yuill, W., Edwards, A., Chowdhury, S., Chowdhury, S.P.: Optimal PMU placement: a comprehensive literature review. In: 2011 IEEE Power and Energy Society General Meeting, July, pp. 1–8 (2011)

Towards a Complexity Dichotomy for Colourful Components Problems on k-caterpillars and Small-Degree Planar Graphs

Janka Chlebíková and Clément Dallard[✉]

School of Computing, University of Portsmouth, Portsmouth, UK
{janka.chlebikova,clement.dallard}@port.ac.uk

Abstract. A connected component of a vertex-coloured graph is said to be *colourful* if all its vertices have different colours, and a graph is colourful if all its connected components are colourful. Given a vertex-coloured graph, the COLOURFUL COMPONENTS problem asks whether there exist at most p edges whose removal makes the graph colourful, and the COLOURFUL PARTITION problem asks whether there exists a partition of the vertex set with at most p parts such that each part induces a colourful component. We study the problems on k-caterpillars (caterpillars with hairs of length at most k) and explore the boundary between polynomial and NP-complete cases. It is known that the problems are NP-complete on 2-caterpillars with unbounded maximum degree. We prove that both problems remain NP-complete on binary 4-caterpillars and on ternary 3-caterpillars. This answers an open question regarding the complexity of the problems on trees with maximum degree at most 5. On the positive side, we give a linear time algorithm for 1-caterpillars with unbounded degree, even if the backbone is a cycle, which outperforms the previous best complexity for paths and widens the class of graphs. Finally, we answer an open question regarding the complexity of COLOURFUL COMPONENTS on graphs with maximum degree at most 5. We show that the problem is NP-complete on 5-coloured planar graphs with maximum degree 4, and on 12-coloured planar graphs with maximum degree 3. Since the problem can be solved in polynomial-time on graphs with maximum degree 2, the results are the best possible with regard to the maximum degree.

Keywords: Colorful component · Caterpillar · Binary tree · Planar subcubic graph

1   Introduction

A *coloured graph* is a graph whose vertices are (not necessarily properly) coloured. A connected component of a coloured graph is a *colourful component*

© Springer Nature Switzerland AG 2019
C. J. Colbourn et al. (Eds.): IWOCA 2019, LNCS 11638, pp. 136–147, 2019.
https://doi.org/10.1007/978-3-030-25005-8_12

if all its vertices have different colours. A graph is said to be colourful if all its connected components are colourful.

In this paper we focus on two closely related problems where a coloured graph and a positive integer p are given as inputs: the COLOURFUL COMPONENTS problem asks if there exist at most p edges whose removal makes the graph colourful; the COLOURFUL PARTITION problem is to decide if there exists a partition of the vertex set with at most p parts such that each part induces a colourful component in the graph.

One key problem in comparative genomics is to partition a set of genes into orthologous genes, which are sets of genes in different species that have evolved through speciation events only, *i.e.* originated by vertical descent from a single gene in the last common ancestor. The problem has been modelled as a graph problem where orthologous genes translate into colourful components in the graph [1,15]. The vertices of the graph represent the genes, and a colour is given to each vertex to symbolise the specie the corresponding gene belongs to. An edge between two vertices is present in the graph if the two corresponding genes are (sufficiently) similar. The quality of a partition of a set of genes into orthologous genes can be expressed in different ways. Minimising the number of similar genes in different subsets of the partition is a well studied variant [4,5,8,13,15], and it corresponds to minimising the number of edges between the colourful components (as in COLOURFUL COMPONENTS). Alternatively, one can ask for a partition of minimum size, *i.e* which contains the minimum number of orthologous genes, or equivalently the minimum number of colourful components [1,5,6] (as in COLOURFUL PARTITION). Another variant, not studied in this paper, considers the objective function that maximises the number of edges in the transitive closure [1,6,13].

Now, we give the formal definitions of the problems considered herein.

COLOURFUL COMPONENTS

Input: A vertex-coloured graph $G = (V, E)$, a positive integer p.
Question: Are there at most p edges in E whose removal makes G colourful?

COLOURFUL PARTITION

Input: A vertex-coloured graph $G = (V, E)$, a positive integer p.
Question: Is there a partition of V with at most p parts s.t. each part induces a colourful component in G?

It is interesting to notice the similarities between COLOURFUL COMPONENTS and the MULTICUT [3,12] and MULTI-MULTIWAY CUT [2] problems. In the MULTICUT problem, a graph and a set of pairs of vertices are given and the goal is to minimise the number of edges to remove in order to disconnect each pair of vertices. In the MULTI-MULTIWAY CUT problem, a graph and sets of vertices are given and the goal is to minimise the number of edges to remove in order to disconnect all paths between vertices from the same vertex set. Thus, COLOURFUL COMPONENTS is a special case where the sets of vertices form a partition.

Both COLOURFUL COMPONENTS and COLOURFUL PARTITION problems can be compared to the GRAPH MOTIF problem [7]. This problem takes a coloured graph and a multiset of colours M (the motif) as input, and the goal is to determine whether there exists a connected subgraph S such that the multiset of colours used by the vertices in S corresponds exactly to M. If M is a set (where each colour appears at most once), then M is said to be colourful.

In this paper, all graphs are simple. We assume that a coloured graph $G = (V, E)$ is always associated with a colouring function c from V to a set of colours, hence for each vertex $u \in V$, $c(u)$ is the colour of the vertex u. The *colour multiplicity* of G corresponds to the maximum number of occurrence of any colour in the graph. If G contains at most ℓ colours we say that G is an ℓ-coloured graph. To simplify the notations, we may say that a vertex u belongs to a path P in G if there exists an edge $e \in P$ with $u \in e$. A path P in G between two vertices u and v is called a *bad path* if $c(u) = c(v) = \gamma$ and u, v are the only two vertices of colour γ in P. Hence, a connected component is colourful if and only if it does not contain a bad path. Lastly, given a set of edges $S \subseteq E$, we denote by $G - S$ the graph $(V, E \setminus S)$, and for a vertex $u \in V$, $N[u]$ is the closed neighbourhood of u.

A *k-caterpillar*, also commonly called a *caterpillar with hairs of length at most k* [10], is a tree in which all the vertices are within distance k of a central path, called the *backbone*. Similarly, we define a *cyclic k-caterpillar* as a k-caterpillar whose backbone is a chordless cycle. Note that 2-caterpillars are also known as *lobster graphs*.

Observe that, on a tree, there is a solution to COLOURFUL COMPONENTS with p edges if and only if there is a solution to COLOURFUL PARTITION with $p + 1$ parts. However, this is not the case on general graphs [5]. Both problems are known to be NP-complete on subdivided stars [6], trees of diameter at most 4 [4], and trees with maximum degree 6 [5]. Trees of diameter at most 4 are in fact a subclass of 2-caterpillars, so both problems are NP-complete on 2-caterpillars when the maximum degree is unbounded. In Sect. 2.1, we prove that COLOURFUL COMPONENTS and COLOURFUL PARTITION are NP-complete on binary 4-caterpillars and on ternary 3-caterpillar, hence with the maximum degree at most 3 or 4. This answers an open question, proposed in [5], regarding the complexity of the problems on trees with maximum degree at most 5. Nonetheless, we propose a linear time algorithm for COLOURFUL COMPONENTS and COLOURFUL PARTITION on 1-caterpillars and cyclic 1-caterpillars with unbounded degree in Sect. 2.2. This result improves the complexity of the known quadratic-time algorithm for paths [6] and widens the class of graphs. We, therefore, obtain a complete complexity dichotomy for the problems on k-caterpillars with regard to k and the maximum degree in the graph.

We also consider the complexity of COLOURFUL COMPONENTS in planar graphs with small degree. It is known that the problem is NP-complete on 3-coloured graphs with maximum degree 6 [4], while COLOURFUL PARTITION is NP-complete on 3-coloured 2-connected planar graphs with maximum degree 3 [5]. However, it was an open question whether COLOURFUL COMPONENTS is NP-complete on ℓ-coloured graphs with maximum degree at most 5. In Sect. 3, we answer that question and show that COLOURFUL COMPONENTS is NP-complete on 5-coloured planar graphs with maximum degree 4 and on 12-coloured planar graphs with maximum degree 3. As COLOURFUL COMPONENTS is polynomial-time solvable on graphs with maximum degree 2, our result is the best possible with regard to the maximum degree.

2   Complexity on k-caterpillars

In this section, we focus on the complexity of COLOURFUL COMPONENTS and COLOURFUL PARTITION on k-caterpillars, depending on the value of k and the maximum degree of the graphs.

2.1   NP-completeness

First, we show that COLOURFUL COMPONENTS and COLOURFUL PARTITION are NP-complete on binary 4-caterpillars and ternary 3-caterpillars. We recall that a binary tree (resp. ternary) is a rooted tree in which each vertex has no more than two children (resp. three children). We propose a reduction from 3-SAT with at most four occurrence of each variable, known as 3, 4-SAT, which is proved NP-complete in [14].

Construction 1. Consider an instance ϕ of 3, 4-SAT, that is, a set of m clauses $C_1, C_2, \ldots, C_m$ on n variables $x_1, x_2, \ldots, x_n$, where each clause contains exactly three literals and where each variable appears at most four times.

For each variable x_i, we define a *variable gadget*. Firstly, we create two vertices labelled x_i and $\bar{x}_i$, which are the roots of two binary trees T_{x_i} and $T_{\bar{x}_i}$, respectively. If a clause C_j contains the literal x_i, then we create a vertex labelled $x_{i,j}$ in T_{x_i}. Similarly, if a clause C_j contains the literal $\bar{x}_i$, then there we create a vertex labelled $\bar{x}_{i,j}$ in $T_{\bar{x}_i}$. All created vertices are connected in such a way that T_{x_i} and $T_{\bar{x}_i}$ are binary trees of depth at most 2. We assume that all the vertices in the trees, except for x_i and $\bar{x}_i$, correspond to one literal in a clause. Finally, we connect x_i and $\bar{x}_i$ to a same new vertex r_{x_i}. Notice that the gadget is a binary tree of depth at most 3 (see Fig. 1).

For each clause C_j, we define a *clause gadget*. Let ℓ_1, ℓ_2 and ℓ_3 be the literals in C_j. We create four vertices y_j, y'_j, z_j and z'_j, three vertices labelled $\ell_{1,j}$, $\ell_{2,j}$ and $\ell_{3,j}$ representing the literals in C_j, and one extra vertex r_{C_j}. Then, we add the edges $\{\ell_{1,j}, y_j\}$, $\{\ell_{2,j}, y'_j\}$, $\{\ell_{3,j}, z'_j\}$, $\{y_j, z_j\}$, $\{y'_j, z_j\}$, and the edges $\{z_j, r_{C_j}\}$ and $\{z'_j, r_{C_j}\}$. Notice that the gadget is a binary tree of depth 3 (see Fig. 1).

Now, we describe two slightly different ways to obtain a tree T containing all the variable and clause gadgets.

- To get T as a binary 4-caterpillar, create a central path with $n + m$ new vertices and connect each of the $n + m$ vertices r_{x_i} and r_{C_j} to a different vertex of the central path.
- To get T as a ternary 3-caterpillar, connect the $n + m$ vertices r_{x_i} and r_{C_j} together (in any order) to create a central path.

In both cases, the central path corresponds to the backbone of T. We set the root r of T such that it belongs to the backbone of T and has minimum degree, hence two or three children, if T is a binary or ternary caterpillar, respectively.

Finally, we assign a colour to each vertex in T. For each gadget representing a variable x_i, let $c(x_i) = c(\bar{x}_i)$ be a new colour. Also, for each vertex $\widetilde{x}_{i,j} \in \{x_{i,j}, \bar{x}_{i,j}\}$, let $c(\widetilde{x}_{i,j})$ be a new colour. Then, for each gadget representing a clause C_j, set the colour of each leaf $\ell_{k,j}$ such that if $\ell_{k,j} = x_i$, then $c(\ell_{k,j}) := c(x_{i,j})$, but if $\ell_{k,j} = \bar{x}_i$, then $c(\ell_{k,j}) := c(\bar{x}_{i,j})$. Furthermore, let $c(y_j) = c(y_j')$ and $c(z_j) = c(z_j')$ be two new colours. Lastly, all the vertices in T which do not belong to any gadget are given new colours. Notice that there are no such vertices if T is a 3-caterpillar. Obviously, in both cases, T is a coloured caterpillar with colour-multiplicity 2.

Note that Construction 1 can be done in polynomial time.

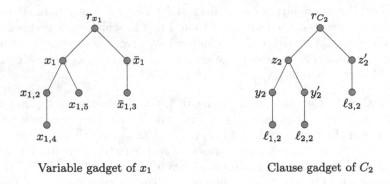

Variable gadget of x_1 · · · · · · · · · · Clause gadget of C_2

Fig. 1. Examples of gadgets used in Construction 1. On the left, the variable gadget of x_1, appearing as a positive literal in C_2, C_4 and C_5, and as a negative literal in C_3. On the right, the clause gadget of C_2.

Theorem 1. COLOURFUL COMPONENTS *and* COLOURFUL PARTITION *are* NP-*complete on coloured ternary 3-caterpillars with colour-multiplicity 2 and on coloured binary 4-caterpillars with colour-multiplicity 2.*

Proof. Obviously, the problem is in NP. Let ϕ be an instance of $3,4$-SAT with m clauses and n variables. We transform ϕ into a coloured tree T as described in Construction 1 such that T has colour multiplicity 2 and is a coloured binary 4-caterpillar or a coloured ternary 3-caterpillar. We claim that there is a solution to $3,4$-SAT on ϕ if an only if there is a set of exactly $n + 2m$ edges in T whose removal makes T colourful.

Let β be a solution to $3,4$-SAT on ϕ. We define the set of edges S as follows:

– For each variable x_i, the set S contains the edge $\{r_{x_i}, x_i\}$ if $x_i = True$ in β, or $\{r_{x_i}, \bar{x}_i\}$ if $x_i = False$ in β.

- For each clause C_j, the set S contains two edges: one from the path between y_j and y'_j, and one from the path between z_j and z'_j. Moreover, in $G - S$, the leaf $\ell_{k,j}$ which belongs to the same connected component as the vertex in r_{C_j} must correspond to (one of) the literal(s) satisfying the clause C_j in β.

Clearly, the set S contains $n + 2m$ edges. We denote by $\mathcal{F}$ the forest $T - S$, and by T' the connected component in $\mathcal{F}$ containing the root r. Obviously, two vertices of the same colour from a same variable gadget do not both belong to a same connected component of $\mathcal{F}$, and the same is true for a clause gadget. Also, note that two vertices of different variable gadgets do not have the same colour, and similarly for vertices of different clause gadgets. Lastly, observe that two vertices of two different gadgets belong to the same connected component if and only if they are connected through the backbone, which is in T'. Thus, if there exist two vertices of the same colour in a same connected component of $\mathcal{F}$, one must be from a variable gadget and the other one from a clause gadget, and they both necessarily belong to T'. Without loss of generality, consider $x_{i,j}$ from the variable gadget of x_i and $\ell_{k,j}$ from the clause gadget of C_j, such that $x_{i,j}, \ell_{k,j} \in T'$. To prove a contradiction, assume that $c(x_{i,j}) = c(\ell_{k,j})$. Note that the literal represented by $\ell_{k,j}$ is $x_{i,j}$, otherwise the two vertices would not have the same colour. Since $\ell_{k,j}$ is in T', it is connected to the vertex r_{C_j} of the clause gadget, hence $\ell_{k,j}$ satisfies the clause C_j. Therefore, x_i satisfies the clause C_j, and the variable $x_i = True$ in β. By construction, this implies that the edge $\{r_{x_i}, x_i\}$ belongs to S, and therefore that the subtree T_{x_i}, containing $x_{i,j}$, is not part of T', which is a contradiction.

Let S be a solution to COLOURFUL COMPONENTS on T such that $|S| = n + 2m$. Observe that one needs to remove at least one edge per variable gadget to put the vertices x_i and $\bar{x}_i$ into different connected components, and at least two edges per clause gadget, the first one to disconnect y_j and y'_j and the second one to disconnect z_j and z'_j. Since $|S| = n + 2m$, S must only contain n edges from variable gadgets and $2m$ edges from clause gadgets. We denote by T' the connected component of $T - S$ containing the root r. Notice that, for each variable gadget, either x_i or $\bar{x}_i$ belongs to T', but not both. Also, for each clause gadget, exactly one leaf $\ell_{k,j}$ belongs to T'. We construct the solution β to 3,4-SAT on ϕ such that, for each variable gadget, if $\{r_{x_i}, x_i\} \in S$, then we set $x_i := True$ in β, and if $\{r_{x_i}, \bar{x}_i\} \in S$, then we set $x_i := False$ in β. To prove a contradiction, assume that there is a clause C_j which is not satisfied in ϕ with regard to β. Consider the leaf $\ell_{k,j} \in T'$ from the gadget clause of C_j, and assume without loss of generality that $\ell_{k,j} = x_i$. If C_j is not satisfied, then the variable $x_i := False$ in β. This means that S contains the edge $\{r_{x_i}, \bar{x}_i\}$, but not the edge $\{r_{x_i}, x_i\}$, and thus $x_{i,j} \in T'$. However, since $c(x_{i,j}) = c(\ell_{k,j})$, then S is not a solution for T, a contradiction.   □

2.2   Polynomiality

Now, we show that COLOURFUL COMPONENTS and COLOURFUL PARTITION can be solved in linear time on 1-caterpillars with unbounded maximum degree, even if the backbone is a chordless cycle. To simplify the notations, we use the term *general caterpillars* to denote both 1-caterpillars and cyclic 1-caterpillars.

We consider the vertices in the backbone as internal vertices of stars, hence vertices of degree 1 are the leaves of a star whose internal vertex belongs to the backbone. We assume that the edges and the vertices in the backbone are either linearly of cyclically ordered, if the backbone is a path or a cycle, respectively.

Remark 1. Consider a coloured general caterpillar. If two vertices of a star have the same colour, then at least one of these vertices is a leaf and it must belong to a different colourful component than the internal vertex of the star. Hence, a general caterpillar can be preprocessed in such a way that, for each such leaf, we add its adjacent edge to a set S_p. This procedure is repeated until there is no such leaf in $G - S_p$. At the end of the preprocessing, each star in $G - S_p$ is a *colourful star*, that is, only contains vertices with different colours.

If a coloured general caterpillar is not colourful, then it contains either one or at least two colours that appear more than once. We deal with these two cases independently in the following lemmas.

Lemma 1. Colourful Components *and* Colourful Partition *can be solved in linear time in coloured general caterpillars where exactly one colour appears at least twice.*

Lemma 2. *Let G be a coloured general caterpillar with only colourful stars such that at least two colours appear at least twice in G. Then there exists an optimal solution S of* Colourful Components *in G such that $S \subseteq B$, where B is the backbone of G.*

Let G be a coloured general caterpillar with backbone B and only colourful stars. We say that a bad path P between two vertices of colour γ in G is a *colour-critical bad* path if and only if there is no other bad path P' between two vertices of colour γ such that $P' \cap B \subseteq P$. Hence, two colour-critical bad paths with endpoints of colour γ do not have any common edge in the backbone B.

Remark 2. Let G be a coloured general caterpillar with only colourful stars such that at least two colours appear twice. We denote by B the backbone of G. Lemma 2 guarantees that there exists an optimal solution S to Colourful Components on G such that $S \subseteq B$. It is clear that if each colour-critical bad path contains an edge in S and $S \subseteq B$, then S is a solution to Colourful Components on G. Hence, there is an optimal solution to Colourful Components that contains only edges in B that also belong to some colour-critical bad path.

Now, the idea is to define a circular-arc graph H (an intersection graph of a collection of arcs on the circle) based on the colour-critical bad paths of G. A minimum clique cover Q of H, which is a partition of the vertex set into a minimum number of cliques, can be obtained in linear time [9]. We show that Q can be translated back into an optimal solution to Colourful Components and Colourful Partition on G in linear time.

Algorithm 1. From coloured general caterpillar to ordered pairs.

Input: $G = (V, E)$, an ℓ-coloured general caterpillar with only colourful stars
and backbone B.
Output: A, a multiset of ordered pairs of vertices.
// Initialisation
1 let $U := \{u \in e \mid e \in B\}$ be an ordered set of vertices, w.r.t. the order on B;
2 let A be an empty multiset of ordered pairs of vertices;
3 let L be an array of length ℓ initialised at $NULL$;
4 let $proceed := True$;
5 let $u \in U$ such that, if U is linearly ordered, then u is minimum in U;
 // if U is cyclically ordered, any $u \in U$ can be taken
6 let $end := NULL$;
 // Main procedure
7 **while** $proceed$ **do**   // $\mathcal{O}(n)$
8 | **foreach** $v \in \{w \in N[u] \mid d(w) = 1 \text{ or } w = u\}$ **do**
9 | | **if** $L[c(v)] = NULL$ **then**
10 | | | $end := u$;
11 | | **else if** $L[c(v)] \neq v$ **then**
12 | | | add $(L[c(v)], u)$ to A;
13 | | | **if** $u = end$ **then**
14 | | | | $proceed := False$;   // U is cyclically ordered
15 | | $L[c(v)] := v$;
16 | **if** U is linearly ordered and u is maximum in U **then**
17 | | $proceed := False$;
18 | $u := v$, such that v is the element after u in U, w.r.t. its order;
19 **return** A;

Lemma 3. *Let G be a coloured general caterpillar with only colourful stars, and A be the multiset of pairs returned by Algorithm 1. Then there is a bijection between the set of colour-critical bad paths in G and the multiset A.*

Proof. Let B be the backbone of $G = (V, E)$. A colour-critical bad path P from a to b is detected in Algorithm 1 at Line 11, when b is found to have the same colour γ as a (the last recorded vertex of colour γ). Let x be the internal vertex of the star to which a belongs, and y for b, respectively. When b is considered in the algorithm, the pair $(L[c(b)], y)$ is added to A at Line 12, and since $L[c(b)] = x$ then $(x, y) \in A$. Thus, the arc with endpoints $(x, y) \in A$ corresponds to the colour-critical bad path P from a to b in G.

An ordered pair (x, y) in A refers to two vertices x and y in V that are internal vertices of two stars. If such a pair exists, then there are two vertices a and b with the same colour γ, such that a belongs to the same star as x and b to the same star as y, and in the path P from a to b, with regard to the order on B, there is no other vertex w with colour γ in a star whose internal vertex is in P (since the last seen vertex of colour γ, before b, is $L[c(b)] = a$). Thus, the path P is a colour-critical bad path and it corresponds to the pair (x, y) in A. □

Lemma 4. *Algorithm 1 runs in linear time.*

Proof. Let G be a coloured general caterpillar with only colourful stars and backbone B, and let A be the multiset of ordered pairs obtained by Algorithm 1 with input G.

In Algorithm 1, when a colour is detected for the first time at Line 9, the internal vertex u of the star is stored in the variable *end*. If the backbone is a chordless cycle, *i.e.* if G is a cyclic caterpillar, the second time that the vertex *end* is considered in the main loop the algorithm sets the variable *proceed* to false at Line 14. If the backbone is a path, *i.e.* if G is a caterpillar, the algorithm considers each vertex exactly once and sets the variable *proceed* to false at Line 17. Thus, Algorithm 1 runs in linear time.   □

Theorem 2. COLOURFUL COMPONENTS *and* COLOURFUL PARTITION *can be solved in linear time on coloured general caterpillars.*

Proof. Let $G = (V, E)$ be a coloured general caterpillar with backbone B. First, we prove that a solution to COLOURFUL COMPONENTS on G can be found in linear time. We apply the preprocessing to G, as defined in Remark 1, and denote by S_p the set of edges that have been removed. Hence, $G - S_p$ contains only colourful stars. If $G - S_p$ is colourful, then S_p is an optimal solution to COLOURFUL COMPONENTS. Otherwise, denote by $G' = (V', E')$ the connected component of $G - S_p$ which contains the backbone B. If G' contains exactly one colour that appears more than once, then according to Lemma 1 COLOURFUL COMPONENTS and COLOURFUL PARTITION can be solved in linear time. Therefore, we assume that G' contains at least two colours that appear at least twice. Let A be the multiset of ordered pairs obtained by Algorithm 1 with input G'.

According to Lemma 4, A can be obtained in linear time. According to Lemma 3, each ordered pair (x, y) in A corresponds to a colour-critical bad path P from x to y in G. These paths can be seen as arcs on the circle, which represent a circular-arc graph $H = (X, F)$. Let $\mathcal{Q}$ be a minimum clique cover of H obtained in linear time [9], and S' be an empty set of edges. Choose a clique $Q_i \in \mathcal{Q}$. From our construction of H, each vertex $u \in Q_i$ corresponds to a colour-critical bad path P_u in G. Let $D_i := \bigcap_{u \in Q_i} P_u$, and notice that $|D_i \cap B| > 0$. Then, choose an edge $e \in D_i \cap B$, and add e to S'. We claim that, once each clique in $\mathcal{Q}$ has been processed, thus once $|S'| = |\mathcal{Q}|$, the set S' is an optimal solution to COLOURFUL COMPONENTS on G. Notice that S' can be computed in linear time.

As stated before, each colour-critical bad path in G' maps to an ordered pair in A, which corresponds to a vertex of H. Hence, a clique Q_i in H corresponds to a set of colour-critical bad paths sharing a common subpath D_i. The set S' contains an edge in $D_i \cap B$ for each $Q_i \in \mathcal{Q}$, hence there is no colour-critical bad path in $G' - S'$. Since $S' \subset B$ and each coloured-minimal bad paths have an edge in S', then following Remark 2 S' is a solution to COLOURFUL COMPONENTS on G'. Moreover, since $\mathcal{Q}$ is optimal, S' is an optimal solution on G'. Let $S := S_p \cup S'$, and note that S is an optimal solution to COLOURFUL COMPONENTS on G.

Since $|E| \in \mathcal{O}(|V|)$, we can detect each connected component of $G - S$ in linear time (for instance, with a breadth-first search). Thus, we can construct the partition π of V such that each part corresponds to a connected component of $G - S$ in linear time. Obviously, π is a solution to COLOURFUL PARTITION on G. Since S is optimal, due to the structure of G, the partition π is optimal.   □

3   COLOURFUL COMPONENTS on Small-Degree Planar Graphs

In [4], the authors prove that COLOURFUL COMPONENTS is NP-complete even when restricted to 3-coloured graphs with maximum degree 6. Using a similar reduction from PLANAR 3-SAT, we show that the vertices of degree 6 can be replaced with gadgets only containing vertices of degree 4, or 3, if we relax the number of colours from 3 to 5, or 12, respectively.

An instance of PLANAR 3-SAT is a 3-CNF formula in which the bipartite graph of variables and clauses is planar. PLANAR 3-SAT has been proved NP-complete in [11].

Construction 2. Given an instance of PLANAR 3-SAT ϕ, that is a set of m clauses $C_1, C_2, \ldots, C_m$ on n variables, we construct the graph $G = (V, E)$ such that:

- For each variable x in ϕ, let m_x denotes the number of clauses in which x appears. We construct a cycle of length $4m_x$ in G with vertices $V_x := \{x_j^1, x_j^2, x_j^3, x_j^4 \mid x \in C_j\}$ with an arbitrary fixed cyclic ordering of the clauses containing x. The vertices are coloured alternatively with two colours c_o and c_e, that is, $c(x_j^1) = c(x_j^3) = c_o$ and $c(x_j^2) = c(x_j^4) = c_e$, for all j such that $x \in C_j$.

- For each clause C_j containing three variables p, q and r, we construct a clause gadget. We propose two types of gadgets:
  - The gadget $\mathcal{A}_j^4$ is made of a cycle of length 3, with vertices a_j^1, a_j^2 and a_j^3 such that each a_j^i is given colour i, different from c_o and c_e. We define how the vertices from V_p are connected to $\mathcal{A}_j^4$. If the variable p appears as a positive literal in C_j, connect the vertices p_j^1 to a_j^1 and p_j^2 to a_j^2. Otherwise, if p occurs as a negative literal, connect the vertices p_j^2 to a_j^1 and p_j^3 to a_j^2. Do the same for the variables q and r by connecting the corresponding vertices in V_q to a_j^2 and a_j^3, and the corresponding vertices in V_r to a_j^3 and a_j^1. Notice that the vertices in $\mathcal{A}_j^4$ have degree 4.
  - The gadget $\mathcal{A}_j^3$ is made of a cycle of length 9 with vertices labelled $a_j^1, \ldots, a_j^9$ and an additional vertex a_j^{10} connected to a_j^2, a_j^5 and a_j^8. We set the colour i to each vertex a_j^i, different from c_o and c_e. We define how the vertices from V_p are connected to $\mathcal{A}_j^3$. If the variable p appears as a positive literal in C_j, connect the vertices p_j^1 to a_j^1 and p_j^2 to a_j^3. Otherwise, the variable p occurs as a negative literal, connect the vertices p_j^2 to a_j^1 and p_j^3 to a_j^3. Do the same for the variables q and r by connecting the corresponding vertices in V_q to a_j^4 and a_j^6, and the corresponding vertices in V_r a_j^7 and a_j^9. Notice that the vertices in $\mathcal{A}_j^3$ have degree 3.

See Fig. 2 for an example of the gadgets.

Since the bipartite graph of variables and clauses of ϕ is planar and each vertex can be replaced by a clause or vertex gadget, with a correct cyclic ordering of the clauses for each variable, the resulting graph G is planar.

Note that Construction 2 can be done in polynomial time.

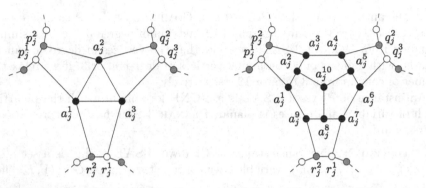

Gadget with vertices of degree 4.   Gadget with vertices of degree 3.

Fig. 2. Two possible clause gadgets of a clause $C_j := (p \vee \bar{q} \vee r)$. White vertices have colour c_o, grey vertices have colour c_e, and each a_j^i is given colour i, different from c_o and c_e.

Theorem 3. COLOURFUL COMPONENTS *is NP-complete on* 5-*coloured planar graphs with maximum degree* 4 *and on* 12-*coloured planar graphs with maximum degree* 3.

4   Conclusion

This paper proposes of complete dichotomy of the computational complexity of COLOURFUL COMPONENTS and COLOURFUL PARTITION on k-caterpillars. The NP-completeness of the problems on 2-caterpillars with unbounded degree demonstrates the inherent complexity of the problems. We prove that both problems remain NP-complete on ternary 3-caterpillars and on binary 4-caterpillars, where both the maximum degree and the hair length are bounded by small constants. Nevertheless, our linear-time algorithm for both problems on general 1-caterpillars, with unbounded degree, generalises the class of paths and cycles, and beats the complexity of the previous best known algorithm for paths. An interesting question is to answer whether the problems remain NP-complete on binary 3-caterpillars and on 2-caterpillars with bounded degree.

We also prove that COLOURFUL COMPONENTS is NP-complete on 5-coloured planar graphs with maximum degree 4 and on 12-coloured planar graphs with maximum degree 3. A natural question is to ask whether the problem remains NP-complete when the number of colours is decreased but the maximum degree is 3 or 4.

Acknowledgements. The authors wish to thank Marthe Bonamy for suggesting these problems, and Paul Ouvrard for our valuable discussions.

References

1. Adamaszek, A., Popa, A.: Algorithmic and hardness results for the colorful components problems. Algorithmica **73**(2), 371–388 (2015)
2. Avidor, A., Langberg, M.: The multi-multiway cut problem. Theoret. Comput. Sci. **377**(1–3), 35–42 (2007)
3. Bousquet, N., Daligault, J., Thomassé, S.: Multicut is FPT. In: Proceedings of the Forty-Third Annual ACM Symposium on Theory of Computing, pp. 459–468. ACM (2011)
4. Bruckner, S., Hüffner, F., Komusiewicz, C., Niedermeier, R., Thiel, S., Uhlmann, J.: Partitioning into colorful components by minimum edge deletions. In: Kärkkäinen, J., Stoye, J. (eds.) CPM 2012. LNCS, vol. 7354, pp. 56–69. Springer, Heidelberg (2012). https://doi.org/10.1007/978-3-642-31265-6_5
5. Bulteau, L., Dabrowski, K.K., Fertin, G., Johnson, M., Paulusma, D., Vialette, S.: Finding a small number of colourful components. arXiv preprint arXiv:1808.03561 (2018)
6. Dondi, R., Sikora, F.: Parameterized complexity and approximation issues for the colorful components problems. Theoret. Comput. Sci. **739**, 1–12 (2018). https://doi.org/10.1016/j.tcs.2018.04.044
7. Fellows, M.R., Fertin, G., Hermelin, D., Vialette, S.: Upper and lower bounds for finding connected motifs in vertex-colored graphs. J. Comput. Syst. Sci. **77**(4), 799–811 (2011)
8. He, G., Liu, J., Zhao, C.: Approximation algorithms for some graph partitioning problems. In: Graph Algorithms and Applications, vol. 2, pp. 21–31. World Scientific (2004)
9. Hsu, W.L., Tsai, K.H.: Linear time algorithms on circular-arc graphs. Inf. Process. Lett. **40**(3), 123–129 (1991)
10. Hung, L.T.Q., Sysło, M.M., Weaver, M.L., West, D.B.: Bandwidth and density for block graphs. Discret. Math. **189**(1–3), 163–176 (1998)
11. Lichtenstein, D.: Planar formulae and their uses. SIAM J. Comput. **11**(2), 329–343 (1982)
12. Marx, D., Razgon, I.: Fixed-parameter tractability of multicut parameterized by the size of the cutset. SIAM J. Comput. **43**(2), 355–388 (2014)
13. Misra, N.: On the parameterized complexity of colorful components and related problems. In: Iliopoulos, C., Leong, H.W., Sung, W.-K. (eds.) IWOCA 2018. LNCS, vol. 10979, pp. 237–249. Springer, Cham (2018). https://doi.org/10.1007/978-3-319-94667-2_20
14. Tovey, C.A.: A simplified NP-complete satisfiability problem. Discret. Appl. Math. **8**(1), 85–89 (1984)
15. Zheng, C., Swenson, K., Lyons, E., Sankoff, D.: OMG! orthologs in multiple genomes – competing graph-theoretical formulations. In: Przytycka, T.M., Sagot, M.-F. (eds.) WABI 2011. LNCS, vol. 6833, pp. 364–375. Springer, Heidelberg (2011). https://doi.org/10.1007/978-3-642-23038-7_30

Maximal Irredundant Set Enumeration in Bounded-Degeneracy and Bounded-Degree Hypergraphs

Alessio Conte[1,2]($\boxtimes$), Mamadou Moustapha Kanté[3], Andrea Marino[4], and Takeaki Uno[1]

[1] National Institute of Informatics, Tokyo, Japan
uno@nii.ac.jp
[2] University of Pisa, Pisa, Italy
conte@di.unipi.it
[3] Université Clermont Auvergne, LIMOS, CNRS, Aubière, France
mamadou.kante@uca.fr
[4] University of Florence, Florence, Italy
andrea.marino@unifi.it

Abstract. An irredundant set of a hypergraph $G = (V, \mathcal{H})$ is a subset S of its nodes such that removing any node from S decreases the number of hyperedges it intersects. The concept is deeply related to that of dominating sets, as the minimal dominating sets of a graph correspond exactly to the dominating sets which are also maximal irredundant sets. In this paper we propose an FPT-delay algorithm for listing maximal irredundant sets, whose delay is $O(2^d \Delta^{d+1} n^2)$, where d is the degeneracy of the hypergraph and Δ the maximum node degree. As $d \leq \Delta$, we immediately obtain an algorithm with delay $O(2^\Delta \Delta^{\Delta+1} n^2)$ that is FPT for bounded-degree hypergraphs. This result opens a gap between known bounds for this problem and those for listing minimal dominating sets in these classes of hypergraphs, as the known running times used to be the same, hinting at the idea that the latter may indeed be harder.

Keywords: Irredundant sets · Enumeration algorithms · FPT · Polynomial delay

1 Introduction

An *enumeration algorithm* is an algorithm that lists all the feasible solutions of a given property, such as subsets of vertices satisfying a given property in graphs. In contrast to optimisation or counting problems, where the size of the output is polynomial in the size of the input, the size of the output of an enumeration problem can be exponential in the size of the input. For instance, any algorithm listing the set of maximal independent sets in any n-vertex graph may output a set of size $3^{n/3}$ as a graph may have such a number of maximal independent sets [18], and then should spend at least such an amount of

© Springer Nature Switzerland AG 2019
C. J. Colbourn et al. (Eds.): IWOCA 2019, LNCS 11638, pp. 148–159, 2019.
https://doi.org/10.1007/978-3-030-25005-8_13

time. It is therefore important in analysing the time complexity of any enumeration algorithm to take into account the size of the output. If the time complexity of an enumeration algorithm depends polynomially in the cumulated sizes of the input and output, we call it an *output-polynomial* algorithm. The existence of output-polynomial algorithms for several enumeration problems has triggered the curiosity in the algorithmic community: for many problems output-polynomial algorithms were proposed (see, e.g., [20] for a survey), while for others the non-existence of output-polynomial algorithms was proved [14,15,17].

Among enumeration problems, the existence of an output-polynomial algorithm for enumerating the *(inclusion-wise) minimal transversals* of a hypergraph,[1] known as HYPERGRAPH TRANSVERSAL ENUMERATION, is arguably the most important open question in this area as a solution would solve the existence of an output-polynomial algorithm for many enumeration problems in many diverse areas ranging from data-mining to integer programming, including biology, databases, game theory, learning theory, etc. (see the many references from [9,20]). Surprisingly, it was proved in [13] that the existence of an output-polynomial algorithm for listing the *(inclusion-wise) minimal dominating sets* in graphs[2] would solve the HYPERGRAPH TRANSVERSAL ENUMERATION problem. This result triggers an interest in the enumeration of (minimal) dominating sets in graphs and for several interesting graph classes output-polynomial algorithms were proposed. Another line of research that emerged also from [13] is the following question:

for which vertex (or edge) set graph property, the enumeration of (inclusion-wise) minimal (or maximal) is equivalent[3] to the HYPERGRAPH TRANSVERSAL ENUMERATION problem?

For some variants of domination like problems an equivalence was proved in [13], while for some others the question remains. The main remarkable open cases are probably the enumeration of *minimal connected dominating sets*, and the enumeration of *maximal irredundant sets*, and we deal with the latter in this paper. A subset I of the vertex set V of a graph is an *irredundant set* if for any $x \in I$, either x has no neighbour in I or there is a vertex $y \in V \setminus I$ having exactly x as neighbour in I. It is folklore to prove that a subset D of the vertex set of a graph is a minimal dominating set if and only if D is an irredundant dominating set. For this reason, irredundant sets in graphs receive a lot of attention in graph theory [12], and recently in algorithmic graph theory, see for instance [3] where the computation of a maximum irredundant set is considered, and [11] where they are interested in an upper-bound on the number of maximal irredundant sets (surprisingly the best known upper bound is the trivial 2^n).

In this paper we are interested in an output-polynomial algorithm for listing the maximal irredundant sets of a hypergraph. Unlike the case of dominating

---

[1] A *transversal* in hypergraph $\mathcal{H} \subseteq 2^V$ is a subset of V intersecting all elements of $\mathcal{H}$.

[2] A *dominating set* in a graph G is a subset D of its vertex set V such that every vertex in $V \setminus D$ has a neighbour in D.

[3] Two enumeration problems are said *equivalent* when one has an output-polynomial algorithm if and only if the other one has.

sets, the set of irredundant sets in a (hyper)graph is an independence set system, and one may expect an equivalence with the HYPERGRAPH TRANSVERSAL ENUMERATION problem. But, unexpectedly the enumeration of maximal irredundant sets in hypergraphs is a hard problem, and is equivalent to the HYPERGRAPH TRANSVERSAL ENUMERATION when restricted to hypergraphs with hyperedges of size 2 (an output-polynomial algorithm is also announced when the maximum degree of the vertices is bounded) [5]. The definition of irredundant sets given in [5] is different from ours, but both are equivalent.

An output-polynomial algorithm has *delay p* if, after a pre-processing running in polynomial time in the input size, the algorithm outputs all the solutions and the time between two outputs is bounded by p. Our main result is the following.

Theorem 1. *For every fixed positive integer d, we can enumerate, with delay bounded by $O(2^d \cdot \Delta^{d+1} \cdot n^2)$ and space $O(nd)$, the set of maximal irredundant sets in any given hypergraph of degeneracy d and maximum degree Δ.*

As the degeneracy d of a hypergraph is at most its maximum node degree (see Sect. 2.1 for the formal definition), this immediately implies an FPT-delay algorithm for bounded-degree hypergraphs:

Corollary 1. *For every fixed positive integer Δ, we can enumerate, with delay bounded by $O(2^\Delta \cdot \Delta^{\Delta+1} \cdot n^2)$ and space $O(n\Delta)$, the set of maximal irredundant sets in any given hypergraph of maximum degree Δ.*

Our result not only improves the announced result in [5], but also considers more general hypergraphs, namely those of bounded degeneracy. Moreover, this result is surprising because the best output-polynomial algorithm for listing the minimal transversals in hypergraphs of degeneracy d has delay $n^{O(d)}$, and it is still open whether we can obtain an output-polynomial algorithm with delay $O(2^d \cdot \Delta^{f(d)} \cdot n^c)$, for some function f and constant c. We observe moreover that we were not able to turn our algorithm into one for listing the minimal dominating sets while keeping the same delay.

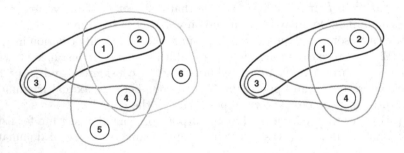

Fig. 1. Left: a hypergraph with 6 nodes, 4 hyperedges, maximum degree 3 and degeneracy 2. Right: its subhypergraph induced by nodes $\{1, 2, 3, 4\}$.

2   Preliminaries

In this paper we consider a hypergraph $G = (V_G, \mathcal{H}_G)$, where V_G is the set of nodes (or vertices) and $\mathcal{H}_G$ the set of hyperedges, where each hyperedge $H \in \mathcal{H}_G$ is a set of nodes of G, i.e., $\mathcal{H}_G \subseteq 2^{V_G}$.

For any $v \in V_G$, $\mathcal{H}_G(v)$ is the set $\{H \in \mathcal{H}_G : v \in H\}$ of hyperedges containing v. The degree of v in G is $|\mathcal{H}_G(v)|$. And we refer as Δ_G to the *maximum degree* of a node in G. A bounded-degree hypergraph is a hypergraph G such that $\Delta_G = O(1)$ (however, hyperedges may be arbitrarily large).

Given a set of nodes $S \subseteq V$ and a node $v \in S$, we say that a hyperedge $H \in \mathcal{H}_G(v)$ is a *private edge* of v for S if v is the only node of S belonging to H, i.e., $H \cap S = \{v\}$. More formally, we define the set of private edges of v for S, in the hypergraph G, as $\text{PRIV}_G(v, S) = \{H \in \mathcal{H}_G : H \cap S = \{v\}\}$.

If every node in S has a private edge for S, i.e., $\forall v \in S$, $\text{PRIV}_G(v, S) \neq \emptyset$, we call S an *irredundant set* (and a *redundant set* otherwise). Furthermore, S is a *maximal irredundant set* (MIRS for short) if no irredundant set $S' \supset S$ exists.

When G is clear from the context, we may remove the subscripts, and for example refer to V_G and $\mathcal{H}_G$ simply as V and $\mathcal{H}$.

We can further assume the hypergraph to be connected, as otherwise the solution set is the Cartesian product of the solution sets of its connected components, which we can compute at no extra cost as in [6].

2.1   Degeneracy of a Hypergraph

The degeneracy of a graph is a well known sparsity measure [1,7,10] and is defined to be the minimum number d such that every subgraph of G has a node of degree d or less. Equivalently, it is the maximum d such that G has a subgraph whose minimum node degree is d (known as a d-core).

Several generalization of degeneracy have been proposed for hypergraphs, such as that in [8] based on CNF-formulas or that in [2] based on Delaunay graphs of hypergraphs. In this work we consider the arguably natural one proposed in [16], which is obtained by simply adapting the concept of subgraph to hypergraphs:

Definition 1 (see [16]). *The degeneracy d_G of a hypergraph G is the minimum integer d such that any subhypergraph of G has a node of degree at most d.*

A *subhypergraph* $G[X] = (X, \mathcal{H}_{G'})$ of G is a hypergraph obtained by removing nodes from G. More formally, we have $X \subseteq V_G$ and $\mathcal{H}_{G[X]} = \{H \cap X | H \in \mathcal{H}_G$ and $H \cap X \neq \emptyset\}$. Note that we are considering hypergraphs and *not* multihypergraphs: for any two edges $H_1, H_2 \in \mathcal{H}_G$, if $H_1 \cap X = H_2 \cap X = H'$, then $G[X]$ will have a single hyperedge corresponding to H' rather than two equal ones.[4] A visual example of subhypergraph is given in Fig. 1.

---

[4] It can be observed that degeneracy would be meaningless otherwise, as the highest degree node would induce by itself a subhypergraph of minimum degree Δ_G.

For a given ordering $v_1, \ldots, v_n$ of the nodes of G, we use G_i as a shorthand for $G[v_1, \ldots, v_i]$.

By iteratively removing the node of smallest degree in G (and taking the subhypergraph induced by the remaining nodes) we may compute a *degeneracy ordering*, that is a node ordering $v_1, \ldots, v_n$ such that each v_i has degree at most d_G in the subhypergraph $G[v_i, \ldots, v_n]$. Similarly, we call *reverse degeneracy ordering* the reverse of a degeneracy ordering: here each v_i has degree at most d_G in $G_i = G[v_1, \ldots, v_i]$. For example, we can observe that $6, 5, 4, 3, 2, 1$ is a degeneracy ordering of the graph in Fig. 1 (left), and $1, 2, 3, 4, 5, 6$ a reverse degeneracy ordering.

3 Algorithm Description

3.1 General Structure

We here describe the operations performed by our incremental algorithm, while its correctness will be proved later in Sect. 3.3.

The structure of the algorithm is simple, in line with the approaches by Tsukiyama et al. [19] and Lawler et al. [17]: given an ordering $v_1, \ldots, v_n$ of the nodes of G,[5] and recalling $G_i = G[v_1, \ldots, v_i]$, we want to compute the solutions of G_i from those of G_{i-1}. Indeed, we can observe that G_1 has only one trivial solution corresponding to $\{v_1\}$ and G_n is G, thus applying this incremental operation is sufficient to generate all the solutions of G.

In the following, we refer to $\mathcal{S}_i$ as the solutions, i.e., the set of all MIRSs, of G_i. We propose an algorithm that computes each $\mathcal{S}_i$ given $\mathcal{S}_{i-1}$, which contains the key algorithmic ideas of our approach. Later, in Sect. 4, we show how to refine this algorithm to run in polynomial delay and use polynomial space.

Firstly, let us observe the differences between G_i and G_{i-1}: recalling the definition of subhypergraph, we have that G_i contains edges of three types:

A All edges in G_{i-1} corresponding to edges that do not contain v_i in G.
B All edges in G_{i-1} corresponding to edges that do contain v_i in G. Note that, by removing v_i, an edge of this type may become equal to a type-A edge, and thus the two will appear as a single edge in G_{i-1}.
C At most one edge containing just v_i, corresponding to all the edges in G containing v_i but no node in $v_1, \ldots, v_{i-1}$.

By keeping this in mind, we can prove the following claim:

Lemma 1. *Let S be a maximal irredundant set of G_i. Then, $T = S \setminus \{v_i\}$ is an (non necessarily maximal) irredundant set of G_{i-1}.*

Proof. Any node $x \in S$ has a private edge H in G_i. We prove the claim showing that for any $x \neq v_i$ the hyperedge corresponding to H in G_{i-1} is still a private edge of x for T. Looking at the types above, note that H is not a type-C edge, it contains $x \in S \setminus \{v_i\}$. We thus prove the claim for the remaining two cases, i.e., whether H is a type-A or type-B edge:

---

[5] This will be a reverse degeneracy ordering, as we later explain.

(A) As H is a private edge of x for S, no other node of $T = S \setminus \{v_i\}$ belongs to H, thus the corresponding edge of G_{i-1} will be a private of x for T in G_{i-1}.

(B) If there is no type-A edge H' equal to H in G_{i-1}, then H is a private of x for T in G_{i-1} by the same logic. Otherwise, if there is an H' equal to H in G_{i-1}, then any node of T except x in H' would also be in H, making H not a private of x in G_{i-1}, a contradiction; thus H' is a private edge of x of type-A and x has a private for T by the above case.   □

Corollary 2. *Let S be a maximal irredundant set of G_i. Then, there exists a maximal irredundant set T of G_{i-1} such that $S \subseteq T \cup \{v_i\}$.*

Using the incremental approach described above, and thanks to Corollary 2, Algorithm 1 shows the general schema to build S_i, where the operation NEIGHBORS(S, v_i) is a generating procedure explained in the next section, which allows us to obtain solutions in S_i starting from solutions $S \in S_{i-1}$.

Algorithm 1. Listing all maximal irredundant sets incrementally.

input : A hypergraph $G = (V, \mathcal{H})$
              A reverse degeneracy ordering $v_1, \ldots, v_n$ of V
output: All maximal irredundant sets of G

```
1 Let G_i be a shorthand for G[v_1, ..., v_i]
2 S_1 ← {{v_1}}                          // only solution of G_1
3 foreach v_i ∈ v_2, ..., v_n do
4     S_i ← ∅                            // solutions of G_i
5     foreach S ∈ S_{i-1} do
6         foreach S' ∈ NEIGHBORS(S, v_i) do
7             add S' to S_i

8 output S_n                             // Maximal irredundant sets of G
```

3.2 Generating New Solutions

Lemma 1, and in particular Corollary 2, tell us that it is possible to find each solution of S_i by just looking at S_{i-1} and v_i. We will show a generating procedure NEIGHBORS(S, v_i) which takes as input an $S \in S_{i-1}$, the node v_i, and generates a set of MIRSs of G_i. The correctness of the algorithm will follow from proving that each solution of G_i is found this way at least once.

The procedure is formalized in Algorithm 2, and we will here describe it in detail:

Firstly, it is possible that $S \cup \{v_i\} \in S_i$, in which case we only generate this unique solution, i.e., NEIGHBORS$(S, v_i) = \{S \cup \{v_i\}\}$ (Lines 2 and 3) and end the procedure. In the rest of the section we consider the opposite case, where adding v_i to S results in a redundant set, and we need to remove some nodes in order to obtain again irredundant sets.

Algorithm 2. Generating new solutions

> **input** : A node v_i in an ordering $v_1, \ldots, v_n$ of the nodes of G.
> A maximal irredundant set S of G_{i-1}
> **output:** A collection of maximal irredundant sets of G_i.

```
 1  Function NEIGHBORS(S, v_i)
 2      if S ∪ {v_i} is a MIRS of G_i then
 3          yield S ∪ {v_i} and return

 4      R ← REDUNDANT(S, v_i)
 5      foreach R' ⊂ R do                                  // case R' ≠ R
 6          foreach E ∈ Π_{r∈R'}(H_{G_i}(r)) do
 7              D ← ⋃_{H∈E}(H ∖ R')
 8              S' ← (S ∪ {v_i}) ∖ ((R ∖ R') ∪ D)
 9              if S' is a MIRS of G_i then yield S'

10      foreach H_i ∈ H(v_i) do                            // case R' = R
11          foreach E ∈ Π_{r∈R}(H_{G_i}(r) ∖ {H_i}) do
12              D ← H_i ∪ ⋃_{H∈E}(H ∖ R)
13              S' ← (S ∪ {v_i}) ∖ D
14              if S' is a MIRS of G_i then yield S'

15      yield S
```

More in detail, there will be a set of nodes of S which do not have a private edge $S \cup \{v_i\}$. We define this set of nodes as

$$\text{REDUNDANT}(S, v_i) = \{x \in S : \text{PRIV}(x, S) \subseteq \mathcal{H}(v_i)\}$$

(v_i may or may not have a private edge in $S \cup \{v_i\}$, but it is *not* included in REDUNDANT(S, v_i) in any case).

For convenience, in the following let $R = \text{REDUNDANT}(S, v_i)$. Our goal is to exploit R to create a collection of irredundant sets of G_i contained in $S \cup \{v_i\}$, among which we aim at selecting the maximal ones.

For example, removing all of R corresponds to $(S \cup \{v_i\}) \setminus R$: every node in $S \setminus R$ has a private edge for $S \cup \{v_i\}$ by definition of R, and v_i has as private edges $\cup_{r \in R}\text{PRIV}(r, S)$, so $(S \cup \{v_i\}) \setminus R$ is an irredundant set.

We can now explain the operations of Algorithm 2 in Lines 4–15 (which, we recall, is the case where $S \cup \{v_i\}$ is not an irredundant set):

- We iterate over *all* possible subsets R' of R.
- For each R', we try to build solutions which include R' and v_i.
- To do so, we forcefully give each $r \in R'$ a private edge H_r, and remove the remaining nodes in $R \setminus R'$.
- We output the irredundant set obtained if it is a MIRS of G_i, and discard it otherwise.
- Finally, we output S as it must also be a MIRS of G_i.

Let us now explain more in detail how this is achieved.

For each r, we select one hyperedge $H_r \in \mathcal{H}_{G_i}(r)$: we force it to become a private edge of r by removing all nodes in H_r (except r) from S. More formally we select a set of hyperedges $E \in \Pi_{r \in R' \cup \{v\}}(\mathcal{H}_{G_i}(r))$.[6] We call D the set $\bigcup_{H \in E}(H) \setminus R'$ of nodes in S covered by these edges: removing D from $S \cup \{v_i\}$ gives each node in R' a private edge. We also remove the remaining nodes in $R \setminus R'$ which may not have private edges at this point.

We now consider v_i: in case $R' \subset R$, i.e., $R' \neq R$, there exists some $w \in R \setminus R'$ which had a private edge H_w for the solution S. Since $w \in R$, this private edge includes v_i, meaning that v_i has H_w as private edge after removing $R \setminus R'$. Thus in this case (Lines 5–9) v_i is guaranteed to have a private edge. After this process, $S' \leftarrow (S \cup \{v_i\}) \setminus ((R \setminus R') \cup D)$ is guaranteed to be an irredundant set as each node in it has a private edge in G_i.

We then separately consider the single case $R' = R$ in Lines 10–14: we perform the same actions as above, but also select an edge $H_i \in \mathcal{H}_{G_i}(v_i)$, and force it to be a private edge of v_i by also adding its contained nodes to D.

In this case S' corresponds to $(S \cup \{v_i\}) \setminus D$ and by the same logic will also be an irredundant set. It is worth remarking that this case covers the (possible) situation where $R = \emptyset$, but $S \cup \{v_i\}$ is not a maximal irredundant set (i.e., v_i is the only node with no private edges).

At this point, we output S' iff it is a MIRS of G_i. We will prove in Sect. 3.3 that each MIRS of G_i is found this way by Algorithm 1.

3.3 Correctness

Since $G_n = G$, to prove the correctness of Algorithm 1 it is sufficient to show that, given $\mathcal{S}_{i-1}$, the algorithm is able to correctly compute $\mathcal{S}_i$, which we do in the following:

Lemma 2. $\bigcup_{S \in \mathcal{S}_{i-1}} \text{NEIGHBORS}(S, v_i) = \mathcal{S}_i$.

Proof. Firstly, looking at Algorithm 2 we can see that only MIRS of G_i are output, so $\bigcup_{S \in \mathcal{S}_{i-1}} \text{NEIGHBORS}(S, v_i) \subseteq \mathcal{S}_i$.

We consider an arbitrary $T \in \mathcal{S}_i$ and show that it is found. Let S be any maximal solution in $\mathcal{S}_{i-1}$ such that $T \subseteq S \cup \{v_i\}$; note that S exists by Corollary 2.

If $T = S \cup \{v_i\}$, it is found in Line 3.

Otherwise, let $R = \text{REDUNDANT}(S, v_i)$ be the set of nodes of S whose private edges are hit by v_i. Some nodes of R (possibly all, or none) will be in T and some will not. As $\text{NEIGHBORS}(S, v_i)$ iterates over all possible subsets $R' \subset S$ (Line 5) and considers the case $R' = R$ (Line 10), at some point it will consider exactly $R' = R \cap T$.

---

[6] As the generated solution must contain v_i, we could avoid choosing edges which contain it, and refine the choice of E to $E \in \Pi_{r \in R' \cup \{v\}}(\mathcal{H}_{G_i}(r) \setminus \mathcal{H}_{G_i}(v))$. However this does not change the complexity analysis.

Furthermore, each node in R' has at least one private edge for T in G_i: let E' be a set containing one of these private edges for each node in R' chosen arbitrarily. Again, as the algorithm iterates over all possible assignment of edges to the nodes in R' (Lines 6 and 11), at some point it will consider $E = E'$.

When the correct R' and E are considered (on either Line 8 or Line 13), we will have that $S' \supseteq T$: indeed, $S \cup \{v_i\} \supseteq T$, and the nodes removed from $S \cup \{v_i\}$ to obtain S' are those in $R \setminus R'$, which are not in T, and those in D. The nodes in D correspond to those not in R' that are in some edge in E; since the edges in $E = E'$ are all private edges of some node in $R' \subseteq T$, they may not contain any other node of T, which implies the claimed $S' \supseteq T$. Finally, since T is by assumption a *maximal* solution of G_i, $S' \supseteq T$ implies $S' = T$, thus T is found. The statement follows. □

Corollary 3. *Algorithm 1 outputs all the maximal irredundant sets of G without duplication.*

Proof. Absence of duplication is trivially guaranteed by $\mathcal{S}_n$, which is a set. As $\mathcal{S}_1 = \{v_1\}$, by Lemma 2 Algorithm 1 correctly computes and outputs $\mathcal{S}_n$ as the set of all MIRSs of $G_n = G$. □

3.4   Complexity

To understand the algorithm's complexity, we need to look at the respective sizes of the $\mathcal{S}_i$ sets, for which we use the following lemma:

Lemma 3. $|\mathcal{S}_i| \geq |\mathcal{S}_{i-1}|$.

Proof. For each maximal solution $S \in \mathcal{S}_{i-1}$ of G_{i-1}, either S is a MIRS of G_i, or only v_i can be added to it: as any other node addition would also be possible in G_{i-1}, meaning S was not a maximal solution in $\mathcal{S}_{i-1}$. In this latter case, $S \cup \{v_i\}$ is a maximal solution of G_i. The statement follows. □

Looking at Algorithm 1 we can observe that the cost is bounded by $O(\sum_{i \in [2,n]} |\mathcal{S}_i|)$ times the worst case cost of the NEIGHBORS(S, v_i) function.[7] By Lemma 3, $O(\sum_{i \in [2,n]} |\mathcal{S}_i|)$ is bounded by $O(n \cdot |\mathcal{S}_n|)$, where $|\mathcal{S}_n|$ is the number of MIRSs of $G_n = G$, i.e., the number of solutions to be output. We thus only need to bound the cost of NEIGHBORS(S, v_i) (Algorithm 2).

Lemma 4. *Function* NEIGHBORS(S, v_i) *in Algorithm 2 takes* $O(2^{d_G} \Delta_G^{d_G+1} n)$ *time.*

Proof. As in Algorithm 2, let $R =$ REDUNDANT(S, v_i), and let P be the set of private edges of nodes in R before the addition of v_i, i.e.:

$$P = \bigcup_{x \in \text{REDUNDANT}(S, v_i)} \text{PRIV}(x, S)$$

---

[7] The cost of adding solutions to $\mathcal{S}_i$ is dominated by that of NEIGHBORS(S, v_i).

Note that all the edges in P are not private anymore for the nodes in R (otherwise the corresponding node would not be in REDUNDANT(S, v_i)), so they must contain v_i. This implies that $P \subseteq \mathcal{H}_{G_i}(v)$. By the fact that Algorithm 1 uses a reverse degeneracy ordering of the nodes of G, and by its properties remarked in Sect. 2, we obtain $|P| \leq |\mathcal{H}_{G_i}(v)| \leq d_G$.

Furthermore, each edge may be the private edge of at most one node (otherwise it would not be a private edge), so there is at most one node in R for each edge in P, meaning that $|R| \leq |P| \leq d_G$.

Consider Lines 5–9. The external **foreach** in Line 5 thus performs $O(2^{|R|}) = O(2^{d_G})$ iterations. The internal one performs $O(\Pi_{r \in R}(|\mathcal{H}_{G_i}(r)|) = O(\Delta^{d_G})$ iterations. For each iteration, computing D can be done in $O(|S| \cdot |E|) = O(|S| \cdot |R|) = O(nd_G)$ time: as we only need to know the nodes of D which belong to S, we can iterate over S, and for each node check its membership to the selected edges in E. Finally, we need to check maximality of S' in G_i: by scanning all edges in $O(\sum_{H \in \mathcal{H}_G} |H|) = O(n\Delta_G)$ we can find, for each $x \in S'$, its private edges, and for each edge the node it is private for. Then, we can test whether any node x can be added to S' in $O(|\mathcal{H}_G(x)|)$ by iterating over each $H \in \mathcal{H}_G(x)$ and checking whether H is the only private edge of some node $y \in S'$ (in which case x would not be addible). Testing addition of all nodes thus takes $O(n\Delta_G)$ time. The total cost of Lines 5–9 is thus $O(2^{d_G} \cdot \Delta^{d_G} n(d_G + \Delta_G)) = O(2^{d_G} \cdot \Delta^{d_G+1} n)$.

The cost of Lines 10–14 is dominated by the earlier, as the loops contain the same operations but the external one only performs $O(|\mathcal{H}_G(v_i)|) = O(d_G)$ iterations rather than $O(2^{d_G})$ as that in Line 5. Line 4 can be executed in $O(n\Delta_G)$ by computing the private edges of each node as above, so its cost is also dominated by $O(2^{d_G} \Delta_G^{d_G+1} n)$. The statement follows. □

We thus obtain a total cost of $O(2^{d_G} \Delta^{d_G+1} n^2 |\mathcal{S}_n|)$ time, i.e., $O(2^{d_G} \Delta^{d_G+1} n^2)$ amortized time per solution.

4   Achieving Polynomial Space and Delay

Algorithm 1 requires storing the sets $\mathcal{S}_i$ to avoid duplication, thus using exponential space. However, we show here how to further refine the space usage of the algorithm, and give bounds to its delay, by simple modifications of Algorithm 1.

By looking at the proof of Lemma 2 we can see that $T \in \mathcal{S}_i$ is found by NEIGHBORS(S, v_i) for any S such that $T \subseteq S \cup \{v_i\}$.

We designate exactly one of these solutions as the *parent* of T, i.e., the solution *in charge* of finding T: for any $T \in \mathcal{S}_i$, let $S = $ PARENT(T, i) be the solution of G_{i-1}, i.e., in $\mathcal{S}_{i-1}$, obtained by taking $T' = T \setminus \{v_i\}$ (which is an irredundant set of G_{i-1} by Lemma 1), and recursively adding to it the node v_j of smallest index in G_{i-1} such that $T' \cup \{v_j\}$ is still an irredundant set.

We replace Lines 9 and 14 of Algorithm 2 as

" **if** S' is a MIRS of G_i and PARENT$(S', i) = S$ **then yield** S' "

Note that Lines 3 and 15 do not need this check as PARENT$(S \cup \{v_i\}, v_i)$ and PARENT(S, v_i), respectively, will both always give S.

The PARENT(T, i) operation can be implemented in $O(n\Delta_G)$ time similarly to the maximality check performed in the proof of Lemma 4: first find the private edges of each node in $T \setminus \{v_i\}$, then test node additions starting from the smallest index node. This does not impact the complexity of the algorithm (as the same cost is paid for the maximality check), but it does ensure that T is found exactly once by Algorithm 1 when considering v_i.

This defines a tree-like structure between solutions, where each solution in S_i is the child of some in S_{i-1} and the leafs are the solutions of S_n, i.e., those to be output. We can thus perform a DFS-like traversal of this tree with a recursive algorithm: i.e., immediately processing NEIGHBORS(S', v_{i+1}) when we find $S' \in S_i$ from its parent $S = $ PARENT(S', i). The delay of the algorithm becomes the cost of a root-to-leaf path plus a leaf-to-root path in this tree, i.e., $O(n)$ times the cost of the NEIGHBORS(S, v_i) function. The modified Algorithm 1 will thus have $O(2^{d_G} \Delta^{d_G+1} n^2)$ time delay, finally proving our main result, Theorem 1.

As for the space usage, a trivial implementation requires storing S, the R, R' and E sets, for each recursion level, taking $O(n^2)$ space. However, we can rebuild S from its child S' by computing PARENT(S', v_i) and just store R, R' and E sets whose size is $O(d_G)$. The total space this way is just $O(nd_G)$.

5   Concluding Remarks

We prove in this paper that we can enumerate the set of MIRSs with delay $O(2^d \cdot \Delta^{d+1} \cdot n^2)$ in any hypergraph of degeneracy d and maximum degree Δ. Therefore, we can enumerate with polynomial delay the set of MIRSs in many sparse graphs, including bounded expansion graphs. We wonder whether we can enumerate the set of MIRSs in nowhere dense graph classes, which would suggest that the difficulty of enumerating MIRSs resides in dense graphs, as for the enumeration of minimal dominating sets.

Despite the announced negative result from [5], we can still wonder whether the enumeration of MIRSs in graphs can be easier than the enumeration of minimal dominating sets as suggested by our result, which would be surprising as maximal irredundancy is commonly believed to be more difficult than domination, which is a local property. We observe however that in hypergraphs the two problems seem to behave differently: in hypergraphs with hyperedges of size 3 MIRSs is coNP-complete while HYPERGRAPH TRANSVERSAL ENUMERATION is output-polynomial in the more general class of hypergraphs of bounded conformality [4]; and in hypergraphs of degeneracy d MIRSs admits an $f(d, \Delta) \cdot n^2$-delay enumeration algorithm (this paper), while the best known delay for HYPERGRAPH TRANSVERSAL ENUMERATION is $n^{O(d)}$.

Acknowledgements. This work was partially supported by JST CREST, grant number JPMJCR1401, Japan, and by MIUR, Italy. Mamadou M. Kanté is supported by the French Agency for Research under the GraphEN project ANR-15-CE40-0009.

References

1. Alon, N., Kahn, J., Seymour, P.D.: Large induced degenerate subgraphs. Graphs Comb. **3**(1), 203–211 (1987)
2. Bar-Noy, A., Cheilaris, P., Olonetsky, S., Smorodinsky, S.: Online conflict-free colouring for hypergraphs. Comb. Probab. Comput. **19**(4), 493–516 (2010)
3. Binkele-Raible, D., et al.: Breaking the 2^n-barrier for irredundance: two lines of attack. J. Discret. Algorithms **9**(3), 214–230 (2011)
4. Boros, E., Elbassioni, K., Gurvich, V., Khachiyan, L.: Generating maximal independent sets for hypergraphs with bounded edge-intersections. In: Farach-Colton, M. (ed.) LATIN 2004. LNCS, vol. 2976, pp. 488–498. Springer, Heidelberg (2004). https://doi.org/10.1007/978-3-540-24698-5_52
5. Boros, E., Makino, K.: Generating maximal irredundant and minimal redundant subfamilies of a given hypergraph. announced at WEPA 2016 (Clermont-Ferrand, November 2016) and Boolean Seminar 2017 (Liblice, March 2017) (2016). http://clp.mff.cuni.cz/booleanseminar/presentations/endre.pdf
6. Conte, A., Grossi, R., Marino, A., Rizzi, R., Versari, L.: Listing subgraphs by cartesian decomposition. In: MFCS 2018, pp. 84:1–84:16 (2018)
7. Conte, A., Kanté, M.M., Otachi, Y., Uno, T., Wasa, K.: Efficient enumeration of maximal k-degenerate induced subgraphs of a chordal graph. Theor. Comput. Sci. (2018)
8. Eiter, T., Gottlob, G., Makino, K.: New results on monotone dualization and generating hypergraph transversals. SIAM J. Comput. **32**(2), 514–537 (2003)
9. Eiter, T., Makino, K., Gottlob, G.: Computational aspects of monotone dualization: a brief survey. Discret. Appl. Math. **156**(11), 2035–2049 (2008)
10. Eppstein, D., Löffler, M., Strash, D.: Listing all maximal cliques in large sparse real-world graphs. J. Exp. Algorithmics **18**, 3.1:3.1–3.1:3.21 (2013). https://doi.org/10.1145/2543629. ISSN 1084-6654, Article no. 3.1. ACM, New York
11. Golovach, P.A., Kratsch, D., Sayadi, M.Y.: Enumeration of maximal irredundant sets for claw-free graphs. Theor. Comput. Sci. **754**, 3–15 (2019)
12. Haynes, T.W., Hedetniemi, S., Slater, P.: Fundamentals of Domination in Graphs. CRC Press, Boca Raton (1998)
13. Kanté, M.M., Limouzy, V., Mary, A., Nourine, L.: On the enumeration of minimal dominating sets and related notions. SIAM J. Discret. Math. **28**(4), 1916–1929 (2014)
14. Khachiyan, L., Boros, E., Borys, K., Elbassioni, K.M., Gurvich, V., Makino, K.: Generating cut conjunctions in graphs and related problems. Algorithmica **51**(3), 239–263 (2008)
15. Khachiyan, L., Boros, E., Borys, K., Gurvich, V., Elbassioni, K.: Generating all vertices of a polyhedron is hard. 1–17 (2009)
16. Kostochka, A.V., Zhu, X.: Adapted list coloring of graphs and hypergraphs. SIAM J. Discret. Math. **22**(1), 398–408 (2008)
17. Lawler, E.L., Lenstra, J.K., Kan, A.H.G.R.: Generating all maximal independent sets: NP-hardness and polynomial-time algorithms. SIAM J. Comput. **9**(3), 558–565 (1980)
18. Moon, J.W., Moser, L.: On cliques in graphs. Isr. J. Math. **3**(1), 23–28 (1965)
19. Tsukiyama, S., Ide, M., Ariyoshi, H., Shirakawa, I.: A new algorithm for generating all the maximal independent sets. SIAM J. Comput. **6**(3), 505–517 (1977)
20. Wasa, K.: Enumeration of enumeration algorithms. arXiv preprint arXiv:1605.05102 (2016)

Dual Domination

Gennaro Cordasco[1](✉), Luisa Gargano[2], and Adele Anna Rescigno[2]

[1] Università della Campania "Luigi Vanvitelli", Caserta, Italy
gennaro.cordasco@unicampania.it
[2] Department of Computer Science, Università di Salerno, Fisciano, Italy
{lgargano,arescigno}@unisa.it

Abstract. Inspired by the feedback scenario, which characterizes online social networks, we introduce a novel domination problem, which we call *Dual Domination (DD)*. We assume that the nodes in the input network are partitioned into two categories: Positive nodes (V^+) and negative nodes (V^-). We are looking for a set $D \subseteq V^+$ that dominates the largest number of positive nodes while avoiding as many negative nodes as possible. In particular, we study the *Maximum Bounded Dual Domination* (MBDD) problem, where given a bound k, the problem is to find a set $D \subseteq V^+$, which maximizes the number of nodes dominated in V^+, dominating at most k nodes in V^-. We show that the MBDD problem is hard to approximate to a factor better than $(1 - 1/e)$. We give a polynomial time approximation algorithm with approximation guaranteed $(1 - e^{-1/\Delta})$, where Δ represents the maximum number of neighbors in V^+ of any node in V^-. Furthermore, we give an $O(|V|k^2)$ time algorithm to solve the problem on trees.

1 Introduction

Let $G = (V, E)$ be an undirected graph modeling a network. We denote by $N_G(v)$ and by $d_G(v) = |N_G(v)|$, respectively, the neighborhood and the degree of the node v in G. In general, for each $S \subseteq V$ we denote by $N_G(S) = \bigcup_{v \in S} N_G(v)$ the neighborhood of the nodes in S. In the rest of the paper we will omit the subscript G whenever the graph G is clear from the context.

A dominating set for $G = (V, E)$ is a subset of the nodes $D \subseteq V$ such that each $v \in V - D$ has at least one neighbor in D. The concept of domination in graphs and its many related problems have been widely studied (see [17] and references therein quoted). Inspired by some scenarios in social networking, which we shall briefly describe in Sect. 1.2, we introduce a new domination problem, which we call *Dual Domination (DD)*. We assume that the nodes in the input network are partitioned into two categories: Positive nodes (V^+) and negative nodes (V^-); i.e., $V = V^+ \cup V^-$. For any $D \subseteq V^+$, we denote by $\Gamma(D)$ (resp. $\Gamma^+(D)$ and $\Gamma^-(D)$) the set of nodes (resp. positive and negative nodes) dominated by D. That is,

$$\Gamma(D) = D \cup N(D), \qquad \Gamma^+(D) = \Gamma(D) \cap V^+ \quad \text{and} \quad \Gamma^-(D) = \Gamma(D) \cap V^-.$$

© Springer Nature Switzerland AG 2019
C. J. Colbourn et al. (Eds.): IWOCA 2019, LNCS 11638, pp. 160–174, 2019.
https://doi.org/10.1007/978-3-030-25005-8_14

For sake of simplicity we pose $\Gamma^+(v) = \Gamma^+(\{v\})$ and $\Gamma^-(v) = \Gamma^-(\{v\})$.

Formally the problem we study in this paper is the following.

Problem 1. (MAXIMUM BOUNDED DUAL DOMINATION (MBDD)) Given a network $G = (V = (V^+ \cup V^-), E)$ and an integer $k \geq 0$, find a set $D \subseteq V^+$ such that $|\Gamma^-(D)| \leq k$, which maximizes $|\Gamma^+(D)|$.

1.1   Our Results

We first show hardness results on the approximability of the MBDD problem, then we give a polynomial time approximation algorithm with approximation guaranteed $(1 - e^{-1/\Delta})$, where Δ represents the maximum number of neighbors in V^+ of any node in V^-. The algorithm uses the fact that $|\Gamma^+(D)|$ is a submodular, nondecreasing set function and is inspired by [27] where an approximation algorithm for maximizing a submodular set function subject to a knapsack constraint has been presented. However, we stress that the constraint, to which a solution of our MBDD problem is subject, is not a knapsack constraint since in our problem two or more positive nodes might share a negative neighbor. In Sect. 4, we depict an $O(|V|k^2)$ time algorithm for the MBDD problem on trees, state some related Dual Domination problems, and give some open problems.

Due to space constraint, most of the proofs are omitted or only sketched.

1.2   The Online Social Networks Context

Online social networks have become an important media for the dissemination of opinions, beliefs, new ideas etc. The increasing popularity of such platforms, together with the availability of large amounts of contents and user profile/behaviour information, has contributed to the rise of viral marketing as an effective advertising strategy. The idea is to exploit the word-of-mouth effect in such a way that an initial set of influential users could influence their friends, friends of friends, and so on, generating a large influence cascade. The key problem is how to select an initial set of users (given a limited budget) so to maximize the influence within the network. This *influence maximization* (IM) problem has been extensively studied in recent years [5–10,16] and a number of approximation algorithms and scalable heuristics have been devised. However, the studies above only look at networks with positive relationships/activities (e.g., positive feedback or influence), where in real scenarios, social actor relationships/activities also include negative ones (e.g., adverse opinion, negative feedback or distrust relationships). For instance in Ebay, buyers and sellers develop trust and distrust relationship; in online review and news forums, such as Slashdot, users comment (positively or negatively) reviews and articles of each other [21].

Research has provided evidence that the benefits of a marketing campaign are not purely increasing in the number of people reached and the exposure to different groups can help or hurt adoption [2,18,19]. As an other example, in a social network composed by individuals with some social problem, people can have both positive and negative impact on each other. In order to implement an intervention programme, it becomes important to target a group of users which allow to reinforce a positive behavior through the network while minimizing the negative reactions (also to maximize the impact of future campaigns) [28]. A somehow similar finding applies to political campaigns where candidates want to reinforce positive messages without promoting resistance to persuasion [25]. Such empirical research suggests that marketing campaigns can suffer negative payoff due to the existence of subsets of the population that will react negatively to the message/product. Hence, the marketing campaign can suffer negative payoff. These can come in the form of harm to the firm's reputation in several ways, as for example through negative reviews on rating sites [3,11,12]. Recently, a variation of the influence maximization problem named *opinion maximization* (OM) has been proposed [4,22]. The goal of opinion maximization is to maximize the number of positive opinions while minimizing the number of negative opinions generated by the activated users during the cascading behavior. A first algorithmic study of an OM problem was done in [1], where the authors propose a theoretical model for the problem of seeding a cascade when there are benefits from reaching positively inclined customers and costs from reaching negatively inclined customers. Namely, the problem studied in [1] is: Given a graph G with node set $V = V^+ \cup V^-$ partitioned into positive and negative nodes, determine a subset of the nodes S that can trigger a cascade which maximize the difference between positive and negative payoff.

1.3 Related Domination Problems

Domination in graphs, and its several variants, is a widely studied problem in graph theory [17]. The variation of the domination problem which we study in this paper is related, but not equivalent, to the concepts of signed and minus domination introduced in [14,15]. For instance, in signed domination the sign of the nodes is not part of the input; namely, given an input graph $G = (V, E)$ one looks for a function of the form $f : V \rightarrow \{-1, 1\}$ such that, $\sum_{u \in N(v) \cup \{v\}} f(u) \geq 1$ for all $v \in V$.

Another recently studied related problem is *domination with required and forbidden nodes* [13]: Given a graph G and two disjoint sets $R, F \subset V$, construct dominating set D of G such that no forbidden node is in D and every required node of R is in D, that is $F \cap D = \emptyset$ and $R \subseteq D$.

2   Hardness Results

Theorem 1. *The MBDD problem is such that:*
(i) There is no polynomial time approximation algorithm with any constant factor better than $(1 - 1/e)$ unless P=NP.
(ii) There is no polynomial time approximation algorithm providing an $n^{-1/\text{polyloglog } n}$-approximation unless the exponential time hypothesis is false.

Proof. (Sketch.) We are going to show that both the k-MaxVD problem [24] and the Densest k-subgraph (DkS) problem [23] are reducible (preserving the approximation factor) in polynomial time, to the MBDD problem. The results (i) and (ii) will then follow from [24] and [23] respectively. For space reasons, the reduction from the DkS problem is omitted.

The k-MaxVD problem is one of the optimization versions of the well known Dominating Set problem [17] and it is defined as follows.

*Problem 2. k–*MAXIMUM VERTEX DOMINATION (k-MaxVD): Given a network $G = (V, E)$ and an integer $k \geq 0$, find a set $D \subseteq V$ with $|D| \leq k$, which maximizes the cardinality of the dominated nodes $|\Gamma(D)|$.

Consider an instance of the k-MaxVD problem, consisting of a graph $G = (V, E)$ having $n = |V|$ nodes and a bound k. Let $V = \{v_1, v_2, \dots, v_n\}$, we build a graph $G' = (V' = (V^+ \cup V^-), E')$ as follows:

Replace each v_i by a gadget G_i' having two nodes v_i^+ and v_i^-. The node v_i^+ plays the role of v_i in G and is also connected to v_i^-. Formally,

$$V' = V^+ \cup V^- \text{ where } V^+ = \{v_i^+ | 1 \leq i \leq n\} \text{ and } V^- = \{v_i^- | 1 \leq i \leq n\}$$

$$E' = \{(v_i^+, v_j^+) | (v_i, v_j) \in E\} \cup \{(v_i^+, v_i^-) | 1 \leq i \leq n\}.$$

Notice that G corresponds to the subgraph of G' induced by V^+. We prove that:

Given an integer t, there exists a set D of nodes in G of size at most k such that $|\Gamma(D)| \geq t$ iff there exists a set $D' \subseteq V^+$ such that $|\Gamma^-(D')| \leq k$ and $|\Gamma^+(D')| \geq t$ in G'.

Assume that there exists a dominating set $D \subseteq V$ in G such that $|D| \leq k$ and $|\Gamma(D)| \geq t$. Then let $D' = \{v_i^+ \in V^+ | v_i \in D\}$, since G is isomorphic to the subgraph of G' induced by V^+, we have that D' dominates the corresponding of all the nodes in $\Gamma(D)$. Hence, $|\Gamma^+(D')| = |\Gamma(D)| \geq t$. Moreover, by construction, in G' each positive node has exactly one connection with a negative one. Hence, $|\Gamma^-(D')| = |D'| = |D| \leq k$.

On the other hand, assume that there exists a set $D' \subseteq V^+$ in G' such that $|\Gamma^-(D')| \leq k$ and $|\Gamma^+(D')| \geq t$. Then, by using exactly the same argument above, the reader can easily see that the set $D = \{v_i \in V | v_i^+ \in D'\}$ satisfies $|D| \leq k$ and $|\Gamma(D)| \geq t$ and this completes the proof.   □

3 An Approximation Algorithm for MBDD

Theorem 2. *Let $G = (V, E)$ be any graph with $V = V^+ \cup V^-$. There exists a polynomial time approximation algorithm for the MBDD problem on G with approximation factor $1 - e^{-1/\Delta}$, where $\Delta = max_{v \in V^-} |\Gamma^+(v)|$ is the maximum degree[1] of any negative node in V^-.*

---

[1] We can assume that no edge exists between two nodes in V^-, since such edges are irrelevant for our problem.

In order to prove Theorem 2, we distinguish two cases on the value of Δ.

If $\Delta = 1$ then two nodes in V^+ cannot share a neighbor in V^-. As a consequence, if we define the weight of a node in V^+ as the number of its neighbors in V^-, then the problem reduces to select a set of nodes in V^+ so that the union of their neighborhood sets in V^+ has maximum size and the sum of their weights is at most k. This is the Budgeted maximum coverage problem and its approximation factor is $(1 - 1/e)$ [20].

Algorithm 1. The Dual Domination algorithm: $\text{DUAL}(G, k)$

Input: A graph $G = (V^+ \cup V^-, E)$ (with $\Delta \geq 2$) and a positive integer k.

1   $P = \emptyset$
2   **forall the** $u \in V^+$ **do**
3   **if** $|\Gamma^-(u)| \leq k$ **then** $P =$ the largest set between P and
      $\{v \in V^+ \mid \Gamma^-(v) \subseteq \Gamma^-(u)\}$
4   **forall the** $v \in V^+ - \{u\}$ **do**
5   **if** $|\Gamma^-(\{u, v\})| \leq k$ **then** $P =$ the largest set between P and
        $\text{DD}(G, \{u, v\}, k)$;

6   **return** P

Algorithm 2. $\text{DD}(G, U, k)$

Input: A graph $G = (V^+ \cup V^-, E)$, a set $U \subseteq V^+$ with $|U| = 2$, a positive integer k.

1   Set $I = V^+$, $S = U$ and $P = \{w \in V^+ \mid \Gamma^-(w) \subseteq \Gamma^-(S)\}$
2   **while** $(I - P \neq \emptyset)$ **do**
3   **forall the** $u \in I - P$ **do** $P_u = \{w \in I - P \mid \Gamma^-(w) \subseteq \Gamma^-(S \cup \{u\})\}$
4      $v = \arg\max_{u \in I - P} \frac{|\Gamma^+(P \cup P_u) - \Gamma^+(P)|}{|\Gamma^-(S \cup \{u\}) - \Gamma^-(S)|}$
5   **if** $|\Gamma^-(S \cup \{v\})| \leq k$ **then** $\{S = S \cup \{v\}; P = P \cup P_v\}$ **else** $I = I - \{v\}$
6   **return** P

The rest of this section is devoted to prove Theorem 2 in the case $\Delta \geq 2$.

The proposed Algorithm $\text{DUAL}(G, k)$ first computes all the feasible solutions of cardinality one, by simply enumerating all nodes in V^+. In order to consider feasible solutions with cardinality two or more, it exploits Algorithm $\text{DD}(G, U, k)$ to greedily enlarge each feasible solution of cardinality two. It is worth noticing that the algorithm $\text{DD}(G, U, k)$ is executed for each couple of nodes in G. This fact will be exploited to obtain the desired approximation factor.

Algorithm $\text{DD}(G, U, k)$, starting from a partial solution $S = U$ of cardinality 2, greedily adds nodes to such a solution, until no feasible node is available: For each node u not in the current solution, the algorithm measures

- the *cost* of u (how many new nodes of V^- are dominated by adding u) and
- the *profit* of u (how many more nodes of V^+ we can dominate by adding u, as well as all the other nodes which become "*cost-free*" because of u, that is, all their neighbors in V^- are already neighbors of the current solution augmented by u).

The algorithm then selects the node, say v, that provides the best *profit/cost* ratio; if the current solution augmented by v is feasible (i.e., $|\Gamma^-(S \cup \{v\})| \leq k$) then v is added to it, otherwise v is definitively discarded (because v will make any solution, that includes the current solution, infeasible).

Define the padding set of $X \subseteq V^+$ as

$$P_X = \{v \in V^+ \mid \Gamma^-(v) \subseteq \Gamma^-(X)\} \tag{1}$$

Notice $\Gamma^-(X) = \Gamma^-(P_X)$. Also, define *the padding set of u with respect to a ground set I and a set X* as

$$P_u(I, X) = \{v \in I - P_X \mid \Gamma^-(v) \subseteq \Gamma^-(\{u\} \cup X)\}. \tag{2}$$

Starting from any set $U \subset V^+$ consisting of two nodes such that $|\Gamma^-(U)| < k$, Algorithm 2 greedily augments U while preserving the constraint. For a given $U = \{w_1, w_2\}$, the algorithm starts with $S = U$ and a padding set $P = P_S$, fixes the initial ground set I to V^+, and iteratively adds feasible nodes to the solution.

At each iteration, Algorithm 2 maintains the relation $P = P_S$ between the sets S and P. For each node $u \in I - P$, the algorithm identifies the set $P_u \subseteq I - P$ whose neighbors in V^- are dominated when we add u to the current set S, namely the algorithm sets

$$P_u = P_u(I, S).$$

The node to be added to the solution is chosen as to maximize the ratio of the number of positive nodes that will be dominated thanks to the contribution of u to the cost of u (i.e., the increment on the number of negative nodes dominated by $\{u\} \cup S$.) Once a node v has been selected:

- If $S \cup \{v\}$ is not feasible (i.e., $|\Gamma^-(S \cup \{v\})| > k$), then v is removed from the ground set I.
- If $|\Gamma^-(S \cup \{v\})| \leq k$, then the algorithm augments S by v and consequently P by P_v thus maintaining the equality $P = P_S$.

The algorithm ends when $I = P = P_S$.

In the following, we analyze Algorithm 2 and derive the desired approximation factor. Let OPT be an optimal solution to the MBDD problem on G. Let u_1, u_2 be respectively, the two nodes in OPT that dominate the maximum number of positive neighbors, namely

$$u_1 = \arg\max_{u \in \text{OPT}} |\Gamma^+(P_{\{u\}})| \quad \text{and} \quad u_2 = \arg\max_{u \in \text{OPT} - P_{\{u_1\}}} |\Gamma^+(P_u(V^+, \{u_1\}))|. \tag{3}$$

Recalling that Algorithm 2 is executed for each pair of nodes in G, from now on we focus on the execution of Algorithm 2 on input $U = \{u_1, u_2\}$. Hence, Algorithm 2 initially (at at line 1) sets $S = \{u_1, u_2\}$ and $P = P_{\{u_1, u_2\}}$.

We denote by S^i the partial set S after exactly i nodes are added to it. Namely,

$$S^0 = \{u_1, u_2\}$$

and for $i \geq 1$, we define S^i as the solution consisting of the initial set $\{u_1, u_2\}$ and the i nodes $v_1, v_2, \ldots, v_i$ (added at line 5 of the algorithm), i.e.,

$$S^i = \{u_1, u_2\} \cup \{v_1, v_2, \ldots, v_i\}.$$

We also set

$$P^0 = P_{S^0} = P_{\{u_1, u_2\}} \text{ and } P^i = P_{S^i} = P_{\{u_1, u_2\} \cup \{v_1, v_2, \ldots, v_i\}} \quad \text{for each } i \geq 1$$

and denote by I^i the ground set I at the end of the iteration in which v_i (cfr. line 5) is added to have S^i. Moreover, for each $i \geq 0$, and $u \in I^i - P^i$, we set

$$c_{i,u} = |\Gamma^-(S^i \cup \{u\}) - \Gamma^-(S^i)|$$

the increment in number of dominated nodes in V^- with respect to $\Gamma^-(S^i)$. Furthermore, recalling that the set P_u at line 3 is $P_u(I^i, S^i)$, we denote by

$$\theta_{i+1} = \max_{u \in I^i - P^i} \frac{|\Gamma^+(P^i \cup P_u(I^i, S^i))| - |\Gamma^+(P^i)|}{c_{i,u}}.$$

Hence, the node v selected at line 4 of Algorithm 2 satisfies the equality

$$|\Gamma^+(P^i \cup P_v(I^i, S^i))| - |\Gamma^+(P^i)| = c_{i,v}\theta_{i+1} \tag{4}$$

while for any other node $u \in (I^i - S^i) - \{v\}$ it holds

$$|\Gamma^+(P^i \cup P_u(I^i, S^i))| - |\Gamma^+(P^i)| \leq c_{i,u}\theta_{i+1}. \tag{5}$$

In the following we use c_u instead of $c_{i,u}$ whenever the index i is clear from the context. Furthermore, we use P_u instead of $P_u(I, S)$ whenever the ground set I and the set S are clear from the context.

We assume that the solution provided by the Algorithm 2 is not the optimal solution OPT. Let $S^t = \{u_1, u_2\} \cup \{v_1, \ldots, v_t\}$, for some $t \geq 0$, be the partial set constructed by Algorithm 2 when, for the first time, the node v selected at line 4 satisfies both the following conditions

1. $v \in$ OPT;
2. v is discarded, i.e. v is removed from the ground set because $|\Gamma^-(S^t \cup \{v\})| > k$.

We notice that,

- it is possible that other nodes have been previously discarded by the algorithm but these nodes do not belong to OPT.

– t is well defined. Indeed, since the solution provided by the Algorithm 2 differs from OPT, there exists at least one node $v \in$ OPT, which is discarded by the Algorithm 2.

Let $I' \subseteq I^t$ denote the ground set when v is selected and let

$$\theta = (|\Gamma^+(P^t \cup P_v(I', S^t))| - |\Gamma^+(P^t)|)/c_v \tag{6}$$

Then for any other node $u \in (I' - S^t) - \{v\}$ it holds

$$|\Gamma^+(P^t \cup P_u(I', S^t))| - |\Gamma^+(P^t)| \leq c_u \theta. \tag{7}$$

We prove now some claims, relating S^t, the discarded node v, and the optimal solution OPT, that will be useful to prove the desired approximation ratio.

Claim 1

$$|\Gamma^-(P^0)| + \sum_{i=1}^{t} c_{v_i} + c_v > k. \tag{8}$$

Proof. It suffices to notice that $S^t \cup \{v\}$ dominates more than k nodes in V^-.

□

Claim 2. For any $i = 0, \dots, t$, it holds

$$\sum_{u \in \text{OPT} - P^i} c_u \leq (k - |\Gamma^-(P^0)|)\Delta \tag{9}$$

Proof. Since Δ is the maximum degree of any node in V^-, a node $x \in \Gamma^-(\text{OPT}) - \Gamma^-(P^i)$ can have at most Δ neighbors in $\text{OPT} - P^i$, hence we have[2]

$$\sum_{u \in \text{OPT} - P^i} c_u = \sum_{u \in \text{OPT} - P^i} |\Gamma^-(S^i \cup \{u\}) - \Gamma^-(S^i)|$$
$$\leq |\Gamma^-(\text{OPT}) - \Gamma^-(P^i)|\Delta$$
$$\leq |\Gamma^-(\text{OPT}) - \Gamma^-(P^0)|\Delta \qquad \text{since } P^0 \subseteq P^i$$
$$= (|\Gamma^-(\text{OPT})| - |\Gamma^-(P^0)|)\Delta \qquad \text{since } P^0 \subseteq \text{OPT}$$
$$\leq (k - |\Gamma^-(P^0)|)\Delta \qquad \text{since } |\Gamma^-(\text{OPT})| \leq k.$$

□

Given a set $A \subseteq V^+$ such that $P^0 \subseteq A$, we define the function

$$g(A) = |\Gamma^+(A)| - |\Gamma^+(P^0)|.$$

In order to obtain the desired bound on the approximation factor of Algorithm 2, we first prove some preliminary results regarding the function $g(\cdot)$ which will be exploited to derive a lower bound on the ratio $g(P^t \cup P_v)/g(\text{OPT})$.

---

[2] (Notice that $S^0 \subseteq$ OPT and since OPT is an optimal solution we have $P_{\text{OPT}} \subseteq$ OPT and then $P^0 \subseteq$ OPT.)

Claim 3. For $i = 0, \ldots, t$, it holds $\quad g(P^i) = \sum_{\ell=1}^i c_{v_\ell} \theta_\ell$.

Proof. Recalling that for $i = 0, \ldots, t$, we have $P^i = P^0 \cup P_{v_1} \cup \cdots \cup P_{v_i}$, we get

$$g(P^i) = |\Gamma^+(P^i)| - |\Gamma^+(P^0)| = |\Gamma^+(P^0 \cup P_{v_1} \cup \cdots \cup P_{v_i})| - |\Gamma^+(P^0)|$$

$$= \sum_{\ell=1}^i (|\Gamma^+(P^0 \cup P_{v_1} \cup \cdots \cup P_{v_\ell})| - |\Gamma^+(P^0 \cup P_{v_1} \cup \cdots \cup P_{v_{\ell-1}})|)$$

$$= \sum_{\ell=1}^i |\Gamma^+(P^{\ell-1} \cup P_{v_\ell})| - |\Gamma^+(P^{\ell-1})| = \sum_{\ell=1}^i c_{v_\ell} \theta_\ell \qquad \text{by (4).}$$

$\square$

Claim 4. Let $P_v = P_v(I', S^t)$, then $\quad g(P^t \cup P_v) = \sum_{\ell=1}^t c_{v_\ell} \theta_\ell + c_v \theta$.

Proof. We have that $P^t = P^0 \cup P_{v_1} \cup \cdots \cup P_{v_t}$ and v was selected when the value of S in Algorithm 2 was S^t. We can then apply Claim 3 and (6) to get the claim.

$\square$

Claim 5

$$g(\text{OPT}) \leq \min_{0 \leq i \leq t} g_i \text{ where } g_i = \begin{cases} \sum_{\ell=1}^i c_{v_\ell} \theta_\ell + \theta_{i+1}(k - |\Gamma^-(P^0)|)\Delta & \text{if } 0 \leq i \leq t-1, \\ \sum_{\ell=1}^t c_{v_\ell} \theta_\ell + \theta(k - |\Gamma^-(P^0)|)\Delta & \text{if } i = t. \end{cases}$$

$$(10)$$

Proof. Fix any $i = 0, \cdots, t$. We notice that the set function g is *non-decreasing*, indeed $g(A) \leq g(A')$ for all $A \subseteq A'$. Moreover, recalling that a set function $f : 2^X \to \mathbb{R}^+$ on the ground set X is *submodular* iff $f(A) + f(A') \geq f(A \cup A') + f(A \cap A')$, for all $A, A' \subseteq X$, it is easy to see that $g(A)$ is also a submodular function on the ground set of the subsets of V^+ that contain P^0. Considering that $P^0 \subseteq \text{OPT}$, we can apply to g a result in [26] and we have that

$$g(\text{OPT}) \leq g(P^i) + \sum_{u \in \text{OPT} - P^i} (g(P^i \cup \{u\}) - g(P^i)) \qquad (11)$$

Hence, for any $i = 0, \cdots, t - 1$ we get

$$g(\text{OPT}) \leq g(P^i) + \sum_{u \in \text{OPT} - P^i} (g(P^i \cup \{u\}) - g(P^i)) \qquad \text{by (11)}$$

$$= g(P^i) + \sum_{u \in \text{OPT} - P^i} (|\Gamma^+(P^i \cup \{u\})| - |\Gamma^+(P^i)|) \quad \text{by the definition of } g(\cdot)$$

$$\leq g(P^i) + \sum_{u \in \text{OPT} - P^i} (|\Gamma^+(P^i \cup P_u)| - |\Gamma^+(P^i)|) \quad \text{since } \{u\} \subseteq P_u$$

$$\leq g(P^i) + \sum_{u \in \text{OPT} - P^i} c_u \theta_{i+1} \qquad \text{by (5), since } u \in \text{OPT} - P^i \subseteq I^i - P^i$$

$$\leq g(P^i) + \theta_{i+1}(k - |\Gamma^-(P^0)|)\Delta \qquad \text{by (9)}$$

$$= \sum_{\ell=1}^i c_{v_\ell} \theta_\ell + \theta_{i+1}(k - |\Gamma^-(P^0)|)\Delta \qquad \text{by Claim 3.}$$

and, following the above reasoning for $i = t$ we get

$$g(\text{OPT}) \leq g(P^t) + \sum_{u \in \text{OPT} - P^t} (|\Gamma^+(P^t \cup P_u)| - |\Gamma^+(P^t)|)$$

$$\leq g(P^t) + \sum_{u \in \text{OPT} - P^t} c_u \theta \qquad \text{by (7), since } u \in \text{OPT} - P^t \subseteq I' - P^t$$

$$= \sum_{\ell=1}^{t} c_{v_\ell} \theta_\ell + \theta(k - |\Gamma^-(P^0)|)\Delta \qquad \text{by Claim 3 and (9) .}$$

$\square$

Lemma 1.

$$\frac{g(P^t \cup P_v)}{g(\text{OPT})} > 1 - e^{-1/\Delta} \tag{12}$$

Proof. We need some definitions. Define

$$B_0 = 0, \qquad B_i = \sum_{\ell=1}^{i} c_{v_\ell} \text{ for } i = 1, \cdots, t, \qquad B_{t+1} = \sum_{\ell=1}^{t} c_{v_\ell} + c_v.$$

By (8) we have

$$\beta = k - |\Gamma^-(P^0)| < B_{t+1} \tag{13}$$

Furthermore, for $i = 0, \cdots, t$ define

$$\rho_j = \begin{cases} \theta_i & \text{if } j = B_{i-1} + 1, \cdots, B_i \\ \theta & \text{if } j = B_t + 1, \cdots, B_{t+1} \end{cases} \tag{14}$$

Hence, for $i = 1, \cdots, t$, we have

$$\sum_{j=1}^{B_i} \rho_j = \sum_{\ell=1}^{i} c_{v_\ell} \theta_\ell \quad \text{and} \quad \sum_{j=1}^{B_{t+1}} \rho_j = \sum_{\ell=1}^{t} c_{v_\ell} \theta_\ell + c_v \theta. \tag{15}$$

We use now the above definitions to bound $g(\text{OPT})$ and $g(P^t \cup P_v)$. By (10)

$$g(\text{OPT}) \leq \min \left\{ \min_{0 \leq i \leq t-1} \left\{ \sum_{\ell=1}^{i} c_{v_\ell} \theta_\ell + \theta_{i+1} \beta \Delta \right\}, \sum_{\ell=1}^{t} c_{v_\ell} \theta_\ell + \theta \beta \Delta \right\}$$

$$\text{by the definition of } \beta \text{ in (13)}$$

$$= \min_{0 \leq i \leq t} \left\{ \sum_{j=1}^{B_i} \rho_j + \rho_{B_{i+1}} \beta \Delta \right\} \qquad \text{by (15) and (14)}$$

$$= \min_{1 \leq s \leq B_{t+1}} \left\{ \sum_{j=1}^{s-1} \rho_j + \rho_s \beta \Delta \right\} \qquad \text{by the definition of } B_{t+1}.$$

By Claim 4 and (15) we have $g(P^t \cup P_v) = \sum_{\ell=1}^{t} c_{v_\ell} \theta_\ell + c_v \theta = \sum_{j=1}^{B_{t+1}} \rho_j$. Hence,

$$\frac{g(P^t \cup P_v)}{g(\text{OPT})} \geq \frac{\sum_{j=1}^{B_{t+1}} \rho_j}{\min_{s=1,\cdots,B_{t+1}} \left\{ \sum_{j=1}^{s-1} \rho_j + \rho_s \beta \Delta \right\}} \tag{16}$$

In order to bound the right end side of (16), we use the following fact.

Fact 1 ([26]). *If a and b are arbitrary positive integers, ρ_j for $j = 1, \cdots, a$ are arbitrary non negative reals and $\rho_0 > 0$*

$$\frac{\sum_{j=1}^{a} \rho_j}{\min_{s=1,\cdots,a} \left\{ \sum_{j=1}^{s-1} \rho_j + b\rho_s \right\}} > 1 - e^{-a/b}$$

Hence, we get $\frac{g(P^t \cup P_w)}{g(\text{OPT})} \geq 1 - e^{-B_{t+1}/(\beta\Delta)} > 1 - e^{-1/\Delta}$ where the last inequality holds since $B_{t+1} > \beta$ by (13). □

We show now that the bound $1 - e^{-1/\Delta}$ also holds for $|\Gamma^+(P^t)|/|\Gamma^+(\text{OPT})|$. Recalling that we are considering Algorithm 2 with input $U = \{u_1, u_2\}$, where u_1 and u_2 are the nodes defined in (3), we are able to prove the following claim.

Claim 6. $|\Gamma^+(P^t \cup P_v)| - |\Gamma^+(P^t)| \leq |\Gamma^+(P^0)|/2$.

Proof. Recalling that the set $P^t = P^0 \cup P_{v_1} \cup \cdots \cup P_{v_t}$ is the union of disjoint sets, and that $|\Gamma^+(\cdot)|$ is a submodular set function we can write

$$|\Gamma^+(P^t \cup P_v)| - |\Gamma^+(P^t)| = |\Gamma^+(P^0 \cup P_{v_1} \cup \cdots \cup P_{v_t} \cup P_v)| - |\Gamma^+(P^0 \cup P_{v_1} \cup \cdots \cup P_{v_t})|$$
$$\leq |\Gamma^+(P_v)| - |\Gamma^+(\emptyset)| = |\Gamma^+(P_v)|$$

Furthermore, recalling that $P_v = P_v(I', S^t) = \{u \in I' - P_{S^t} \mid \Gamma^-(u) \subseteq \Gamma^-(\{v\} \cup S^t)\}$ and that $P_{\{v\}} = \{u \in V^+ \mid \Gamma^-(u) \subseteq \Gamma^-(v)\}$ we have $P_v \subseteq P_{\{v\}}$. Since $|\Gamma^+(\cdot)|$ is not decreasing then $|\Gamma^+(P_v)| \leq |\Gamma^+(P_{\{v\}})|$. From this and using the definition of u_1 in (3) we have

$$|\Gamma^+(P^t \cup P_v)| - |\Gamma^+(P^t)| \leq |\Gamma^+(P_{\{v\}})| \leq |\Gamma^+(P_{\{u_1\}})|. \tag{17}$$

We now derive a further bound on $|\Gamma^+(P^t \cup P_v)| - |\Gamma^+(P^t)|$. To this aim, we notice that $P^0 = P_{\{u_1\}} \cup P_{u_2}(V^+, \{u_1\})$ is the union of disjoint sets and that $P_v \subseteq P_v(V^+, \{u_1\}))$; using this and the definition of u_2 in (3), we have

$$|\Gamma^+(P^t \cup P_v)| - |\Gamma^+(P^t)| =$$
$$\leq |\Gamma^+(P_{\{u_1\}} \cup P_v)| - |\Gamma^+(P_{\{u_1\}})| \leq |\Gamma^+(P_{\{u_1\}} \cup P_v(V^+, \{u_1\}))| - |\Gamma^+(P_{\{u_1\}})|$$
$$\leq |\Gamma^+(P_{\{u_1\}} \cup P_{u_2}(V^+, \{u_1\}))| - |\Gamma^+(P_{\{u_1\}})| = |\Gamma^+(P^0)| - |\Gamma^+(P_{\{u_1\}})| \tag{18}$$

The claim follows by summing up (17) and (18). □

We are now ready to conclude the proof of Theorem 2. We have

$$
\begin{aligned}
|\Gamma^+(P^t)| &= |\Gamma^+(P^0)| + g(P^t) \\
&= |\Gamma^+(P^0)| + g(P^t \cup P_v) - (g(P^t \cup P_v) + g(P^t)) \\
&= |\Gamma^+(P^0)| + g(P^t \cup P_v) - (|\Gamma^+(P^t \cup P_v)| - |\Gamma^+(P^t)|) \\
&\geq |\Gamma^+(P^0)| + (1 - e^{-1/\Delta})g(\text{OPT}) - (|\Gamma^+(P^t \cup P_v)| - |\Gamma^+(P^t)|) && \text{by Lemma 1} \\
&\geq |\Gamma^+(P^0)| + (1 - e^{-1/\Delta})g(\text{OPT}) - |\Gamma^+(P^0)|/2 && \text{by Claim 6} \\
&= |\Gamma^+(P^0)|/2 + (1 - e^{-1/\Delta})|\Gamma^+(\text{OPT})| - (1 - e^{-1/\Delta})|\Gamma^+(P^0)| && \text{by def. of } g(\cdot) \\
&= |\Gamma^+(P^0)|(1/2 - (1 - e^{-1/\Delta})) + (1 - e^{-1/\Delta})|\Gamma^+(\text{OPT})| \\
&\geq (1 - e^{-1/\Delta})|\Gamma^+(\text{OPT})| && \text{since } \Delta \geq 2.
\end{aligned}
$$

Hence, after the first iteration in which the algorithm eliminates (at line 5) an element of the optimal solution OPT, it holds that $\frac{|\Gamma^+(P^t)|}{|\Gamma^+(\text{OPT})|} \geq 1 - e^{-1/\Delta}$.

Noticing that subsequent iterations of Algorithm 2 can only improve the ratio, we can conclude that Theorem 2 holds.

4   Concluding Remarks: Extensions and Open Problems

In this section we summarize some additional results and problems related to the MBDD problem. Namely, we consider the following Problems 3 and 4.

Problem 3. (MAXIMUM DUAL DOMINATION (MDD)) Given a network $G = (V = (V^+ \cup V^-), E)$, find $D \subseteq V^+$ which maximizes $|\Gamma^+(D)| - |\Gamma^-(D)|$.

Problem 4. (MINIMUM NEGATIVE DUAL DOMINATION (mNDD)) Given $G = (V = (V^+ \cup V^-), E)$, find $D \subseteq V^+$ which dominates all positive nodes ($\Gamma^+(D) = V^+$) and minimizes the number of dominated negative nodes $|\Gamma^-(D)|$.

First of all, we mention that the MBDD problem is at least as hard as solving any of the Problems 3 and 4. Indeed any optimal strategy OPT that solves the MBDD problem can be used to solve with an extra polynomial time both the Problems 3 and 4. Indeed for the Problem 3 it is sufficient to run the OPT strategy for any budget $i = 1, \ldots, |V^-|$ and than choose the value that maximizes the difference $|\Gamma^+(S_i)| - |\Gamma^-(S_i)|$, where S_i denotes the output of the OPT strategy with budget i. Similarly for the Problem 4 it is sufficient to run the OPT strategy increasing the value of the budget until $\Gamma^+(S_i) = V^+$.

4.1   Trees

The MBDD problem, defined in Sect. 1, can be solved in polynomial time when the graph G is a tree. Let $T = (V = (V^+ \cup V^-), E)$ be a tree network and k be an integer that represents our budget. Without loss of generality, we can root the tree at a node $r \in V^+$. The idea is then that, considering a node v and one of its children u, there are three possibilities: v dominates u (i.e., $v \in S$),

v is dominated by u (i.e., $v \notin S, u \in S$); they do not dominate each other (i.e., $u, v \notin S$). Taking this into account, one can design a dynamic programming algorithm that traverses the input tree T bottom–up, in such a way that each node is considered after all of its children have been processed.

Such an algorithm can be easily adapted to deal with Problems 3 and 4. Summarizing, we have the following results, whose proof is omitted.

Theorem 3. *The MBDD, mNDD, and MDD problems are solvable in linear time on trees.*

4.2 Hardness

By the same construction of the graph G' as in the proof of Theorem 1, it is possible to show that:

- There exists a dominating set D in G of size at most k iff there exists a set $D' \subseteq V^+$ such that $|\Gamma^+(D')| - |\Gamma^-(D')| \geq n - k$ in G'.
- There exists a dominating set D in G of size at most k iff there exists a set $D' \subseteq V^+$ such that $\Gamma^+(D') = V^+$ and $|\Gamma^-(D')| \leq k$.

Hence, DS is reducible in polynomial time to both Problems 3 and 4 and the following result holds.

Theorem 4. *The MDD problem is NP-hard.*

For the mNDD problem, noticing that the above reduction is gap preserving, we have the following result.

Theorem 5. *There is no polynomial time approximation algorithm with any constant factor better than $\log |V|$ for the mNDD problem unless P=NP.*

4.3 Open Problem

From the above, we have that it is a natural question to ask if a logarithmic approximation algorithm can be devised for the $mNDD$ problem.

References

1. Abebe, R., Adamic, L., Kleinberg, J.: Mitigating overexposure in viral marketing. In: The Thirty-Second AAAI Conference on Artificial Intelligence (AAAI 2018), pp. 241–248 (2018)
2. Berger, J., Heath, C.: Where consumers diverge from others: identity signaling and product domains. J. Consum. Res. **34**(2), 121–134 (2007)
3. Byers, J.W., Mitzenmacher, M., Zervas, G.: The groupon effect on yelp ratings: a root cause analysis. In: Proceedings of 13th ACM Conference on Electronic Commerce, EC 2012, pp. 248–265 (2012)
4. Chen, Y., Li, H., Qu, Q.: Negative-aware influence maximization on social networks. In: Proceedings of AAAI Conference on Artificial Intelligence (2018)

5. Cordasco, G., Gargano, L., Rescigno, A.A., Vaccaro, U.: Optimizing spread of influence in social networks via partial incentives. In: Scheideler, C. (ed.) Structural Information and Communication Complexity. LNCS, vol. 9439, pp. 119–134. Springer, Cham (2015). https://doi.org/10.1007/978-3-319-25258-2_9

6. Cordasco, G., Gargano, L., Rescigno, A.A.: Influence propagation over large scale social networks. In: Proceedings of ASONAM 2015, pp. 1531–1538 (2015)

7. Cordasco, G., et al.: Whom to befriend to influence people. Theoret. Comput. Sci. (to appear)

8. Cordasco, G., Gargano, L., Mecchia, M., Rescigno, A.A., Vaccaro, U.: Discovering small target sets in social networks: a fast and effective algorithm. Algorithmica 80(6), 1804–1833 (2018)

9. Cordasco, G., Gargano, L., Rescigno, A.A.: Active influence spreading in social networks. Theoret. Comput. Sci. 764, 15–29 (2019)

10. Cordasco, G., Gargano, L., Rescigno, A.A., Vaccaro, U.: Evangelism in social networks: algorithms and complexity. Networks 71(4), 346–357 (2018)

11. Cravens, K., Goad Oliver, E., Ramamoorti, S.: The reputation index: measuring and managing corporate reputation. Eur. Manag. J. 21(2), 201–212 (2003)

12. Cretu, A.E., Brodie, R.J.: The influence of brand image and company reputation where manufacturers market to small firms: a customer value perspective. Ind. Mark. Manag. 36(2), 230–240 (2007)

13. Delbot, F., Laforest, C., Phan, R.: Graphs with forbidden and required vertices. In: 17th Rencontres Francophones sur les Aspects Algorithmiques des Télécom (ALGOTEL 2015) (2015)

14. Dunbar, J.E., Hedetniemi, S.T., Henning, M.A., Slater, P.J.: Signed domination in graphs. In: Proceedings of the Seventh International Conference in Graph Theory, Combinatorics, Algorithms, and Applications, pp. 311–321 (1995)

15. Dunbar, J., Hedetniemi, S., Henning, M.A., McRae, A.: Minus domination in graphs. Discret. Math. 199(1), 35–47 (1999)

16. Easley, D., Kleinberg, J.: Networks, Crowds, and Markets: Reasoning About a Highly Connected World. Cambridge University Press, New York (2010)

17. Haynes, T.W., Hedetniemi, S., Slater, P.: Fundamentals of Domination in Graphs (Pure and Applied Mathematics (Marcel Dekker)). CRC, Boca Raton (1998)

18. Hu, Y., Van den Bulte, C.: Nonmonotonic status effects in new product adoption. Mark. Sci. 33(4), 509–533 (2014)

19. Joshi, Y.V., Reibstein, D.J., Zhang, Z.J.: Turf wars: product line strategies in competitive markets. Mark. Sci. 35(1), 128–141 (2016)

20. Khuller, S., Moss, A., Naor, J.: The budgeted maximum coverage problem. Inf. Process. Lett. 70(1), 39–45 (1999)

21. Li, Y., Chen, W., Wang, Y., Zhang, Z.-L.: Influence diffusion dynamics and influence maximization in social networks with friend and foe relationships. In: Proc. of WSDM 2013, pp. 657–666 (2013)

22. Liu, X., Kong, X., Yu, P.S.: Active opinion maximization in social networks. In: Proceedings of the 24th ACM SIGKDD International Conference on Knowledge Discovery & Data Mining, (KDD 2018) (2018)

23. Manurangsi, P.: Almost-polynomial ratio ETH-hardness of approximating densest K-subgraph. In: Proceedings of the 49th ACM SIGACT Symposium on Theory of Computing (STOC 2017), pp. 954–961 (2017)

24. Miyano, E., Ono, H.: Maximum domination problem. In: Proceedings of the 17th Computing on The Australasian Theory Symposium (CATS 2011), vol. 119, pp. 55–62 (2011)

25. Neiheisel, J.R., Niebler, S.: On the limits of persuasion: campaign ads and the structure of voters' interpersonal discussion networks. Polit. Commun. **32**(3), 434–452 (2015)
26. Nemhauser, G.L., Wolsey, L.A., Fisher, M.L.: An analysis of approximations for maximizing submodular set functions-I. Math. Prog. **14**(1), 265–294 (1978)
27. Sviridenko, M.: A note on maximizing a submodular set function subject to a knapsack constraint. Oper. Res. Lett. **32**(1), 41–43 (2004)
28. Wakefield, M., Flay, B., Nichter, M., Giovino, G.: Effects of anti-smoking advertising on youth smoking: a review. J. Health Comm. **8**(3), 229–247 (2003)

Reaching 3-Connectivity
via Edge-Edge Additions

Giordano Da Lozzo[1]([⊠]) and Ignaz Rutter[2]

[1] Roma Tre University, Rome, Italy
giordano.dalozzo@uniroma3.it
[2] University of Passau, Passau, Germany
rutter@fim.uni-passau.de

Abstract. Given a graph G and a pair $\langle e', e'' \rangle$ of distinct edges of G, an *edge-edge addition on* $\langle e', e'' \rangle$ is an operation that turns G into a new graph G' by subdividing edges e' and e'' with a dummy vertex v' and v'', respectively, and by adding the edge (v', v''). In this paper, we show that any 2-connected simple planar graph with minimum degree $\delta(G) \geq 3$ and maximum degree $\Delta(G)$ can be augmented by means of edge-edge additions to a 3-connected planar graph G' with $\Delta(G') = \Delta(G)$, where each edge of G participates in *at most one* edge-edge addition. This result is based on decomposing the input graph into its 3-connected components via SPQR-trees and on showing the existence of a planar embedding in which edge pairs from a special set share a common face. Our proof is constructive and yields a linear-time algorithm to compute the augmented graph.

As a relevant application, we show how to exploit this augmentation technique to extend some classical NP-hardness results for bounded-degree 2-connected planar graphs to bounded-degree 3-connected planar graphs.

1 Introduction

Many computational problems on planar graphs drastically change their complexity when the degree of connectivity increases. This turns out to be particularly relevant for graph embedding problems, where connectivity plays a key role. In fact, while simply-connected and 2-connected planar graphs may admit exponentially-many different combinatorial embeddings, a celebrated result by Whitney [15] states that a 3-connected planar graph admits a unique embedding (up to a flip). As an example, while testing whether a planar digraph admits an *upward-planar drawing*, i.e., a planar drawing in which edges are drawn as y-monotone curves, or a planar graph admits a *rectilinear-planar drawing*, i.e.,

This research was supported in part by MIUR Project "MODE" under PRIN 20157EFM5C, by MIUR Project "AHeAD" under PRIN 20174LF3T8, by H2020-MSCA-RISE project 734922 – "CONNECT", by MIUR-DAAD JMP N° 34120, and by Ru 1903/3-1 of the German Science Foundation (DFG).

C. J. Colbourn et al. (Eds.): IWOCA 2019, LNCS 11638, pp. 175–187, 2019.
https://doi.org/10.1007/978-3-030-25005-8_15

a planar drawing in which edges are drawn has either horizontal or vertical segments, are NP-complete problems for 2-connected planar graphs [11], these problems become polynomial-time solvable for 3-connected planar graphs [4,14].

When proving NP-hardness results for graph embedding problems on planar graphs, it is often convenient to restrict the embedding choices of parts of a construction by building a rigid structure or a structure with a fixed embedding. To this aim, the most frequent approach is to reduce from problems that are NP-complete for subclasses of 3-connected planar graphs, due to the uniqueness of their combinatorial embedding.

Bounds on the maximum degree have also been frequently considered when studying the complexity of graphs visualization problems. For instance, while testing for the existence of an *orthogonal drawing*, i.e., a planar drawing where vertices are mapped to grid points and edges are mapped to paths on the grid, with the minimum number of bends is NP-complete for maximum degree-4 graphs [11], it is instead polynomial-time solvable for maximum degree-3 graphs [9]. Clearly, when devising reductions to prove NP-hardness results for problems in bounded-degree planar graphs, it is very helpful to reduce from graph problems that are hard even when the maximum degree is bounded.

Contributions. In this paper, we introduce a novel augmentation technique for 2-connected planar graphs based on the existence of a special set of pairwise-disjoint edge pairs (Theorem 1). We illustrate how to exploit this technique to obtain strong NP-hardness results for planar graphs that are both 3-connected and have bounded maximum degree, by giving simple proofs for some classical NP-hard problems for these classes of graphs. Namely, we show that MAXIMUM INDEPENDENT SET (Theorem 2) and STEINER TREE (Theorem 5) are NP-complete even for 3-connected cubic planar graphs, and that 3-COLORING is NP-complete even for 3-connected planar graphs of maximum degree 8 (Theorem 4).

For space limits, proofs marked with ($\star$) can be found in the full version [6].

2   Preliminaries

The graphs considered in this paper are finite, simple, and undirected. We assume familiarity with basic concepts about vertex connectivity, graph embeddings, and planarity. We also provide all the required definitions in the full version [6].

Graph Operations. Let $G = (V, E)$ be a planar graph with two distinct designated edges $e', e'' \in E$. An *expansion* on the pair $\langle e' = (u, v), e'' = (x, y) \rangle$ turns G into a new graph G' by replacing edges e' and e'' with a given connected graph $\mathcal{A}$, called *gadget*, containing four vertices each of which is identified with one of $\{u, v, x, y\}$. The expansion is *planar* if G' is planar; see Fig. 1 for some examples. Clearly, for an expansion to be planar $\mathcal{A}$ must also be planar and G must admit an embedding in which e' and e'' are incident to the same face. Gadget $\mathcal{A}$ is k-*stable* if G' is k-connected and G' has no k-cut $X \subset V(\mathcal{A})$ such that $G' - X$ contains vertices of $\mathcal{A}$. We also say that the planar expansion based on $\mathcal{A}$ is k-*stable*.

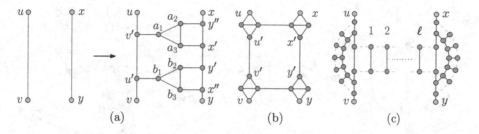

Fig. 1. Planar expansions: Gadget $\mathcal{A}_0$ (a), $\mathcal{A}_1$ (b), and $\mathcal{L}_\ell$ (c), resp., for the NP-hardness of MIS (Theorem 2), 3-COLORING (Theorem 4), and ST (Theorem 5).

Barnette and Grünbaum gave a complete set of operations to construct any 3-connected graph from K_4 [3,13]. We are going to exploit one of these operations, which is also a simple planar expansion, that is defined as follows: An *edge-edge addition* on the pair $\langle e', e'' \rangle$ is a planar expansion that turns G into a new planar graph G' by subdividing edges e' and e'' with a dummy vertex v' and v'', respectively, and by adding the edge (v', v''). Clearly, an edge-edge addition is 2-stable.

SPQR-Trees. We consider uv-graphs with two special *pole* vertices u and v, which can be constructed in a fashion very similar to series-parallel graphs. Namely, an edge (u, v) is a uv-graph with poles u and v. Now, let G_i be a uv-graph with poles u_i, v_i, for $i = 1, \ldots, k$, and let H be a planar graph with two designated vertices u and v and $k + 1$ edges $uv, e_1, \ldots, e_k$. We call H the *skeleton* of the composition and its edges are called *virtual edges*; the edge uv is the *parent edge* and u and v are the poles of the skeleton H. To compose the G_i's into a uv-graph with poles u and v, we remove the edge uv and replace each e_i by G_i, for $i = 1, \ldots, k$, by removing e_i and identifying the poles of G_i with the endpoints of e_i. In fact, we only allow three types of compositions: in a *series composition* the skeleton H is a cycle of length $k + 1$, in a *parallel composition* H consists of two vertices connected by $k + 1$ parallel edges, and in a *rigid composition* H is 3-connected.

It is known that for every 2-connected graph G with an edge uv, the graph $G - uv$ is a uv-graph with poles u and v. Much in the same way as series-parallel graphs, the uv-graph $G - uv$ gives rise to a (de-)composition tree T describing how it can be obtained from single edges. The nodes of T corresponding to edges, series, parallel, and rigid compositions of the graph are Q-, S-, P-, and R-nodes, respectively. To obtain a composition tree for G, we add an additional root Q-node representing the edge uv. To fully describe the composition, we associate with each node μ its skeleton denoted by $\mathrm{skel}(\mu)$. For a node μ of T, the *pertinent graph* $\mathrm{pert}(\mu)$ is the subgraph represented by the subtree with root μ.

Let μ be a node of T, we denote the poles of μ by u_μ and v_μ. In this paper, we will assume the edge (u_μ, v_μ) to be part of both $\mathrm{skel}(\mu)$ and $\mathrm{pert}(\mu)$. Let $\mathcal{E}_\mu$ be an embedding of $\mathrm{pert}(\mu)$. The *left* (resp., *right*) *outer face* of $\mathcal{E}_\mu$ is the face that is to the right (resp., to the left) of the edge (u_μ, v_μ) when traversing this edge from u_μ to v_μ. The *outer face* of $\mathcal{E}_\mu$ is the one obtained from $\mathcal{E}_\mu$ after removing

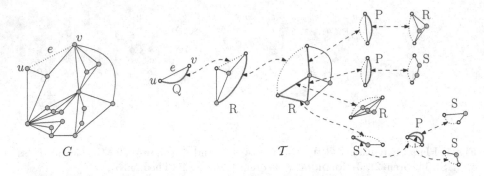

Fig. 2. (left) A 2-connected planar graph G and (right) the SPQR-tree $\mathcal{T}$ of G rooted at the edge $e = uv$. Skeletons of non-leaf nodes of $\mathcal{T}$ are depicted; virtual edges corresponding to edges of G are thin, whereas virtual edges corresponding to S-, P-, and R-nodes are thick blue curves. Dashed arrowed curves connect the (dotted) parent edge in the skeleton of a child node with the virtual edge representing the child node in the skeleton of its parent. (Color figure online)

the edge (u_μ, v_μ), i.e., the open region obtained as the union of the left outer face of $\mathcal{E}_\mu$, of the right outer face of $\mathcal{E}_\mu$, and of the edge (u_μ, v_μ).

The *SPQR-tree* of G with respect to the edge uv, originally introduced by Di Battista and Tamassia [8], is the (unique) smallest decomposition tree $\mathcal{T}$ for G; refer to Fig. 2 for an example. Using a different edge $u'v'$ of G and a composition of $G - u'v'$ corresponds to re-rooting $\mathcal{T}$ at the node representing $u'v'$. It thus makes sense to say that $\mathcal{T}$ is the SPQR-tree of G. The SPQR-tree of G has size linear in the size of G and can be computed in linear time [12]. Planar embeddings of G correspond bijectively to planar embeddings of all skeletons of $\mathcal{T}$; the choices are the orderings of the parallel edges in P-nodes and the embeddings of the R-node skeletons, which are unique up to a flip [15]. When considering rooted SPQR-trees, we assume that the embedding of G is such that the root edge is incident to the outer face, which is equivalent to the parent edge being incident to the outer face in each skeleton. Hence, we only consider embeddings of the skeletons and of the pertinent graphs of each node with their poles lying on the outer face.

Canonical Ordering. Let $G = (V, E)$ be a 3-connected plane graph with vertices v_2, v_1, and v_n in this clockwise order along the outer face of G. Let $\pi = (P_1, \ldots, P_k)$ be an ordered partition of V into paths, where $P_1 = (v_1, v_2)$ and $P_k = (v_n)$. Define G_i to be the subgraph of G induced by $P_1 \cup \ldots \cup P_i$, and denote by C_i the boundary of the outer face of G_i. We say that π is a *canonical ordering* [10] for G if:

- each C_i with $i > 1$ is a cycle containing the edge (v_1, v_2);
- each G_i is 2-connected and *internally 3-connected*, that is, removing two interior vertices of G_i does not disconnect it; and
- for each $i \in \{2, \ldots, k-1\}$, one of the two following conditions holds:

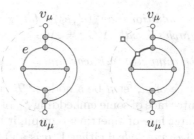

Fig. 3. (left) Pertinent graph pert(μ) of an e-externally 3-connectible S-node μ and (right) the subdivision of a 3-connected planar graph obtained by performing an edge-edge addition on $\langle e, (u_\mu, v_\mu) \rangle$ in pert(μ).

(a) P_i is a singleton $\{z\}$, where z belongs to C_i and has at least one neighbor in $G \setminus G_i$.

(b) P_i is a chain $\{z_1, \ldots, z_l\}$, where each z_j has at least one neighbor in $G \setminus G_i$, and where z_1 and z_l each have one neighbor on C_{i-1}, and these are the only two neighbors of P_i in G_{i-1}.

Observe that, if P_i is a chain, then the two neighbors of z_1 and z_l on C_{i-1} are adjacent in C_{i-1}.

NP-Hard Problems. An *independent set* in a graph $G = (V, E)$ is a subset $V' \subseteq V$ of pairwise non-adjacent vertices. The MAXIMUM INDEPENDENT SET (MIS) problem asks for a maximum-size independent set of a graph.

A 3-*coloring* of a graph $G = (V, E)$ is an assignment $c: V \to \{1, 2, 3\}$ such that $c(u) \neq c(v)$, for every edge $(u, v) \in E$. The 3-COLORING problem asks whether a given graph admits a 3-coloring.

Let (G, T, k) be a triple where $G = (V, E)$ is a graph, $T \subseteq V$ is a subset of the vertices of G, called *terminals*, and k is a positive integer. The STEINER TREE (ST) problem asks whether (G, T) admits a *Steiner tree*, i.e., a subtree of G containing all the terminals in T, consisting of at most k edges.

3 Augmentation Technique

In this section, we present an algorithm to augment in linear time any 2-connected planar graph G with minimum degree $\delta(G) \geq 3$ and maximum degree $\Delta(G)$ to a 3-connected planar graph G' with $\Delta(G') = \Delta(G)$, by means of edge-edge additions on edge pairs of G, where each edge participates in *at most one* edge-edge addition. More formally, we prove the following.

Theorem 1. *Let $G = (V, E)$ be a 2-connected planar graph with minimum degree $\delta(G) \geq 3$ and maximum degree $\Delta(G)$. There exist pairs $\langle e'_1, e''_1 \rangle$, $\langle e'_2, e''_2 \rangle, \ldots, \langle e'_k, e''_k \rangle$ of distinct edges in E such that:*

(i) any two pairs $\langle e'_i, e''_i \rangle$ and $\langle e'_j, e''_j \rangle$, with $1 \leq i < j \leq k$, are pairwise disjoint, i.e., it holds $\{e'_i, e''_i\} \cap \{e'_j, e''_j\} = \emptyset$, and

(ii) *performing edge-edge additions on* $\langle e'_i, e''_i \rangle$, *for* $i = 1, \ldots, k$, *yields a 3-connected planar graph* G' *with* $\Delta(G') = \Delta(G)$.

Further, such pairs can be computed in linear time.

Let $\mathcal{T}$ be the SPQR-tree of G, let μ be a node of $\mathcal{T}$, and let e be an edge of pert(μ) incident to the outer face of some embedding $\mathcal{E}_\mu$ of pert(μ). Observe that, by the definition of the outer face of a pertinent graph, it holds that $e \neq (u_\mu, v_\mu)$. Then, $\mathcal{E}_\mu$ is *e-externally 3-connectible* if either 1. μ is a Q-node, i.e., pert$(\mu) = e$, or 2. the graph obtained from pert(μ) by performing an edge-edge addition on $\langle e, (u_\mu, v_\mu) \rangle$ is a subdivision of a 3-connected planar graph whose only degree-2 vertices, if any, are the poles u_μ and v_μ of μ; refer to Fig. 3. Also, we say that μ is *e-externally 3-connectible* (or, simply, *externally 3-connectible*) if pert(μ) admits an e-externally 3-connectible embedding for some edge e of pert(μ).

Let μ be a non-Q node of $\mathcal{T}$, let e be an edge of pert(μ), and let pert$^*(\mu)$ be a planar graph obtained by means of edge-edge additions on pairwise-disjoint edge pairs of pert(μ). If e has not been used in any edge-edge addition, i.e., $e \in E(\text{pert}(\mu)) \setminus E(\text{pert}^*(\mu))$, we say that e is a *free edge*.

Definition 1. *Let* $L_\mu = [e_1, e_2]$ *and* $R_\mu = [e_3]$ *be two lists of free edges of* pert$^*(\mu)$ *and let* $\mathcal{E}^*_\mu$ *be an embedding of* pert$^*(\mu)$. *We say that the 4-tuple* $\langle \text{pert}^*(\mu), \mathcal{E}^*_\mu, L_\mu, R_\mu \rangle$ *is extensible if (i)* $L_\mu \cap R_\mu = \emptyset$, *(ii) the free edges in* L_μ *and in* R_μ *belong to distinct faces of* $\mathcal{E}^*_\mu$ *that are incident to the parent edge* (u_μ, v_μ) *of* μ, *i.e., the left and the right outer face of* $\mathcal{E}^*_\mu$, *and (iii)* $\mathcal{E}^*_\mu$ *is e-externally 3-connectible for any* $e \in L_\mu$.

Without loss of generality, by possibly flipping the embedding, we may further assume that the free edges in L_μ and in R_μ are incident to the left and to the right outer face of $\mathcal{E}^*_\mu$, respectively. Observe that, once pert$^*(\mu)$, L_μ, and R_μ have been fixed, there exists a unique (up to a flip) embedding $\mathcal{E}^*_\mu$ of pert$^*(\mu)$ such that $\langle \text{pert}^*(\mu), \mathcal{E}^*_\mu, L_\mu, R_\mu \rangle$ is extensible. Hence, to simplify the notation, in the following we will omit to specify the embedding of pert$^*(\mu)$ in an extensible tuple.

If μ is a Q-node representing edge $e = (u_\mu, v_\mu)$, then we also say that the triple $\langle \text{pert}^*(\mu) = e, L_\mu = [e], R_\mu = [e] \rangle$ is *extensible*, thus allowing $|L_\mu| = 1$ and $L_\mu \cap R_\mu \neq \emptyset$ in this case.

Let μ be an internal node of $\mathcal{T}$ and let f be a face of an embedding of skel(μ). Consider the counter-clockwise sequence $(e_{\mu_1}, e_{\mu_2}, \ldots, e_{\mu_{|f|}})$ of virtual edges of skel(μ) incident to f. This sequence induces a natural circular order $\mathcal{O}_f$ for the free edges in $\bigcup_{i=1}^{|f|} L_{\mu_i}$. Further, order $\mathcal{O}_f$ induces a unique linear order for each list L_{μ_i} such that the free edges in the list are consecutive in $\mathcal{O}_f$. In the following, we assume the lists L_{μ_i}'s be ordered according to such a linear order; see, e.g., Fig. 4a.

Our strategy to prove Theorem 1 is as follows. We root the SPQR-tree $\mathcal{T}$ of G at an arbitrary Q-node ρ whose unique child ξ is an R-node. Observe that such a Q-node exists, since $\delta(G) \geq 3$ and G is simple, and that the poles of ξ have degree at least 2 in pert(ξ). We process the nodes of $\mathcal{T}$ bottom-up and show how to compute, for each non-root node μ with children $\mu_1, \ldots, \mu_k$, an extensible

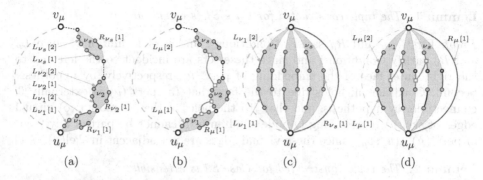

Fig. 4. Illustration for the proof of Theorem 1 when μ is an S-node with three non-Q-node children ν_1, ν_2, and ν_3 ((a),(b)), and a P-node with a Q-node child ((c),(d)); free edges are thick red. ((a),(c)) Graphs $\text{pert}^+(\mu)$. ((b),(d)) Obtaining $\text{pert}^*(\mu)$ from $\text{pert}^+(\mu)$ via edge-edge additions.

triple $\langle \text{pert}^*(\mu), L_\mu, R_\mu \rangle$, starting from the extensible triples $\langle \text{pert}^*(\mu_i), L_{\mu_i}, R_{\mu_i} \rangle$ of its children. When we reach the root ρ, since the triple $\langle \text{pert}^*(\xi), L_\xi, R_\xi \rangle$ is extensible and since the poles of ξ have degree at least 2 in $\text{pert}^*(\xi)$, performing the edge-edge addition on $\langle (u_\rho, v_\rho), e \in L_\xi \rangle$ in the graph $\text{pert}^*(\xi)$ clearly yields a 3-connected planar graph G' obtained via edge-edge additions on pairwise-disjoint edge pairs of G.

If μ is a leaf Q-node representing edge $e = (u_\mu, v_\mu)$, then there is nothing to be done, as the triple $\langle pert^*(\mu) = e, L_\mu = [e], R_\mu = [e] \rangle$ is extensible by definition.

We now show how to compute extensible triples for each internal node $\mu \in \mathcal{T}$.

S-nodes. Suppose that μ is an S-node with children $\mu_1, \ldots, \mu_k$. Recall that, since μ is an internal node of $\mathcal{T}$, it has at least two children; also, since $\delta(G) \geq 3$, no two virtual edges corresponding to Q-node children of μ are adjacent in $\text{skel}(\mu)$. Hence, μ has at least a non-Q-node child ν. We distinguish two cases based on whether μ has exactly one (Case S1) or more than one (Case S2) non-Q-node children.

Case S1. Let ν be the unique non-Q-node child of μ and let e_ν be the virtual edge representing ν in $\text{skel}(\mu)$. We set $\text{pert}^*(\mu)$ to be the graph obtained from $\text{skel}(\mu)$ by replacing e_ν with $\text{pert}^*(\nu)$, and by setting $L_\mu = L_\nu$ and $R_\mu = R_\nu$.

Case S2. Let $\nu_1, \ldots, \nu_s$ be the non-Q-node children of μ ordered as the corresponding virtual edges appear in $\text{skel}(\mu)$ from u_μ to v_μ. First, we construct an auxiliary embedded graph $\text{pert}^+(\mu)$ starting from $\text{skel}(\mu)$, by replacing each virtual edge e_{ν_i} in $\text{skel}(\mu)$ with $\text{pert}^*(\nu_i)$, for $i = 1, \ldots, s$; where the replacement is performed in such a way that the free edges in lists $L[\nu_i]$'s (resp. $R[\nu_i]$'s) are incident to the left outer face (resp. the right outer face) of the embedding (see Fig. 4a). Then, we obtain $\text{pert}^*(\mu)$ from $\text{pert}^+(\mu)$ by performing edge-edge additions on $\langle L_{\nu_i}[2], L_{\nu_{i+1}}[1] \rangle$, for $i = 1, \ldots, s - 1$ (see Fig. 4b). Finally, we set $L_\mu = [L_{\nu_1}[1], L_{\nu_s}[2]]$ and $R_\mu = R_{\nu_1}$.

Lemma 1. *The tuple constructed for Case S1 is extensible.*

Proof. First, L_μ and R_μ satisfy Conditions i and ii of Definition 1, since L_ν and R_ν satisfy Condition i and since these sets are incident to the left and to the right outer face of the embedding of pert*(ν), respectively, by hypothesis. Second, Condition iii holds due to the fact that (a) pert*(ν) is externally 3-connectible, by hypothesis, and that (b) the poles of the (at most two) virtual edges of skel(μ) representing Q-node children of μ cannot be part of any 2-cut of pert*$(\mu) \setminus (u_\mu, v_\mu)$, since these virtual edges are not adjacent in skel(μ). □

Lemma 2. *The tuple constructed for Case S2 is extensible.*

Proof. Condition i of Definition 1 holds since $L_{\nu_1} \cap R_{\nu_1} = \emptyset$, by hypothesis, and since $(L_{\nu_1} \cup R_{\nu_1}) \cap L_{\nu_s} = \emptyset$, due to the fact that the children of μ are edge-disjoint. Condition ii holds since edges in L_{ν_1} and L_{ν_s} are incident to the same face, by construction. To see that Condition iii holds, observe that each edge-edge addition on $\langle L_{\nu_i}[2], L_{\nu_{i+1}}[1]\rangle$ has the effect of turning two non-Q-node children ν_i and ν_{i+1} of μ, together with the unique Q-node child of μ possibly separating them in skel(μ), into an externally 3-connectible node; see, e.g., nodes ν_2 and $\nu_3 = \nu_s$ in Fig. 4b. Hence, at the end of the augmentation, there might exist at most two non-adjacent virtual edges representing Q-node children of μ incident to the poles of μ, whose poles however do not contribute to any 2-cut of pert*$(\mu) \setminus (u_\mu, v_\mu)$. □

P-nodes. Suppose that μ is a P-node. Note that, since G is simple, μ has at most one Q-node child. First, we select an arbitrary embedding $\mathcal{H}$ of skel(μ) such that the unique Q-node child of μ, if any, is incident to the right outer face of $\mathcal{H}$. Let $\nu_1, \ldots, \nu_s$ be the left-to-right ordering of the non-Q-node children of μ in $\mathcal{H}$, where ν_1 is the child of μ whose corresponding virtual edge is incident to the left outer face of $\mathcal{H}$. Second, we construct an auxiliary embedded graph pert$^+(\mu)$ starting from $\mathcal{H}$, by replacing each virtual edge e_{ν_i} in skel(μ) with pert*(e_{ν_i}), for $i = 1, \ldots, s$; where the replacement is performed in such a way that the free edges in $L[\nu_1]$ are incident to the left outer face of the constructed embedding and that, for any two consecutive children ν_i and ν_{i+1}, it holds that $R[\nu_i]$ and $L[\nu_{i+1}]$ are incident to a common face (see Fig. 4c). Third, we obtain pert*(μ) from pert$^+(\mu)$ by performing edge-edge additions on $\langle R_{\nu_i}[1], L_{\nu_{i+1}}[2]\rangle$, for $i = 1, \ldots, s-1$ (see Fig. 4d). Finally, we set $L_\mu = L_{\nu_1}$ and $R_\mu = R_{\nu_s}$, if there exists no Q-node child of μ, or $R_\mu = [(u_\mu, v_\mu)]$, otherwise.

Lemma 3. *The tuple constructed in the case in which μ is a P-node is extensible.*

Proof. If μ has no Q-node child, then Condition i of Definition 1 holds since $L_{\nu_1} \cap R_{\nu_s} = \emptyset$ (due to the fact that the children of μ are edge-disjoint); otherwise, it holds since $(u_\mu, v_\mu) \notin$ pert*(ν_1). Condition ii holds since, in the constructed embedding, we have that (a) the free edges in L_{ν_1} are incident to the left outer face, and that (b) the edges in R_{ν_s} are incident to the right outer face, if μ has no Q-node child, or edge (u_μ, v_μ) is incident to the right outer face, otherwise.

To see that also Condition iii holds, it suffices to observe that the performed edge-edge additions merge each two consecutive externally 3-connectible children into a new externally 3-connectible node. Thus, the constructed triple is extensible.   □

R-nodes. Suppose that μ is an R-node. Assume that μ has at least a non-Q-node child, as otherwise $\text{pert}(\mu)$ is a 3-connected planar graph and it is trivial to define an extensible triple. To simplify the description of this case, we remove the parent edge from $\text{skel}(\mu)$, which is hence only internally 3-connected. Let $\mathcal{H}$ be the unique (up to a flip) embedding of $\text{skel}(\mu)$ and let $\pi = (P_1 = (u_\mu, v_2), P_2, \dots, P_k = v_\mu)$ be a canonical ordering of $\text{skel}(\mu)$, where v_2 is the neighbor of u_μ different from v_μ that is incident to the outer face of $\mathcal{H}$. Also, let $\text{skel}_i(\mu)$, with $i = 2, \dots, k$, be the embedded subgraph of $\text{skel}(\mu)$ induced by vertices $\bigcup_{j=1}^i P_j$ and C_i be the cycle bounding the outer face of $\text{skel}_i(\mu)$. For $i = 2, \dots, k$, we denote by P_i' the path $(v_{i,1}, P_i, v_{i,s})$, where $v_{i,1}$ and $v_{i,s}$ are the neighbors of the first and of the last vertex of P_i in $\text{skel}_{i-1}(\mu)$, respectively. Recall that, since π is a canonical ordering, $\text{skel}_i(\mu)$ is internally 3-connected. Further, let $\text{pert}_i^+(\mu)$, with $i = 2, \dots, k$, be the embedded graph inductively defined as follows. Graph $\text{pert}_2^+(\mu)$ is the embedded planar graph obtained from cycle $\text{skel}_2(\mu)$, by replacing each virtual edge e_h corresponding to a non-Q-node child ν_h of μ with $\text{pert}^*(\mu_h)$; where the replacement is performed in such a way that the free edges in R_{ν_h} are incident to the outer face of the resulting embedding (see Figs. 5a and b). For $i = 3, \dots, k$, graph $\text{pert}_i^+(\mu)$ is the embedded planar graph obtained from $\text{pert}_{i-1}^+(\mu) \cup P_i'$, by replacing each virtual edge $e_h \in P_i'$ corresponding to a non-Q-node child ν_h of μ with $\text{pert}^*(\mu_h)$.

Our proof strategy is as follows. We show how to augment $\text{pert}_i^+(\mu)$ to an embedded planar graph $\text{pert}_i^*(\mu)$ via edge-edge additions so that the following two properties hold: (a) the outer face of $\text{pert}_i^*(\mu)$ contains a set F_i of $|C_i|$ free edges, each of which is separated by two consecutive vertices of $\text{skel}_i(\mu)$, and (b) $\text{pert}_i^*(\mu)$ is internally 3-connected and its 2-cuts, if any, are 2-cuts of $\text{skel}_i(\mu)$ as well. This allows us to construct an extensible tuple for μ as follows. We set $\text{pert}^*(\mu) = \text{pert}_k^*(\mu)$, and initialize L_μ and R_μ by selecting edges in F_k incident to the left and to the right outer face of $\text{pert}_k^*(\mu)$, respectively. Since the paths incident to the left and to the right outer face of $\text{skel}_k(\mu)$ are edge-disjoint and have length at least 2 (due to the fact that μ is an R-node), then Property (a) allows us to initialize L_μ and R_μ so to satisfy Conditions i and ii of Definition 1. Also, Conditions iii of Definition 1 is immediately satisfied, as adding the edge (u_μ, v_μ) to $\text{pert}^*(\mu)$ yields a 3-connected planar graph, due to Property (b).

The augmentation is done by induction on i. We first assume that the child of μ corresponding to the virtual edge (u_μ, v_2) is a Q-node. Then, we show how to drop this assumption by, possibly, performing an extra edge-edge addition.

The base case is $i = 2$; refer to Fig. 5. To obtain $\text{pert}_2^*(\mu)$ we proceed as follows. We initialize $\text{pert}_2^*(\mu) = \text{pert}^+(\mu)$. Then, we perform edge-edge additions on $\langle L_{\nu_i}[2], L_{\nu_{i+1}}[1] \rangle$, for $i = 1, \dots, s - 1$, where we assume that the virtual edge corresponding to ν_i precedes the virtual edge corresponding to ν_{i+1} when traversing P_2' from $v_{2,1}$ to $v_{2,s}$ (see Fig. 5c). Clearly, $\text{pert}_2^*(\mu)$ satisfies Properties (a) and (b).

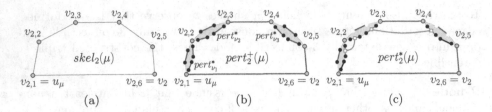

Fig. 5. Graphs for the base case of the proof of Theorem 1 when μ is an R-node.

In the inductive case $2 < i \leq k$, and we assume to have already computed graph $\text{pert}^*_{i-1}(\mu)$. Recall that, $\text{pert}^*_{i-1}(\mu)$ is obtained from $\text{pert}^+_{i-1}(\mu)$ via edge-edge additions so that Properties (a) and (b) are satisfied, by the inductive hypothesis. Let P_i be the i-th path in π and suppose $|P_i| > 1$, i.e., P_i is not a single vertex; the case $|P_i| = 1$ can be treated similarly. Let $P'_i = (v_{i,1}, \dots, v_{i,s})$ be the path in $\text{skel}_i(\mu)$ associated with P_i. Note that, since π is a canonical ordering, there exists a virtual edge $e_\ell = (v_{i,1}, v_{i,s})$ that is incident to the outer face of $\text{skel}_{i-1}(\mu)$. Let ν_ℓ be the child of μ corresponding to the virtual edge e_ℓ in $\text{skel}(\mu)$. Let e be a free edge along the outer face of $\text{pert}^*_{i-1}(\mu)$ belonging to $\text{pert}^*(\nu_\ell)$ (which exists by Property (a)). We construct $\text{pert}^*_i(\mu)$ as follows. First, initialize $\text{pert}^*_i(\mu) = \text{pert}^*_{i-1}(\mu)$. Second, we add to $\text{pert}^*_i(\mu)$ all the vertices and edges of P'_i and embed them planarly in the outer face of $\text{pert}^*_i(\mu)$ (see Fig. 6a). Third, we replace each virtual edge in P'_i corresponding to a non-Q-node child ν_h of μ with $\text{pert}^*(\nu_h)$ (see Fig. 6b). Finally, we perform an edge-edge addition on $\langle L_{\nu_1}[1], e \rangle$ and edge-edge additions on $\langle L_{\nu_i}[2], L_{\nu_{i+1}}[1] \rangle$, for $i = 1, \dots, s-1$ (see Fig. 6c). Thus, $\text{pert}^*_i(\mu)$ is obtained from $\text{pert}^+_i(\mu)$ via edge-edge additions, and satisfies Properties (a) and (b).

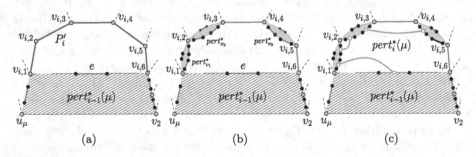

Fig. 6. Graphs for the inductive case of Theorem 1 when μ is an R-node.

To complete the proof of Theorem 1, we need to show that the child σ of μ corresponding to the virtual edge (u_μ, v_2) need not be a Q-node. Observe that in the construction, we did not make use of the free edges of $\text{pert}^*(\sigma)$ in any edge-edge addition. Thus, performing an edge-edge addition on $\langle L_\sigma[2], R_\tau[1] \rangle$, where τ is the child of μ whose corresponding virtual edge is incident to u_μ and to the right outer face of $\text{skel}(\mu)$ yields a graph $\text{pert}^*(\mu)$ satisfying Properties (a) and (b).

Lemma 4 ($\star$). *For $2 < i \leq k$, graph $\text{pert}_i^*(\mu)$ constructed in the case in which μ is an R-node satisfies Properties (a) and (b).*

By Lemmas 1 to 4, we can therefore compute extensible triples for all non-root nodes of $\mathcal{T}$. We now argue about the running time of the presented algorithm. First, the tree $\mathcal{T}$ has size linear in the size of G and can be computed in linear time [12]. Second, a canonical ordering of $\text{skel}(\mu)$ can be computed in $O(|\text{skel}(\mu)|)$ time [10]. Third, by simply storing the free edges involved in each edge-edge addition without actually performing such a planar expansion, we can process each node μ of $\mathcal{T}$ in $O(|\text{skel}(\mu)|)$ time. This concludes the proof of Theorem 1.

4   NP-Hardness Results via 2-Stable Planar Expansions

We now show three examples of how to exploit Theorem 1 to extend NP-hardness results from the class of 2-connected planar graphs with minimum degree 3 and bounded maximum degree to the class of 3-connected planar graphs with bounded maximum degree.

The general strategy is as follows. Given an NP-hard problem $\mathcal{P}$ that takes as input a planar graph G and, possibly, a parameter k, we show how to construct a new planar graph G' by performing 2-stable planar augmentations on edge pairs of G, computed by applying Theorem 1, based on a problem-specific gadget $\mathcal{A}$ of constant size. This is done so to obtain a graph G' such that (G, k) is a **yes** instance for problem $\mathcal{P}$ if and only if (G', k') is a **yes** instance for problem $\mathcal{P}$, where $k' = f(k)$ and f is a computable polynomial function. While the correctness of the transformation will depend on the choice of gadget $\mathcal{A}$, it is easy to see that G' will be 3-connected by Condition ii of Theorem 1, assuming that $\mathcal{A}$ is 2-stable. Moreover, since no two edge pairs computed by Theorem 1 share an edge, by Condition i of Theorem 1, and since $\mathcal{A}$ has constant size, we have that G' has bounded maximum degree if and only if G has bounded maximum degree.

MIS. For a graph G, we denote by $\alpha(G)$ the size of a largest independent set in G.

Lemma 5 ($\star$). *Let G be a 2-connected cubic planar graph and let e and e' be two edges in $E(G)$ incident to the same face of a planar embedding of G. Let G' be the 2-connected cubic planar graph obtained from G by performing the planar expansion illustrated in Fig. 1a on $\langle e, e' \rangle$. Then, $\alpha(G') = \alpha(G) + 5$.*

By applying Lemma 5 to the distinct edge pairs of a 2-connected cubic planar graph determined by Theorem 1, we can give a new, and arguably simpler proof, of the following result by Biedl, Kant, and Kaufmann [5, Theorem 4.2].

Theorem 2 ($\star$). *MIS is NP-complete for 3-connected cubic planar graphs.*

3-Coloring. We are going to use the following known result.

Theorem 3. ([7], **Theorem 4**). *3-COLORING is NP-complete for 2-connected 4-regular planar graphs.*

Given an edge pair $\langle (u, v), (x, y) \rangle$, we define the gadget $\mathcal{A}_1$ as the 2-connected planar graph containing u, v, x, and y, illustrated in Fig. 1b. Clearly, gadget $\mathcal{A}_1$ is 2-stable. The key property of gadget $\mathcal{A}_1$ is that in any of its 3-colorings the vertices in each pair $\{u, u'\}$, $\{v, v'\}$, $\{x, x'\}$, and $\{y, y'\}$ need to have the same color.

We can thus exploit Theorem 3, Theorem 1, and the fact that gadget $\mathcal{A}_1$ is 2-stable to obtain the following theorem, where the bound on the maximum degree derives from the fact that (i) a planar expansion using gadget $\mathcal{A}_1$ increases the degree of the end-vertices of the corresponding edge pair by 1, that (ii) each edge is involved in at most one planar expansion, by Theorem 1, and that (iii) the original graph is assumed to be 4-regular, by Theorem 3.

Theorem 4. 3-COLORING *is NP-complete for 3-connected planar graphs of maximum degree* 8.

Steiner Tree. Given an edge pair $\langle (u, v), (x, y) \rangle$ and an integer $\ell \geq 1$, we define the *ladder gadget* $\mathcal{L}_\ell$ *of length* ℓ as the planar graph containing u, v, x, and y, illustrated in Fig. 1c; the *safe edges* of $\mathcal{L}_\ell$ are the dashed edges. Note that $\mathcal{L}_\ell$ is 2-stable. We are going to use the following.

Lemma 6 ($\star$). *ST is NP-complete for 2-connected cubic planar graphs.*

Let (G, T, k) be an instance of ST, where G is a 2-connected cubic planar graph. We construct a graph G' from G by performing a planar expansion on each of the edge pairs computed by applying Theorem 1 on G, using a ladder gadget $\mathcal{L}_{7k}$. By Theorem 1 and since ladder gadgets are 2-stable, we have that G' is cubic, 3-connected, and planar. Let $\mathcal{E}$ be a planar embedding of G'. As long as there exists $e \in E(G) \setminus E(G')$, we select an edge e' incident to a face of $\mathcal{E}$ the edge e is incident to and perform a planar expansion on the pair $\langle e, e' \rangle$, using a ladder gadget $\mathcal{L}_{7k}$. The edge e' is selected such that either $e' \in E(G) \setminus E(G')$ or e' is a safe edge of some $\mathcal{L}_{7k}$. Clearly, G' is still planar, cubic, and 3-connected. The expansions can be performed until $E(G) \setminus E(G') = \emptyset$ due to the fact that the ladder gadgets contribute with safe edges on all the faces edges e and e' are incident to. Since each edge of G participates in exactly one planar expansion, since none of the ladder gadgets can be used as a shortcut, and since the shortest path in $\mathcal{L}_{7k}$ between pairs (u, v) and (x, y) has length 7, we have that instances $(G', T' = T, k' = 7k)$ and (G, T, k) are equivalent. Thus, Lemma 6 and the fact that (G', T', k') can easily be constructed in polynomial time imply the following.

Theorem 5. *ST is NP-complete for 3-connected cubic planar graphs.*

Theorem 5 strengthens, for bounded-degree graphs, previous NP-hardness results for ST in graphs with a unique (up to a flip) combinatorial embedding [1,2].

5   Conclusions

In this paper, we introduced a new augmentation technique for 2-connected planar graphs. As a notable application, we showed how this technique can be exploited to obtain strong NP-hardness results for planar graphs that are both 3-connected and have bounded maximum degree. We are confident that this technique may turn useful in other contexts, such as in proving open conjectures about planar graphs. For instance, consider a graph property $\mathcal{H}$ that is invariant under edge subdivision and *monotone*, i.e., every subgraph of a graph satisfying property $\mathcal{H}$ also satisfies property $\mathcal{H}$. Then, our technique implies that in order to prove that bounded-degree planar graphs satisfy property $\mathcal{H}$, it suffices to show that the property holds for 3-connected bounded-degree planar graphs.

References

1. Angelini, P., Da Lozzo, G., Di Battista, G., Frati, F., Patrignani, M., Roselli, V.: Relaxing the constraints of clustered planarity. Comput. Geom. **48**(2), 42–75 (2015)
2. Angelini, P., Da Lozzo, G., Neuwirth, D.: Advancements on SEFE and partitioned book embedding problems. Theor. Comput. Sci. **575**, 71–89 (2015)
3. Barnette, D.W.: On Steinitz's theorem concerning convex 3-polytopes and on some properties of planar graphs. In: Chartrand, G., Kapoor, S.F. (eds.) The Many Facets of Graph Theory. LNM, vol. 110, pp. 27–40. Springer, Heidelberg (1969). https://doi.org/10.1007/BFb0060102
4. Bertolazzi, P., Di Battista, G., Liotta, G., Mannino, C.: Upward drawings of triconnected digraphs. Algorithmica **12**(6), 476–497 (1994)
5. Biedl, T.C., Kant, G., Kaufmann, M.: On triangulating planar graphs under the four-connectivity constraint. Algorithmica **19**(4), 427–446 (1997)
6. Da Lozzo, G., Rutter, I.: Strengthening hardness results to 3-connected planar graphs. CoRR abs/1607.02346 (2016). http://arxiv.org/abs/1607.02346
7. Dailey, D.P.: Uniqueness of colorability and colorability of planar 4-regular graphs are NP-complete. Discret. Math. **30**(3), 289–293 (1980)
8. Di Battista, G., Tamassia, R.: On-line graph algorithms with SPQR-trees. In: Paterson, M.S. (ed.) ICALP 1990. LNCS, vol. 443, pp. 598–611. Springer, Heidelberg (1990). https://doi.org/10.1007/BFb0032061
9. Didimo, W., Liotta, G., Patrignani, M.: Bend-minimum orthogonal drawings in quadratic time. In: Biedl, T., Kerren, A. (eds.) GD 2018. LNCS, vol. 11282, pp. 481–494. Springer, Cham (2018). https://doi.org/10.1007/978-3-030-04414-5_34
10. de Fraysseix, H., Pach, J., Pollack, R.: How to draw a planar graph on a grid. Combinatorica **10**(1), 41–51 (1990)
11. Garg, A., Tamassia, R.: On the computational complexity of upward and rectilinear planarity testing. SIAM J. Comput. **31**(2), 601–625 (2001)
12. Gutwenger, C., Mutzel, P.: A linear time implementation of SPQR-trees. In: Marks, J. (ed.) GD 2000. LNCS, vol. 1984, pp. 77–90. Springer, Heidelberg (2001). https://doi.org/10.1007/3-540-44541-2_8
13. Schmidt, J.M.: Construction sequences and certifying 3-connectedness. In: Marion, J., Schwentick, T. (eds.) STACS 2010. LIPIcs, vol. 5, pp. 633–644. Schloss Dagstuhl - Leibniz-Zentrum fuer Informatik (2010)
14. Tamassia, R.: On embedding a graph in the grid with the minimum number of bends. SIAM J. Comput. **16**(3), 421–444 (1987)
15. Whitney, H.: Congruent graphs and the connectivity of graphs. Am. J. Math. **54**(1), 150–168 (1932)

Cops and Robber on Some Families
of Oriented Graphs

Sandip Das[1], Harmender Gahlawat[1(✉)], Uma Kant Sahoo[1], and Sagnik Sen[2]

[1] Indian Statistical Institute, Kolkata, India
harmendergahlawat@gmail.com
[2] Indian Institute of Technology, Dharwad, India

Abstract. Cops and robber game on a directed graph $\overrightarrow{D}$ initiates by
Player 1 placing k cops and then Player 2 placing one robber on the
vertices of $\overrightarrow{D}$. After that, starting with Player 1, alternately the players
may move each of their tokens to the adjacent vertices. Player 1 wins if,
after a finite number of moves, a cop and the robber end up on the same
vertex and Player 2 wins otherwise. However, depending on the type of
moves the players make, there are three different models, namely, *the
normal cop model*: both cops and robber move along the direction of the
arcs; *the strong cop model*: cops can move along or against the direction
of the arcs while the robber moves along them; and *the weak cop model*:
the robber can move along or against the direction of the arcs while
the cops move along them. A graph is *cop-win* if Player 1 playing with
a single cop has a winning strategy. In this article, we study the three
models on some families of oriented graphs and characterize the cop-win
directed graphs for the third model.

1 Introduction

Cops and Robber is a popular two-player game introduced by Nowakowski and
Winkler [22] in 1983 having applications in artificial intelligence, graph search,
game development etc. [3,15,16] as well as significant implications in theory [23].
The game is extensively studied since its introduction giving rise to deep theory
(see [4] for details) and its numerous variations contributed to its extent.

Recently, Nicolas Nisse[1] [8] introduced one of the variants (the strong cop
model) in directed graphs (digraphs) and asked to characterize the "cop-win"
graphs in two variants (the normal cop model and the strong cop model). In a
seminar (summer 2018) at Simon Fraser University, another natural variant was
discussed where the first author was present. In this article, we study all three
variants, starting by presenting their precise definitions.

This work is partially supported by the IFCAM project Applications of graph homo-
morphisms (MA/IFCAM/18/39).
[1] GRASTA 2014: http://www-sop.inria.fr/coati/events/grasta2014/.

© Springer Nature Switzerland AG 2019
C. J. Colbourn et al. (Eds.): IWOCA 2019, LNCS 11638, pp. 188–200, 2019.
https://doi.org/10.1007/978-3-030-25005-8_16

Setup and initiation: We start with an oriented[2] graph $\overrightarrow{G}$ and Player 1 places k cops on its vertices (multiple cops can be on the same vertex). After that Player 2 places the robber on one vertex of the graph.

Play: After the setup, Player 1 and 2 take turns to move their cops and robber, respectively, with Player 1 taking the first turn.

Winning: Player 1 wins if after finitely many turns the robber and a cop are on the same vertex. In this case, we say that the cop *captures* the robber. Player 2 wins if Player 1 does not win in finite number of moves.

Normal Move: Suppose uv is an arc. In a normal move, the cop/robber can move only from u to v.

Strong Move: Suppose uv is an arc. In a strong move, the cop/robber can move from u to v as well as from v to u.

Normal Cop Model: In their respective turns, Player 1 and Player 2 can perform at most one normal move on each of its cops, and the robber respectively.

Strong Cop Model: In their respective turns, Player 1 can perform at most one strong move on each of its cops, whereas Player 2 can perform at most one normal move on the robber.

Weak Cop Model: In their respective turns, Player 1 can perform at most one normal move on each of its cops, whereas Player 2 can perform at most one strong move on the robber.

Now that we have described the three models, we define a few necessary parameters. The *normal (resp., strong, weak) cop number* $c_n(\overrightarrow{G})$ (resp., $c_s(\overrightarrow{G})$, $c_w(\overrightarrow{G})$) of an oriented graph $\overrightarrow{G}$ is the minimum number of cops needed by Player 1 to have a winning strategy in the normal (resp., strong, weak) cop model. Furthermore, for a family $\mathcal{F}$ of oriented graphs

$$c_x(\mathcal{F}) = \max\{c_x(\overrightarrow{G}) | \overrightarrow{G} \in \mathcal{F}\}$$

where $x \in \{n, s, w\}$. Given a fixed model, an oriented graph is *cop-win* if Player 1 has a winning strategy playing with a single cop.

Below we give a brief survey of the literature concerning the normal cop model, followed by a summary of our results.

Survey: Hamidoune [12] considered the game on Cayley digraphs. Frieze et al. [9], studied the game on digraphs and gave an upper bound of $O\left(\frac{n(\log\log n)^2}{\log n}\right)$ for cop number in digraphs. Along these lines, Loh and Oh [21] constructively proved the existence of a strongly connected planar digraph with cop number greater than three. They also prove that every n-vertex strongly connected planar digraph has cop number at most $O(\sqrt{n})$.

---

[2] An oriented graph is a directed graph without 2-cycles i.e. each edge has a direction. For the purposes of this article, they are the same.

Goldstein and Reingold [10] proved that deciding if k cops can capture a robber is EXPTIME-complete if k is not fixed and either the initial positions are given or the graph is directed. Later Kinnersley [19] proved that determining the cop number of a graph or digraph is EXPTIME-complete. Kinnersley [20] also showed that n-vertex strongly connected cop-win digraphs can have capture time $\Omega(n^2)$.

Hahn and MacGillivray [11] gave an algorithmic characterization of the cop-win finite reflexive digraphs. They also showed that any k-cop game can be reduced to 1-cop game (resulting in an algorithmic characterization for k-cop-win finite reflexive digraphs). However, these results do not give a structural characterization of such graphs. Later Darlington et al. [6] tried to structurally characterize cop-win oriented graphs and gave a conjecture which was later disproved by Khatri et al. [17], who also study the game in oriented outerplanar graphs and line digraphs.

Recently, Hosseini and Mohar [13] (also see [14]) studied whether cop number of planar Eulerian digraphs is bounded or not, and point to evidence of the former.

Organization and Results: In Sect. 2, we compare the parameters $c_n(\cdot), c_s(\cdot),$ $c_w(\cdot)$. The normal, strong and weak cop models are studied in Sects. 3, 4 and 5, respectively. We give an outline of our results.

1. Normal Cop Model
   - Prove a Mycielski-type result by constructing oriented graphs with high normal cop number and girth.
   - Characterize oriented triangle-free and outerplanar normal cop-win graphs.
2. Strong Cop Model
   - Find strong cop number of oriented planar graphs, oriented outerplanar graphs and oriented series-parallel graphs.
   - Prove that a specific class of oriented outerplanar graphs (whose weak dual is a collection of paths) and oriented grids are strong cop-win.
3. Weak Cop Model
   - Characterize weak cop-win oriented graphs.

Now we look into some relations between the parameters $c_n(\cdot), c_s(\cdot),$ $c_w(\cdot)$ and some definitions.

2   Basic Results and Preliminaries

The first result follows directly from the definitions.

Proposition 1. *For any oriented graph $\overrightarrow{G}$ we have $c_s(\overrightarrow{G}) \leq c_n(\overrightarrow{G}) \leq c_w(\overrightarrow{G})$.*

Observe that there are plenty of oriented graphs, the transitive tournament for instance, where equality hold in each of the cases. However, it is interesting to study the gap between these parameters. But first we will introduce some notations and terminologies. Let uv be an arc of an oriented graph $\overrightarrow{G}$. We say that u is an *in-neighbor* of v and v is an *out-neighbor* of u. Let $N^-(u)$ and $N^+(u)$ denote the set of in-neighbors and out-neighbors of u respectively. A vertex without any in-neighbor is a *source* and a vertex without any out-neighbor is a *sink*.

Proposition 2. *Given any* $m, n \in \mathbb{N}$, *there exists an oriented graph* $\overrightarrow{G}$ *such that* $c_n(\overrightarrow{G}) - c_s(\overrightarrow{G}) = n$ *and* $c_w(\overrightarrow{G}) - c_n(\overrightarrow{G}) \geq m$.

Proof. The oriented graph $\overrightarrow{G} = \overrightarrow{G}_{m,n}$ is composed of two oriented graphs $\overrightarrow{A}_n$ and $\overrightarrow{B}_m$. The oriented graph $\overrightarrow{A}_n$ is an orientation of the star graph such that its central vertex v is a sink having degree $n + 1$.

We know that there exist graphs with arbitrarily high cop number in undirected case [1]. Let B_m be a connected undirected graph with cop number at least m. Let $\overrightarrow{B}_n$ be such an orientation of B_m that it is a directed acyclic graph having a single source u. The graph $\overrightarrow{G}_{m,n}$ is obtained by merging vertices u and v (call this vertex v_{merge}). Note that $c_s(\overrightarrow{G}_{m,n}) = 1$ as Player 1 can place one cop on v_{merge} and capture the robber in one move if it is in $\overrightarrow{A}_n$ or capture the robber in a finite number of moves if it is in $\overrightarrow{B}_m$.

On the other hand, $c_n(\overrightarrow{G}_{m,n}) = n + 1$, as Player 1 must keep a cop on each source to win, and since $\overrightarrow{B}_m$ is a directed acyclic graph, one of the cops reaches v_{merge} and then captures the robber in $\overrightarrow{B}_m$. Also, $c_w(\overrightarrow{G}_{m,n}) \geq m + n + 1$ as Player 1 needs to place $n + 1$ cops at sources in $\overrightarrow{A}_n$ and it needs at least as many cops as the cop number of B_m.   □

We end this section with some general notations and terminologies. The *out-degree* of v is $d^+(v) = |N^+(v)|$ and its *in-degree* is $d^-(v) = |N^-(v)|$. Let $N^+[v] = N^+(v) \cup \{v\}$ denote the *closed out-neighbourhood* of v.

In the rest of this article, we refer to the robber as $\mathcal{R}$; and to the cop, only in case of cop-win graphs, as $\mathcal{C}$.

If a cop moves to an in-neighbour of the robber $\mathcal{R}$, then we say that the cop *attacks* the robber. The robber is on a *safe* vertex from a cop if it cannot be captured by the cop in the next turn of Player 1. The robber *evades* capture if every time the cop attacks it, $\mathcal{R}$ can move to a safe vertex.

3   Normal Cop Model

In the context of cops and robbers on oriented graphs, the weakly connected case reduces to solving the strongly connected case [9]. Hence it suffices to consider strongly connected oriented graphs. We begin by constructing strongly connected oriented graphs with arbitrarily high normal cop number and girth (length of a smallest cycle in the graph).

Theorem 1. *Given any* $g \geq 5$ *and* $c \geq 3$, *there exists a strongly connected oriented graph* $\overrightarrow{G}_{g,c}$ *with girth at least* g *having* $c_n(\overrightarrow{G}_{g,c}) \geq c + 1$.

Proof. We borrow a construction to form *regular expander graphs* with high girth [18]. For sake of completeness, we present their complete construction. (Also see [2].) Let G be a simple graph and let L be a set. Define the graph G^L to be a *L-lift* of G if $V(G^L) = V(G) \times L$, and for every edge $uv \in E(G)$, the sets $F_u = \{(u, l_i)\}_{l_i \in L}$ and $F_v = \{(v, l_i)\}_{l_i \in L}$ induce a perfect matching in G^L. Here G is the base graph and G^L depends on the matching between F_u and F_v assigned to each edge uv. Observe that the lifts of $k-$regular graphs are also $k-$regular. From Amit and Nilial [2], it follows that there are lifts of G which are $\delta-$connected for $\delta \geq 3$, where δ is the minimum degree of G. For path $u - v - \cdots - w$ in G, G^L will have a unique path $(u, l_{i_1}) - (v, l_{i_2}) - \cdots - (w, l_{i_k})$, for some $l_{i_1}, l_{i_2}, \ldots, l_{i_k} \in L$. The path $u - v - \cdots - w$ is called as the *projection* of $(u, l_{i_1}) - (v, l_{i_2}) - \cdots - (w, l_{i_k})$.

Now consider a graph G with a fixed ordering π of $m = |E(G)|$ edges, and let $L = \{0, 1\}^m$, that is, the set of all possible m-tuples of 0's and 1's. For $u, v \in V(G)$ and $l_i, l_j \in L$, we connect (u, l_i) with (v, l_j) in G^L if $uv \in E(G)$ and l_i and l_j differ only at the index of edge uv in the ordering π. So (u, l_i) is adjacent to (v, l_j) and (u, l_j) is adjacent to (v, l_i). Hence for edge $uv \in E(G)$, the sets $F_u(= \{(u, l_i)\}_{l_i \in L})$ and $F_v(= \{(v, l_i)\}_{l_i \in L})$ induce a perfect matching in G^L. Thus G^L is a $L-$lift of G.

Now we pick a shortest cycle C_0 in G^L. Its projection in G is also a cycle C. We claim that for every edge $uv \in C$ there are at least two edges in C_0 between F_u and F_v. Start at point (u, l_i) of C_0. Let the next vertex in C_0 be (v, l_j); so l_i and l_j differ only at the index of uv in π. Now to reach (u, l_i) we need to flip the value at the index of uv in π. This happens only if we traverse uv once again. So $|C_0| \geq 2|C|$. Hence girth of G^L is at least twice the girth of G.

To construct the oriented graph with arbitrarily high cop number and girth, do the following. Take a K_{2c+1} and go on applying the above-mentioned lift construction repeatedly until the girth is at least g. The resulting graph $G_{g,c}$ is Eulerian as degree of v in G is even and is the same as the degree of (v, l_i) in $G_{g,c}$. Make the Eulerian circuit a directed circuit by assigning orientations to the edges. This results in a strongly connected oriented graph with girth at least g. Observe that the out-degree of each vertex is c. Thus its normal cop number is at least $c + 1$ as we know that a strongly connected oriented graph with girth at least 5 have normal cop number $c_n(\overrightarrow{G}) \geq \delta^+(\overrightarrow{G}) + 1$, where $\delta^+(\overrightarrow{G})$ is the minimum out-degree of $\overrightarrow{G}$ [21]. $\qquad\square$

Darlington et al. [6] characterized cop-win oriented paths and trees in the normal cop model. We are also going to do so for some other families of oriented graphs.

A *transitive-triangle-free* oriented graph is an oriented graph with no transitive triangles. The following theorem characterizes cop-win transitive-triangle-free oriented graphs, a superclass of triangle-free oriented graphs.

Proposition 3. *A transitive-triangle-free oriented graph $\overrightarrow{G}$ is cop-win if and only if it is a directed acyclic graph with one source.*

Proof. Observe that any directed acyclic graph with one source is cop-win and that every cop-win oriented graph has exactly one source. So it suffices to prove that if a transitive triangle-free oriented graph $\overrightarrow{G}$ is cop-win, then it is a directed acyclic graph.

Suppose $\overrightarrow{G}$ has a directed cycle[3] $\overrightarrow{C}$ on at least 3 vertices. We will now give a strategy for the robber $\mathcal{R}$ to escape. Note that the cop c must be placed at the source initially, as otherwise Player 2 places $\mathcal{R}$ on the source and wins. $\mathcal{R}$ initially places himself at some safe vertex of $\overrightarrow{C}$. Such a vertex exists, as any vertex in $\overrightarrow{G}$ cannot dominate two consecutive vertices in $\overrightarrow{C}$, else a transitive triangle is created. $\mathcal{R}$ moves to the next vertex in $\overrightarrow{C}$ whenever $\mathcal{R}$ lies in the out-neighbour of c. Whenever c attacks $\mathcal{R}$, the robber moves to the next vertex in $\overrightarrow{C}$ and evades the attack. Since $\overrightarrow{C}$ is a directed cycle, c cannot capture $\mathcal{R}$. This contradicts that $\overrightarrow{G}$ is a cop-win graph; hence the result.   □

As bipartite graphs are triangle-free, we have the following corollary.

Corollary 1. *An oriented bipartite graph is cop-win if and only if it is a directed acyclic graph with one source.*

Next, we characterize cop-win oriented outerplanar graphs.

Proposition 4. *An oriented outerplanar graph $\overrightarrow{G}$ is cop-win if and only if it is a directed acyclic graph with one source.*

Proof. The 'if' part is obvious.

For proving the 'only if' part, first note that a graph cannot be cop-win if it has no source or at least two sources. Thus suppose that there exists an oriented outerplanar cop-win graph $\overrightarrow{G}$ containing a directed cycle $\overrightarrow{C}$ with exactly one source v. The cop c must be initially placed on the source v.

Note that at most two vertices of $\overrightarrow{C}$ can have a path made up of vertices from outside $\overrightarrow{C}$ connecting v in order to avoid a K_4-minor. So there is at least one safe vertex u in $\overrightarrow{C}$ such that any directed path connecting v to u must go through some vertex of $\overrightarrow{C}$ other than u. Thus if the robber $\mathcal{R}$ places itself on u and does not move until c comes on a vertex of $\overrightarrow{C}$, it cannot be captured.

If c is on a vertex of $\overrightarrow{C}$ and starts moving towards $\mathcal{R}$ following the direction of the arcs of $\overrightarrow{C}$, then $\mathcal{R}$ also moves forward and evades c.

Thus c must go out of $\overrightarrow{C}$ in order to try and capture $\mathcal{R}$. The moment c goes out to some vertex w outside $\overrightarrow{C}$, $\mathcal{R}$ either is on a safe vertex or it can move to a safe vertex on $\overrightarrow{C}$ in its next move as w can be adjacent to at most two vertices of $\overrightarrow{C}$ in order to avoid a K_4-minor.

This brings us to a situation similar to the initial situation. Thus, the robber will always evade the cop, a contradiction.   □

---

[3] We use the term *directed cycle* instead of *oriented cycle* as it is commonly used.

4   Strong Cop Model

The strong cop number of an oriented graph is upper bounded by cop number in classical version of the game on the underlying undirected graph. We begin this section by finding strong cop number of planar graphs, outerplanar graphs, and series-parallel graphs. But first, we find a lower bound of the strong cop number of a specific oriented graph.

Construction: Given an undirected graph G on vertex set $\{v_1, \ldots, v_n\}$, we form an oriented graph $\overrightarrow{H}$ from G by replacing each edge $v_i v_j$ in G by a directed 4-cycle $v_i u_{ij} v_j u_{ji} v_i$. We have the following lemma relating strong cop number of $\overrightarrow{H}$ with cop number of G. Here $c(G)$ is the cop number of the undirected graph G.

Lemma 1. $c_s(\overrightarrow{H}) \geq c(G)$.

Proof. Each vertex $v_i \in V(G)$ corresponds to the set $N^+[v_i]$ in $V(\overrightarrow{H})$. Note the sets $N^+[v_i]$ partition $V(\overrightarrow{H})$. For each $v \in N^+[v_i]$ in $\overrightarrow{H}$, define its image in G as $I(v) = v_i$. We know that $c_s(\overrightarrow{H})$ cops have a strategy to capture the robber $\mathcal{R}$ in $\overrightarrow{H}$. We will show that $c_s(\overrightarrow{H})$ cops have a winning strategy in G.

We use the winning strategy of $c_s(\overrightarrow{H})$ cops in $\overrightarrow{H}$ to obtain a winning strategy in G. As the game is played in G, we also (sort of) play it in $\overrightarrow{H}$ by following $\mathcal{R}$'s move in G. The move of the cops in $\overrightarrow{H}$ following the winning strategy is translated to G using the images. This procedure is done as follows.

Initially in G, place the cops and then $\mathcal{R}$ is placed. In $\overrightarrow{H}$, place the cops and $\mathcal{R}$ at the vertices with same labels as in the occupied vertices in G. The cops in G pass their first move and then $\mathcal{R}$ moves or passes its move. For each move of $\mathcal{R}$ in G (say from v_i to v_j), we play two turns in $\overrightarrow{H}$: in the first turn $\mathcal{R}$ moves from v_i to u_{ij} and then to v_j in the second turn. In each of these two turns in $\overrightarrow{H}$, the cops move following their winning strategy. After two turns in $\overrightarrow{H}$, the cops in G move to the images of cops in $\overrightarrow{H}$ (this is always possible). Following the winning strategy, when $\mathcal{R}$ is captured in $\overrightarrow{H}$, $\mathcal{R}$ is also captured in G. $\square$

As a result of Lemma 1, we find the strong cop number of oriented planar graphs and then form oriented graphs with arbitrarily high strong cop number.

Corollary 2. *The strong cop number of the family of oriented planar graphs is three.*

Proof. Recall that the strong cop number of an oriented graph is upper bounded by cop number in classical version of the game on the underlying undirected graph. The cop number of planar graphs is three [1]. Apply the construction used in Lemma 1 to a planar graph with cop number 3 to get an oriented planar graph with $c_s \geq 3$ (the construction used in Lemma 1 maintains planarity). $\square$

Corollary 3. *For every $k \in \mathbb{N}$, there exists an oriented graph $\overrightarrow{H}$ such that $c_s(\overrightarrow{H}) \geq k$.*

Proof. Apply the construction used in Lemma 1 to a graph with cop number at least k (whose existence is given in [1]). $\qquad\square$

Next, we find strong cop numbers of the family of oriented outerplanar and series-parallel graphs.

Theorem 2. *The strong cop number of the family of oriented outerplanar graphs is two.*

Proof. The cop number of outerplanar graphs in the classical game on undirected graphs is two [5]. Hence it suffices to construct an oriented graph which is not strong cop-win.

Consider an outerplanar graph on 47 vertices formed by joining 2 copies of the following biconnected outerplanar graph at a vertex, say v_0. Take a cycle C_1 (see Fig. 1) on $v_0, v_1, \ldots, v_{23}$ arranged in counterclockwise manner with arcs $v_i v_{i-1}$ and $v_{2i} v_{2i+2}$ (under modulo 24). Let the other copy C_2 of the cycle be on vertices $v_0 = u_0, u_1, \ldots, u_{23}$ with arcs $u_i u_{i-1}$ and $u_{2i} u_{2i+2}$ (under modulo 24). The arcs of the form $v_i v_{i-1}$ are called *cycle arcs* and the arcs of the form $v_{2i} v_{2i+2}$ are called *chord arcs*. The vertices v_i and u_j with even indices are called *even vertices* and with odd indices are called *odd vertices*.

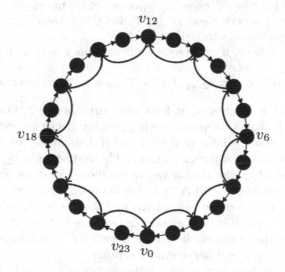

Fig. 1. The biconnected outerplanar graph C_1.

If the cop c is placed at v_0, then robber $\mathcal{R}$ enters at v_4; else $\mathcal{R}$ starts at v_4 or u_4 depending on whether c starts at C_2 or C_1 respectively. In the latter case (that is c is not placed at v_0), $\mathcal{R}$ passes his moves until c is at v_0 (in order to catch $\mathcal{R}$, c has to go through v_0). Once c reaches v_0, $\mathcal{R}$ passes his move once more; reducing this case to the former case. Hence, without loss of generality, assume that c and $\mathcal{R}$ start at v_0 and v_4 respectively.

In the rest of this proof, we show that if c (at v_0) tries to capture $\mathcal{R}$ (at v_4), then $\mathcal{R}$ reaches the *initial configuration* (c at v_0 and $\mathcal{R}$ at v_4) or its *equivalent configuration* (c at v_0 and $\mathcal{R}$ at u_4) without being captured. Precisely, we show that if c pursues $\mathcal{R}$, then $\mathcal{R}$ reaches v_0 two turns before c. So after two more turns $\mathcal{R}$ can be at u_4 or v_4 depending on whether c is in C_1 or C_2 respectively; and then passes its moves until c is at v_0. So $\mathcal{R}$ evades capture indefinitely; thereby proving that the graph constructed above is not strong cop-win.

To simplify our presentation, we use the following notations. Let variable $X = \{c, \mathcal{R}\}$; variables U, V denote two adjacent vertices; and symbol $*$ denote $\circlearrowright$ for clockwise or $\circlearrowleft$ for counter-clockwise. Read $X(U * V)$ as "X moves from U to V in $*$ sense". Read $X(*)$ as "X moves in $*$ sense to an adjacent vertex". Let d_c denote the distance between c and $\mathcal{R}$ at the given instant in the underlying undirected graph. Note that $\mathcal{R}(*)$ results in a fixed final position, where as $c(*)$ results in two possible final positions.

$\mathcal{R}$ moves according to the following rules. All operations are performed under modulo 24.

R0: At any turn, if c passes its move then $\mathcal{R}$ passes its move.
R1: For $i = 1$ to 7, $\mathcal{R}(v_{2i} \circlearrowright v_{2i+2})$ only if $c(v_{2i-4} \circlearrowleft v_{2i-2})$ or $c(v_{2i-3} \circlearrowleft v_{2i-2})$; else it passes its move.
R2: For $i = 8$ to 11, $\mathcal{R}(v_{2i} \circlearrowleft v_{2i+2})$ irrespective of c's move.
R3: If $\mathcal{R}$ is at an even vertex v_{2i}, for $i \leq 7$, and $C(\circlearrowleft)$, then
- $\mathcal{R}(\circlearrowright)$, if d_c increases to at least 4.
- $\mathcal{R}$ passes its move, if d_c increases but remains less than 4.
- $\mathcal{R}(\circlearrowright)$, if d_c decreases.
R4: If $\mathcal{R}$ is at an odd vertex v_{2i+1}, for $i < 7$, then $\mathcal{R}(\circlearrowright)$ irrespective of c's move.

We claim that $\mathcal{R}$ reaches v_0 at least two turns before c. Once $\mathcal{R}$ is at v_{14} and $c(v_{10} \circlearrowleft v_{12})$, then $\mathcal{R}$ keeps on moving counter-clockwise and reaches v_0 at least two turns before c. However if $c(\circlearrowleft)$ and if d_c increases to at least 4, then $\mathcal{R}(\circlearrowleft)$, else if $d_c < 4$ then $\mathcal{R}$ passes its move. The restriction $d_c \geq 4$ ensures that if c moves counter-clockwise then $\mathcal{R}$ can safely move clockwise to the next even vertex. For subsequent steps, if $c(\circlearrowleft)$ and $\mathcal{R}$ is on an even vertex v_{2i}, for $i = 1$ to 7, then $\mathcal{R}(\circlearrowleft)$, provided the restrictions in *R3* are met. In any intermediate step if $c(\circlearrowleft)$, then $\mathcal{R}(\circlearrowleft)$ if it is at an odd vertex; else $\mathcal{R}(\circlearrowleft)$ or $\mathcal{R}$ passes its move depending on whether c attacks it or not. In such a case $\mathcal{R}$ always stays at least two moves away from c and hence evades capture.

The only way left for c to capture $\mathcal{R}$ is if c continues moving counter-clockwise along the chord arcs and then tries to capture $\mathcal{R}$ which now moves counter-clockwise along the cycle arcs. However in such a case also it is easy to see that $\mathcal{R}$ reaches v_0 at least two moves before c. Hence, either $\mathcal{R}$ evades capture indefinitely or reaches the initial or its equivalent configuration; which implies that $\mathcal{R}$ is never captured. Hence the constructed graph is not strong cop-win. $\square$

It is known that the cop number of series-parallel graphs in the classical game on undirected graphs is two [24]. Since outerplanar graphs are also series-parallel graphs, we have the following corollary.

Corollary 4. *The strong cop number of oriented series-parallel graphs is two.*

As mentioned earlier all the oriented graphs whose underlying graphs are cop-win graphs in the classical (undirected graph) version are strong cop-win. Next, we find some families of oriented graphs which are strong cop-win but whose underlying undirected graphs are not cop-win in the classical version. We begin with a specific class of outerplanar graphs.

We need the following definitions. For a plane graph G (i.e. the planar embedding of G), its *dual graph* has vertices that represent faces of G and edges represent the adjacency of faces in G separated by an edge. The *weak dual* of G is the induced subgraph of the dual graph whose vertices correspond to the bounded faces of G.

Theorem 3. *Oriented outerplanar graphs whose weak dual is a collection of paths are strong cop-win.*

Proof. Let G be an outerplanar graph on n vertices, whose weak dual is a collection of paths and $\overrightarrow{G}$ denote the oriented outerplanar graph on G. We call the edges in the outer face of G as *cycle edges*. For a cycle C in $\overrightarrow{G}$, *image* $I_C(\mathcal{R})$ of the robber $\mathcal{R}$, is the set of vertices in C that are closest to $\mathcal{R}$.

First we claim that $|I_C(\mathcal{R})| \leq 2$. Suppose $|I_C(\mathcal{R})| > 2$; then let $u_1, u_2, u_3 \in I_C(\mathcal{R})$ be three vertices arranged in a cyclic order in C. Since distance from $\mathcal{R}$ to u_1, u_2, u_3 are same, the paths from the robber to u_i does not contain u_j, for $i \neq j$ and $i, j \leq 3$. So u_2 does not lie in the outer face; a contradiction.

Furthermore, if $|I_C(\mathcal{R})| = \{u_1, u_2\}$, then u_1 and u_2 are adjacent, else the internal vertices on a $u_1 u_2$ path does not lie in the outer face. If $\mathcal{R}$ is in the cycle then $I_C(\mathcal{R})$ contains the vertex occupied by $\mathcal{R}$. So if $|I_C(\mathcal{R})| = 2$, then $\mathcal{R}$ does not lie in the cycle C.

Now we prove the theorem by induction on the order of G. The base case is easy to verify. Now assume every outerplanar graph of order less than n, whose weak dual is a collection of paths is strong cop-win.

Now consider an outerplanar graph $\overrightarrow{G}$ of order n, whose weak dual is a collection of paths. Select a cycle C in $\overrightarrow{G}$ and place the cop c in some vertex of C. After $\mathcal{R}$ is placed in $\overrightarrow{G}$, we find $I_C(\mathcal{R})$ and capture it in subsequent moves. If $|I_C(\mathcal{R})| = 2$, then we capture any one of them. This can always be done in a cycle that is not directed. If C is directed, c moves against the orientations. So a vertex in $I_C(\mathcal{R})$ can be captured by c. If $\mathcal{R}$ is in the cycle then it is captured. If $|I_C(\mathcal{R})| = 2$, then $\mathcal{R}$ is not in the cycle when some vertex of $I_C(\mathcal{R})$ is captured by c. Once c captures a vertex in $I_C(\mathcal{R})$, the robber cannot enter C. So $\mathcal{R}$ is now trapped in one component of $\overrightarrow{G}$ obtained after deleting the cycle edges of C. By our inductive hypothesis, $\mathcal{R}$ can be captured in this component. Hence oriented outerplanar graphs whose weak dual is a collection of paths are cop-win in the strong cop model. □

Our next class of strong cop-win graphs are oriented grids.

Theorem 4. *Oriented grids are strong cop-win.*

Outline of the Proof. Fix a $m \times n$ grid with points $\{(i,j)|0 \le i \le m-1, 0 \le j \le n-1\}$. The cop c starts at $(0,0)$. If the robber $\mathcal{R}$ is at (x_r, y_r), then define the vertices $(x_r \pm 1, y_r \pm 1)$ (when they exist) as *guard positions*. The reader can check that c can *guard* $\mathcal{R}$ at $(x_r - 1, y_r - 1)$ (see [7, Step 1 in Thm. 1]). So $\mathcal{R}$ can only move either up or to the right if the orientations allow (else $\mathcal{R}$ is caught). Note that once it reaches $(m-1, n-1)$, it gets captured. If $\mathcal{R}$ stays stagnant at a vertex, then c can force it to move. Although the guard position is lost, after a few steps c can regain the guard position (or capture $\mathcal{R}$). So the Y-coordinate of $\mathcal{R}$ gradually increases. Eventually $\mathcal{R}$ ends up at $(m-1, n-1)$ or gets captured. □

5   Weak Cop Model

A vertex u in a directed graph is said to be a *corner vertex*, if there exists a vertex v such that $N^+[u] \cup N^-(u) \subseteq N^+[v]$ where $N^\alpha[v] = N^\alpha(v) \cup \{v\}$ for each $\alpha \in \{+, -\}$. We also say that v *dominates* u.

Now we characterize all cop-win directed graphs in this model, which is adapted from the cop-win characterization of undirected graphs (whose proof follows from a couple of lemmas).

Theorem 5. *A directed graph is cop-win in the weak cop model if and only if by successively removing corner vertices, it can be reduced to a single vertex.*

Lemma 2. *If a directed graph has no corner vertex, then it is not weak cop-win.*

Proof. Let $\overrightarrow{G}$ have no corner vertex. The robber $\mathcal{R}$ starts from a vertex that is not an out-neighbour of the cop c. The robber does not move unless c attacks it. Whenever $\mathcal{R}$ is under attack, it can move to a vertex that in not an out-neighbour of c (as there are no corner vertices in $\overrightarrow{G}$). Hence $\mathcal{R}$ never gets caught.   □

Lemma 3. *A directed graph $\overrightarrow{G}$ with a corner u is weak cop-win if and only if $\overrightarrow{H} = \overrightarrow{G} \setminus \{u\}$ is weak cop-win.*

Proof. Let vertex v dominate u in G. Suppose $\overrightarrow{H}$ is cop-win. Define the image $I_\mathcal{R}$ of the robber $\mathcal{R}$ as follows: $I_\mathcal{R}(u) = v$ and $I_\mathcal{R}(x) = x$ for all $x \in V(\overrightarrow{H})$. So $I_\mathcal{R}$ is restricted to $\overrightarrow{H}$ and it can be captured by the cop c. If $\mathcal{R}$ is not on u, then it is captured. If $\mathcal{R}$ is on u, then c is on v and will capture $\mathcal{R}$ in its next move.

Suppose, on the other hand, $\overrightarrow{H}$ is not weak cop-win. Define the image I_C of the cop c as follows: $I_C(u) = v$ and $I_C(x) = x$ for all $x \in V(\overrightarrow{H})$. So I_C is restricted to $\overrightarrow{H}$ and $\mathcal{R}$ has a winning strategy against I_C. If c is not on u, then $\mathcal{R}$ follows its winning strategy and does not get captured in c's next move. If c is on u, then $\mathcal{R}$ follows its winning strategy assuming c is on $I_C(u) = v$. Since $\mathcal{R}$ has a winning strategy against c if c were at v instead, $\mathcal{R}$ does not get captured in c's next move (as v dominates u). So $\mathcal{R}$ evades capture; hence $\overrightarrow{G}$ is not weak cop-win.   □

Finally, we are ready to prove Theorem 5.

Proof of Theorem 5. Lemma 3 implies that upon removing the corner vertices, the weak cop-win property of the graph remains the same. Now remove all possible corner vertices successively in the directed graph. If we end up with a single vertex, then it is weak cop-win. Otherwise we end up with some other graph that has no corner vertices, Lemma 2 implies that it is not weak cop-win.   □

References

1. Aigner, M., Fromme, M.: A game of cops and robbers. Discret. Appl. Math. **8**, 1–12 (1984)
2. Amit, A., Linial, N.: Random graph coverings I: general theory and graph connectivity. Combinatorica, **22**(1), 1–18 (2002)
3. Alspach, B.: Sweeping and searching in graphs: a brief survey. Matematiche **59**, 5–37 (2006)
4. Bonato, A., Nowakowski, R.: The Game of Cops and Robbers on Graphs. American Mathematical Society, Providence (2011)
5. Clarke, N.E.: Constrained cops and robber. Ph.D. thesis, Dalhousie University (2002)
6. Darlington, E., Gibbons, C., Guy, K., Hauswald, J.: Cops and robbers on oriented graphs. Rose-Hulman Undergrad. Math. J. **17**(1), Article 14 (2016)
7. Das, S., Gahlawat, H.: Variations of cops and robbers game on grids. In: Panda, B.S., Goswami, P.P. (eds.) CALDAM 2018. LNCS, vol. 10743, pp. 249–259. Springer, Cham (2018). https://doi.org/10.1007/978-3-319-74180-2_21
8. Fomin, F., Fraigniaud, P., Nisse, N., Thilikos, D.M.: Report on GRASTA 2014, pp. 14 (hal-01084230) (2014)
9. Frieze, A., Krivelevich, M., Loh, P.: Variations on cops and robbers. J. Graph Theory **69**, 383–402 (2012)
10. Goldstein, A.S., Reingold, E.M.: The complexity of pursuit on a graph. Theoret. Comput. Sci. **143**, 93–112 (1995)
11. Hahn, G., MacGillivray, G.: A note on k-cop, l-robber games on graphs. Discret. Math. **306**, 2492–2497 (2006). Creation and Recreation: A Tribute to the Memory of Claude Berge
12. Hamidoune, Y.O.: On a pursuit game on Cayley digraphs. Eur. J. Comb. **8**, 289–295 (1987)
13. Hosseini, S.A., Mohar, B.: Game of cops and robbers in oriented quotients of the integer grid. Discret. Math. **341**, 439–450 (2018)
14. Hosseini, S.A.: Game of Cops and Robbers on Eulerian Digraphs. Ph.D. thesis, Simon Fraser University (2018)
15. Isaza, A., Lu, J., Bulitko, V., Greiner, R.: A cover-based approach to multi-agent moving target pursuit. In: Proceedings of The 4th Conference on Artificial Intelligence and Interactive Digital Entertainment (2008)
16. Isler, V., Kannan, S., Khanna, S.: Randomized pursuit-evasion with local visibility. SIAM J. Discret. Math. **1**, 26–41 (2006)
17. Khatri, D., et al.: A study of cops and robbers in oriented graphs. ArXiv e-prints, arXiv:1811.06155 (2019)
18. Kilbane, J.: Notes on Graphs of Large Girth

19. Kinnersley, W.B.: Cops and robbers is EXPTIME-complete. J. Comb. Theory Ser. B **111**, 201–220 (2015)
20. Kinnersley, W.B.: Bounds on the length of a game of cops and robbers. Discret. Math. **341**, 2508–2518 (2018)
21. Loh, P., Oh, S.: Cops and robbers on planar directed graphs. ArXiv e-prints, arXiv:1507.01023 (2015)
22. Nowakowski, R., Winkler, P.: Vertex-to-vertex pursuit in a graph. Discret. Math. **43**, 253–259 (1983)
23. Seymour, P.D., Thomas, R.: Graph searching and a min-max theorem for treewidth. J. Comb. Theory Ser. B **58**, 22–33 (1993)
24. Theis, D.O.: The cops and robber game on series-parallel graphs. ArXiv e-prints, arXiv:0712.2908 (2008)

Disjoint Clustering in Combinatorial Circuits

Zola Donovan[1], K. Subramani[1(✉)], and Vahan Mkrtchyan[2]

[1] West Virginia University, Morgantown, WV, USA
K.Subramani@mail.wvu.edu
[2] Gran Sasso Science Institute, L'Aquila, AQ, Italy

Abstract. As the modern integrated circuit continues to grow in complexity, the design of very large-scale integrated (VLSI) circuits involves massive teams employing state-of-the-art computer-aided design (CAD) tools. An old, yet significant CAD problem for VLSI circuits is physical design automation. In this problem, one needs to compute the best physical layout of millions to billions of circuit components on a tiny silicon surface. The process of mapping an electronic design to a chip involves several physical design stages, one of which is clustering. Even for combinatorial circuits, there exists several models for the clustering problem. In particular, our primary consideration is the problem of disjoint clustering in combinatorial circuits for delay minimization (CN). The problem of clustering with replication for delay minimization has been well-studied and known to be solvable in polynomial time. However, replication can become expensive when it is unbounded. Consequently, CN is a problem worth investigating. We establish the computational complexities of several variants of CN. We also present a 2-approximation algorithm for an **NP-hard** variant of CN.

1 Introduction

In this paper, we focus on the problem of disjoint clustering in combinatorial circuits for delay minimization (CN). Generally, it is not possible to place every circuit element in one chip because of various requirements and constraints. As a result, the circuit is partitioned into clusters, where each cluster represents a chip in the overall circuit design. While satisfying specific design constraints (e.g., cluster capacity), the circuit elements are assigned to clusters [11].

Gates and their interconnections usually have delays. The delays of the interconnections are determined by the way the circuit is clustered. Intra-cluster delays, d, are associated with the interconnections between gates in the same cluster. Inter-cluster delays, D, are associated with the interconnections between

This research was supported in part by the Air Force Research Laboratory Information Directorate (AFRL/RI), through the Air Force Office of Scientific Research (AFOSR's) Summer Faculty Fellowship Program, and the AFRL/RI Information Institute®, contract numbers FA9550-15-F-0001 and FA8750-16-3-6003.

C. J. Colbourn et al. (Eds.): IWOCA 2019, LNCS 11638, pp. 201–213, 2019.
https://doi.org/10.1007/978-3-030-25005-8_17

gates in different clusters. The delay along a path from an input to an output is the sum of the delays of the gates and interconnections on the path. The delay of the overall circuit, induced by a clustering, is the longest delay among all paths connecting an input to an output.

The problem of clustering combinatorial circuits for delay minimization when logic replication is allowed (CA) is well-studied [6,11] and frequently arises in VLSI design. In CA, the goal is to find a clustering of a circuit that minimizes the delay of the overall circuit. CA is known to be solvable in polynomial time [6,11]. With replication, circuit elements may be assigned to more than one cluster. Therefore, unbounded replication can be quite expensive. As systems grow in complexity, disjoint clustering (i.e., clustering without logic replication) becomes more necessary. It follows that there is a pressing need to study CN in VLSI design. In this paper, we consider several variants of CN and discuss their computational complexities. A more detailed discussion of related work can be found in the extended version of this paper.

The rest of this paper is organized as follows: The problems that we study are formally described in Sect. 2. In Sect. 3, we give some computational complexity results. In Sect. 4, we propose an approximation algorithm for an **NP-hard** variant of CN. We conclude the paper with Sect. 5, by summarizing our main results and identifying avenues for future work.

2   Statement of Problems

In this section, we define the main graph-theoretic concepts that are used in this paper.

Graphs considered in this paper do not contain loops or parallel edges. The *degree* of a vertex v of an undirected graph G is the number of edges of G that are incident with v. The *maximum degree* of G is denoted by $\Delta(G)$ or simply Δ when G is known from the context.

A *directed path* (or, just a *path*) of a directed graph G is a sequence $Q = v_0 e_1 v_1 \ldots e_l v_l$, where $v_0, v_1, \ldots, v_l$ are vertices of G, $e_1, \ldots, e_l$ are edges (also called *arcs*) of G, and $e_j = (v_{j-1}, v_j)$, $1 \leq j \leq l$. We call l the *length* of the path Q, and sometimes we say that Q is an *l-path* of G. If $v_0 = v_l$, then Q is called a *directed cycle* (or, just *cycle*). G is said to be a *directed acyclic graph (DAG)*, if it contains no directed cycles. For further terminology on graphs and directed graphs, one may consult [1,13].

A *cluster* is an arbitrary subset of the vertices of a DAG, and it does not have to be strongly connected. If C is a cluster in a DAG G, then an edge is said to be a *cut-edge* if it connects a vertex of C to a vertex from $V(G)\backslash C$. The *degree* of C is the number of cut-edges incident with a vertex in C.

The *indegree* and *outdegree* of a vertex are the number of arcs that enter and leave the vertex, respectively. A *source* (*sink*, resp.) is a vertex with indegree zero (outdegree zero, resp.). It is well-known that every DAG has a source and a sink [1].

2.1 Formulation of CN Using Combinatorial Circuits

A combinatorial circuit can be represented as a DAG $G = (V, E)$. In G, each vertex $v \in V$ represents a gate, and each edge $(u, v) \in E$ represents an interconnection between gates u and v. In general, each gate in a circuit has an associated delay [9]. In the model that we consider in this paper, each interconnection has one of the following types of delays: (1) an intra-cluster delay, d, when there is an interconnection between two gates in the same cluster, or (2) an inter-cluster delay, D, when there is an interconnection between two gates in different clusters.

The delay along a path from an input to an output is the sum of the delays of the gates and interconnections that lie on the path. The delay of the overall circuit is the maximum delay among all source to sink paths in the circuit.

A clustering partitions the circuit into disjoint subsets. A clustering algorithm tries to achieve one or both of the following goals, subject to one or more constraints:

(1) The delay minimization through the circuit [3, 6, 9, 11].
(2) The minimization of the total number of cut-edges [2, 4, 7, 8, 12].

In this paper, we study CN under the delay model described as follows:

1. Associated with every gate v of the circuit, there is a delay $\delta(v)$ and a size $w(v)$.
2. The delay of an interconnection between two gates within a single cluster is d.
3. The delay of an interconnection between two gates in different clusters is D, where $D \gg d$.

The size of a cluster is the sum of the sizes of the gates in the cluster. The precise formulation of CN is as follows:

Given a combinatorial circuit, with each gate having a size and a delay, maximum degree Δ, intra- and inter-cluster delays d and D, respectively, and a positive integer M called cluster capacity, the goal is to partition the circuit into clusters such that

1. *The size of each cluster is bounded by M,*
2. *The delay of the circuit is minimized.*

2.2 Graph-Theoretic Formulation of CN

In the rest of the paper, we focus on a graph-theoretic formulation of CN. Given a clustering of a combinatorial circuit represented as a DAG $G = (V, E)$, the delays on the interconnections between gates induce an *edge-delay function* $\delta : E \rightarrow \{d, D\}$ of G. The *weight* of a cluster is the sum of the weights of the vertices in the cluster. The *delay-length* of a directed path $P = v_0 e_1 v_1 \ldots e_l v_l$ of G is $\sum_{i=0}^{l} \delta(v_i) + \sum_{i=1}^{l} \delta(e_i)$, where $\delta(e_i)$ is equal to d if v_{i-1} and v_i are inside the same cluster, or D, otherwise.

CN$\langle X, M, \Delta \rangle$ is formulated (graph-theoretically) as follows: *Given a DAG $G = (V, E)$, with vertex-weight function $w : V \rightarrow \mathbb{N}$, delay function $\delta : V \rightarrow \mathbb{N}$, maximum degree Δ, constants d and D, and a cluster capacity M, the goal is to partition V into clusters such that*

1. *The weight of each cluster is bounded by M,*
2. *The maximum delay-length of any path from a source to a sink of G is minimized.*

The symbol X in our 3-tuple notation may represent some weighted set W of vertices or some unweighted set N of vertices. For some sets, $W = [n]$, where $[n] = \{1, 2, 3, \ldots, n\}$ with $n \in \mathbb{N}$. The symbol M is the cluster capacity.

A clustering of G, such that the weight of each cluster is bounded by M, is called *feasible*. Given a feasible clustering of G, one can consider the corresponding edge-length function $\delta : E \rightarrow \{d, D\}$ of G. A clustering of G is *optimal* if the maximum delay-length of any path from a source to a sink is the minimum among all clusterings.

The main contributions of this paper are as follows:

1. Establishing the computational complexities of several variants of CN$\langle X, M, \Delta \rangle$ (Sect. 3).
2. Design and analysis of a 2-approximation algorithm for an **NP-hard** variant of CN$\langle X, M, \Delta \rangle$ (Sect. 4).

3   Computational Complexities of Clustering Variants

In this section, we establish the computational complexities of several variants of CN.

In [5], CN is considered under area constraints and pin constraints, separately. The decision version of the area-constrained problem is formulated by them as follows: Given a directed acyclic graph $G = (V, E)$ representing a combinatorial circuit, a delay $\delta(v)$ and area $\alpha(v)$ for each $v \in V$, an inter-cluster delay constant $D \geq 0$, a cluster area bound M, and a maximum delay bound B, determine whether there exists a clustering with no replication so that in each cluster C, $\sum_{v \in C} \alpha(v) \leq M$, and for any path $P = (p_1, p_2, \ldots, p_n)$ from a primary input to a primary output, $\sum_{i=1}^{n} \delta(p_i) + k \cdot D \leq B$, where $k = |\{(p_i, p_{i+1}) : (p_i, p_{i+1}) \in P) \wedge (p_i, p_{i+1}$ appear in different clusters)$\}|$. Note that primary inputs and primary outputs represent sources and sinks of the DAG, respectively. The decision version of the pin-constrained problem has an analogous formulation. However, the area of each cluster C is not restricted, while the total number of I/O pins of each cluster must not exceed a given constant Q.

We observe that the decision version of $CN\langle W, M, \Delta\rangle$ belongs to **NP**. This follows from the observation that if we have an edge-weighted DAG, then we can compute a path of maximum edge-weight in polynomial time. Below, we will consider several restrictions of $CN\langle W, M, \Delta\rangle$, which also belong to **NP**.

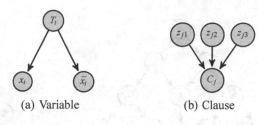

(a) Variable   (b) Clause

Fig. 1. Gadgets used to represent variables and clauses.

Our first result establishes **NP-hardness** and inapproximability of $CN\langle[4], 5, \Delta\rangle$.

Theorem 1. $CN\langle[4], 5, \Delta\rangle$ *is* **NP-hard**.

Proof. We recall $CN\langle[4], 5, \Delta\rangle$ as follows: Given a DAG $G = (V, E)$, with vertex-weight function $w : V \to \{1, 2, 3, 4\}$, $\delta(v) = 0 \ \forall v \in V$, maximum degree Δ, constants d and D, and cluster capacity $M = 5$, the goal is to partition V into clusters such that the weight of each cluster is bounded by M, and the maximum delay-length of any path from a source to a sink of G is minimized.

To show that $CN\langle[4], 5, \Delta\rangle$ is **NP-hard**, we reduce from 3SAT (cf. Theorem 2.1 in [5]). For that purpose, we recall 3SAT as follows: Given a 3-CNF formula ϕ with n variables $x_1, \ldots, x_n$ and m clauses $C_1, \ldots, C_m$, the goal is to check whether ϕ has a satisfying assignment.

Let each variable x_i $(1 \le i \le n)$ be represented by a variable gadget as shown in Fig. 1(a). Let each clause C_j $(1 \le j \le m)$ be represented by a clause gadget as shown in Fig. 1(b). If a variable x_i or its complement $\bar{x}_i$ is the pth literal of a clause C_j, where $p \in \{1, 2, 3\}$, then we add edges (x_i, z_{jp}) or $(\bar{x}_i, z_{jp})$, respectively. The resulting DAG G represents a combinatorial circuit. Let U denote the set of all vertices labeled x_i or $\bar{x}_i$ $(1 \le i \le n)$. There are n sources labeled T_i $(1 \le i \le n)$ and m sinks labeled C_j $(1 \le j \le m)$. They are connected through some vertices in U and $3 \cdot m$ vertices labeled z_{jp} $(1 \le j \le m, 1 \le p \le 3)$. Each z_{jp} is connected to exactly one variable gadget. For every j, no two vertices in the set $\{z_{j1}, z_{j2}, z_{j3}\}$ are adjacent to both x_i and $\bar{x}_i$ of the same variable gadget. In other words, x_i and $\bar{x}_i$ cannot both be connected to the same clause gadget. Every T_i and C_j has a weight of 1, every $x_i, \bar{x}_i \in U$ has a weight of 4, and every z_{jp} has a weight of 2. Let $d = 0$ and let D be any positive integer. All vertices are given a delay of 0. The cluster capacity M is set to 5, and set $k = 2 \cdot D$. It is shown that an instance I of 3SAT is a "yes" instance if and only if an instance I' of $CN\langle[4], 5, \Delta\rangle$ is a "yes" instance.

Theorem 2. $CN\langle N, 2, 3\rangle$ *is* **NP-hard**.

Proof. We recall $CN\langle N, 2, 3\rangle$ as follows: Given a DAG $G = (V, E)$, with $w(v) = 1 \ \forall v \in V$, $\delta(v) = 0 \ \forall v \in V$, maximum degree $\Delta = 3$, constants d and D, cluster capacity $M = 2$, and a positive integer k, the goal is to partition V into

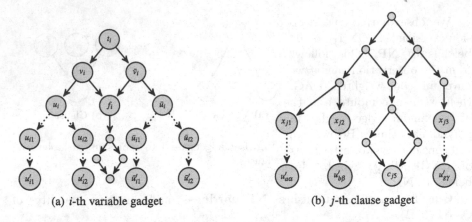

(a) i-th variable gadget   (b) j-th clause gadget

Fig. 2. Gadgets used to represent variables and clauses.

clusters such that the weight of each cluster is bounded by M, and the maximum
delay-length of any path from a source to a sink of G is minimized.

In order to establish **NP-hardness** of $CN\langle N, 2, 3\rangle$, we present a reduction
from $3SAT_{\leq 3, \leq 2}$. For that purpose, we recall $3SAT_{\leq 3, \leq 2}$ as follows: Given a 3-
CNF formula ϕ with n variables $u_1, \ldots, u_n$ and m clauses $C_1, \ldots, C_m$, such that
each variable occurs at most three times and each literal occurs at most twice, the
goal is to check whether ϕ has a satisfying assignment. Note that the requirement
that each clause has exactly three literals is relaxed in this restriction of 3SAT.
Any variable, say u_i, with q occurrences (for some $q > 3$) can be replaced with q
new variables $w_1, \ldots, w_q$. The clauses $(\bar{w}_1 \vee w_2) \wedge (\bar{w}_2 \vee w_3) \wedge (\bar{w}_q \vee w_1)$ can then
be added to ϕ to ensure that the q new variables retain the truth assignment of
the original variable u_i [10].

Given an instance I of $3SAT_{\leq 3, \leq 2}$, we construct an instance I' of $CN\langle N, 2, 3\rangle$.
Let each variable u_i $(1 \leq i \leq n)$ be represented by a variable gadget as shown in
Fig. 2(a), where the dashed arrows indicate possible successors. Note that since
each variable u_i occurs at most three times, then the size of the neighborhood
of $\{u_i, \bar{u}_i\}$ is at most three. Let each clause C_j $(1 \leq j \leq m)$ be represented by a
clause gadget as shown in Fig. 2(b). A set of edges also connects clause gadgets
to variable gadgets. For example, if the p-th literal of clause C_j is the α-th
occurrence of some literal u_a, where $p \in \{1, 2, 3\}$, $\alpha \in \{1, 2\}$ and $a \in \{1, \ldots, n\}$,
then we add edge $(x_{jp}, u'_{a\alpha})$. Every vertex has a weight of 1. We set $d = 0$ and
let D be any positive integer. All vertices are given a delay of 0. The cluster
capacity M is set to 2, and we set $k = 3 \cdot D$. The description of I' is complete.

Observe that I' can be constructed from I in polynomial time. To complete
the proof of the theorem, we show that I is a "yes" instance of $3SAT_{\leq 3, \leq 2}$ if
and only if I' is a "yes" instance of $CN\langle N, 2, 3\rangle$.

Suppose that I is a "yes" instance of $3SAT_{\leq 3, \leq 2}$. This means that there
exists an assignment of ϕ such that every clause has at least one *true* literal. If
literal u_i (or $\bar{u}_i$) is set to *true*, then we cluster the vertices as follows:

1. t_i is clustered with v_i and f_i is clustered with $\bar{v}_i$ (or t_i is clustered with $\bar{v}_i$ and f_i is clustered with v_i).
2. For each $r \in \{1, 2\}$, $\bar{u}_{ir}$ is clustered with $\bar{u}'_{ir}$ (or u_{ir} is clustered with u'_{ir}).
3. If the r-th occurrence of literal u_i (or $\bar{u}_i$) is the p-th literal of clause C_j, then u'_{ir} (or $\bar{u}'_{ir}$) is clustered with clause gadget vertex x_{jp}, where $p \in \{1, 2, 3\}$.
4. For any p-th literal of clause C_j that is set to *true*, then x_{jp} is clustered with its successor.
5. The successors of the variable gadget vertex f_i, say V_{f_i}, are clustered in such a way that the edges of the underlying undirected graph of $G[V_{f_i}]$ form a perfect matching.
6. The clause gadget vertex c_{j5} and its predecessors, say $V_{c_{j5}}$, are clustered in such a way that the edges of the underlying undirected graph of $G[V_{c_{j5}} \cup c_{j5}]$ form a perfect matching.
7. All other vertices are clustered alone.

Observe that the cluster capacity constraint is satisfied, and the maximum delay-length of any path from a source to a sink is $3 \cdot D$. This means that I' is a "yes" instance of $CN\langle N, 2, 3\rangle$.

Conversely, suppose that I' is a "yes" instance of $CN\langle N, 2, 3\rangle$. This means that there is a way of partitioning the vertices of G into clusters of capacity $M = 2$, such that the delay-length of any path from a source to a sink is at most $3 \cdot D$. Observe that under any partitioning, the delay-length of any path from a source to a sink is at least $3 \cdot D$. In any partitioning with delay-length equal to $3 \cdot D$, we have that either t_i is clustered with v_i or t_i is clustered with $\bar{v}_i$, for every $i \in \{1, \dots, n\}$. Furthermore, in any partitioning with delay-length equal to $3 \cdot D$, there is at least one x_{jp} that must be clustered with its successor. If t_i is clustered with v_i, then for each $r \in \{1, 2\}$, $\bar{u}_{ir}$ must be clustered with $\bar{u}'_{ir}$. Set literal u_i to *true* and consider each u'_{ir} *free*. Otherwise, if t_i is clustered with $\bar{v}_i$, then for each $r \in \{1, 2\}$, u_{ir} must be clustered with u'_{ir}. Set literal $\bar{v}_i$ to *true* and consider each $\bar{u}'_{ir}$ *free*. At least one x_{jp} is clustered with its successor, namely some *free* vertex. This means that at least one *true* literal appears in every clause. Thus, a satisfying clustering for G yields a satisfying assignment for ϕ. Hence, I is a "yes" instance of $3SAT_{\leq 3, \leq 2}$.

In order to present our next results, we will need some definitions. We say that two edges of G are independent if they are not incident with the same vertex. A matching of G is a set of pairwise independent edges of G. A matching is maximal if it is not a subset of a larger matching.

Proposition 1. *For any instance of $CN\langle N, 2, \Delta\rangle$ there exists an optimal clustering, such that the edges of G with delay d form a maximal matching of G.*

Proof. Consider an optimal clustering of G. Since $M = 2$, we have that any two edge with delay d are independent. Thus, they form a matching I. Now, if I is not maximal, then there is an edge e, such that $I \cup \{e\}$ is a matching. Put the end-vertices of e to the same cluster. Observe that the resulting clustering is feasible, moreover, its delay does not exceed the delay of the original clustering.

Thus, the resulting clustering is again optimal. By continuing this process, we will end-up with a clustering such that the edges of delay d form a maximal matching. The proof is complete.

Our next proposition states that the clustering problem remains difficult even if we assume that the input DAG G contains a path which contains sufficiently many edges.

Proposition 2. *For each fixed integer t, $CN\langle N, 2, 3\rangle$ remains* **NP-hard** *even if we assume that G contains a path with at least t edges.*

Proof. We present a reduction from $CN\langle N, 2, 3\rangle$ to its restriction stated in the statement. Assume that the input DAG G of maximum degree 3 is given. Consider a component which is a directed 4-cycle, whose edges are directed from the left to right. Let c_1 be the source in it. Assume that c_1 is adjacent to c_2 and c_3, which are neither source nor a sink. Finally, let them be adjacent to the sink c_4 (Fig. 3).

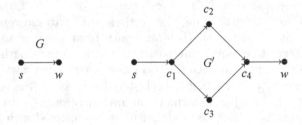

Fig. 3. The reduction with the directed 4-cycle.

Now, let the input DAG G be given which is of maximum degree three. Consider all edges of G which are incident to a source, and replace them with a 4-cycle (one 4-cycle per every edge), that is, if sw is an edge of G, then we replace it with a new 4-cycle, and connect s to c_1 and c_4 to w (Fig. 3). Let G' be the resulting DAG. Observe that this procedure increases the length of any longest path by three. Thus, applying it sufficiently many times, we can get a DAG with desired lower bound for the length of the longest path. Moreover, observe that this process does not increase the maximum degree of the vertex, that is, the resulting graph is still of maximum degree three.

Let $OPT(H)$ denote the optimal delay in a DAG H. We claim that

$$OPT(G') = OPT(G) + (d + 2 \cdot D).$$

Consider an optimal clustering in G. Let $e = sw$ be an edge of G. If it is a d-edge in G, then declare the edges sc_1 and c_4w of G' as a d-edge. The rest of the edges of G' in the part corresponding to e are declared as D-edges. On the other hand, if e is a D-edge, then declare the edges c_1c_2 and c_3c_4 as d-edges, and the rest of edges corresponding to e as D-edges. Now consider a critical path P. If it starts with a d-edge on P, then its delay d is replaced with $2 \cdot (d + D)$ in G'. Thus the increase of the delay is $d + 2 \cdot D$. If it started with a D-edge, then its delay D becomes $d + 3 \cdot D$ in G'. Thus, the increase in the delay is again $d + 2 \cdot D$. Hence, we have the same increase in the delay. Clearly, this implies that

$$OPT(G') \leq OPT(G) + (d + 2 \cdot D).$$

To prove the converse inequality, let us show that we can always find an optimal clustering such that sc_1 and c_4w are d-edges or D-edges at the same time. If sc_1 and c_4w are d-edges then we are done. Thus, we can assume that one of them is a D-edge. Then we show that we can assume that the other one is also a D-edge. First, assume that sc_1 is a d-edge and c_4w is a D-edge. Since we can always assume that the optimal clustering is a maximal matching (Proposition 1), we have that one of the edges of the 4-cycle incident to c_4 is a d-edge f. Now, if we consider a new clustering of G' by replacing the edge sc_1 with the edge opposite to f (the unique edge of the 4-cycle that is not adjacent to f) in the 4-cycle as a d-edge. Clearly, the resulting clustering has the same delay as the original one. However, sc_1 now is a D-edge. Similarly, assume that c_4w is a d-edge and sc_1 is D-edge. Since we can always assume that the optimal clustering is a maximal matching (Proposition 1), we have that one of the edges of the 4-cycle incident to c_1 is a d-edge f. Now, if we consider a new clustering of G' by replacing the edge c_4w with the edge opposite to f in the 4-cycle as a d-edge. Clearly, the resulting clustering has the same delay as the original one. However, c_4w now is a D-edge. Thus, we can always find an optimal clustering such that sc_1 and c_4w are d-edges or D-edges at the same time.

The proved property allows us to get a clustering in G simply by looking at sc_1 and c_4w. If they are d-edges at the same time, we declare sw as a d-edge in G. On the other hand, if they are D-edges, then we declare sw as a D-edge in G. Observe that the resulting clustering of G will have delay at most $OPT(G') - (d + 2 \cdot D)$. Thus,

$$OPT(G) \leq OPT(G') - (d + 2 \cdot D)$$

or

$$OPT(G') \geq OPT(G) + (d + 2 \cdot D).$$

This completes the proof of the equality. The proved equality implies that optimizing the delay in G' is the same that optimizing in G. Thus the above process is a reduction.

Let us say that a graph is cubic if any vertex is of degree 3. We are ready to obtain the final result of this section.

Theorem 3. *For each fixed integer t, $CN\langle N, 2, 3\rangle$ remains **NP-hard** even if we assume that G is a cubic graph that contains a path with at least t edges.*

Proof. We get a reduction from the restriction of $CN\langle N, 2, 3\rangle$ where we assume that G contains a path with at least 6 edges. Since $M = 2$, we have that $OPT(G) \geq 3 \cdot D$. As $\Delta = 3$, we have that the vertices of G are of degree 0, 1, 2 or 3. Now, we will show how to get rid of the vertices which have degree less than three. First, if we have a vertex of degree 0, we can remove it from G. In order to overcome the vertices of degree 1 and 2, we will make use of the following orientation of the complete graph K_4 on 4 vertices by removing an edge $e = uv$. Let the other two vertices of K_4 be u' and v'. Direct the edges starting from u' towards u and v. Direct the edges starting from v' towards u and v. Finally, direct the edge $u'v'$ from v' to u' (Fig. 4).

Fig. 4. The orientation of the edges of $K_4 - e$.

Now, assume that our input DAG G contains a vertex w of degree one. Assume that w is a source. Take a copy of the above orientation of K_4 minus an edge, and join w to u and v with directed edges, so that w is a source, and u and v are sinks in resulting DAG G'. Observe that in the added part the maximum delay will be at most $2 \cdot D$, which is less than $OPT(G)$. Since w is a source, the added part will play no role. Similarly, one can overcome the case when w is a sink. One just needs to reverse the orientations of edges in the copy of K_4 minus an edge.

Next let us assume that we have a vertex w of degree two. First, let us assume that w is a source. We add a new vertex w' and join w to w' with a directed edge. Observe that w is of degree 3 and w' is of degree one. Thus we can apply the trick from the previous paragraph. Similarly, if w is a sink, we will join w' to the vertex w with a directed edge and again apply the trick from the previous paragraph.

Thus, we are left with the case when w-the degree two vertex, is neither a source nor a sink. Let x and z be the neighbors of w such that xw and wz are directed edges. We consider two cases. If z is a sink in G, then we add a new vertex w' of degree 1 and join it to w with a directed edge $w'w$. Observe that in the optimal clustering w' will not be on a path of optimal delay as it reaches only z and the delay of this path is at most $2 \cdot D$. On the other hand, if we assume that G contains a path of length at least six, the optimal delay will be at least $3 \cdot D$. Finally, if we assume that z is not a sink, then we add a new vertex w' and join w to w' with a directed edge ww'. Since z is not a sink and $M = 2$, in any clustering z will be incident to at least one D-edge. Hence, there will be a path of optimal delay in G' that will not terminate at w'. Thus, the addition of w' will not play a role. Since w' is a degree one vertex, we can overcome it via the trick mentioned above. Thus, without loss of generality, we can assume that the input graph in $CN\langle N, 2, 3\rangle$ is a cubic graph.

4   A 2-approximation Algorithm

We now provide an integer program (IP) for $CN\langle W, M, \Delta\rangle$.

4.1   An IP for $CN\langle W, M, \Delta\rangle$

Let w_j be the weight of vertex j. Define x_{ij} to be an integer variable that is set to 1 if vertices i and j are in the same cluster, and 0 otherwise. We present the following IP:

Packing constraints

$$x_{ii} = 1, \quad \forall i \in V \tag{1}$$

$$\sum_{j=1}^{n} w_j \cdot x_{ij} \leq M, \quad \forall i \in V \tag{2}$$

Consistency constraints

$$x_{ij} = x_{ji}, \quad \forall i, j \subset V \tag{3}$$

$$x_{ik} \geq x_{ij} + x_{jk} - 1, \quad \forall i, j, k \in V \tag{4}$$

Condition (1) ensures that every vertex is clustered. Condition (2) ensures that every cluster has weight at most M. Condition (3) ensures that either i and j are in the same cluster or they are in different clusters. Likewise, condition (4) ensures that if i and j are in one cluster, and j and k are in one cluster, then i and k must be in the same cluster and all clusters are disjoint.

We now come to the objective function. For any vertex j, let δ_j be the delay at j in a clustering. This delay is completely dependent upon its predecessors. We can write

$$\delta_j = \max_{i:(i,j)\in E} \{\delta_i + d \cdot x_{ij} + D \cdot (1 - x_{ij})\}. \tag{5}$$

Hence the function to be minimized is δ_t, where t is the sink of the circuit. Note that condition (5) can be easily linearized.

The correctness of reduction will be shown in the journal version of this paper.

4.2   An LP-rounding Algorithm for $CN\langle N, 2, \Delta\rangle$

In this section, we present an LP-rounding algorithm for $CN\langle N, 2, \Delta\rangle$.

Let $\mathrm{LP}_{CN\langle N,2,\Delta\rangle}$ be the linear programming relaxation obtained from $\mathrm{IP}_{CN\langle W,M,\Delta\rangle}$ when vertices are unweighted and $M = 2$ (i.e., the problem restricted to $CN\langle N, 2, \Delta\rangle$), by replacing its 0-1 integrality constraints for x_{ij} with $x_{ij} \in [0, 1]$.

Algorithm 1. An LP rounding algorithm for $CN\langle N, 2, \Delta\rangle$

---

   input : A DAG $G = (V, E)$, where $|V| = n$ and $|E| = m$.
   output: A clustering Γ of G.

1 Solve $LP_{CN\langle N,2,\Delta\rangle}$. Let each $\hat{x}_{ij}$ denote the delay on the edge connecting vertices i and j and let $\hat{\delta}_j$ denote the delay at vertex j.

2 Let $G' = G$.

3 **while** $G' \neq \emptyset$ **do**

4 | Consider a source s of G' such that the delay-length of the path from s to the sink t is maximum.

5 | Let vertex $v \in N^+(s)$ be such that $\hat{\delta}_v = \min_{j \in N^+(s)}\{\hat{\delta}_j\}$.

6 | We round some $\hat{x}_{ij}$ to 0-1 values $\bar{x}_{ij}$ as follows: set $\bar{x}_{sv} = 1$, set $\bar{x}_{sj} = 0 \ \forall j \in N^+(s) \setminus v$, and set $\bar{x}_{vj} = 0 \ \forall j \in (N^-(v) \cup N^+(v)) \setminus s$.

7 | Let $G' = G'[V \setminus \{s, v\}]$.

8 Cluster together all vertices i and j such that $\bar{x}_{ij} = 1$, where $i \neq j$. Put the remaining vertices into singleton clusters.

9 **return** Γ.

Theorem 4. *Algorithm 1 is a 2-approximation algorithm.*

Proof. Let Q be a path of G from a source to the sink t with maximum delay-length. Let OPT be the delay of an optimal clustering of G. This means that OPT is the sum of the fractional intra- and inter-cluster delays of the edges along Q. Let ALG be the delay of the clustering of G returned by Algorithm 1. Since the algorithm returns a solution with an integral delay, notice that for each intra-cluster edge $(i, j) \in Q$, the delay is increased by $d \cdot (1 - \hat{x}_{ij})$. Moreover, for each inter-cluster edge $(i, j) \in Q$, the delay is decreased by $D \cdot (1 - \hat{x}_{ij})$. Hence,

$$
\frac{ALG}{OPT} \leq \frac{\sum_{(i,j)\in Q} d \cdot \hat{x}_{ij} + D \cdot (1 - \hat{x}_{ij}) + \sum_{(i,j)\in Q} d \cdot (1 - \hat{x}_{ij}) - D \cdot (1 - \hat{x}_{ij})}{\sum_{(i,j)\in Q} d \cdot \hat{x}_{ij} + D \cdot (1 - \hat{x}_{ij})}
$$

$$
= 1 + \frac{\sum_{(i,j)\in Q} d \cdot (1 - \hat{x}_{ij}) - D \cdot (1 - \hat{x}_{ij})}{\sum_{(i,j)\in Q} d \cdot \hat{x}_{ij} + D \cdot (1 - \hat{x}_{ij})}
$$

$$
\leq 1 + \frac{\sum_{(i,j)\in Q} d \cdot (1 - \hat{x}_{ij})}{\sum_{(i,j)\in Q} d \cdot \hat{x}_{ij} + D \cdot (1 - \hat{x}_{ij})}
$$

$$
\leq 1 + \frac{\sum_{(i,j)\in Q} D \cdot (1 - \hat{x}_{ij})}{\sum_{(i,j)\in Q} d \cdot \hat{x}_{ij} + D \cdot (1 - \hat{x}_{ij})}
$$

$$
= 2 - \frac{\sum_{(i,j)\in Q} d \cdot \hat{x}_{ij}}{\sum_{(i,j)\in Q} d \cdot \hat{x}_{ij} + D \cdot (1 - \hat{x}_{ij})}
$$

5   Conclusion

In this paper, we studied the problem of disjoint clustering in combinatorial circuits for delay minimization (CN). We obtained the computational complexities of several variants of CN. We also proposed an approximation algorithm for a variant of CN and analyzed it.

We are interested in the following open problems:

1. Finding inapproximability bounds for variants of $CN\langle X, M, \Delta \rangle$ using other assumptions.
2. Finding approximation, parameterized and exact exponential algorithms for other variants of $CN\langle X, M, \Delta \rangle$.

References

1. Bang-Jensen, J., Gutin, G.: Digraphs: Theory, Algorithms and Applications. Springer, London (2010). https://doi.org/10.1007/978-1-84800-998-1
2. Behrens, D., Hebrich, E.B.K.: Heirarchical partitioning. In: Proceedings of the IEEE International Conference on CAD, pp. 171–191 (1997)
3. Cong, J., Ding, Y.: FlowMap: an optimal technology mapping algorithm for delay optimization in lookup-table based FPGA designs. IEEE Trans. Comput. Aided Des. Integr. Circuits Syst. 13(1), 1–12 (1994)
4. Hwang, L.J., El Gamal, A.: Min-cut replication in partitioned networks. IEEE Trans. Comput. Aided Des. Integr. Circuits Syst. 14(1), 96–106 (1995)
5. Kagaris, D.: On minimum delay clustering without replication. Integ. VLSI J. 36(1), 27–39 (2003)
6. Lawler, E.L., Levitt, K.N., Turner, J.: Module clustering to minimize delay in digital networks. IEEE Trans. Comput. 18(1), 47–57 (1969)
7. Liu, L.-T., Kuo, M.-T., Cheng, C.-K., Hu, T.C.: A replication cut for two-way partitioning. IEEE Trans. Comput. Aided Des. Integr. Circuits Syst. 14(5), 623–630 (1995)
8. Mak, W.-K., Wong, D.F.: Minimum replication min-cut partitioning. In: Proceedings of International Conference on Computer Aided Design, pp. 205–210 (1996)
9. Murgai, R., Brayton, R.K., Sangiovanni-Vincentelli, A.: On clustering for minimum delay/area. In: 1991 IEEE International Conference on Computer-Aided Design Digest of Technical Papers, pp. 6–9 (1991)
10. Papadimitriou, C.H.: Computational Complexity. Addison-Wesley, Reading (1994)
11. Rajaraman, R., Wong, D.F.: Optimal clustering for delay minimization. In: 30th ACM/IEEE Design Automation Conference, pp. 309–314 (1993)
12. Shih, M., Kuh, E.S.: Circuit partitioning under capacity and I/O constraints. In: Proceedings of the IEEE on Custom Integrated Circuits Conference, pp. 659–662. IEEE, May 1994
13. West, D.B.: Introduction to Graph Theory, 2nd edn. Prentice Hall, Upper Saddle River (2001)

The Hull Number in the Convexity of Induced Paths of Order 3

Mitre C. Dourado[1] ⓘ, Lucia D. Penso[2(✉)], and Dieter Rautenbach[2]

[1] Instituto de Matemática, Universidade Federal do Rio de Janeiro,
Rio de Janeiro, Brazil
mitre@dcc.ufrj.br
[2] Institute of Optimization and Operations Research, Ulm University, Ulm, Germany
{lucia.penso,dieter.rautenbach}@uni-ulm.de

Abstract. A set S of vertices of a graph G is P_3^*-*convex* if there is no vertex outside S having two non-adjacent neighbors in S. The P_3^*-*convex hull of S is the minimum P_3^*-convex set containing S. If the P_3^*-convex hull of S is $V(G)$, then S is a P_3^*-*hull set*. The minimum size of a P_3^*-hull set is the P_3^*-*hull number of G. In this paper, we show that the problem of deciding whether the P_3^*-hull number of a chordal graph is at most k is NP-complete and present a linear-time algorithm to determine this parameter and provide a minimum P_3^*-hull set for unit interval graphs.

Keywords: Graph convexity · Hull number · Unit interval graph. · 2-distance shortest path

1   Introduction

We consider finite, undirected, and simple graphs. The path with k vertices is denoted by P_k and an *induced path* is a path having no chords. Given a set S of vertices of a graph G, the *interval of S in the convexity of induced paths of order 3*, also known as the P_3^* *convexity*, is the set $[S]_3^* = S \cup \{u : u$ belongs to an induced P_3 between two vertices of $S\}$. The set S is P_3^*-*convex* if $S = [S]_3^*$ and is P_3^*-*concave* if $V(G) \setminus S$ is P_3^*-convex. The P_3^*-*convex hull of S is the minimum P_3^*-convex set containing S and it is denoted by $\langle S \rangle_3^*$. If $\langle S \rangle_3^* = V(G)$, then S is a P_3^*-*hull set*. The minimum size of a P_3^*-hull set is the P_3^*-*hull number* $h_3^*(G)$ of G.

Recently, the P_3^* convexity has attracted attention as an alternative to other quite known convexities with different behavior despite a similar definition. It is particularly interesting in spreading dynamics which forbid the same influence by two neighbors to get spread to a common neighbor. For instance, in [4], it is shown that the problem of deciding whether the P_3^*-hull number of a bipartite graph is at most k is NP-complete, while polynomial-time algorithms for determining this parameter for P_4-sparse graphs and cographs are presented. Apart from these results very little is known, as results of quite similarly defined

© Springer Nature Switzerland AG 2019
C. J. Colbourn et al. (Eds.): IWOCA 2019, LNCS 11638, pp. 214–228, 2019.
https://doi.org/10.1007/978-3-030-25005-8_18

well-known convexities do not help, since the proofs depend on the existence of longer shortest paths or a non-induced P_3.

In the well-known *geodetic convexity*, the *geodetic interval of S* is $[S]_g = S \cup \{u : u$ belongs to some shortest path between two vertices of $S\}$. The terms *geodesically convex*, *geodesically concave*, *geodetic convex hull* $\langle S \rangle_g$, *geodetic hull set*, and *geodetic hull number* $h_g(G)$ are defined in a similar way to the P_3^* convexity. The literature concerning the geodetic hull number is large. It is known that this problem is NP-complete for chordal graphs [5], P_9-free graphs [10], and partial cubes [1]; and that one can find a minimum geodetic hull set in polynomial time if the input graph is unit interval [9], $(q, q - 4)$-graph [2], cobipartite [2], cactus [2], (paw, P_5)-free [10], $(C_3, \ldots, C_{k-2}, P_k)$-free [10], P_5-free bipartite [3], or planar partial cube quadrangulation [1]. However, as already remarked, though an induced path of order 3 is a shortest path of length 2 between a pair of nodes, those results do not directly apply to the P_3^* convexity due to the use of longer shortest paths in proofs.

Unlike the P_3^* convexity, the P_3*convexity* considers all paths of order 3. In this convexity, also known as irreversible 2-conversion, the problem of computing the hull number is APX-hard for bipartite graphs with maximum degree at most 4 and NP-complete for planar graphs with maximum degree at most 4 [7,15], and can be found in polynomial time for chordal graphs [6] as well as for cubic or subcubic graphs [15]. Finally, in the convexity that considers all induced paths, the *monophonic convexity*, the hull number can be computed in polynomial time for any graph [11].

The main result of this paper is a linear-time algorithm to determine both the P_3^*-hull number and a minimum P_3^*-hull set of a unit interval graph (Sect. 2). We point out that Theorem 1 is not only an explicit formula for the P_3^*-hull number $h_3^*(G)$ but also an almost explicit one for the minimum P_3^*-hull set, since one needs to compute the necessary labels before. We also show that the problem of deciding whether this parameter is at most k for a chordal graph is NP-complete (Sect. 3). Remember that unit interval graphs have a variety of applications in operations research, including resource allocation problems in scheduling [13] and in genetic modeling such as DNA mapping in bioinformatics [14], where an overall agreement (on a value, a signal, a failure, a disease, a characteristic, etc.) might get forced by a minimum key set under some convexity such as the P_3^*.

We conclude this section giving some definitions. The distance between vertices u and v is denoted by $d(u, v)$ and the neighborhood of a vertex v is denoted by $N(v)$. The set $\{1, \ldots, k\}$ for an integer $k \geq 1$ is denoted by $[k]$. A subgraph of G induced by vertex set S is denoted by $G[S]$. A vertex u is *simplicial* if its neighborhood induces a complete graph. Note that every P_3^*-hull set contains all simplicial vertices and at least one vertex of each P_3^*-concave set of the graph. Given an ordering $\alpha = (v_1, \ldots, v_n)$ of $V(G)$ and a set $I \subseteq [n]$, the subordering $\alpha' = (v_{i_1}, \ldots, v_{i_{|I|}})$*of αinduced by I* is the ordering of the set $\{v_{i_j} : i_j \in I\} \subseteq V(G)$ such that v_{i_j} appears before v_{i_k} if and only if $i_j < i_k$. If $I = [j] \setminus [i - 1]$ for $0 \leq i < j \leq n$, then the suborderingof α induced by I is denoted by $\alpha_{i,j}$.

2   Unit Interval Graphs

A *unit interval graph* G is the intersection graph of a collection of intervals of the same size on the real line. Since one can assume that all left endpoints of the intervals of such a collection are distinct, a *canonical ordering* $\Gamma = (v_1, v_2, \ldots, v_n)$ of $V(G)$ is defined as the one such that $i < j$ if and only if the left endpoint of the interval of v_i is smaller than the left endpoint of the one of v_j. This ordering has the property that if $v_i v_j \in E(G)$ for $i < j$, then $\{v_i, \ldots, v_j\}$ is a clique [8,16]. It is easy to see that if $v_i \in [v_j, v_k]_3^*$ for $j < k$, then $j \le i \le k$. In the next, we consider a canonical ordering $\Gamma = (v_1, v_2, \ldots, v_n)$ of a given unit interval graph G.

Lemma 1. *If v_j is simplicial, then $h_3^*(G) = h_3^*(G[\{v_1, \ldots, v_j\}]) + h_3^*(G[\{v_j, \ldots, v_n\}]) - 1$.*

Proof. Let S, S_1, and S_2 be minimum hull sets of $G, G_1 = G[\{v_1 \ldots v_j\}]$, and $G_2 = G[\{v_j \ldots v_n\}]$, respectively. Since $\Gamma = (v_1, v_2, \ldots, v_n)$ is a canonical ordering and v_j is a simplicial vertex of G, G_1, and G_2, it holds $S \supseteq \{v_j\} = S_1 \cap S_2$. It is clear that $S_1 \cup S_2$ is a hull set of G, and hence $h_3^*(G) \le h_3^*(G[\{v_1, \ldots, v_j\}]) + h_3^*(G[\{v_j, \ldots, v_n\}]) - 1$. Now, consider an induced path $v_i v_k v_\ell$ with $v_i \in V(G_1) \setminus \{v_j\}$ and $v_\ell \in V(G_2) \setminus \{v_j\}$. If $v_k \in V(G_2)$, then $i < j < k < \ell$ and $\{v_i, \ldots, v_k\}$ is a clique containing v_j. Since v_j is simplicial, if $v_j v_\ell \in E(G)$, then $v_i v_\ell \in E(G)$, which would contradict the assumption that $v_i v_k v_\ell$ is an induced path. Therefore $v_j v_\ell \notin E(G)$ and $v_k \in [v_j, v_\ell]_3^*$. The case for $v_k \in V(G_1)$ is analogue. Hence, $S \cap V(G_i)$ is a minimum hull set of G_i for $i \in [2]$, which means that $h_3^*(G) \ge h_3^*(G[v_1, \ldots, v_j]) + h_3^*(G[v_j, \ldots, v_n]) - 1$. □

Due to Lemma 1, we can assume that G has only two simplicial vertices, namely, v_1 and v_n. In the sequel, we use the geodetic interval to obtain a useful partition of the vertices of G. We say that v_i is a *black vertex* if v_i lies in a shortest (v_1, v_n)-path and that it is a *red vertex* otherwise. The black vertices v such that $d(v_1, v) = i$ form the *black region B_i*. Note that all vertices of B_i are consecutive in Γ. The set of vertices between the black regions B_{i-1} and B_i in Γ form the *red region R_i*. These definitions induce a partition of Γ into black and red regions $B_0, R_1, B_1, \ldots, R_q, B_q$. Note that a red region can be empty, $B_0 = \{v_1\}$, $B_q = \{v_n\}$, and $d(v_1, v_n) = q = d(G)$, where $d(G)$ stands for the diameter of G. Besides, the black (and red) regions are precisely defined by the following two (not necessarily distinct) shortest (v_1, v_n)-paths, namely, the one whose internal vertices have all the highest possible indexes in Γ and the one whose internal vertices have all the lowest possible indexes in Γ. Each black region B_i contains precisely the vertices of those two paths having distance i to v_1 as well as all vertices between them in the canonical ordering Γ. (See Fig. 1.)

For $i \in [q]$, denote by r_i^ℓ and r_i^r the leftmost and rightmost vertices of R_i in Γ, respectively. For $i \in [q] \cup \{0\}$, denote by b_i^ℓ and b_i^r the leftmost and rightmost vertices of B_i in Γ, respectively. If, for $i \in [q]$, $b_{i-1}^\ell b_i^r \in E(G)$, then we say that $b_{i-1}^\ell b_i^r$ is a *long edge*. For $i \in [q]$, denote by R_i^r the vertices of R_i having edges to R_{i+1} and by R_i^ℓ the vertices of R_i having edges to R_{i-1}. We say that a red vertex $v \in R_i$ is a *right vertex* if $v \notin R_i^\ell$ and that it is a *left vertex* if $v \notin R_i^r$.

Let $\Theta = (u_{k_1}, u_{k_2}, \ldots, u_{k_{|R|}})$ be the subordering of Γ induced by the set of all red vertices $R = R_1 \cup \ldots \cup R_q$. A subordering $\Theta_{i,j} = (u_{k_i}, \ldots, u_{k_j})$ for $i \leq j$ is a *component* of G if u_{k_i} is a right vertex, u_{k_j} is a left vertex, there is no $i' \in [j-1] \setminus [i-1]$ such that $u_{k_{i'}}$ is a left vertex and $u_{k_{i'+1}}$ is a right vertex, and it is maximal (that is, it must hold that either $u_{k_{i-1}}$ is a left vertex or $i = 1$, and additionally, that either $u_{k_{j+1}}$ is a right vertex or $j = |R|$). If, in addition, for every $v_{k'} \in \Theta_{i,j}$ there exists a long edge $v_{k''}v_{k'''} \in E(G)$ such that $k'' < k' < k'''$, then $\Theta_{i,j}$ is said to be a *covered component*. Note that a component can be a singleton. However, since we are assuming that G has only two simplicial vertices v_1 and v_n, a covered component can neither be a singleton nor intersect only one red region as its elements would be simplicial vertices different from v_1 and v_n. Thus, every covered component must contain vertices of at least two red regions. (See Fig. 1.) Finally, the components of a unit interval graph can be determined in linear time [9].

Now we present some structural results. In the next, we characterize some P_3^*-concave sets.

Lemma 2. *It holds that:*

(a) for $i \in [q-1]$, $R_i^r \cup B_i \cup R_{i+1} \cup B_{i+1}$ is a P_3^*-concave set;
(b) for $i \in [q-1]$, $R_i \cup B_i \cup R_{i+1} \cup B_{i+1}$ is a P_3^*-concave set;
(c) for $i \in \{0\} \cup [q-2]$, $B_i \cup R_{i+1} \cup B_{i+1} \cup R_{i+2}$ is a P_3^*-concave set; and
(d) for $i \in \{0\} \cup [q-1]$, if $R_{i+1}^\ell \cap R_{i+1}^r = \varnothing$, then $B_i \cup R_{i+1} \cup B_{i+1}$ is a P_3^*-concave set.

Proof. First note that if $\Gamma_{j,k} = (v_j, \ldots, v_k)$ for $j < k$ is the ordering of a set $S = \{v_j, \ldots, v_k\}$ in Γ, then all vertices in $\Gamma_{1,j-1}$ having a common neighbor in S are adjacent, and the same is true for all vertices in $\Gamma_{k+1,n}$ having a common neighbor in S. Thus, S is always a P_3^*-concave set for $j = 1$. Besides, S is a P_3^*-concave set for $j > 1$ if the distance of any vertex of $\Gamma_{1,j-1}$ to any vertex of $\Gamma_{k+1,n}$ is at least 3.

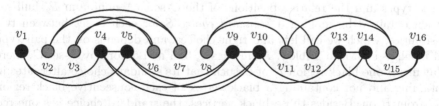

Fig. 1. Here there are two components: $\{v_2, v_3, v_6, v_7\}$ (covered) and $\{v_8, v_{11}, v_{12}, v_{15}\}$ (not covered), as both red vertices v_{11} and v_{12} are not covered by a long edge such as v_1v_5, v_4v_{10} and $v_{13}v_{16}$. Note that only edges between black vertices and between red vertices of distinct red regions are being depicted, and that the two shortest paths v_1 v_4 v_9 v_{13} v_{16} and v_1 v_5 v_{10} v_{14} v_{16} define precisely the black (and red) regions. (Color figure online)

(a) Take $S = R_i^r \cup B_i \cup R_{i+1} \cup B_{i+1}$. For $j > 1$, the fact that v_{j-1} has no edges to R_{i+1} implies that the distance of any vertex of $\Gamma_{1,j-1}$ to any vertex of $\Gamma_{k+1,n}$ is at least 3.

(b) By (a) and the fact that if $\Gamma_{j,k}$ is P_3^*-concave, then $\Gamma_{j',k'}$ is P_3^*-concave for $j' \leq j$ and $k' \geq k$.

(c) By symmetry, this case is equivalent to (b).

(d) Now take $S = B_i \cup R_{i+1} \cup B_{i+1}$. For $j > 1$, $d(v_{j-1}, v_{k+1}) \geq 3$ if $R_{i+1}^\ell \cap R_{i+1}^r = \varnothing$. $\qquad\square$

Next, we present a partition of Γ into parts called C-sets and classify them into 4 types.

– If $\Theta_{i,j}(u_{k_i}, \ldots, u_{k_j})$ is a covered component, then $\Gamma_{i',j'}$ is a C-set where $v_{i'} = u_{k_i}$ and $v_{j'} = u_{k_j}$. If $\Gamma_{i',j'}$ contains an odd number of black regions, then $\Gamma_{i',j'}$ has type 1. Otherwise $\Gamma_{i',j'}$ has type 2.

– If $\Gamma_{i,j}$ is maximal having no vertex belonging to a C-set of type 1 or 2, then $\Gamma_{i,j}$ is a C-set as well. If $\Gamma_{i,j}$ contains an odd number of black regions, then $\Gamma_{i,j}$ has type 3. Otherwise, $\Gamma_{i,j}$ has type 4.

For example, the unit interval graph of Fig. 1 gets partitioned into exactly three C-sets: $\Gamma_{1,1}$ (of type 3), $\Gamma_{2,7}$ (of type 1), and $\Gamma_{8,16}$ (of type 3). It is clear that the C-sets always form a partition of Γ. In fact, they also induce a partition of the black regions of G, i.e., all vertices of any black region are contained in a unique C-set and every C-set contains at least the vertices of one black region. These facts allow us to denote by $C_{i,j}$ the C-set whose set of black regions is $\{B_i, \ldots, B_j\}$. If $C_{i,j} \cap R_i \neq \emptyset$, then $C_{i,j} \cap R_i = R_i^r$. Similarly, if $C_{i,j} \cap R_{j+1} \neq \emptyset$, then $C_{i,j} \cap R_{j+1} = R_{j+1}^\ell$. Therefore, from now on, if not empty, both R_i^r and R_{j+1}^ℓ as well as R_k for $i < k < j+1$ will be called the *red regions of* $C_{i,j}$. Finally, note that those red regions might be empty if $C_{i,j}$ has type 3 or 4. However, if $C_{i,j}$ has type 1 or 2, then its red regions form a covered component, and consequently, not only both $R_i^r \neq \emptyset$ and $R_{j+1}^\ell \neq \emptyset$, but also both $R_k \neq \emptyset$ and $R_k^\ell \cap R_k^r \neq \emptyset$ for $i < k < j+1$, as a covered component does not contain a left vertex followed by a right vertex.

The proposed algorithm finds a $\{0, -1\}$ labeling of the C-sets, which depends on the types and the relative positions of the C-sets. A minimum P_3^*-hull set is then obtained by making a *standard choice* S for each C-set between two possibilities, which we call left and right choices, and depends on the pair type and label, as shown in Table 1. As in Figs. 2, 3, 4 and 5, for any C-set C, there is (at most) one black region B of C such that its standard choice alternates in containing and not containing a black vertex of each consecutive black region of C from B on. Besides those black vertices, the standard choice has one red vertex if C contains a covered component and no red vertex otherwise.

The *left choice* S for a C-set $C_{i,j}$ with type $t \in [4]$ and $k = j - i + 1$ black regions is defined as

$$S = \begin{cases} \{r_i^r\} \cup \{b_{i+1+2k'}^r : 0 \leq k' < \lfloor \frac{k}{2} \rfloor\} & \text{, if } t \in [2]; \\ \{b_{i+2k'}^r : 0 \leq k' < \lceil \frac{k}{2} \rceil\} & \text{, if } t \in \{3,4\}. \end{cases}$$

Algorithm 1. Finding a minimum P_3^*-hull set.

input: A unit interval graph G having exactly 2 simplicial vertices
1 Γ canonical ordering of $V(G)$
2 $\mathcal{C} \leftarrow$ the partition $(C_1, \ldots, C_t)$ of Γ into C-sets, where C_1 and C_t have types in $\{3,4\}$
3 **if** C_1 *has type* 3 **then**
4 $\quad \lfloor$ label$(C_1) \leftarrow 0$ (left choice for type 3)
5 **else**
6 $\quad \lfloor$ label$(C_1) \leftarrow -1$ (left choice for type 4)
7 **for** j *from* 2 *to* q **do**
8 $\quad$ **if** C_j *has type* 2 *or* 3 **then**
9 $\quad\quad \lfloor$ $label(C_j) \leftarrow -1 - label(C_{j-1})$
10 $\quad$ **else**
11 $\quad\quad \lfloor$ $label(C_j) \leftarrow label(C_{j-1})$
12 $S \leftarrow \{v_n\}$
13 **for** $C_i \in \mathcal{C}$ **do**
14 $\quad \lfloor$ $S \leftarrow S \cup S'$ where S' is a standard choice for $C_{i,j}$ according to Table 1
15 **return** S

Table 1. Standard choices.

Type	Label 0	Label -1
1	Right choice	Left choice
2	Left choice	Right choice
3	Left choice	Right choice
4	Right choice	Left choice

The *right choice* S for a C-set $C_{i,j}$ with type $t \in [4]$ and $k = j - i + 1$ black regions is defined as

$$S = \begin{cases} \{r\} \text{ for some } r \in R_{i+1}^\ell & \text{, if } t = 1 \text{ and } j = i; \\ \{r_{i+1}\} \cup \{b_{i+2k'}^r : 0 < k' < \lceil \frac{k}{2}\rceil\} \text{ for some } r_{i+1} \in R_{i+1}^\ell \cap R_{i+1}^r & \text{, if } t \in [2] \text{ and } j > i; \\ \{b_{i+1+2k'}^r : 0 \le k' < \lfloor \frac{k}{2}\rfloor\} & \text{, if } t \in \{3,4\}. \end{cases}$$

The idea of the linear algorithm is to give the left choice for the first C-set, and then alternate the standard choice from left to right and vice-versa if and only if the preceding C-set had type 2 or 3. Note that label -1 means "missing" and indicates that no vertex of the last black region B_j of $C_{i,j}$ belongs to its standard choice S. Note also that the first and the last C-sets in Γ are not a covered component, and therefore, always have types in $\{3,4\}$. Figures 2, 3, 4 and 5 depict the left and right choices for a C-set depending on its type.

In the next lemma we show that, for any $j \in [q]$, if B_{j-1} and B_j belong to distinct C-sets, then there is no vertex of R_j having neighbors in both R_{j-1} and R_{j+1}.

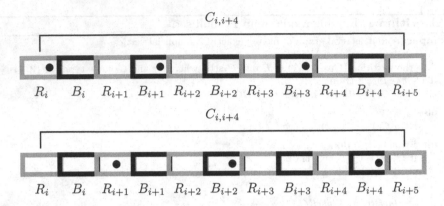

Fig. 2. Scheme representing the left (on top) and right (on bottom) choices of C-set $C_{i,i+4}$ with type 1. (Color figure online)

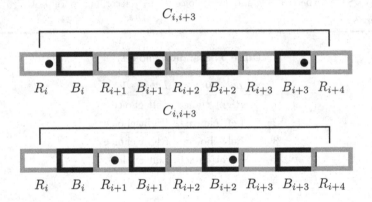

Fig. 3. Scheme representing the left (on top) and right (on bottom) choices of C-set $C_{i,i+3}$ with type 2. (Color figure online)

Lemma 3. *If $C_{i,j-1}$ and $C_{j,k}$ are C-sets, then $R_j^\ell \cap R_j^r = \varnothing$.*

Proof. By definition, if one of these C-sets has type 3 or 4, then the other one has type 1 or 2. This means that exactly one of these two C-sets has type 1 or 2, and therefore the red vertices of this C-set form a covered component. However, if $R_j^\ell \cap R_j^r \neq \varnothing$, then there would be a contradiction, as no vertex in R_j could be neither a left vertex nor a right vertex, i.e., neither the first nor the last red vertex of this C-set of type 1 or 2. $\qquad\square$

Lemma 4. *Every covered component of a unit interval graph G is P_3^*-concave.*

Proof. Let $C_{i,j}$ be the C-set of type either 1 or 2 containing the covered component, therefore, the red regions of $C_{i,j}$ form the covered component. Lemma 3 implies that, for $i < k < j + 1$, each red vertex of R_k neighbors only red vertices of R_{k-1}, R_k and R_{k+1}, each red vertex of R_i^r neighbors only red vertices

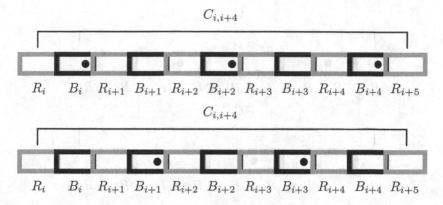

Fig. 4. Scheme representing the left (on top) and right (on bottom) choices of C-set $C_{i,i+4}$ with type 3. (Color figure online)

of R_i and R_{i+1} and each red vertex of R_{j+1}^ℓ neighbors only red vertices of R_j and R_{j+1}. Finally, for $i \leq k \leq j+1$, since $B_{k-1} \cup R_k \cup B_k$ form a clique and R_k neighbors only black vertices of $B_{k-1} \cup B_k$, each vertex of the covered component is such that all its neighbors not belonging to the covered component form a clique, meaning that the covered component is P_3^*-concave, and therefore, every covered component of G must intersect with every P_3^*-hull set of G. (As an alternative proof, the fact that $\langle S \rangle_3^* \subseteq \langle S \rangle_g$ for every set $S \subseteq V(G)$ also implies the claim, as from [9] it is known that every covered component of G is geodesically concave.) □

Remember that b_{i-1}^r is the rightmost vertex of B_{i-1}, and that b_{i+1}^ℓ is the leftmost vertex of B_{i+1}, being both b_{i-1}^r and b_{i+1}^ℓ adjacent to every vertex in B_i, but not to one another. Note also that b_{i+1}^ℓ has no neighbor in R_i^r, as no vertex in that set belongs to a shortest path between v_1 and v_n, and that b_{i-1}^r has no neighbor in R_{i+1}^ℓ, as no vertex in that set has distance i to v_1. The following property will be very useful.

Lemma 5. *If* $r \in R_i^r \cup R_{i+1}^\ell$, *then* $R_i^r \cup B_i \cup R_{i+1}^\ell \subseteq \langle \{b_{i-1}^r, r, b_{i+1}^\ell\} \rangle_3^*$.

Proof. Note that $B_i \subseteq [b_{i-1}^r, b_{i+1}^\ell]_3^*$ and that $R_i^r \neq \varnothing$ if and only if $R_{i+1}^\ell \neq \varnothing$. If $r \in R_i^r$ and v_k is the vertex with maximum index k in Γ belonging to $N(r) \cap R_{i+1}^\ell$, then $R_{i+1}^\ell \cap \Gamma_{1,k} \subseteq [\{r, b_{i+1}^\ell\}]_3^*$, $R_i^r \subseteq [\{b_{i-1}^r\} \cup (R_{i+1}^\ell \cap \Gamma_{1,k})]_3^*$, and $R_{i+1}^\ell \subseteq [R_i^r \cup \{b_{i+1}^\ell\}]_3^*$ as well. Similarly, in a symmetric way, if $r \in R_{i+1}^\ell$ and $v_{k'}$ is the vertex with minimum index k' in Γ belonging to $N(r) \cap R_i^r$, then $R_i^r \cap \Gamma_{k',n} \subseteq [\{r, b_{i-1}^r\}]_3^*$, $R_{i+1}^\ell \subseteq [\{b_{i+1}^\ell\} \cup (R_i^r \cap \Gamma_{k',n})]_3^*$ and $R_i^r \subseteq [R_{i+1}^\ell \cup \{b_{i-1}^r\}]_3^*$ as well. □

Now we are ready to understand why the linear algorithm presented, which starts with a left choice for the first C-set and then flips the standard choice from left to right and vice-versa if and only if the preceding C-set has type 2 or

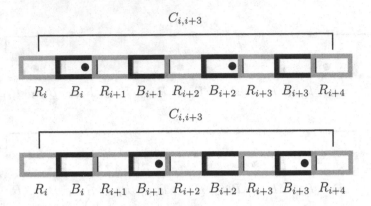

Fig. 5. Scheme representing the left (on top) and right (on bottom) choices of C-set $C_{i,i+3}$ with type 4. (Color figure online)

3, indeed provides a minimum P_3^*-hull set of G when the choices of the C-sets get united together with v_n. The following lemma throws light on that.

Lemma 6. *If $S_{i,j}$ is a standard choice for a C-set $C_{i,j}$, then the following holds:*

(a) $C_{i,j} \subseteq \langle S_{i,j} \cup \{b_{i-1}^r, b_{j+1}^\ell\} \rangle_3^*$;

(b) $C_{i,j} \setminus (R_i^r \cup R_{i+1}^\ell) \subseteq \langle S_{i,j} \cup B_i \cup \{b_{j+1}^\ell\} \rangle_3^*$;

(c) $C_{i,j} \setminus (R_j^r \cup R_{j+1}^\ell) \subseteq \langle S_{i,j} \cup B_j \cup \{b_{i-1}^r\} \rangle_3^*$;

(d) $C_{i,j} \setminus (R_i^r \cup R_{i+1}^\ell \cup R_j^r \cup R_{j+1}^\ell) \subseteq \langle S_{i,j} \cup B_i \cup B_j \rangle_3^*$.

Proof. Write $S_a = S_{i,j} \cup \{b_{i-1}^r, b_{j+1}^\ell\}$, $S_b = S_{i,j} \cup B_i \cup \{b_{j+1}^\ell\}$, $S_c = S_{i,j} \cup B_j \cup \{b_{i-1}^r\}$, and $S_d = S_{i,j} \cup B_i \cup B_j$. We give only one proof for all four cases, thus let $x \in \{a, b, c, d\}$.

First consider $C_{i,j}$ with type in $\{1, 2\}$, that is, its red regions form a covered component. It is clear that $B_i \subseteq [S_x]_3^*$ for $x \in \{a, b, c, d\}$. Since $b_k^\ell b_{k+1}^r \in E(G)$ for $i - 1 \le k \le j$, we have $B_k \subseteq [[S_x]_3^*]_3^* \subseteq \langle S_x \rangle_3^*$ for $x \in \{a, b, c, d\}$ and $i \le k \le j$. Besides, for each choice $S_{i,j}$, note that there is $r \in S_{i,j}$ such that either $r = r_i^r$ (left choice) or $r \in R_{i+1}^\ell \cap R_{i+1}^r$ (right choice), and thus, not only $[S_x]_3^* \cap R_{i+1}^\ell \cap R_{i+1}^r \ne \varnothing$ for $x \in \{a, b, c, d\}$ and $i < j$, but also by Lemma 5 we have that $R_i^r \cup R_{i+1}^\ell \subseteq \langle S_x \rangle_3^*$ for either $x \in \{a, c\}$ and $i < j$, or $x = a$ and $i = j$. Now, since $R_{k+1}^\ell \cap R_{k+1}^r \ne \varnothing$ for $i \le k \le j - 1$ as the red regions of $C_{i,j}$ form a covered component, due to Lemma 5 we have for $i < j$ by forwards induction starting on $[S_x]_3^* \cap R_{i+1}^\ell \cap R_{i+1}^r \ne \varnothing$ that $R_{k+1}^r \cup R_{k+2}^\ell \subseteq \langle S_x \rangle_3^*$ for either $x \in \{a, b\}$ and $i \le k \le j - 1$, or $x \in \{c, d\}$ and $i \le k \le j - 2$.

Now, consider that $C_{i,j}$ has type 3 or 4. Note that $\{b_k^r | i \le k \le j\} \subseteq [S_x]_3^*$ and $B_j \subseteq [[S_x]_3^*]_3^*$ for $x \in \{a, b, c, d\}$, which implies, by backwards induction starting on B_j, that $B_k \subseteq \langle S_x \rangle_3^*$ for $x \in \{a, b, c, d\}$ and $i \le k \le j$. Therefore, if some red region R_k for $i < k \le j$ is not covered by a long edge, then $R_k \subseteq [b_{k-1}^r, b_k^\ell]_3^* \subseteq \langle S_x \rangle_3^*$ for $x \in \{a, b, c, d\}$ as well. Thus, suppose that $(R_{i'}, \ldots, R_{j'})$ is a maximal sequence of covered non-empty red regions for $i + 1 \le i' \le k \le j' \le j$. Since

$C_{i,j}$ does not contain a covered component, $R_{i'-1} \supseteq R_{i'-1}^r \neq \emptyset$ with $i' > i+1$ or $R_{j'+1} \supseteq R_{j'+1}^\ell \neq \emptyset$ with $j' < j$ is not a covered red region. Without loss of generality, assume that $R_{i'-1} \supseteq R_{i'-1}^r \neq \emptyset$ with $i' > i+1$ is not a covered red region, meaning that $R_{i'-1} \subseteq \langle S_x \rangle_3^*$ for $x \in \{a, b, c, d\}$. (Otherwise, an analogous argument using Lemma 5 works with a backwards induction instead of a forwards one.) Note that Lemma 5 applied on $b_{i'-2}^r, r_{i'-1}^r, b_{i'}^\ell$ yields $R_{i'}^\ell \subseteq \langle S_x \rangle_3^*$. Now, as $C_{i,j}$ does not contain a covered component, $R_k^\ell \cap R_k^r \neq \emptyset$ for $i' \leq k < j'$, implying by forwards induction that Lemma 5 applied on $b_{k-1}^r, r, b_{k+1}^\ell$ with $r \in R_k^\ell \cap R_k^r$ yields $R_k^r \cup R_{k+1}^\ell \subseteq \langle S_x \rangle_3^*$ for either $i' \leq k \leq j'$ (if $R_{j'+1} \neq \emptyset$) or $i' \leq k < j'$ (if $R_{j'+1} = \emptyset$), that is, $R_k \subseteq \langle S_x \rangle_3^*$ for $i' \leq k \leq j'$, as either $R_{j'}^\ell \cap R_{j'}^r \neq \emptyset$ with $j' < j$ when $R_{j'+1} \neq \emptyset$ or $R_{j'} = R_{j'}^\ell$ when $R_{j'+1} = \emptyset$. Finally, it remains to show that $R_i^r \subseteq \langle S_x \rangle_3^*$ for $x \in \{a, c\}$ and that $R_{j+1}^\ell \subseteq \langle S_x \rangle_3^*$ for $x \in \{a, b\}$, but these facts are directly derived from Lemma 5, as in this case both R_i and R_{j+1} are covered red regions. $\qquad \square$

Let $(C_1, \ldots, C_t)$ be the C-sets ordered according to Γ. The set S returned by Algorithm 1 is a minimum P_3^*-hull set of G containing v_n as well as the standard choices selected by the algorithm for the C-sets, based on both the types and the received labels. Remark that the label is applied in such a way that the algorithm gives the left choice for C_1, and then consecutively alternates the standard choice from left to right and vice-versa if and only if the preceding C-set had type 2 or 3, maintaining it otherwise. In Lemma 8 we prove that S is in fact a P_3^*-hull set of G, whereas in Lemmas 9 to 11 we prove that there is no P_3^*-hull set of G with less than $|S|$ vertices. Define $f(C_i)$ as the cardinality of the standard choice that the algorithm associated with C_i and $f'(G)$ as the number of times that the labeling changes from -1 to 0, plus 1 if C_1 has type 3, and again plus 1 if C_t received label -1. In Theorem 1 we show that $|S| = f'(G) + \sum_{1 \leq i \leq t} f(C_i)$.

The next lemma combined with the previous one is key to comprehend the correctness.

Lemma 7. *If $C_{i,j}$ is a C-set of G and S is the set returned by Algorithm 1, then $B_i \cup \ldots \cup B_j \subseteq \langle S \rangle_3^*$. Hence, $b_{i-1}^r \in \langle S \rangle_3^*$ for $1 \leq i \leq q$ and $b_{j+1}^\ell \in \langle S \rangle_3^*$ for $0 \leq j \leq q - 1$.*

Proof. First, consider the case where $C_{i,j}$ has type 3 or 4. Let $S_{i,j} = S \cap C_{i,j}$. We begin assuming that $i \geq 1$ and $j \leq q - 1$. Observe that if $b_i^r \in S_{i,j}$, then there is $v \in S \cap (\{b_{i-2}^r\} \cup R_{i-1})$; otherwise there is $v \in S \cap (\{b_{i-1}^r\} \cup R_i)$. Observe also that if $b_j^r \in B$, then there is $w \in S \cap (\{b_{j+1}^r, b_{j+2}^r\} \cup R_{j+2})$; otherwise there is $w \in S \cap (\{b_{j+1}^r\} \cup R_{j+1})$. In all cases, it holds that $b_k^r \in [S_{i,j} \cup \{v, w\}]_3^*$ for $i - 1 \leq k \leq j + 1$. Since $C_{i,j}$ has type 3 or 4, the C-set containing B_{j+1} has type 1 or 2, which means that the edge $b_j^\ell b_{j+1}^r$ exists. Hence $b_j^\ell \in \langle S \rangle_3^*$, which implies that $B_i \cup \ldots \cup B_j \subset \langle S \rangle_3^*$. Now, if $i = 0$, then $B_i = \{v_1\}$ and $b_i^r = v_1 \in S$; and if $j = q$, then $B_j = \{v_n\}$ and $b_j^\ell = v_n \in S$, which means that $B_i \cup \ldots \cup B_j \subset \langle S \rangle_3^*$ even if for $i = 0$ or $j = q$.

Now, consider the case where $C_{i,j}$ has type 1 or 2. Note that the first C-set C_1 as well as the last C-set C_t have both types in $\{3, 4\}$. Thus, a C-set $C_{i,j}$ of type in $\{1, 2\}$ is such that not only $0 < i \leq j < q$, but also both its preceding and subsequent C-sets have types in $\{3, 4\}$. This fact jointly with both the previous case and (a) of Lemma 6 imply that $B_i \cup \ldots \cup B_j \subseteq \langle S \rangle_3^*$. $\qquad\square$

Lemma 8. *Algorithm 1 returns a P_3^*-hull set of G.*

Proof. Recall that G has exactly 2 simplicial vertices. Let S be the set returned by Algorithm 1 and $C_{i,j}$ be a C-set of G having type t. Consider first $i = 0$. If $i = 0$, $B_i = \{v_1\} \subseteq S$ and clearly by definition $R_0^r \cup R_1^\ell = \emptyset$. If $j = q$, $B_j = \{v_n\} \subseteq S$ and clearly by definition $R_q^r \cup R_{q+1}^\ell = \emptyset$. By (d) of Lemma 6, $V(G) = C_{i,j} = \langle S \rangle_3^*$. Now, consider that $j < q$. By Lemma 7, it holds $b_{j+1}^\ell \in \langle S \rangle_3^*$. Thus, by (b) of Lemma 6, $C_{i,j} \subseteq \langle S \rangle_3^*$. Next, consider $j = q$ and $i > 0$. By Lemma 7, $b_{i-1}^r \in \langle S \rangle_3^*$. By (c) of Lemma 6, $C_{i,j} \subseteq \langle S \rangle_3^*$. Finally, suppose $i > 0$ and $j < q$. By Lemma 7, $b_{i-1}^r, b_{j+1}^\ell \in \langle S \rangle_3^*$. By (a) of Lemma 6, $C_{i,j} \subseteq \langle S \rangle_3^*$. $\qquad\square$

We now define a lower bound, proved in Lemma 9, for the number of vertices that any P_3^*-hull set contains from a C-set $C_{i,j}$ as a function of its type t.

$$f(C_{i,j}) = \begin{cases} \frac{j-i+1}{2} & \text{, if } t \in \{2, 4\}; \\ \frac{j-i+2}{2} & \text{, if } t = 1; \\ \frac{j-i}{2} & \text{, if } t = 3. \end{cases}$$

Let $S_{i,j}$ be a standard choice of $C_{i,j}$. Note that $f(C_{i,j}) = |S_{i,j}|$ if $t \in \{1, 4\}$ or $S_{i,j}$ is a right choice; otherwise, $f(C_{i,j}) = |S_{i,j}| - 1$.

Lemma 9. *If S is a P_3^*-hull set and $C_{i,j}$ is a C-set of a unit interval graph G, then $|S \cap C_{i,j}| \geq f(C_{i,j})$.*

Proof. The number of black regions contained in $C_{i,j}$ is $j-i+1$. By Lemma 2 (a), $R_i^r \cup B_i \cup R_{i+1} \cup B_{i+1}$ is a P_3^*-concave set and $R_k \cup B_k \cup R_{k+1} \cup B_{k+1}$ is a P_3^*-concave set for $i + 1 \leq k \leq j - 1$. Therefore, $C_{i,j}$ contains at least $\lfloor \frac{j-i+1}{2} \rfloor$ disjoint P_3^*-concave sets, which implies the result if the type of $C_{i,j}$ is 2 or 4 as $j - i + 1$ is even or if its type is 3 as $j - i$ is not only even but also smaller than $j - i + 1$.

Now, consider that $C_{i,j}$ has type 1 and let S be a P_3^*-hull set of G. By definition, $j - i + 1$ is odd. Since $C_{i,j}$ contains a covered component, Lemma 4 implies that at least one vertex of a red region of $C_{i,j}$ belongs to S. Now, due to this fact, if $|S \cap C_{i,j}| < \lceil \frac{j-i+1}{2} \rceil$, then there are four consecutive regions of Γ, w.l.o.g. say $V' = R_{i'} \cup B_{i'} \cup R_{i'+1} \cup B_{i'+1}$ for $i \leq i' < j$ such that $V' \cap S = \emptyset$. By Lemma 2, V' is a P_3^*-concave set, which is a contradiction. Thus, the result also holds for type 1. $\qquad\square$

Lemma 10. *Let S be a P_3^*-hull set and let $C_{i,j}$ be a C-set of G such that $|S \cap C_i| = f(C_i)$.*

(a) If $C_{i,j}$ has type 2, then $S \cap (R_i^r \cup B_i \cup B_j \cup R_{j+1}^\ell) = \varnothing$;

(b) If $C_{i,j}$ has type 1 or 4, then $S \cap (R_i^r \cup B_i) = \varnothing$ or $S \cap (B_j \cup R_{j+1}^\ell) = \varnothing$;

(c) If $C_{i,j}$ has type 3, then $S \cap (R_i^r \cup B_i \cup R_{i+1} \cup R_j \cup B_j \cup R_{j+1}^\ell) = \varnothing$.

Proof. We first count the number of regions of $C_{i,j}$ in terms of $f(C_{i,j})$. (a) Suppose for contradiction that $v \in S \cap (R_i^r \cup B_i \cup B_j \cup R_{j+1}^\ell)$. By symmetry, assume that $S \cap (R_i^r \cup B_i) \neq \varnothing$. The number of black regions of $C_{i,j}$ is $j - i + 1$, which means that $C_{i,j}$ has $2(j - i + 1) + 1 = 4f(C_{i,j}) + 1$ regions, namely, $R_i^r, B_i, R_{i+1}, B_{i+1}, \ldots, R_j, B_j, R_{j+1}^\ell$. (b) First, consider that $C_{i,j}$ has type 1. Suppose for contradiction that $S \cap (R_i^r \cup B_i) \neq \varnothing$ and $S \cap (B_j \cup R_{j+1}^\ell) \neq \varnothing$. Then, $C_{i,j}$ has $2(j-i+1)+1 = 2(j-i+2)-1 = 4f(C_{i,j})-1$ regions. Now, consider that $C_{i,j}$ has type 4. Suppose for contradiction that $S \cap (R_i^r \cup B_i) \neq \varnothing$ and $S \cap (B_j \cup R_{j+1}^\ell) \neq \varnothing$. Then, $C_{i,j}$ has $2(j-i+1)+1 = 4f(C_{i,j})+1$ regions. (c) Suppose for contradiction that $v \in S \cap (R_i^r \cup B_i \cup R_{i+1} \cup R_j \cup B_j \cup B_{j+1}^\ell)$. By symmetry, assume $v \in R_{i-1}^r \cup B_i \cup R_i$. Then, $C_{i,j}$ has $2(j-i+1)+1 = 2(j-i)+3 = 4f(C_{i,j})+3$ regions.

Besides, by Lemma 4, S contains a vertex of a red region of $C_{i,j}$ if its type is either 1 or 2 (that is, if it contains a covered component). Now, using the pigeonhole principle in all (a), (b) and (c) items, we conclude that in all cases there are four consecutive regions of $C_{i,j}$ having no vertices of S. By Lemma 2, these four regions form a P_3^* concave set, which implies that S is not a P_3^*-hull set of G, a contradiction. $\qquad\square$

Lemma 11. *Consider the labeling obtained by Algorithm 1 and let S be a minimum P_3^*-hull set of G. The following sentences hold:*

(a) *If $(C_i, \ldots, C_j)$ is a maximal sequence of C-sets such that $label(C_j) = 0$ and $label(C_k) = -1$ for $i \leq k < j$, then $|S \cap (C_i \cup \ldots \cup C_j)| \geq f(C_i) + \ldots + f(C_j) + 1$; and*

(b) *If C_{j-1} and $C_j = C_{\ell_j, d(G)}$ are C-sets and $label(C_j) = -1$, then $|S \cap (C_{j-1} \cup C_j)| \geq f(C_{j-1}) + f(C_j) + 1$.*

Proof. (a) Suppose that $|S \cap (C_i \cup \ldots \cup C_j)| \leq f(C_i) + \ldots + f(C_j)$. If $j = 1$, then C_j has type 3. By Lemma 10, $S \cap B_0 = \varnothing$, which is a contradiction since $B_0 = \{v_1\}$ and v_1 is a simplicial vertex. Then consider $j > 1$. Remember that a C-set has label different of its predecessor if and only if its type is 2 or 3. Hence, $C_j = C_{j', j''}$ has type 2 or 3. By Lemma 10, it holds $S \cap (R_{j'}^r \cup B_{j'}) = \varnothing$. If $i = 1$, then $C_1 = C_{0, \ell_2 - 1}$ has type 4. Since v_1 is a simplicial vertex, $v_1 \in S$, then, by Lemma 10, $S \cap (B_{\ell_2 - 1} \cup R_{\ell_2}^\ell) = \varnothing$. If $i \geq 2$, then $C_i = C_{\ell_i, \ell_{i+1} - 1}$ has type 2 or 3. By Lemma 10, $S \cap (B_{\ell_2 - 1} \cup R_{\ell_2}^\ell) = \varnothing$. In both cases, $C_k = C_{\ell_k, \ell_{k+1} - 1}$ has type 1 or 4 for $i + 1 \leq k < j$. This means by Lemma 10 that $S \cap (R_{\ell_k}^r \cup B_{\ell_k}) = \varnothing$ or $S \cap (B_{\ell_{k+1} - 1} \cup R_{\ell_{k+1}}^\ell) = \varnothing$ for $i + 1 \leq k < j$. Therefore, by the pigeonhole principle, there is some $p \in \{i + 1, \ldots, j\}$ such that $B_{\ell_p} \cup R_{\ell_p + 1}^r \subset C_{\ell_p - 1, \ell_p - 1}$ and $R_{p+1}^\ell \cup B_{p+1} \subset C_{\ell_p, \ell_p + 1 - 1}$ such that $S \cap (B_p \cup R_{p+1} \cup B_{p+1}) = \varnothing$. By Lemmas 2 (d) and 3, $B_p \cup R_{p+1} \cup B_{p+1}$ is a P_3^*-concave set, which contradicts the assumption that S is a P_3^*-hull set.

(b) Suppose that $|S \cap (C_{j-1} \cup C_j)| \leq f(C_{j-1}) + f(C_j)$. We know that C_j has type $t \in \{3,4\}$, C_{j-1} has type 1 or 2, $B_{d(G)} = \{v_n\}$, and $R_{d(G)+1} = \varnothing$. If $t = 3$, then Lemma 10 (c) implies that $S \cap (R_i^r \cup B_i \cup R_{i+1} \cup R_{d(G)} \cup B_{d(G)}) = \varnothing$. But this is a contradiction because $v_n \in S$. Then consider $t = 4$. By Lemma 10 (b), $S \cap (R_{\ell_j}^r \cup B_{\ell_j}) = \varnothing$ or $S \cap B_{d(G)} = \varnothing$. Since $v_n \in S$, it holds $S \cap (R_{\ell_j}^r \cup B_{\ell_j}) = \varnothing$. Note that $label(C_{j-1}) = -1$. Let $(C_i, \ldots, C_j)$ be the maximal sequence of C-sets such that $label(C_k) = -1$ for $i \leq k \leq j$. Note that C_k has type 1 or 4 for $i < k \leq j$. Write $C_k = C_{\ell_k, \ell_{k+1}-1}$ for $i \leq k < j$. Consider first $i = 1$. By the algorithm, $C_1 = C_{0,\ell_2-1}$ has type 4. By Lemma 10 (b), $B_0 \cap S = \varnothing$ or $(B_{\ell_2-1} \cup R_{\ell_2}^\ell) \cap S = \varnothing$. Since $v_1 \in B_0$ is a simplicial vertex, it holds $(B_{\ell_2-1} \cup R_{\ell_2}^\ell) \cap S = \varnothing$. Now consider $i > 1$. The algorithm implies that C_i has type 2 or 3. Lemmas 10 (a) and (c) imply that $(B_{\ell_{i+1}-1} \cup R_{\ell_{i+1}}^\ell) \cap S = \varnothing$. Thus, in any case, Lemma 10 (b) implies that $S \cap (R_{\ell_k}^r \cup B_{\ell_k}) = \varnothing$ or $S \cap (B_{\ell_{k+1}-1} \cup R_{\ell_{k+1}}^\ell) = \varnothing$ for $i < k < j$. This means that there is some $p \in \{i+1, \ldots, j\}$ such that $S \cap (B_p \cup R_{p+1} \cup B_{p+1}) = \varnothing$, which is a contradiction by Lemmas 2 and 3. □

Theorem 1. *If G is a unit interval graph with exactly two simplicial vertices, then $h_3^*(G) = f'(G) + \sum_{1 \leq i \leq t} f(C_i)$. Besides, the P_3^*-hull number of a unit interval graph G can be found in linear time.*

Proof. Consequence of Lemmas 8, 9, 10, and 11. Besides, a canonical ordering of a unit interval graph can be found in linear time [8,16], and thus, its simplicial vertices as well. Since the components of a unit interval graph can be determined in linear time [9], the result follows due to Lemma 1. □

3   Chordal Graphs

We conclude by pointing out the succeeding NP-completeness for the superclass of chordal graphs.

Theorem 2. *Given a chordal graph G and an integer k, it is NP-complete to decide whether $h_3^*(G) \leq k$.*

The main idea behind the NP-completeness proof (omitted here due to lack of space) is a polynomial reduction from a restricted version of SATISFIABILITY which is NP-complete [10,12] . Let $\mathcal{C}$ be an instance of SATISFIABILITY consisting of m clauses $C_1, \ldots, C_m$ over n boolean variables $x_1, \ldots, x_n$ such that every clause in $\mathcal{C}$ contains at most three literals and, for every variable x_i, there are exactly two clauses in $\mathcal{C}$, say $C_{j_i^1}$ and $C_{j_i^2}$, that contain the literal x_i, and exactly one clause in $\mathcal{C}$, say $C_{j_i^3}$, that contains the literal $\bar{x}_i$, and these three clauses are distinct.

Let the graph G be constructed as follows starting with the empty graph. For every $j \in [m]$, add a vertex c_j. For every $i \in [n]$, add 10 vertices $x_i, y_i, z_i, x_i^1, x_i^2, w_i^1, w_i^2, \bar{x}_i, \bar{y}_i, \bar{w}_i$ and 17 edges to obtain the subgraph indicated in Fig. 6. Add a vertex z and the edges to make a clique of $C \cup Z \cup \{z\}$, where

$C = \{c_j | j \in [m]\}$ and $Z = \{z_i | i \in [n]\}$. Setting $k = 4n + 1$, we show in the full version of the paper that C is satisfiable if and only if G contains a P_3^*-hull set of order at most k.

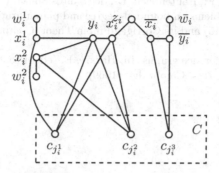

Fig. 6. When the construction of G ends, z_i will belong to the clique $C \cup \{z_1, \ldots, z_n\} \cup \{z\}$.

References

1. Albenque, M., Knauer, K.: Convexity in partial cubes: the hull number. Discret. Math. **339**(2), 866–876 (2016)
2. Araujo, J., Campos, V., Giroire, F., Nisse, N., Sampaio, L., Soares, R.: On the hull number of some graph classes. Theoret. Comput. Sci. **475**, 1–12 (2013)
3. Araujo, J., Morel, G., Sampaio, L., Soares, R., Weber, V.: Hull number: P_5-free graphs and reduction rules. Discret. Appl. Math. **210**, 171–175 (2016)
4. Araújo, R.T., Sampaio, R.M., dos Santos, V.F., Szwarcfiter, J.L.: The convexity of induced paths of order three and applications: Complexity aspects. Discret. Appl. Math. **237**, 33–42 (2018)
5. Bessy, S., Dourado, M.C., Penso, L.D., Rautenbach, D.: The geodetic hull number is hard for chordal graphs. SIAM J. Discret. Math. **32**(1), 543–547 (2018) and LAGOS'17 - ENDM 62 (2017) 291–296
6. Centeno, C.C., Dourado, M.C., Penso, L.D., Rautenbach, D., Szwarcfiter, J.L.: Irreversible conversion of graphs. Theoret. Comput. Sci. **412**, 3693–3700 (2011)
7. Coelho, E.M.M., Dourado, M.C., Sampaio, R.M.: Inapproximability results for graph convexity parameters. Theoret. Comput. Sci. **600**, 49–58 (2015)
8. Deng, X., Hell, P., Huang, J.: Linear-time representation algorithms for proper circular-arc graphs and proper interval graphs. SIAM J. Comput. **25**(2), 390–403 (1996)
9. Dourado, M.C., Gimbel, J.G., Kratochvíl, J., Protti, F., Szwarcfiter, J.L.: On the computation of the hull number of a graph. Discret. Math. **309**, 5668–5674 (2009)
10. Dourado, M.C., Penso, L.D., Rautenbach, D.: On the geodetic hull number of P_k-free graphs. Theoret. Comput. Sci. **640**, 52–60 (2016)
11. Dourado, M.C., Protti, F., Szwarcfiter, J.L.: Complexity results related to monophonic convexity. Discret. Appl. Math. **158**, 1268–1274 (2010)
12. Garey, M.R., Johnson, D.S.: Computers and Intractability: A Guide to the Theory of NP-Completeness. WH Freeman & Co., New York (1979)

13. Gavruskin, A., Khoussainov, B., Kokho, M., Liu, J.: Dynamising interval scheduling: the monotonic case. Theoret. Comput. Sci. **562**, 227–242 (2015)
14. McConnell, R.M., Nussbaum, Y.: Linear-time recognition of probe interval graphs. SIAM J. Discret. Math. **29**(4), 2006–2046 (2015)
15. Penso, L.D., Protti, F., Rautenbach, D., dos Santos Souza, U.: Complexity analysis of P_3-convexity problems on bounded-degree and planar graphs. Theoret. Comput. Sci. **607**, 83–95 (2015) and its Corrigendum in Theor. Comput. Sci. 704 (2017) 92–93
16. Roberts, F.S.: Indifference graphs. In: Harary, F. (ed.) Proof Techniques in Graph Theory. Academic Press, Cambridge (1969)

Supermagic Graphs with Many
Odd Degrees

Dalibor Froncek[✉][iD] and Jiangyi Qiu

University of Minnesota Duluth, Duluth, USA
{dalibor,qiuxx284}@d.umn.edu

Abstract. A graph $G = (V, E)$ is called *supermagic* if there exists a
bijection $f : E \to \{1, 2, \ldots, |E|\}$ such that the *weight* of every vertex
$x \in V$ defined as the sum of labels $f(xy)$ of all edges xy incident with x
is equal to the same number m, called the *supermagic constant*.

Recently, Kovář et al. affirmatively answered a question by Madaras
about existence of supermagic graphs with arbitrarily many different
degrees. Their construction provided graphs with all degrees even. Therefore, they asked if there exists a supermagic graph with d different odd
degrees for any positive integer d.

We answer this question in the affirmative by providing a construction
based on the use of 3-dimensional magic rectangles.

Keywords: Supermagic graphs · Magic-type labeling · Edge labeling

1 Introduction

A finite simple graph $G = (V, E)$ is called *supermagic* if there exists a bijection
$f : E \to \{1, 2, \ldots, |E|\}$ called *supermagic labeling* such that the *weight* of every
vertex $x \in V$ defined as the sum of labels $f(xy)$ of all edges xy incident with x
is equal to the same number m, called the *supermagic constant*. That is,

$$\exists m \in \mathbb{N} \ \forall x \in V : w(x) = \sum_{xy \in E} f(xy) = m.$$

Most often, graphs studied in this context are vertex-regular or even vertex-
transitive.

Recently, Kovář, Kravčenko, Krbeček, and Silber, [2] affirmatively answered
a question by Madaras: Does there exist a supermagic graph with arbitrarily
many different degrees?

Because their construction provided only graphs where all degrees were even,
they asked the following more specific question: Does there exist a supermagic
graph with d different odd degrees for any positive integer d?

We construct a class of disconnected graphs based on lexicographic products
(also called compositions) of graphs with many different odd degrees and complements of complete graphs. We also present a modification of this construction
that provides connected graphs with the same required properties.

© Springer Nature Switzerland AG 2019
C. J. Colbourn et al. (Eds.): IWOCA 2019, LNCS 11638, pp. 229–236, 2019.
https://doi.org/10.1007/978-3-030-25005-8_19

The labeling used in the construction is based on the existence of three-dimensional magic rectangles.

2   Tools

To construct the graphs, we will use one of the standard graph products, called the lexicographic product or also composition.

A graph $G \circ H$ called the *lexicographic product* or *composition* of graphs G and H arises from G by replacing each vertex of G by a copy of H, and every edge of G by the complete bipartite graph $K_{t,t}$, where t is the number of vertices of H.

More formally, let $V(G) = \{g_1, g_2, \ldots, g_s\}$ and $V(H) = \{h_1, h_2, \ldots, h_t\}$. Then $V(G \circ H) = V(G) \times V(H)$ and $(g_a, h_b)(g_c, h_d) \in E(G \circ H)$ if and only if $g_a g_c \in E(G)$ or $g_a = g_c$ and $h_b h_d \in E(H)$.

If the graph H is isomorphic to $\overline{K_t} = tK_1$, that is, consists of t independent vertices, then we say that $G \circ t K_1$ is a *blown up G*, or that we have *blown up G by tK_1*. In this case, the graph $G \circ t K_1$ will be denoted simply by $G[t]$.

An important ingredient of our construction is a 3-dimensional magic rectangle. We start with a more general definition introduced by Hagedorn [1].

Definition 1. *An n-dimensional magic rectangle n-MR$(a_1, a_2, \ldots, a_n)$ is an $a_1 \times a_2 \times \cdots \times a_n$ array with entries $d_{i_1, i_2, \ldots, i_n}$ which are elements of the set $\{1, 2, \ldots, a_1 a_2 \ldots a_n\}$, each appearing once, such that all sums in the k-th direction are equal to a constant σ_k. That is, for every k, $1 \leq k \leq n$, and every selection of indices $i_1, i_2, \ldots, i_{k-1}, i_{k+1}, \ldots, i_n$, we have*

$$\sum_{j=1}^{a_k} d_{i_1, i_2, \ldots, i_{k-1}, j, i_{k+1}, \ldots, i_n} = \sigma_k,$$

where $\sigma_k = a_k(a_1 a_2 \ldots a_n + 1)/2$.

The following existence results were also proved by Hagedorn in [1].

Theorem 1 [1].   *If there exists an n-dimensional magic rectangle n-MR$(a_1, a_2, \ldots, a_n)$, then $a_1 \equiv a_2 \equiv \cdots \equiv a_n \pmod 2$.*

For $a_1 \equiv a_2 \equiv \cdots \equiv a_n \equiv 0 \pmod 2$, Hagedorn found a complete existence characterization.

Theorem 2 [1]. *An n-dimensional magic rectangle n-MR$(a_1, a_2, \ldots, a_n)$ with $a_1 \leq a_2 \leq \cdots \leq a_n$ and all a_i even exists if and only if $2 \leq a_1$ and $4 \leq a_2 \leq \cdots \leq a_n$.*

For odd dimensions, Hagedorn proved the following.

Theorem 3 [1]. *A 3-dimensional magic rectangle 3-MR(a_1, a_2, a_3) with $3 \leq a_1 \leq a_2 \leq a_3$ exists whenever $\gcd(a_i, a_j) > 1$ for some $i, j \in \{1, 2, 3\}$.*

We will use a special case of his result in our construction.

Corollary 1 [1]. *A 3-dimensional magic rectangle* 3-MR$(3,3,a)$ *exists for every odd* a, $a \geq 3$.

Recently, Zhou, Li, Zhang, and Su [3] proved that for 3-dimensional magic rectangles the necessary conditions are also sufficient.

Theorem 4 [3]. *A 3-dimensional magic rectangle* 3-MR(a_1, a_2, a_3) *with* $3 \leq a_1 \leq a_2 \leq a_3$ *exists whenever* $a_1 \equiv a_2 \equiv a_3 \equiv 1$ (mod 2).

3 Construction

We build our graphs in two steps. First we build a graph G with the required number of different odd degrees, and label the edges with just two different labels so that the sum of labels at every vertex is constant. Then we blow up G into $G[3]$ and label the edges of each $K_{3,3}$ using entries of one 3×3 rectangle from a 3-MR$(3,3,a)$, or from a slightly modified 3-MR$(3,3,b)$ with $b = 8a + 1$.

We call a graph G 2-*pseudomagic* if there are positive integers l and h and a mapping $\tilde{g} : E(G) \rightarrow \{l, h\}$ called 2-*pseudomagic labeling* such that the weight of every vertex x, defined as the sum of labels of all edges incident with x, is equal to the same constant $\tilde{m}$. The edges labeled l will be called *light*, and those labeled h will be called *heavy*.

Let 3-MR(a_1, a_2, a_3) be a 3-dimensional magic rectangle as defined above, with entries d_{j_1, j_2, j_3} and magic constants $\sigma_i, i = 1, 2, 3$. For 3-MR$(3, 3, a)$ we have $\sigma_1 = \sigma_2 = 3(9a + 1)/2$. Recall that a must be odd.

Then the 3-dimensional $a_1 \times a_2 \times a_3$ array with entries $d^+_{j_1, j_2, j_3} = c + d_{j_1, j_2, j_3}$ is called a 3-*dimensional c-lifted magic rectangle*. It should be obvious that the magic constants here are equal to $a_i c + \sigma_i$ for each $i = 1, 2, 3$. Such a rectangle will be denoted 3-MR$^+(a_1, a_2, a_3; c)$ with magic constants σ_i^+, where by the reasoning above we have $\sigma_i^+ = a_i c + \sigma_i$.

In order to use consecutive integers in the labeling of $G[3]$, we want to find 3-MR$(3, 3, a)$ and 3-MR$^+(3, 3, b; c)$ such that $c = 9a$ and $l\sigma_i^+ = h\sigma_i$ for some positive integers l and h. Note that because 3-MR$^+(3, 3, b; c)$ is constructed from a 3-dimensional magic rectangle 3-MR$(3, 3, b)$, we must have b odd.

We present one such pair in the following. For simplicity, we choose $l = 1$.

Lemma 1. *Let* $c = 9a$, $b = 8a + 1$, *and* $\sigma_1 = \sigma_2$ *and* $\sigma_1^+ = \sigma_2^+$ *be magic constants of* 3-MR$(3, 3, a)$ *and* 3-MR$^+(3, 3, b; c)$, *respectively. Then* $\sigma_1^+ = 10\sigma_1$.

Proof. We have

$$\sigma_1 = \frac{9a(9a + 1)}{2 \cdot 3a} = \frac{3(9a + 1)}{2}$$

Since the labels in 3-MR$^+(3, 3, b; c)$ start with $9a + 1$, we have

$$\sigma_1^+ = \frac{(9a + 1 + 9(a + b))9b}{2 \cdot 3b} = \frac{3(9a + 1 + 9(a + b))}{2}$$

Substitute $b = 8a + 1$ to get

$$\sigma_1^+ = \frac{3(9a + 1 + 9(a + 8a + 1))}{2} = \frac{3(9a + 1)(1 + 9))}{2} = 10\frac{3(9a + 1)}{2} = 10\sigma_1$$

as desired.

Now we construct a 2-pseudomagic host graph G with d different odd degrees. The labels we use are $l = 1$ and $h = 10$. We will say that a vertex x has a *light degree* $\deg_l(x)$ if it is incident with $\deg_l(x)$ light edges, that is, edges labeled 1. Similarly, vertex x has a *heavy degree* $\deg_h(x)$ if it is incident with $\deg_h(x)$ heavy edges, that is, edges labeled 10. The weight of a vertex x in G will be denoted $\widetilde{w}(x)$. From the above it follows that

$$\widetilde{w}(x) = \deg_l(x) + 10\deg_h(x). \tag{1}$$

Construction 1. Let p be a prime, $p \geq 2d + 1$, and $d > 1$. Our host graph G will consist of $d + 2$ components. A light component G_0, mixed components $G_1, G_2, \ldots, G_d$, and heavy component G_{d+1}. To use the magic rectangles from Lemma 1 for blowing up G to $G[3]$, we will need a light and $b = 8a + 1$ heavy edges. We denote by a_i and b_i the number of light and heavy edges in G_i, respectively.

We start with $G_0 \cong K_{10p+1}$ with all light edges. Notice that the number of edges in G_0 is $(10p + 1)5p$, which is odd, since p is a prime. We have $\deg(x_{0,j}) = \deg_l(x_{0,j}) = 10p$ and from (1) it follows that $\widetilde{w}(x_{0,j}) = 10p$ for every $x_{0,j} \in G_0$.

Then for $i = 1, 2, \ldots, d - 1$ we first take $G_i' \cong K_{10p+2} - M_{10p+2}$, where M_{10p+2} is a perfect matching. This is a $10p$-regular graph, so we have used an even number of light edges in each $G_i', i > 0$.

Next we build $G_i, i = 1, 2, \ldots, d - 1$ by removing $10(2i - 1)$ light one-factors and adding back $(2i - 1)$ heavy one-factors. The number of light edges in G_i is still even, since for every $x_{i,j} \in G_i$ we have

$$\deg_l(x_{i,j}) = 10p - 10(2i - 1) \tag{2}$$

and the order of G_i is $10p + 2$. Because we added back $(2i - 1)$ heavy one-factors, we have

$$\deg_h(x_{i,j}) = 2i - 1. \tag{3}$$

Therefore, by adding (2) and (3), the degree of $x_{i,j}$ is

$$\deg(x_{i,j}) = \deg_l(x_{i,j}) + \deg_h(x_{i,j}) = 10p - 9(2i - 1) = 10p - 18i + 9, \tag{4}$$

which is indeed odd. From (1) it follows that for every $x_{i,j} \in G_i$ we have

$$\widetilde{w}(x_{i,j}) = \deg_l(x_{i,j}) + 10\deg_h(x_{i,j}) = 10p - 10(2i - 1) + 10(2i - 1) = 10p. \tag{5}$$

All components except G_0 contain an even number of light edges, making a odd as needed. It should be obvious that

$$b_1 + b_2 + \cdots + b_d < 8(a_1 + a_2 + \cdots + a_d). \tag{6}$$

Since $b = 8a + 1$ and $b_0 = 0$, we must add some heavy edges. We denote the number of lacking heavy edges by b^*. If b^* is a multiple of p, say $b^* = ps$, we can build a p-regular component G_d with $2s$ vertices, and we are done. Notice that $2s > p$, because we still need to compensate for $a_0 = (10p + 1)5p$ light edges in G_0, that is, we need to add at least $8a_0 = 40(10p + 1)p$ heavy edges.

If b^* is not a multiple of p, we keep adding copies of G_1 to G until the number of lacking heavy edges is divisible by p.

For each copy of G_1, we will add

$$b_1 = \frac{10p + 2}{2} = 5p + 1$$

heavy edges and

$$a_1 = \frac{(10p - 10)(10p + 2)}{2} = 10(p - 1)(5p + 1)$$

light edges. Suppose we are adding q copies, and recall that the number of lacking heavy edges is b^*. Denote the number of light and heavy edges in G by $\bar{a}$ and $\bar{b}$, respectively. Hence, we have

$$b = \bar{b} + qb_1 + b^*,$$

and

$$a = \bar{a} + qa_1.$$

Because $b = 8a + 1$, we obtain

$$\bar{b} + qb_1 + b^* = 8(\bar{a} + qa_1) + 1,$$

or

$$b^* = (8\bar{a} + 1 - \bar{b}) + q(8a_1 - b_1), \tag{7}$$

where

$$\begin{aligned}
8a_1 - b_1 &= 8(10(p - 1)(5p + 1)) - (5p + 1) \\
&= (5p + 1)(80(p - 1) - 1) \\
&= (5p + 1)(80p - 81) \tag{8}
\end{aligned}$$

We assumed that $8\bar{a} + 1 - \bar{b}$ is not divisible by p, so suppose

$$8\bar{a} + 1 - \bar{b} \equiv k \pmod{p}.$$

If we want to have b^* a multiple of p, it follows from (7) that we must have

$$q(8a_1 - b_1) \equiv -k \pmod{p}. \tag{9}$$

Note that the congruence has a solution for q if and only if

$$\gcd(8a_1 - b_1, p) \mid k$$

and moreover, whenever there is a solution, then there is always one such that $1 \leq q \leq p - 1$.

Recall that we assumed $d > 1$ and $p \geq 2d + 1$. Hence, $p > 3$ and we clearly have $\gcd(81, p) = 1$. Then from Eq. (8) we will have $\gcd(8a_1 - b_1, p) = 1$ because $p \nmid 5p + 1$ and $p \nmid (80p - 81)$ when $\gcd(81, p) = 1$. Therefore, we can always solve congruence (9) for q.

Hence, we can always construct a 2-pseudomagic graph G with edge labels 1 and 10 and d different odd degrees, where the number of light edges is a and the number of heavy edges is $b = 8a + 1$.

4   Main Result

We start with an easy observation, tying magic rectangles to supermagic labelings of complete bipartite graphs.

Lemma 2. *Let* $3\text{-MR}(a_1, a_1, a_3)$ *be a 3-dimensional magic rectangle and* $H = a_3 K_{a_1, a_1}$ *a disjoint union of* a_3 *copies of the complete bipartite graph* K_{a_1, a_1}. *Then there exists a supermagic labeling of* H.

Proof. Denote the k-th copy of K_{a_1, a_1} by H^k and vertices in its respective partite sets by $u_1^k, u_2^k, \ldots, u_{a_1}^k$ and $v_1^k, v_2^k, \ldots, v_{a_1}^k$. Let

$$f(u_i^k v_j^k) = d_{i,j,k}$$

for every $i, j = 1, 2, \ldots, a_1$ and every $k = 1, 2, \ldots, a_3$. Then

$$w(u_i^k) = \sum_{j=1}^{a_1} f(u_i^k v_j^k) = \sum_{j=1}^{a_1} d_{i,j,k}$$

and

$$w(v_j^k) = \sum_{i=1}^{a_1} f(u_i^k v_j^k) = \sum_{i=1}^{a_1} d_{i,j,k}.$$

Because we have $a_1 = a_2$, from Definition 1 it follows that

$$\sum_{j=1}^{a_1} d_{i,j,k} = \sum_{i=1}^{a_1} d_{i,j,k} = \frac{a_1(a_1^2 a_3 + 1)}{2} = \sigma_1,$$

which implies

$$w(u_i^k) = w(v_j^k) = \sigma_1$$

for every $i, j = 1, 2, \ldots, a_1$ and every $k = 1, 2, \ldots, a_3$, and f is the desired supermagic labeling.

Now we are ready to prove our main result.

Theorem 5. *For every positive integer d there exists a supermagic graph with d different odd vertex degrees.*

Proof. For $d = 1$ we take $G \cong K_2$ and label the only edge by 1. For $d > 1$, we take the graph G with d different odd degrees from Construction 1 and blow it up to $G[3]$. Each vertex with an odd degree $\deg_G(x)$ in G is now of odd degree $3 \deg_G(x)$. Then we label each $K_{3,3}$ arising from a light edge by entries of a 3-MR$(3, 3, a)$ and each $K_{3,3}$ arising from a heavy edge by entries of a 3-MR$^+(3, 3, b; 9a)$. We will call the graphs $K_{3,3}$ in $G[3]$ arising from light and heavy edges in G also *light* and *heavy* graphs, respectively. The triple of vertices arising from a vertex x will be denoted by $x^{[1]}, x^{[2]}, x^{[3]}$.

In particular, for $a_1 = a_2 = 3$ and $a_3 = a$ in Lemma 2 we have $\sigma_1 = 3(9a + 1)/2$. From Lemma 1 it follows that $\sigma_1^+ = 10\sigma_1 = 15(9a + 1)$.

We observe that each light $K_{3,3}$ contributes to every $w(x^{[i]})$ for $i = 1, 2, 3$ by σ_1 and a heavy $K_{3,3}$ contributes by $\sigma_1^+ = 10\sigma_1$. Therefore, the total contribution of all light graphs at a vertex $x^{[i]}$ is $\sigma_1 \deg_l(x)$ and the contribution of all heavy graphs is $\sigma_1^+ \deg_h(x) = 10\sigma_1 deg_h(x)$. It follows that

$$w(x^{[i]}) = \sigma_1 \deg_l(x) + 10\sigma_1 deg_h(x) = \sigma_1 \left(\deg_l(x) + 10 deg_h(x)\right). \qquad (10)$$

But from (5) we have $\widetilde{w}(x) = \deg_l(x) + 10 \deg_h(x) = 10p$, and thus we immediately obtain from (10) that

$$w(x) = \sigma_1 \widetilde{w}(x) = 10p\sigma_1 \qquad (11)$$

for every vertex $x^{[i]}$ in $G[3]$, which concludes our proof. $\qquad \square$

It is not difficult to modify the graphs constructed above to make them connected.

Corollary 2. *For every positive integer d there exists a connected supermagic graph with d different odd vertex degrees.*

Proof. For $d = 1$, we again have $G \cong K_2$. For $d > 1$, we construct a connected graph H from G first, and find the blown-up graph $H[3]$ exactly the same way as for $G[3]$ in Theorem 5.

Recall that the graph G may contain more than one copy of G_1. We denoted the number of these components by $q + 1$ and observed that $q + 1 \leq p$. Denote the copies of G_1 by G_1^k for $k = 0, 1, \ldots, q$. First we observe that each component $G_1^k, G_2, \ldots, G_d$ is large enough to contain two independent heavy edges, say $x_{i,1}y_{i,1}$ and $x_{i,2}y_{i,2}$ in component G_i.

For $i = 1, 2, \ldots, d$ we replace edges $x_{i,1}y_{i,1}$ and $x_{i+1,2}y_{i+1,2}$ by $x_{i,1}y_{i+1,2}$ and $x_{i+1,2}y_{i,1}$ (here $x_{1,1}y_{1,1}$ and $x_{1,2}y_{1,2}$ belong to G_1^0). This connects all graphs $G_1^0, G_2, \ldots, G_d$ into one component without changing any vertex degrees.

Now we connect G_0 to each G_1^k for $k = 0, 1, \ldots, q$ in a similar manner. Clearly, $G_0 \cong K_{10p+1}$ contains $p \geq q + 1$ independent light edges. Select $q + 1$ and call them $u_i v_i$ for $i = 0, 1, \ldots, q$. Then select a light edge $t_i z_i$ in the i-th copy of G_1 and replace the pair of edges $u_i v_i$ and $t_i z_i$ by edges $u_i t_i$ and $v_i z_i$.

As before, this connects all copies of G_1 to G_0 without changing the degree of any vertex.

We have thus constructed a connected graph H with d different odd degrees, a light edges, $b = 8a + 1$ heavy edges, and a constant weight $\widetilde{w}(x) = 10p$ for every vertex x. Blowing it up the same way as in the proof of Theorem 5 concludes this proof.

References

1. Hagedorn, T.R.: On the existence of magic n-dimensional rectangles. Discrete Math. **207**(1–3), 53–63 (1999)
2. Kovář, P., Kravčenko, M., Krbeček, M., Silber, A.: Supermagic graphs with many different degrees. Discuss. Math. Graph Theory (accepted)
3. Zhou, C., Li, W., Zhang, Y., Su, R.: Existence of magic 3-dimensional rectangles. J. Combin. Des. **26**(6), 280–309 (2018)

Incremental Algorithm for Minimum Cut and Edge Connectivity in Hypergraph

Rahul Raj Gupta and Sushanta Karmakar[✉]

Department of Computer Science and Engineering,
IIT Guwahati, Guwahati, Assam, India
rahulrg.raj@gmail.com, sushantak@iitg.ac.in

Abstract. For an uncapacitated hypergraph $H = (V, E)$ with $n = |V|$, $m = |E|$ and $p = \Sigma_{e \in E}|e|$, and edge connectivity λ, this paper presents an insertion-only algorithm which updates minimum cut and edge connectivity incrementally on addition of a set of hyperedges to an existing hypergraph. The algorithm is deterministic and takes $O(\lambda n)$ amortized time per insertion of a hyperedge. The algorithm can answer queries on edge-connectivity in $O(1)$ time and returns a cut of size λ in $O(n)$ time. First we propose a method to maintain a hypercactus [3] under the addition of a set of hyperedges. It is observed that the time for maintaining a hypercactus on addition of a set U of hyperdeges is $O(n + p_u)$ where $p_u = \Sigma_{e \in U}|e|$. This method is then used as a subroutine in our incremental algorithm for maintaining minimum cut and edge connectivity.

Keywords: Hypergraph · Minimum cut · Edge connectivity · Hypercactus

1 Introduction

Computing the minimum cut (or edge connectivity) is a fundamental problem in graph algorithms. There are many algorithms to compute the edge connectivity in a simple graph [6,12,16,19]. There are also a few algorithms that maintain the edge connectivity in a simple graph under the addition of a few edges and vertices [9,10]. Computing the minimum cut or the edge connectivity for a hypergraph is also an important problem. It has applications in various fields e.g., circuit and chip design, network communication, planning in transportation, circuit partitioning and cluster analysis. There are a few algorithms to compute the edge connectivity for a static hypergraph [2,13,15]. The best algorithm known so far to compute the minimum cut for a static hypergraph is given by Chekuri and Xu [2]. However, to the best of our knowledge, no algorithm exists in the literature that maintains the minimum cut or the edge connectivity for dynamic hypergraphs where a few hyperedges are added to or deleted from a given hypergraph. For dynamic hypergraphs, one straightforward approach is to apply a known algorithm [2,13,15] to compute the minimum cut whenever there is a change in the hypergraph (due to addition or deletion of hyperedges). However this simple

C. J. Colbourn et al. (Eds.): IWOCA 2019, LNCS 11638, pp. 237–250, 2019.
https://doi.org/10.1007/978-3-030-25005-8_20

approach has many drawbacks. For example, even for a single hyperedge insertion, the application of the algorithm by Chekuri and Xu [2] requires $O(p+\lambda n^2)$ time to update the minimum cut and the edge connectivity. In a dynamic environment usually the number of hyperedges being added or deleted is relatively small compared to the overall size of the hypergraph. It is therefore desirable to tolerate a small change (edge or node) in an efficient manner. Also the approach helps answer queries on the minimum cut and the edge connectivity in an online setting. The existing algorithms for computing the minimum cut in a hypergraph suffer from the fact that they do not work efficiently under dynamic or online setting.

In this paper, we present an incremental algorithm that updates the minimum cut and the edge connectivity of an existing hypergraph on addition of a set of hyperedges. The algorithm takes $O(n + p_u)$ time if the edge connectivity does not change in spite of the addition of new hyperedges. Otherwise the algorithm recomputes the edge connectivity and the hypercactus using the method of Chekuri and Xu [2] that requires $O(p + \lambda n^2)$ time. Our algorithm is deterministic and takes $O(\lambda n)$ amortized time per insertion of a hyperedge. The algorithm can answer queries on edge-connectivity in $O(1)$ time and returns a cut of size λ in $O(n)$ time. Note that the claimed bound is a significant improvement over the trivial algorithm that computes everything from scratch. This is because any static algorithm must take $\Omega(p)$ time, where p could be exponential in n.

Our Contribution: In this paper, our contributions are the following:

(i) We present a method to maintain a given hypercactus efficiently under dynamic addition of a few hyperedges. Our method takes $O(n + p_u)$ time compared to $O(p + \lambda n^2)$ time taken by the current best approach (Chekuri and Xu [2]) to update the hypercactus given that the edge connectivity does not change. Here p_u, p, λ and n have usual meanings.

(ii) We present an incremental algorithm that updates the minimum cut and the edge connectivity of an existing hypergraph on addition of a set of hyperedges. We use the aforesaid method of maintaining a hypercactus as a subroutine in this incremental algorithm. We show that our algorithm has an amortized time of $O(\lambda n)$ per insertion of a hyperedge.

Organization: In Sect. 2, we discuss the related work. Section 3 contains some basic preliminaries related to the problem. In Sect. 4 we discuss our algorithm. The proof of correctness of the proposed algorithm is given in Sect. 5. We conclude the work in Sect. 6.

2   Related Work

There exist many algorithms for computing the minimum cut and the edge connectivity in a simple graph [6,12,16,19] and in the last few decades, dynamically updating the edge connectivity in a simple graph has been addressed by many researchers. In 2016, Goranci et al. [9] gave a deterministic incremental algorithm that maintains the edge connectivity in $\widetilde{O}(1)$ amortized time per edge insertion in undirected and unweighted graph. Henzinger [10] has also given an incremental algorithm that maintains the minimum cut and the edge connectivity of a graph G under dynamic addition of a set of edges. It takes $O(\lambda \log n)$ amortized time per insertion of a simple edge. Dinitz and Westbrook proposed a method [5] to maintain a cactus representation [4] which stores all possible minimum cuts in a graph and this method is used as a subroutine in the incremental algorithm given by Henzinger [10]. The current best deterministic algorithm to compute a cactus representation of a graph G is given by Gabow [7] which requires $O(m + \lambda^2 n \log(n/\lambda))$ time.

In case of hypergraphs there are many static algorithms to compute different hypergraph properties like minimum cut, minimum weight hyperpath, transitive closure, rank, independent sets, etc. A careful implementation of the method given by Queyranne [17] to compute the minimum cut in a hypergraph takes $O(np + n^2 \log n)$ time for capacitated hypergraph and $O(np)$ time for an uncapacitated hypergraph. Klimmek and Wagner [13], Mak and Wong [15] independently gave algorithms for computing the minimum cut in a hypergraph having same time bound. The current best algorithm to compute the edge connectivity and the hypercactus is given by Chekuri and Xu [2] that requires $O(p + \lambda n^2)$ time. Ausiello et al. [1] proposed an algorithm to maintain transitive closure for a hypergraph under dynamic addition of hyperedges. Italiano and Nanni [11] proposed an algorithm to maintain minimum rank and minimum gap hyperpath over a batch of hyperedge insertions. In [18], a Dijkstra-like procedure has been proposed for maintaining a weighted shortest path in a *fully-dynamic hypergraph*.

3   Preliminaries

Let $H = (V, E)$ be an uncapacitated hypergraph where V is the set of vertices and E is the set of hyperedges where each hyperedge $e \in E$ is a subset of vertices, $n = |V|$, $m = |E|$ and $p = \Sigma_{e \in E}|e|$ where $|e|$ is the number of vertices in a hyperedge e. A *cut* is the partitioning of V into two non-empty sets A and $V \backslash A$. The set of hyperedges connecting the two sets (also called cut-edges) contribute to the value of the cut. Out of all possible cuts in a hypergraph, any cut whose value is minimum is known as a minimum cut.

We denote the set of hyperedges intersecting both A and $V \backslash A$ with $\delta_H(A)$ and call it a *cut-edge set of H*. A hypergraph $H' = (V', E')$ is said to be a *subhypergraph* of a hypergraph $H = (V, E)$ if $V' \subseteq V$ and there is a bijective mapping $\phi : E \to E'$ where $\phi(e) \subseteq e$ for each $e \in E$. For vertices $u, v \in V$, a (u, v)-walk of length k in H is a sequence $v_0 e_1 v_1 e_2 v_2 \ldots v_{k-1} e_k v_k$ of vertices and hyperedges (possibly repeated) such that $v_0, v_1, \ldots, v_k \in V$, $e_1, \ldots, e_k \in E$, $v_0 = u$, $v_k = v$, and for all $i = 1, 2, \ldots, k$, the vertices v_{i-1} and v_i are adjacent in H via the hyperedge e_i. The vertices $u, v \in V$ are said to be connected in H if there exists a (u, v)-walk in H. The hypergraph H is said to be connected if every pair of distinct vertices is connected in H. The minimum number of hyperedges whose removal disconnects H is called the *edge connectivity* of H, which is denoted by λ. A cut is k-*cut* if $|\delta_H(A)| = k$. The vertices s and t are said to be k-*edge connected* if there exists no k'-cut, $k' < k$ that disconnects the pair $\{s, t\}$.

A *cactus tree* [4] is an $O(n)$ sized data structure which compactly represents all possible minimum cuts of size k in a k-edge connected simple graph G. A cactus tree $\tau(G)$ can be represented as (G^*, Φ) where G^* is a simple weighted graph and $\Phi : V(G) \longrightarrow V(G^*)$ such that Φ is a many to one mapping. The properties of $\tau(G)$ are:

(i) $\Phi(u) = \Phi(v)$ if and only if the vertex u and v are not separated by a λ-cut in G.
(ii) If $(X, \bar{X})$ is a λ-cut of G^* for $X \subset V(G^*)$ then $(Y, \bar{Y})$ is a λ-cut of G where $Y = \{\Phi^{-1}(u)|u \in X\}$.
(iii) If λ is odd, G^* is a tree and every edge of G^* has a weight of λ. If λ is even, two simple cycles of G^* have at most one common node, every edge that belongs to a cycle has weight $\lambda/2$ and every edge that does not belong to a cycle has weight λ.

For a given hypergraph H and edge connectivity λ, a *hypercactus* is an $O(n)$ sized data structure which compactly represents all possible minimum cuts of size λ. The hypercactus $\tau(H)$ [3] can be represented as (H^*, Φ) where H^* is a hypergraph and $\Phi : V(H) \rightarrow V(H^*)$ such that Φ is a many to one mapping. The properties of $\tau(H)$ are as follows:

(i) $\Phi(u) = \Phi(v)$ if and only if the vertex u and v are not separated by a λ-cut in H.
(ii) If $(X, \bar{X})$ is a λ-cut of H^* for $X \subset V(H^*)$ then $(Y, \bar{Y})$ is a λ-cut of H where $Y = \{\Phi^{-1}(u)|u \in X\}$.

The main difference between a cactus and a hypercactus is that a hypercactus can have hyperedges (see Fig. 2 for example) in addition to simple edges and cycles whereas a cactus can have only simple edges and cycles. The weight of each non cycle edge and each hyperedge in $\tau(H)$ is λ whereas the weight of each cycle edge in $\tau(H)$ is $\lambda/2$. Figure 2 represents a hypercactus $\tau(H)$ for a hypergraph H as shown in Fig. 1 whose edge connectivity is 2. In Fig. 2, $x1 = x2 = \phi$ and $\Phi(1) = \Phi(2) = \Phi(3) = \Phi(4) = a$. For every other vertex u of H, $\Phi(u) = u$ in $\tau(H)$ as shown in the figure. The edge weights are also shown in the figures. In addition to structural differences between a cactus and a hypercactus, there are also operational differences between them. For example, we can apply *shrinking* and *squeezing* operations on a cactus structure whereas we can apply *tuning* operation in addition to *shrinking* and *squeezing* operations in a hypercactus.

Dinitz and Westbrook [5] defined *shrinking* of nodes and *squeezing* of cycles as follows. In *shrinking* a subset of vertices $W \subseteq V$, the operation replaces all vertices in W by a single vertex w, deletes all edges whose both end points lie in W. For any edge (x, y) where $x \in W$ and $y \notin W$, the operation replaces (x, y) with (w, y).

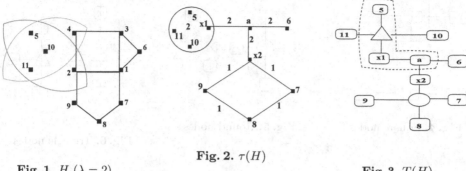

Fig. 2. $\tau(H)$

Fig. 1. H ($\lambda = 2$)   **Fig. 3.** $T(H)$

Let $C = (v_1, v_2, ..., v_k, v_1)$, $k \geq 2$ be a cycle. Then *squeezing* C at v_i and v_j, $i < j$, results in *shrinking* v_i and v_j. The *squeezing* results in two new cycles: $(w, v_{j+1}, ..., v_k, v_1, ..., v_{i-1}, w)$ and $(w, v_{i+1}, ..., v_{j-1}, w)$. If the length of a resulting cycle is 2, the cycle gets replaced by a simple edge. In this paper we introduce a new operation named *tuning* of a hyperedge which is defined in Subsect. 4.2.

4   The Proposal

In this section we first briefly discuss the work of Dinitz and Westbrook [5] for maintaining the cactus tree $\tau(G)$ of a graph G. This is used as a subroutine for maintaining the minimum cut and the edge connectivity in a simple graph under dynamic addition of edges (Henzinger [10]). Next we propose our method to maintain a hypercactus $\tau(H)$ of a hypergraph H on dynamic addition of hyperedges. We use this hypercactus maintenance method as a subroutine for designing the incremental algorithm for maintaining the minimum cut and the edge connectivity in a hypergraph, which is described in Subsect. 4.3.

4.1   Cactus Maintenance

Let n be the number of vertices in G. The data structure used by Dinitz and Westbrook takes $O(n + m + q)$ time to perform m number of *Insert-Edge(u,v)* operations and q number of *Same-k-Class(u,v)* queries. The *Insert-Edge(u,v)* operation inserts a new edge (u, v) dynamically to a graph G. The *Same-k-Class(u,v)* query returns *true* if vertex u and v are k-edge connected and returns *false* otherwise. The algorithm takes $O(m + k^2 n \log(n/k))$ preprocessing time in order to construct the initial data structure where m is the number of edges in G. Previously the algorithms for maintaining all possible minimum cuts of size 1, 2 and 3 in a graph G has been described in [8,14]. The algorithm by Dinitz and Westbrook [5] is a generalization for an increasing value of k.

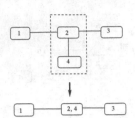

Fig. 4. Square nodes

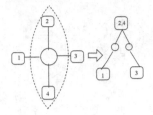

Fig. 5. Round nodes

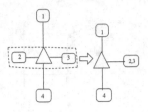

Fig. 6. Triangle nodes

The author introduced the definition of the *auxiliary tree* which is an extension of the *cactus tree*. The auxiliary tree $T(G)$ consists of two types of nodes: *square node*, one for each node in $\tau(G)$ and *round node*, one for each cycle in $\tau(G)$. Two square nodes in $T(G)$ are connected by a simple edge if corresponding nodes in $\tau(G)$ are connected by a simple edge. The square nodes in $T(G)$ which are a part of a cycle C in $\tau(G)$ are made adjacent to the round node in $T(G)$ representing C. The square nodes connected to a round node in $T(G)$ follow the order in which the corresponding nodes in $\tau(G)$ are connected in a cycle. A leaf node is created for each vertex v in G. These leaf nodes are made children of the corresponding square nodes in $T(G)$. The query *Same-(k+1)-Class(u,v)* returns *true* only if vertices u and v have same parent in $T(G)$.

On inserting a new edge, the algorithm first finds a unique path between the corresponding square nodes in $T(G)$. It modifies the path such that it correctly reflects the effects of *squeezing* cycles and *shrinking* nodes in $\tau(G)$. If two square nodes in the path are connected by a simple edge in $T(G)$ then the square nodes are merged in $T(G)$ as shown in Fig. 4 and *shrinking* of corresponding nodes in $\tau(G)$ are applied. If two square nodes in the path are connected to a round node in $T(G)$ then the modification of $T(G)$ is done as shown in Fig. 5 and *squeezing* of corresponding cycle in $\tau(G)$ is applied. There will be leaf nodes connected to the square nodes in $T(G)$, however we omit them for the clarity of the figures.

4.2 Hypercactus Maintenance

In this subsection we propose a method to maintain a hypercactus on dynamic addition of a set U of hyperedges which takes $O(n + p_u)$ time. This method inherits the ideas proposed by Dinitz and Westbrook [5] described in Subsect. 4.1. The method is described using the following cases:

(i) $\lambda = 0$: In this case the given hypergraph H is disconnected. On adding a new hyperedge $e = \{u_1, u_2, ..., u_l\}$, the method updates H using a fast disjoint set-union data structure [20] which takes $O(q\alpha(q, n))$ time to perform any sequence of q number of *union* and *find* operations. Here α is the functional inverse of the Ackermann's function (practically, α is a constant). The method first creates a set of each connected component of H.

The *find(v)* operation returns the name of a component containing the vertex v. The *union* operation takes two connected components as the input, merges the two components and returns the merged component. The query *Same-($\lambda + 1$)-Class(u,v)* returns true if $find(u) = find(v)$. The method iterates through each vertex in e. If $find(u_i) = find(u_{i+1})$ for $1 \leq i < l$ then the method does not change anything. Otherwise *union* operation on disjoint sets $find(u_i)$ and $find(u_{i+1})$ is applied. The method follows the same procedure for each hyperedge $e \in U$. Note that total $O(p_u)$ number of *find* and *union* operations are called in this case and thus the time complexity is $O(p_u \alpha(p_u, n))$. After a certain number of hyperedge insertions if the hypergraph has a single component then the hypergraph is no more disconnected. In this situation the method computes the edge connectivity and the hypercactus using the algorithm of Chekuri and Xu [2] and applies the technique given in the following case to maintain the hypercactus.

(ii) $\lambda \geq 1$: In this case the hypergraph is connected. In case of a graph, a cactus consists of simple edges and cycles. In case of a hypergraph, a hypercactus can have hyperedges in addition to simple edges and cycles. The method extends the definition of the auxiliary tree $T(H)$ used by Dinitz and Westbrook for the case of hypercactus. This auxiliary tree is used in the proposed method for hypercactus maintenance. In addition to *square* nodes and *round* nodes, another type of node called *triangle* node is introduced in $T(H)$, one for each hyperedge in $\tau(H)$. Square nodes and round nodes follow the same rules as described earlier in the case of cactus maintenance. Here the rules for triangle nodes are described. The square nodes in $T(H)$ which are the part of a hyperedge e in $\tau(H)$ are made adjacent to the triangle node in $T(H)$ corresponding to e. Unlike the case of round nodes, the square nodes can be connected to a triangle node in any order. A leaf node is created for each vertex v in H. These leaf nodes are made children of the corresponding square nodes in $T(H)$. Figure 3 represents the auxiliary tree $T(H)$ for a hypercactus $\tau(H)$ shown in Fig. 2. The query *Same-($\lambda+1$)-Class(u,v)* returns *true* if vertices u and v belong to same parent in $T(H)$. The *find(v)* operation for a vertex $v \in e$ returns the corresponding square node in $T(H)$. Similarly, the *union(x,y)* operation merges the nodes x and y in $T(H)$.

On inserting a hyperedge $e = \{u_1, u_2, ..., u_k\}$, the method iterates through each vertex in e. Let $x = find(u_i)$ and $y = find(u_{i+1})$. If $x = y$ then the method does not change anything. Otherwise the method finds the unique path between the nodes x and y in $T(H)$. There can be three types of nodes in the path: square nodes, round nodes and triangle nodes. This path is then modified in such a way that it correctly reflects the effects of *shrinking* nodes, *squeezing* cycles and *tuning* hyperedges in $\tau(H)$. The *tuning* of a hyperedge is defined as follows.

Tuning Operation: Let $e = (v_1, v_2, ..., v_k)$, $k > 2$ be a hyperedge (for $k = 2$, e is a simple edge). Then *tuning* e at v_i and v_j, $i < j$, results in *shrinking* v_i and v_j. This results in a new hyperedge: $(v_1, v_2, ..., v_{i-1}, w, v_{i+1}, ..., v_{j-1}, v_{j+1}, ..., v_k)$. Here w denotes the supervertex obtained after merging nodes v_i and v_j throughout the entire hypergraph. If the size of the resulting hyperedge is 2, the hyperedge is replaced by a simple edge.

If two square nodes in the path are connected by a simple edge in $T(H)$ then the *union* operation is applied on the square nodes in $T(H)$ as shown in Fig. 4 and the *shrinking* operation on corresponding nodes in $\tau(H)$ is applied. If two square nodes in the path are connected to a round node in $T(H)$ then the nodes are modified as shown in Fig. 5 and the *squeezing* operation is applied on the corresponding cycle in $\tau(H)$. Similarly, if two square nodes in the path are connected to a triangle node in $T(H)$ then the nodes are modified as described below and the *tuning* operation is applied on the corresponding hyperedge in $\tau(H)$.

Triangle Node Modification: Let s_1 and s_2 be the square nodes connected to a triangle node t i.e., (s_1, t) and (t, s_2) are the edges in the path between nodes x and y in $T(H)$. The method merges s_1 and s_2 into a supernode w and connects this w with t. All the edges connected to s_1 and s_2 get connected to w. Rest other edges connected to t remain the same. After the modification, if t has 2 square nodes connected to it then t is deleted and these two square nodes get connected with a direct edge. An example of triangle node modification is shown in Fig. 6.

Theorem 1. *Let $\tau(H) = (H^*, \Phi)$ be the hypercactus representation of a hypergraph H whose edge connectivity is λ. Under dynamic addition of a set U of hyperedges to H, $\tau(H)$ can be maintained in $O(n + p_u)$ time, where n is the number of vertices in H and $p_u = \Sigma_{e \in U} |e|$.*

Proof. The method described in Subsect. 4.2 to maintain $\tau(H)$ under dynamic addition of a set of hyperedges uses the method similar to Dinitz and Westbrook [5] described in Subsect. 4.1 with additional case of handling hyperedges. From the construction of the auxiliary tree $T(H)$ it is clear that each hyperedge of size k has $k + 1$ number of nodes and k number of edges in $T(H)$. From Cheng [3], we know that $|V(H^*)| = O(|V(H)|)$ and $|E(H^*)| = O(|V(H)|)$. Thus, the construction of $T(H)$ corresponding to $\tau(H)$ can be done in linear time. Under dynamic addition of a set of hyperedges, the method applies $find$ and $union$ operations for each hyperedge iteratively. The merging of two square nodes in the path between two nodes in the auxiliary tree takes $O(1)$ time. Correspondingly, *shrinking* an edge and *squeezing* a cycle in $\tau(H)$ takes $O(1)$ time. The method introduces a new technique called *tuning* which modifies a hyperedge in $\tau(H)$. In *tuning*, the method *shrinks* the two nodes and as a result a new hyperedge forms. The operation can be done in $O(1)$ time using the method of disjoint set union-find operation as used for *shrinking* and *squeezing* technique. On inserting a hyperedge e, the method uses $O(|e|)$ number of find and union operations to update $T(H)$ and $\tau(H)$. Thus, total $O(p_u)$ number of $find$ and $union$ operations are used under the insertion of a set U of hyperedges. The total time taken by $find$ and $union$ operation is $O(p_u \alpha(p_u, n))$. The total running time to maintain $\tau(H)$ is $O(n + p_u \alpha(p_u, n))$ time. In practical scenario, $\alpha \leq 6$. Thus, the total cost to maintain a hypercactus under dynamic addition of a set U of hyperedges is $O(n + p_u)$. □

Algorithm 1. The Incremental Algorithm

1: Compute λ, $\tau(H)$ using Chekuri and Xu [2]
2: $N \leftarrow \phi$
3: **while** there is ≥ 1 mincut of size λ **do**
4:     **if** next operation is **query-size()**
5:         print λ
6:     **if** next operation is **query-mincut()**
7:         print the mincut   //Refer to point 2 in Subsection 4.3
8:     **if** next operation is **Same-(λ+1)-Class**(u, v)
9:         print "true" or "false"
10:    **if** next operation is **InsertHyperedges(U)**
11:        update $\tau(H)$ //Refer to Subsection 4.2
12:        $N \leftarrow N \cup U$
13: Recompute λ and $\tau(H)$ with $H = (V, E')$ where $E' = E \cup N$
14: Goto step 2

4.3   The Incremental Algorithm

In this subsection we describe an incremental approach to maintain the minimum cut and the edge connectivity of an uncapacitated hypergraph H under dynamic addition of a set of hyperedges. The psuedocode of the proposed algorithm is given in Algorithm 1. In line 1, the algorithm first computes λ and $\tau(H)$ using the algorithm of Chekuri and Xu [2]. The algorithm checks in $O(1)$ time if there exist at least one minimum cut of size λ in $\tau(H)$ by asserting that $\tau(H)$ has more than one node. In the algorithm we discuss four queries considering that the hypercactus is modified and then these queries come:

 (i) **query-size()**: It returns the current value of λ. Since λ is always known, we can return the result in $O(1)$ time.
 (ii) **query-mincut()**: It returns a minimum cut of the form $(A, \bar{A})$ such that $A \subset V$. The auxiliary tree $T(H)$ is first split by any edge (u, v). The DFS (Depth First Search) method is then applied on nodes u and v to get two connected components $(X, \bar{X})$ such that $u \in X$ and $v \in \bar{X}$. Then $(\Phi^{-1}(X), \Phi^{-1}(\bar{X}))$ is the resulting minimum cut. The number of vertices and edges in $T(H)$ are of size $O(n)$ (due to Cheng [3]), therefore the query takes $O(n)$ time.
 (iii) **Same-$(\lambda + 1)$-Class**(u, v): It returns true if vertices u and v have edge connectivity greater than λ.; here both u and v have same parent node in $T(H)$. If $find(u) = find(v)$ the algorithm returns true otherwise it returns false. The time complexity for this query is $O(1)$.
 (iv) **InsertHyperedges(U)**: This query inserts a set U of hyperedges into H. Due to the addition of new hyperedges some cuts of size λ in H may no more remain a minimum cut. In order to update all the minimum cuts of size λ the algorithm applies the method described in Subsect. 4.2 that maintains hypercactus $\tau(H)$ on insertion of a set U of hyperedges in $O(n + p_u)$ time.

Let N denote the set of all hyperedges added so far such that no minimum cut of size λ exists in $\tau(H)$. If the algorithm keeps receiving single edge insertions at any point of time i.e, $U = \{e\}$ then the algorithm sets $\lambda = \lambda + 1$ at line 13 of Algorithm 1 instead of actually recomputing λ. Otherwise it recomputes λ. The algorithm recomputes the hypercactus $\tau(H)$ for updated λ using the method of Chekuri and Xu [2] that takes $O(p + \lambda n^2)$ time. The algorithm goes to line 2 to continue the incremental process.

Theorem 2. *The amortized time to maintain the minimum cut and the edge connectivity for a dynamic hypergraph H is $O(\lambda n)$ per hyperedge insertions.*

Proof. Let λ_0 be the initial edge connectivity in line 1 of Algorithm 1. It takes $O(p_0 + \lambda_0 n^2)$ time to compute λ_0 and $\tau(H)$ in line 1 of Algorithm 1. During the execution of Algorithm 1, let λ assume the values $\lambda_0, \ldots, \lambda_f$ in an increasing order. Phase i consists of all steps executed while $\lambda = \lambda_i$. Let U_i denote the set of hyperedges inserted in Phase i. In Phase i, we compute the new edge connectivity λ_i and $\tau(H)$ in line 13 and maintain $\tau(H)$ in line 11. The time to compute λ_i and the corresponding hyercactus $\tau(H)$ in line 13 is $O(p + \lambda_i n^2)$ where p is calculated in the modified hypergraph. From Theorem 1, the time taken to maintain $\tau(H)$ is $O(n + p_i)$ where $p_i = \Sigma_{e \in U_i} |e|$. The total time spent in executing Phase i is $O(n + p_i + p + \lambda_i n^2)$. The maximum number of phases can be λ. Thus the total time to execute all phases is asymptotically $O(\lambda p + \lambda^2 n^2)$ where $p = \Sigma_{e \in E \cup U} |e|$. The amortized cost of a hyperedge insertion is $O(\lambda + \lambda^2 n^2 / p)$. For a hypergraph H with edge connectivity λ, $p = \Omega(\lambda n)$ Thus the amortized insertion time is $O(\lambda + \lambda n) = O(\lambda n)$. □

4.4 Analysis

Let H be the given uncapacitated hypergraph whose edge connectivity is λ. Let $\tau(H)$ be the corresponding hypercactus. Let U be the set of hyperedges that is dynamically inserted to H. In order to compute λ and $\tau(H)$, we can apply the static algorithm (Chekuri and Xu) which takes $O(p_0 + p_1 + \lambda' n^2)$ time where $p_0 = \Sigma_{e \in E} |e|$, $p_1 = \Sigma_{e \in U} |e|$ and λ' is the new egde connectivity. With our proposed incremental algorithm, if the value of λ does not change then updating $\tau(H)$ takes $O(n + p_u)$ time. Otherwise, computing the new edge connectivity λ' and the corresponding hypercactus takes $O(\lambda' n)$ amortized time (Theorem 2). Hence the cost of the proposed algorithm is better than the static algorithm.

Now we show the probabilistic approach to compute the cost of updating λ and $\tau(H)$ using the proposed incremental algorithm. Let f denote the probability that the value of λ gets changed under dynamic addition of a set U of hyperedges to H. Then the total cost to update λ and $\tau(H)$ is asymptotically $O(f(p_0 + p_1 + \lambda n^2) + (1 - f)(n + p_1))$. Thus, if the value of f is very small, then the proposed incremental algorithm requires less computation time than the static method.

5   Proof of Correctness

Lemma 1. *A hypercactus has the following properties: (a) No two hyperedges can have more than one node in common. (b) A hyperedge and a cycle can have* *at most one node in common. (c) No two cycles can have more than one node in common.*

Proof. Let there be two hyperedges that have more than one node in common in a hypercactus. Each hyperedge in the hypercactus has a weight of λ. From the hypercactus construction, we have one minimum cut of the form $(A, \bar{A})$ such that at least one common node belongs to A and the other common nodes belong to $\bar{A}$. But to get such a minimum cut, we need to remove two hyperedges since the common nodes belong to both the hyperedges. Thus, the size of the cut becomes 2λ which is not minimum. This is a contradiction. Hence two hyperedges in $\tau(H)$ can not have more than one node in common. Using similar arguments we can proof properties **(b)** and **(c)**. □

Lemma 2. *On introducing a triangle node into the auxiliary tree $T(H)$, it remains a tree.*

Proof. We first claim that $T(H)$ constructed after introducing a triangle node is connected i.e., there is a path between any two nodes in $T(H)$. Let us assume that there exists two nodes in $T(H)$ between which no path exists. This can be possible only if $T(H)$ is disconnected. It means each square node connected to the triangle node has no edge with other square nodes or round nodes in $T(H)$. But $T(H)$ is constructed from a hypercactus, it implies that the hypercactus is disconnected. This is a contradiction. Thus the auxiliary tree $T(H)$ is always connected.

Now we claim that $T(H)$ has no cycles. Let us assume that $T(H)$ have a cycle after introducing a triangle node. It means that the square nodes connected to a triangle node forms a cycle with other square nodes or round nodes in $T(H)$. But if this is the case then in the hypercactus the corresponding hyperedge forms a cycle with the cycle edges or simple edges which is a contradiction as per Lemma 1. Thus, the auxiliary tree can not have any cycle after introduction of a triangle node. Hence the auxiliary tree remains a tree on introducing a triangle node into it. □

Lemma 3. *On applying a tuning operation, the hypercactus preserves it's properties, i.e., it preserves all the minimum cuts of size λ.*

Proof. Dinitz and Westbrook [5] gave the definition of *shrinking* and *squeezing* operations. On applying a *shrinking* or *squeezing* operation to the hypercactus, the updated hypercactus preserves it's properties. In this paper, the *tuning* operation is introduced. Let $\tau(H)$ be the hypercactus corresponding to the hypergraph H. Let $Z = \{a_1, a_2, ..., a_k\}$ be a hyperedge in $\tau(H)$. Let $\tau(H')$ be the updated hypercactus after applying a tuning operation on nodes a_i and a_j in Z, $i \neq j$. We prove that the $\tau(H')$ preserves it's properties using three cases:

- Case $|Z| = 2$: As per the definition of tuning operation, the hyperedge of size 2 is treated as a simple edge. This simple edge gets the same weight as of hyperedge Z i.e., λ. The shrinking operation between the nodes a_i and a_j is applied in this case and thus preserves the properties of $\tau(H')$. This follows from the work of Dinitz and Westbrook.

- Case $|Z| = 3$: In this case, the shrinking operation on nodes a_i and a_j is first applied. The size of updated hyperedge Z' becomes 2. This Z' is replaced with a simple edge with same weight λ. In a hypercactus the weights of each hyperedge and each simple edge is always λ. Thus from the construction, all the mincuts of size λ are preserved in the updated hypercactus.
- Case $|Z| \geq 4$: In this case, the shrinking operation on nodes a_i and a_j is first applied. The size of hyperedge Z gets reduced by 1. The updated hyperedge Z' is still a hyperedge with same weight λ. Thus, all the mincuts of size λ are preserved in the updated hypercactus. $\square$

Lemma 4. *An update on $T(H)$ eventually leads to a corresponding update on $\tau(H)$.*

Proof. On insertion of a new hyperedge $e = \{v_1, v_2, .., v_k\}$ to the given hypergraph H, the method first finds a unique path P in $T(H)$ between the two square nodes s_1 and s_2 which corresponds to the nodes v_i and v_{i+1} in e respectively. In the path P the two consecutive square nodes can either be directly connected, connected by a round node or connected by a triangle node. If in the path P, no triangle node is involved i.e, every consecutive square nodes are either directly connected or connected to a round node then the modification technique to update $T(H)$ and $\tau(H)$ is exactly the same as Dinitz and Westbrook [5]. Hence for this case the updated $T(H)$ corresponds to the updated $\tau(H)$.

If two consecutive square nodes in the path P are connected by a triangle node then we update $T(H)$ as described above in **Triangle Node Modification** and apply *tuning* operation between the two corresponding vertices in $\tau(H)$. We show that the updated $T(H)$ corresponds to the updated $\tau(H)$. Let (s_1, t) and (t, s_2) be the consecutive edges in the path P. Here, s_1 and s_2 denote the two consecutive square nodes connected to the triangle node t. After modifying the path P in $T(H)$, let s denotes the square node in $T(H)$ after merging s_1 and s_2. Similarly, let w_1 and w_2 denote the vertices in $\tau(H)$ corresponding to the square nodes s_1 and s_2 in $T(H)$ respectively. After applying the tuning operation on w_1 and w_2, let w denotes the merged node in $\tau(H)$. This w in $\tau(H)$ should correspond to s in $T(H)$. For the sake of contradiction, let us assume that w and s do not correspond to each other. It means that w either maps to some other square node $s' \neq s$ in $T(H)$ or it maps to empty. From the construction of $T(H)$ we create one square node for each vertex in $\tau(H)$. Therefore w can not map to empty. Let us consider the case where w maps to s'. In the modification of path P, s' is not touched. Thus there must exist a vertex w' in $\tau(H)$ that maps to s'. From our assumption w maps to s' which means $w = w'$. But w is formed after merging w_1 and w_2. This leads to a contradiction that $w = w'$. Thus w in $\tau(H)$ maps to s in $T(H)$. This proves that the updated $T(H)$ corresponds to the updated $\tau(H)$. $\square$

6   Conclusion

Under dynamic addition of hyperedges, when the edge connectivity changes the proposed incremental algorithm relies on Chekuri and Xu's approach to recompute hypercactus. It may be worth investigating a method to recompute the hypercactus efficiently using the structures behind the Chekuri and Xu's static algorithm, instead of recomputing everything from scratch. Similarly it would be interesting to recompute or update the hypercactus efficiently in the deletion case. It may help in designing an efficient decremental algorithm to maintain the edge connectivity and the minimum cut under dynamic deletion of hyperedges. All the contributions made in this paper are for uncapacitated hypergraph. It will be worth investigating a method for the capacitated case.

References

1. Ausiello, G., Nanni, U., Italiano, G.F.: Dynamic maintenance of directed hypergraphs. Theor. Comput. Sci. **72**(2–3), 97–117 (1990)
2. Chekuri, C., Xu, C.: Computing minimum cuts in hypergraphs. In: Proceedings of the Twenty-Eighth Annual ACM-SIAM Symposium on Discrete Algorithms, pp. 1085–1100. Society for Industrial and Applied Mathematics (2017)
3. Cheng, E.: Edge-augmentation of hypergraphs. Math. Program. **84**(3), 443–465 (1999)
4. Dinits, E.A., Karzanov, A.V., Lomonosov, M.V.: On the structure of the system of minimum edge cuts in a graph. In: Investigations in Discrete Optimization, pp. 290–306 (1976) (in Russian)
5. Dinitz, Y., Westbrook, J.: Maintaining the classes of 4-edge-connectivity in a graph on-line. Algorithmica **20**(3), 242–276 (1998)
6. Gabow, H.N.: A matroid approach to finding edge connectivity and packing arborescences. In: Proceedings of the Twenty-third Annual ACM Symposium on Theory of Computing, pp. 112–122. ACM (1991)
7. Gabow, H.N.: The minset-poset approach to representations of graph connectivity. ACM Trans. Algorithms **12**(2), 24:1–24:73 (2016)
8. Galil, Z., Italiano, G.F.: Maintaining the 3-edge-connected components of a graph on-line. SIAM J. Comput. **22**(1), 11–28 (1993)
9. Goranci, G., Henzinger, M., Thorup, M.: Incremental exact min-cut in polylogarithmic amortized update time. In: 24th Annual European Symposium on Algorithms, ESA 2016, 22–24 August 2016, Aarhus, Denmark, pp. 46:1–46:17 (2016)
10. Henzinger, M.R.: A static 2-approximation algorithm for vertex connectivity and incremental approximation algorithms for edge and vertex connectivity. J. Algorithms **24**(1), 194–220 (1997)
11. Italiano, G.F., Nanni, U.: On-line maintenance of minimal directed hypergraphs. In: Italian Conference on Theoretical Computer Science, pp. 335–349 (1989)
12. Kawarabayashi, K.I., Thorup, M.: Deterministic global minimum cut of a simple graph in near-linear time. In: Proceedings of the Forty-seventh Annual ACM Symposium on Theory of Computing, pp. 665–674. ACM (2015)
13. Kilmmek, R., Wagner, F.: A simple hypergraph min cut algorithm. Technical Report B 96–02, Department of Mathematics and Computer Science, Freie Universität Berlin (1996)
14. La Poutré, J.A., van Leeuwen, J., Overmars, M.H.: Maintenance of 2-and 3-edge-connected components of graphs. Discrete Math. **114**(1–3), 329–359 (1993)
15. Mak, W.K., Wong, D.: A fast hypergraph min-cut algorithm for circuit partitioning. Integr. VLSI J. **30**(1), 1–11 (2000)

16. Nagamochi, H., Ibaraki, T.: Computing edge-connectivity in multigraphs and capacitated graphs. SIAM J. Discrete Math. **5**(1), 54–66 (1992)
17. Queyranne, M.: Minimizing symmetric submodular functions. Math. Program. **82**(1–2), 3–12 (1998)
18. Ramalingam, G., Reps, T.: An incremental algorithm for a generalization of the shortest-path problem. J. Algorithms **21**(2), 267–305 (1996)
19. Stoer, M., Wagner, F.: A simple min-cut algorithm. J. ACM (JACM) **44**(4), 585–591 (1997)
20. Tarjan, R.E., Van Leeuwen, J.: Worst-case analysis of set union algorithms. J. ACM (JACM) **31**(2), 245–281 (1984)

A General Algorithmic Scheme for Modular Decompositions of Hypergraphs and Applications

Michel Habib[1,3]([⊠]), Fabien de Montgolfier[1,3], Lalla Mouatadid[2], and Mengchuan Zou[1,3]

[1] IRIF, UMR 8243 CNRS & Université de Paris, Paris, France
habib@irif.fr
[2] Department of Computer Science, University of Toronto, Toronto, ON, Canada
[3] Gang Project, Inria, Paris, France

Abstract. We study here algorithmic aspects of modular decomposition of hypergraphs. In the literature one can find three different definitions of modules, namely: the standard one [19], the k-subset modules [6] and the Courcelle's one [11]. Using the fundamental tools defined for combinatorial decompositions such as partitive and orthogonal families, we directly derive a linear time algorithm for Courcelle's decomposition. Then we introduce a general algorithmic tool for partitive families and apply it for the other two definitions of modules to derive polynomial algorithms. For standard modules it leads to an algorithm in $O(n^3 \cdot l)$ time (where n is the number of vertices and l is the sum of the size of the edges). For k-subset modules we obtain $O(n^3 \cdot m \cdot l)$ (where m is the number of edges). This is an improvement from the best known algorithms for k-subset modular decomposition, which was not polynomial w.r.t. n and m, and is in $O(n^{3k-5})$ time [6] where k denotes the maximal size of an edge. Finally we focus on applications of orthogonality to modular decompositions of tournaments, simplifying the algorithm from [18]. The question of designing a linear time algorithms for the standard modular decomposition of hypergraphs remains open.

1 Introduction

In this paper we study hypergraph modular decomposition; an important generalization of graph modular decomposition. Hypergraph modular decomposition is equivalent to the modular decomposition of both set systems [19] as well as monotone Boolean functions [20], while that of general Boolean functions was shown to be NP-hard [5]. We study here algorithmic aspects of modular decomposition of hypergraphs. In the literature one can find three different definitions of modules, namely: the standard one [19], the k-subset modules [6] and Courcelle's one [11]. In the following we recall the fundamental tools defined for combinatorial decompositions such as partitive and orthogonal families. This directly yields a linear time algorithm for Courcelle's decomposition.

© Springer Nature Switzerland AG 2019
C. J. Colbourn et al. (Eds.): IWOCA 2019, LNCS 11638, pp. 251–264, 2019.
https://doi.org/10.1007/978-3-030-25005-8_21

In Sect. 3 we propose a general algorithmic tool for partitive families and apply it for the other two definitions of modules to derive polynomial algorithms. For standard modules it leads to an algorithm in $O(n^3 \cdot l)$ time (where n is the number of vertices and l the sum of the size of the m edges). For k-subset modules we obtain $O(n^3 \cdot m \cdot l)$ algorithm, improving the previous known $O(n^{3k-5})$ time algorithm [6]. In Sect. 5 we show that the orthogonality may also bring some new insights to graph modular decomposition, i.e. application to factorizing permutations and simplify decomposition algorithm for tournaments.

1.1 Definitions

Following Berge's definition of hypergraphs [1], a hypergraph H over a finite ground set $V(H)$ is made by a family of subsets of $V(H)$, denoted by $\mathcal{E}(H)$ such that (i) $\forall e \in \mathcal{E}(H)$, $e \neq \emptyset$ and (ii) $\cup_{e \in \mathcal{E}(H)} e = V(H)$. In other words, a hypergraph admits no empty edge and no isolated vertex. Furthermore we deal only with **simple** hypergraphs, where $\mathcal{E}(H) \subseteq 2^{V(H)}$ (no multiple edges). When analyzing algorithms, we use the standard notations: $|V(H)| = n$, $|\mathcal{E}(H)| = m$ and $l = \Sigma_{e \in \mathcal{E}(H)} |e|$. For every edge $e \in \mathcal{E}(H)$, we denote by $H(e) = \{x \in V(H)$ such that $x \in e\}$, and for every vertex $x \in V(H)$, we denote by $N(x) = \{e \in \mathcal{E}(H)$ such that $x \in H(e)\}$. To each hypergraph one can associate a bipartite graph G, namely its **incidence bipartite graph**, such that: $V(G) = V(H) \cup \mathcal{E}(H)$ and $E(G) = \{xe$ with $x \in V(H)$ and $e \in \mathcal{E}(H)$ such that $x \in H(e)\}$. For a hypergraph H and a subset $M \subseteq V(H)$, let $H(M)$ denote the **hypergraph induced by** M, where $V(H(M)) = M$ and $\mathcal{E}_{H(M)} = \{e \cap M \in \mathcal{E}(H)$, for $e \cap M \neq \emptyset\}$. Similarly, let H_M denote the **reduced hypergraph** where $V(H_M) = (V \setminus M) \cup \{m\}$ with $m \notin V$, and $\mathcal{E}(H_M) = \{e \in \mathcal{E}(H)$ with $e \cap M = \emptyset\} \cup \{(e \setminus M) \cup \{m\}$ with $e \in \mathcal{E}(H)$ and $e \cap M \neq \emptyset\}$. By convention in case of multiple occurrences of a similar edge, only one edge is kept and so H_M is a simple hypergraph.

Two non-empty sets A and B **overlap** if $A \cap B \neq \emptyset, A \setminus B \neq \emptyset$, and $B \setminus A \neq \emptyset$. Sets that do not overlap are said to be **orthogonal**, which is denoted by $A \perp B$. Let $\mathcal{F}$ be a family of subsets of a ground set V. We denote by $\mathcal{F}^{\perp}$ the family of subsets of V which are orthogonal to **every** element of $\mathcal{F}$. A set $S \in \mathcal{F}$ is called **strong** if $\forall S' \neq S \in \mathcal{F} : S \perp S'$. Let Δ denote the symmetric difference operation.

Definition 1 [10]. *A family of subsets $\mathcal{F}$ over a ground set V is **partitive** if it satisfies the following properties:*

(i) $\emptyset$, V and all singletons $\{x\}$ for $x \in V$ belong to $\mathcal{F}$.
(ii) $\forall A, B \in \mathcal{F}$ that overlap, $A \cap B, A \cup B, A \setminus B$ and $A \Delta B \in \mathcal{F}$

Both orthogonal and partitive families play fundamental roles in combinatorial decompositions [10,17]. Every partitive family admits a unique decomposition tree, with only two types of nodes: **complete** and **prime**. As for graphs, a node in a decomposition tree is said to be complete if the subgraph rooted at that node is either a clique or an independent set, and said to prime if the subgraph rooted at that node cannot be decomposed any further. It is well known that the strong elements of $\mathcal{F}$ form a tree ordered by the inclusion relation [10]. In this decomposition tree, every node corresponds to a set of the elements of the ground set V of $\mathcal{F}$, and the leaves of the tree are single elements of V. For a **complete** (resp. **prime**) node, every union of its child nodes (resp. no union of its child nodes other than itself) belongs to the partitive family.

Here we introduce some properties on orthogonality that will be useful for Courcelle's module and applications on graphs.

Property 1 [16]. Given a family $\mathcal{F}$ of subsets over a ground set V, $\mathcal{F}^\perp$ is partitive. Furthermore $\mathcal{F}$ is partitive iff $\mathcal{F} = (\mathcal{F}^\perp)^\perp$.

Moreover, if $\mathcal{F}$ and $\mathcal{F}'$ are two partitive families on the same ground set V, then $\mathcal{F} \cap \mathcal{F}'$ is also partitive and thus we can search for the smallest partitive family that contains a given family.

Definition 2. *Let $\mathcal{F}$ be a family of subsets over a ground set V. Let $\mathcal{P}(\mathcal{F})$ denote the smallest – by inclusion – completion of $\mathcal{F}$ that admits a unique tree decomposition with nodes labeled prime and complete.*

Property 2. For every subset family $\mathcal{F}$, $\mathcal{P}(\mathcal{F}) = (\mathcal{F}^\perp)^\perp$.

We denote by $\mathcal{O}(\mathcal{F})$ the **overlap graph** of $\mathcal{F}$ constructed as follows: The vertices are the elements of $\mathcal{F}$, and two vertices are adjacent if their corresponding subsets overlap. A pair of vertices is said to be **twins** if the vertices appear in exactly the same members of $\mathcal{F}$. The **block of twins** are the equivalence classes of the twin relation. Putting together Theorems 3.3 and 5.1 of [16] we get:

Theorem 1. *Let $\mathcal{F}$ be a subset family on V. A subset N is a node of the decomposition tree of $\mathcal{F}^\perp$ if and only if N is either:*

1. *V (the ground set), or $\{v\}$ (a one-subset element), or a block of twins, or*
2. *$\cup C$ for some connected component C of the overlap graph of $\mathcal{O}(\mathcal{F})$*

An internal node N is labeled prime if there exists a component C of $\mathcal{O}(\mathcal{F})$ with at least two members of $\mathcal{F}$ such that $N = \cup C$. Otherwise N is labeled complete.

Non-trivial nodes are mostly given by Case 2, since in Case 1 we either get the root of the tree, or a leaf, and all the siblings of any block-of-twins node are leaves. In [16] the following theorem is proposed, using as a blackbox Dahlhaus's algorithm [12] plus some post-treatment. But it has been largely simplified by [9].

Theorem 2. *Let $\mathcal{F}$ be a subset family on V. The decomposition tree of $\mathcal{F}^\perp$ can be computed in $O(n + l)$ time.*

Corollary 1. *$\mathcal{P}(\mathcal{F})$ can be computed in $O(n + l)$ time.*

2 Hypergraph Modular Decomposition

Hypergraph Substitution: *Substitution* in general is the action of replacing a vertex v in a graph G by a graph $H(V', E')$ while preserving the same neighborhood properties. To apply this concept to hypergraphs, we use the following definition presented in [19,20]:

Definition 3. *Given two hypergraphs* H, H_1, *and a vertex* $v \in V(H)$ *the **substitution** of vertex* $v \in V(H)$ *by hypergraph* H_1 *is an hypergraph, denoted* $H' = H_v^{H_1}$, *which satisfies* $V(H') = \{V \setminus v\} \cup V(H_1)$, *and* $\mathcal{E}(H') = \{e \in \mathcal{E}(H)$ *s.t.* $v \notin e\} \cup \{f \setminus v \cup e_1$ *s.t.* $f \in \mathcal{E}(H)$ *and* $v \in f$ *and* $e_1 \in \mathcal{E}(H_1)\}$.

Let us consider the example in Fig. 1 where hypergraphs are described using their incidence matrices. In this example, we substitute vertex v_3 in H by the hypergraph H_1 to create H'. Note that even if H, H_1 are undirected graphs, the substitution operation may create edges of size 3, and therefore the resulting hypergraph H' is no longer a graph.

Definition 4 (Standard Hypergraph Module) [19,20]. *Given a hypergraph* H, *a **module** $M \subseteq V(H)$ satisfies:* $\forall A, B \in \mathcal{E}(H)$ *s.t.* $A \cap M \neq \emptyset$, $B \cap M \neq \emptyset$ *then* $(A \setminus M) \cup (B \cap M) \in \mathcal{E}(H)$.

When M is a module of H then $H = (H_M)_m^{H(M)}$. On the previous example: let us take $A, B \in \mathcal{E}$ (resp. 1st and 6th columns of H') then $(A \setminus M) \cup B \cap M = 2^{nd}$ column of H' and therefore belongs to $\mathcal{E}$. If M is a module of H then $\forall e \in \mathcal{E}(H_M)$, the edges of H that strictly contain e and are not included in M are the same. In other words, **all edges in** $\mathcal{E}(H_M)$ **behave the same with respect to the outside**, which is an equivalence relation between edges.

Property 3 [20]. The family of modules of a simple hypergraph H is partitive.

Since every partitive family has a unique decomposition tree [10], it follows that the family of the modules of a simple hypergraph admits a uniqueness decomposition theorem and a unique hypermodular decomposition tree. For hypergraphs, as for graphs we have 2 types of complete node, namely series and parallel. Therefore the modular decomposition tree for hypergraphs has three types of nodes: series, parallel and prime. If $\mathcal{E}(H)$ is the set of all singletons of $V(H)$, then every subset of $V(H)$ is a module; this corresponds to the parallel case. On the other hand if $\mathcal{E}(H) = 2^{|V(H)|}$, then also every subset of $V(H)$ is a module, which corresponds to the series case.

Modular decomposition, applied to bipartite graphs, just leads to the computation of sets of *false twins* in the bipartite graphs (vertices sharing the same neighborhood) and connected components. As Fig. 1 example shows, hypergraph modules are not always set of twins of the associated incidence bipartite.

Some authors [3,4] defined **clutters** hypergraphs, in which no edge is included into another one. In this case, clutters modules are called **committees** [3]. Trivial clutters are closed under hypergraph substitution. The committees of a simple clutter also yields a partitive family which implies a uniqueness

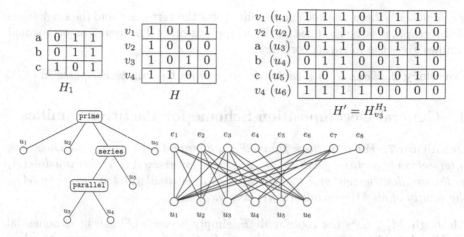

Fig. 1. An exp. of substitution, its decomposition tree, and its incidence bipartite graph. $\{u_3, u_4, u_5\}$ is a module, but only u_3, u_4 are false twins in the incidence bipartite.

decomposition theorem. From this one can recover a well-known Shapley's theorem on the modular decomposition of monotone Boolean functions. It should be noted however that finding the modular decomposition of a Boolean function is NP-hard [5]. It was shown in [7], that computing clutters in linear time would contradict the SETH conjecture.

2.1 Variants of Modular Decomposition of Hypergraphs

Often when generalizing graph concepts to hypergraphs there are several potential generalizations. In fact we found in the literature two variations on the hypergraph module definition: the k-subset modules defined in [6] and the Courcelle's modules defined in [11]. In this section we will first recall them and study their relationships to the standard one (Definition 4).

Definition 5 (k-subset module [6]). *Given a hypergraph H, we call **k-subset module** $M \subseteq V(H)$ satisfies: $\forall A, B \subseteq V(H)$ s.t. $2 \le |A|, |B| \le k$ and $A \cap M \neq \emptyset$, $B \cap M \neq \emptyset$ and $A \setminus M = B \setminus M \neq \emptyset$ then $A \in \mathcal{E}(H) \Leftrightarrow B \in \mathcal{E}(H)$.*

If H is a 2-uniform hypergraph (i.e., an undirected graph) the 2-subset modules are simply the usual graph modules. Families of k-subset modules also yield a partitive family [6].

Definition 6 (Courcelle's module [11]). *Given a hypergraph H, we call* ***Courcelle's module*** *a set $M \subseteq V(H)$ that satisfies $\forall A \in \mathcal{E}(H), A \perp M$.*

Courcelle's modules using our notations of Sect. 1 just correspond to $\mathcal{E}(H)^\perp$. Using Property 1 these modules yield a partitive family. This notion seems to be far from the standard hypergraph module definition [19, 20], this is why we called them Courcelle's modules. Indeed, applied to graphs, the orthogonal of the edge-set is the connected components (plus the vertex-set and the singletons) of the graph, not the modules. A direct application of Theorem 2 on orthogonal families gives the following corollary:

Corollary 2. *Courcelle's modular decomposition tree can be computed in $O(l)$.*

3   General Decomposition Scheme for Partitive Families

Definition 7. *For a partitive family $\mathcal{F}$ on a ground set V, using the closure by intersection of partitive families, we can define for every $A \subseteq V$, $Minmodule(A)$ as the smallest element of $\mathcal{F}$ that contains A. In particular, let us denote by $\mathcal{M}_{x,y}$ the family of all $Minmodule(\{x, y\})$ and $\forall x, y \in V$.*

Although $\mathcal{M}_{x,y}$ does not contain all $\mathcal{F}$, simply because $|\mathcal{F}|$ can be exponential in $|V|$ while $|\mathcal{M}_{x,y}|$ is always quadratic. In this section we propose an algorithm scheme to compute the decomposition tree of a partitive family if the only access to the family is a call of a function that computes: for every $A \subseteq V$, $Minmodule(A)$. Thus designing an efficient algorithm is to minimize the total number of calls. We will now show a simple way to extract the decomposition tree, i.e., the strong elements from of $\mathcal{M}_{x,y}$.

According to the definition, if a node has only two children, we cannot distinguish whether this node is prime or complete. We take the convention that the node is prime in this case. After constructing the tree, we can easily transform it into another convention just by labeling all nodes with only two children as complete nodes.

Theorem 3. *For every partitive family $\mathcal{F}$ over a ground set V, its decomposition tree can be computed using $O(|V|^2)$ calls to $Minmodule(\{x, y\})$, with $x, y \in V$.*

Proof. First choose an initial vertex x_0 and compute $Minmodule(\{x_0, x\}), \forall x \neq x_0 \in V$ and add them to a set $\mathcal{M}$. Then we add all singletons to $\mathcal{M}$. Let μ the unique path from x_0 to the root in the decomposition tree.

Claim 1: Every prime node of μ, belongs to $\mathcal{M}$.

Proof. Consider a prime strong element $A \in \mu$, it corresponds to some node of the tree which admits children $A_0, A_1, \ldots A_k$, with $k \geq 1$ in the decomposition tree. If $x_0 \in A_0$, and take $y \in A_1$, then $Minmodule(\{x_0, y\} = A$, since A is the least common ancestor in the decomposition tree. $\square$

Claim 2: For a complete node $A \in \mu$, with children $A_0, \ldots A_k$, if $x_0 \in A_0$, then
   (i) for every $1 \leq i \leq k$ the set $A_0 \cup A_i$ belongs to $\mathcal{M}$
   (ii) when elements of $\mathcal{M}$ are sorted by their size, $A_0 \cup A_i$ appear consecutively.

Proof. (i) In fact for every $y \in A_i$, $Minmodule(\{x_0, y\}) = A_0 \cup A_i$. Note that it may be possible that $A_0 = \{x_0\}$.
   (ii) If there exists a prime node P such that $\exists i, j$ such that $|A_0 \cup A_i| < |P| < |A_0 \cup A_j|$. Since $x_0 \in P$ and $x_0 \in A_0$ then P must overlap with $A_0 \cup A_i$ or

$A_0 \cup A_j$, which contradicts the fact that P is a prime node that overlaps no other element in the family.   □

The above arguments also show that any $Minmodule(\{x_0, y\})$ corresponds to either a prime node, or the union of two children of a complete node. So the family $\mathcal{M}$ is made up with prime nodes that overlap no other subsets and some daisies, and they all contain x_0, where daisies are these subsets $A_0 \cup A_i$, all containing A_0, the A_i's being the petals of the daisy and A_0 its center. Note that a daisy is a simple particular case of overlap component.

Now to find the decomposition tree we can apply the following algorithm:

1. Sort elements in $\mathcal{M}$ by size, eliminate multiple occurrences of a subset in $\mathcal{M}$.
2. Scan this list in increasing order and checking if the new considered subset overlaps the previous, else merge it to the previous it with **complete** (label both the two as complete) and continue. After the iteration, there is no unlabeled set that overlaps with another unlabeled set, then mark every unlabeled set with the label **prime**.
   Then labeled sets $X_0, X_1, \ldots, X_h$ provide the path from x_0 to the root of the modular decomposition tree, namely: $\mu = [\{x_0\} = X_0, X_1, \ldots, X_h = V]$.
3. Let us consider the partition $\{V_0, \ldots, V_h\}$ of V defined as follows:
   $V_i = \{x \in V | Minmodule(x_0, x) = X_i\}$ for $0 \leq i \leq h$.
   For every $0 \leq i \leq h$ recurse on the partitive family over the ground set V_i by computing the path from a vertex $x \in V_i$ that haven't been computed and attach its tree to X_i.

Claim 3: Every node constructed is a strong module.

Proof. Assume node X, $x_0 \in X$ overlapping with some module X'. We take any element $x' \in X' \setminus X$, then $X \Delta X'$ is a module and thus $Minmodule(x_0, x') \subseteq X \Delta X'$, which overlaps with X. Thus $Minmodule(x_0, x')$ must have been merged into X, contradiction.   □

The validity of the claim directly follows from Claims 1 to 3. For Step 1 we can use any linear sorting by value algorithm, since the size of the subsets are bounded by $n = |V|$. Clearly Step 2 can be done linear time in the size of $\mathcal{M}$. So the bottleneck of complexity is the number of calls of $Minmodule(\{x, y\})$, which is bounded by n^2.   □

Consequently, if computing the function $Minimal$ of a given partitive family can be done $O(p(n))$time, then the computation of the decomposition tree can be done in $O(n^2 \cdot p(n))$. Applied to graphs it yields an $O(n^2 m)$ algorithm, far from being optimal. Such an approach was already used for graphs in [15]. Let us now consider how to compute this function for the three variations of hypergraph modules defined previously.

4   Computing Minimal-Modules for Hypergraphs

For undirected graphs, computing Minimal-modules can be done via a graph search in linear time. We generalize this to hypergraphs for two out of the three

---

Algorithm 1. Modular-closure

Data: H a simple hypergraph and $W \subsetneq V(H)$
Result: The minimal module of H that contains W
1 Compute a lexicographic ordering τ of $\mathcal{E}(H)$ w.r.t an arbitrary ordering of V,
2 $C \leftarrow W, X \leftarrow W,$
3 Compute the induced hypergraph $H(C)$,
4 **for** $1 \leq i \leq |\mathcal{E}(H(C))|$ **do**
5 | Compute the ordered lists L_i of the restriction to $V(G) \setminus C$ of the edges in
| $\mathcal{E}(H)$ that contain $f_i \in \mathcal{E}(H(C))$
6 $\mathcal{Q}(C) \leftarrow \{L_1, \ldots, L_{|\mathcal{E}(H(C))|}\}$ the ordered partition made up with these lists
7 **if** $|\mathcal{Q}(C)| = 1$ **then**
8 | C is a module, STOP
9 **else**
10 | $L \leftarrow First(\mathcal{Q}(C))$, {% the first class in the ordered partition%}
11 | **while** $Next(L) \neq NIL$ {% the next element of L in the ordered partition%}
| **do**
12 | | $X \leftarrow Comparison(L, Next(L))$
13 | | **if** $X = \emptyset$ **then**
14 | | | $L \leftarrow Next(L)$
15 | | **else** $C \leftarrow C \cup X$, update $\mathcal{Q}(C)$ via partition refinement with X,
| | $L \leftarrow First(L)$
16 | | {%if L has been split during the update we take its first part%}
17 |
18 $RESULT \leftarrow C$ {%C is a minimal module that contains W%}

---

definitions of modules: the standard and the k-subset module. For efficiency purposes, we represent our hypergraphs using for each vertex x a list to represent $N(x)$ i.e., the edges its belongs to, and for each edge e a list to represent $H(e)$ i.e., the vertices it contains. For a hypergraph this yields a representation using $O(n + m + l)$ memory. If the hypergraph is simple then $O(n + m + l) = O(l)$.

4.1   Standard Modules

Definition 8. *For a set $C \subsetneq V(H)$, an edge $A \in \mathcal{E}(H)$ is a **edge-splitter** for C, if $A \setminus C \neq \emptyset$ and $A \cap C \neq \emptyset$ and if there exists $B \in \mathcal{E}(H)$ s.t. $B \cap C \neq \emptyset$, and $(A \setminus C) \cup (B \cap C) \notin \mathcal{E}(H)$.*

In other words, a set of vertices is a module iff it admits no edge-splitter.

Property 4. If $X \subseteq V(H)$ is a splitter of $C \subseteq V(H)$ respect to A, B as above. Let $B' \in \mathcal{E}(H)$ be the edge such that $B' \cap C = B \cap C$ and with $|(B' \setminus C) \Delta (A \setminus C)|$ minimum.

Let $X' = (B' \setminus C) \Delta (A \setminus C)$, then there is no module Y of H such that $C \subsetneq Y$ but $X' \nsubseteq Y$.

Theorem 4. *If H is a simple hypergraph, for every set $W \subseteq V(H)$, Algorithm Modular-closure computes $Minmodule(W)$ in $O(n \cdot l)$. And its modular decomposition tree can be computed in $O(n^3 \cdot l)$.*

Algorithm 2. Procedure Comparison

Data: 2 lists L', L'' of the restriction to $V(G) \setminus C$ of the edges in $\mathcal{E}(H)$ that
contain some $f \in \mathcal{E}(H(C))$. They are supposed to be lexicographically
increasingly ordered using τ

Result: X a set of vertices forced be contained in Minmodule(C)

1 if $L' = L''$ then
2 | $X \leftarrow \emptyset$, STOP
3 else
4 | Let $e \in L'$ and $f \in L''$ be the first lexicographically difference,
5 | if $(e <_\tau f)or(e \neq \emptyset \ and \ f = \emptyset)$ then
6 | | {% $e \notin L''$ is a edge-splitter %}
7 | | compute $f' \in L''$ that minimizes $|h(e)\Delta h(f')|$ with $f' \in L''$
   | | $X \leftarrow h(e)\Delta h(f')$
8 | else
9 | | {% $(e >_\tau f)or(e = \emptyset \ and \ f \neq \emptyset)$, i.e. $f \notin L'$ is a edge-splitter %}
10 | | compute $e' \in L'$ that minimizes $|h(f)\Delta h(e')|$ with $e' \in L'$,
11 | | $X \leftarrow h(f)\Delta h(e')$
12
13 {%Note that $e = \emptyset$, $f \neq \emptyset$ (resp. $e \neq \emptyset$, $f = \emptyset$) corresponds to the case
$|L| < |Next(L)|$ (resp. $|L| > |Next(L)|)$%}

Proof. (i) **Correctness:** First we notice that C is a module of H iff all the lexicographically sorted lists L_i are equal. At each step of the lexicographic process a list can only be cut into parts, no lists are merged. If at some step of the algorithm two lists L_i, L_j are equal, and if afterwards they are cut into sublists via the refinement process, equality between sublists is preserved since the refinement act similarly on the lists. Thus the algorithm scan the lists form left to right using a single sweep and the following invariant: at each step of the while loop all the lists before the current list L are all equal to L.

Using the procedure Comparison either the lists are equal and then we proceed else using Property 4 we know that we can add this set of vertices. At the end of the algorithm either all lists are equals and $C \neq V(H)$ and therefore C is the non trivial minimal module containing W or $C = V(H)$ and there is no other module between W and $V(H)$.

(ii) **Complexity Analysis:** To implement the first step (line 1) we can use an ordered partition refinement technique on $\mathcal{E}(H)$ (see [13]) using the sets $N(x)$ for every $x \in V(H)$ as pivot sets. This provides a total ordering τ of $\mathcal{E}(H)$. This can be done in $O(n + m + l)$.

To compute $\mathcal{Q}(C)$, we can use the same ordered partition refinement technique using the sets $N(x)$ for every $x \in C$ as pivot sets we can compute the ordered partition of $\mathcal{E}(H)$. Starting from the partition $P_0 = \{\mathcal{E}(H)\}$, we refine this partition successively using $N(x_i)$ for every $x_i \in C$. Let us denote by P_f the partition obtained after this round of refinements. Each part of P_f can be ordered using τ, since partition refinement can maintain an initial ordering of its elements within the same complexity. So if we start with the initial ordering τ in the unique part of P_0. And the parts are lexicographically ordered with respect to their intersection to C. This can be done in $O(|C| + \Sigma_{x \in C} |N(x)|)$.

In fact after line 6 we can ignore the vertices of C, a similar remark holds when C is updated.

Now we have to check if all edges lists L_i are identical or not and stop at the first difference. Since the lists are ordered lexicographically using an ordering τ of the vertices, a simple scan of these ordered lists is enough to compute (Comparison procedure) of Algorithm 1.

When C and $\mathcal{Q}(C)$ are updated, the algorithm goes on with the first part of the previous current list L. First means that if L has been split during the update we take its first part. Therefore in the worst case some list can be analyzed several times (at most n times) and therefore the overall complexity of the list scan is bounded by $O(n \cdot l)$.

When a difference is found between two lists we have to search for an edge that minimizes the symmetric distance with respect to the differentiating edge. Even though it can be done several times for a given edge, but every time we launch this search, at least one vertex will be added into C, thus at most search for n times. So the overall complexity of these searches is $O(n \cdot l)$.

Therefore the whole process is in $O(n+m+l+n \cdot l) = O(n \cdot l)$. Using Theorem 3 we obtain the decomposition tree in $O(n^3 \cdot l)$.   □

Up to our knowledge, [20] states there is a polynomial time decomposition algorithm for clutters based on its $O(n^4 m^3)$ modular closure algorithm without precising the complexity, our algorithm is an improvement because our total decomposition time is already smaller than $O(n^4 m^3)$ s.

4.2   Decomposition into k-Subset Modules

Definition 9 [6]. *A subset $X \neq \emptyset$ is a **k-subset splitter** of the set C if there exist $A, B \subseteq V$ s.t. $2 \leq |A|, |B| \leq k$ and $A \cap C \neq \emptyset$, $B \cap C \neq \emptyset$ and $A \setminus C = B \setminus C = X$, $A \in \mathcal{E}(H)$ but $B \notin \mathcal{E}(H)$.*

Lemma 1. *Given a set $C \subseteq V(H)$, any k-subset splitter of C is in the form of $H(e) \setminus C$ for some $e \in \mathcal{E}(H)$, $|H(e)| \leq k$.*

Such an edge will be called an **k-edge-splitter** of C. Let $D(k, h) = \Sigma_{i=1}^{i=k} \binom{h}{i}$ for $1 \leq k \leq h$, where $\binom{h}{i}$ denotes the binomial coefficient. All values of $D(k, h)$ strictly greater that $|\mathcal{E}(H)|$ will be set as **Out-of-Range**, a huge number.

Lemma 2. *For a simple hypergraph H, given a set $C \subseteq V(H)$ and an edge $e \in \mathcal{E}(H)$ s.t. $|H(e)| \leq k$, $e \cap C \neq \emptyset$ and $X = e \setminus C \neq \emptyset$, let L be the list of edges in $\mathcal{E}(H)$ with size $\leq k$ and whose intersection with $V(H) \setminus C$ are identical to X and intersection with C is not empty, i.e. $L = \{e' \in \mathcal{E}(H) \mid e' \setminus C = X, e' \cap C \neq \emptyset \text{ and } |H(e')| \leq k\}$. If $|L| < D(k - |X|, |C|)$ then e is an edge-splitter of C.*

Proof. It is equivalent to check for such an e given above and $X = e \setminus C$, whether every non empty subset B of size $\leq k - |X|$ in C has $X \cup B \in \mathcal{E}(H)$. Since H is simple and there are no identical elements in L, a counting argument captures the condition. Moreover, the number of subsets checked this way is $\leq |\mathcal{E}(H)|$. □

Algorithm 3. k-subset modular-closure

---

1 **Algorithm:** k-subset modular-closure

   Data: H a simple hypergraph, k an integer such that $1 \leq k \leq |V(H)|$ and
   $\quad\quad W \subsetneq V(H)$
   Result: The minimal k-module of H that contains W

2 Compute all $D(k,h)$ for $|W| \leq h \leq |V(H)|$,

3 Compute a lexicographic ordering τ of $\mathcal{E}(H)$ with respect to some ordering of
   the vertices,

4 $C \leftarrow W, X \leftarrow W$,

5 **while** $X \neq \emptyset$ **do**

6 $\quad$ For every $e_i \in \mathcal{E}(H)$ overlap C, $|H(e_i)| \leq k$, create the lists L_i of edges in
   $\quad$ $\mathcal{E}(H)$ with size $\leq k$ whose intersection with $V(H) \setminus C$ are identical to $e_i \setminus C$
   $\quad$ and of size h and intersection with C is not empty

7 $\quad$ **if** *For some i, $|L_i| < D(k-h,|C|)$* **then**

8 $\quad\quad$ $X \leftarrow X \cup (H(e_i) \setminus C)$,

9 $\quad\quad$ **if** $V(G) = C \cup X$ **then**

10 $\quad\quad\quad$ % there is no non-trivial module between W and $V(H)$ %

11 $\quad\quad\quad$ $RESULT \leftarrow V(H)$, STOP

12 $\quad\quad$ **else**

13 $\quad\quad\quad$ %X is a splitter for C%,

14 $\quad\quad\quad$ $C \leftarrow C \cup X$

15 $\quad\quad$ |

16 $\quad$ **else** $X \leftarrow \emptyset$

17 $RESULT \leftarrow C$ %C is a non trivial module containing W%

Theorem 5. *For a simple hypergraph H and $A \subsetneq V(H)$, for any fixed integer $k \leq |V(H)|$, Algorithm 3 (k-subset modular-closure) can compute the minimal k-subset module that contains A in $O(n \cdot m \cdot l)$ time, which gives a $O(n^3 \cdot m \cdot l)$ decomposition algorithm.*

5 Using Orthogonality for Graph Modular Decomposition

5.1 Factorizing Permutations and Fractures

Given a permutation σ of a set V, let $\sigma(i)$ denote the ith element of V. An interval $[l,r]$ of σ is a set of elements that follow consecutively ($1 \leq l \leq r \leq n$).

Definition 10. *Let $\mathcal{F}$ be a family of subsets of V. A permutation σ of V is a **factorizing permutation** if every strong set of $\mathcal{F}$ is an interval of σ. Furthermore, it is **perfect** if every set of $\mathcal{F}$ is an interval.*

Factorizing permutations were defined in the context of modular decomposition [8] but can be generalized to any subset family, where a permutation can be obtained by a traversal of its decomposition tree. We adapt a definition from [8]:

Definition 11. *Let $G = (V, E)$ be a graph and σ a permutation of V. Let us consider a pair $\{\sigma(i), \sigma(i+1)\}$ of two consecutive elements. The **left fracture** (resp. **right fracture**) of i is the largest interval $[s, i]$ (resp. $[i+1, s]$) where $\sigma(s)$ is a splitter of $\{\sigma(i), \sigma(i+1)\}$. If that pair admits no splitter on its left (resp. right) then i has no left (resp. right) fracture. The **fracture family**, denoted $\mathcal{F}rac(\sigma)$, of a given permutation of the vertices of a graph is the set of all (left and right) fractures for all $1 \leq i < n$.*

Lemma 3. *Given a graph G and a permutation σ of its vertices, an interval I of σ is a module iff it does not overlap any fracture of $\mathcal{F}rac(\sigma)$.*

Lemma 4. *Let σ be a factorizing permutation of the modules family of G, F be a fracture of σ, and M the smallest module containing F. M is a strong module.*

Theorem 6. *Given a graph G, the family $\mathcal{M}$ of its modules and a factorizing permutation σ of $\mathcal{M}$ we have: $\mathcal{P}(\mathcal{M}) = \mathcal{F}rac(\sigma)^{\perp}$. Furthermore, if the graph is undirected then $\mathcal{M} = \mathcal{F}rac(\sigma)^{\perp}$.*

5.2   Modular Decomposition of Tournaments

We can apply the theorem above to undirected graphs where we get a new algorithm but no improvement of the existing algorithms, or to tournaments (orientation of the complete graph), and we get an simple (much simpler than the existing algorithm in [18]) and optimal modular decomposition algorithm. Among the families admitting a perfect factorizing permutation are the **anti-symmetric-partitive families**, families where Axiom ii of Definition 1 is replaced with: $\forall A, B \in \mathcal{F}$ that overlap, $A \cap B, A \cup B, A \setminus B \in \mathcal{F}$ and $A \triangle B \notin \mathcal{F}$. Their decomposition tree is often called a **PQ-tree**, whose nodes are labeled P (prime) and Q (having a linear ordering of the siblings so that any union of siblings that follow consecutively belong to the family, and no other union). Well-known antisymmetric-partitive families are the intervals of the real line (their intersection model being an interval graph), the common intervals of two permutations [2], or the modules of a tournament.

Theorem 7. *Let G be a tournament. The modular decomposition tree of G can be computed in $O(n^2)$ time.*

6   Conclusions

In this paper, using a general framework for decomposition of partitive families or tools of orthogonality, we have proposed 3 polynomial algorithms to compute hypergraph modular decomposition trees under 3 different definitions of modules. Our general framework yields a $O(n^3 \cdot l)$ algorithm for the standard decomposition of hypergraphs, and a $O(n^3 \cdot m \cdot l)$ time for k-subset modules, an improvement to the previously known non-polynomial $O(n^{3k-5})$ time algorithm [6], where k denotes the maximal size of an edge. Since our approach is brute force, there may exists linear time (in $O(l)$) algorithms for the standard hypergraph decomposition, as for graphs [14]. One would have to develop new hypergraph algorithms, for example one that computes in linear time some factoring permutation which always exists for every partitive family and use some orthogonality.

Conjecture 1. Simple hypergraphs admit a linear time $O(l)$ modular decomposition algorithm.

References

1. Berge, C.: Graphes et hypergraphes. Dunod, Paris (1970)
2. Bergeron, A., Chauve, C., de Montgolfier, F., Raffinot, M.: Computing common intervals of K permutations, with applications to modular decomposition of graphs. SIAM J. Discrete Math. **22**(3), 1022–1039 (2008)
3. Billera, L.J.: Clutter decomposition and monotonic boolean functions. Ann. N.-Y. Acad. Sci. **175**, 41–48 (1970)
4. Billera, L.J.: On the composition and decomposition of clutters. J. Combin. Theory, Ser. B **11**(3), 234–245 (1971)
5. Bioch, J.C.: The complexity of modular decomposition of boolean functions. Discrete Appl. Math. **149**(1–3), 1–13 (2005)
6. Bonizzoni, P., Vedova, G.D.: An algorithm for the modular decomposition of hypergraphs. J. Algorithms **32**(2), 65–86 (1999)
7. Borassi, M., Crescenzi, P., Habib, M.: Into the square: on the complexity of some quadratic-time solvable problems. Electr. Notes Theor. Comput. Sci. **322**, 51–67 (2016)
8. Capelle, C., Habib, M., de Montgolfier, F.: Graph decompositions and factorizing permutations. Discrete Math. Theor. Comput. Sci. **5**(1), 55–70 (2002)
9. Charbit, P., Habib, M., Limouzy, V., de Montgolfier, F., Raffinot, M., Rao, M.: A note on computing set overlap classes. Inf. Process. Lett. **108**(4), 186–191 (2008)
10. Chein, M., Habib, M., Maurer, M.C.: Partitive hypergraphs. Discrete Math. **37**(1), 35–50 (1981)
11. Courcelle, B.: A monadic second-order definition of the structure of convex hypergraphs. Inf. Comput. **178**(2), 391–411 (2002)
12. Dahlhaus, E.: Parallel algorithms for hierarchical clustering and applications to split decomposition and parity graph recognition. J. Algorithms **36**(2), 205–240 (2000)
13. Habib, M., McConnell, R.M., Paul, C., Viennot, L.: Lex-BFS and partition refinement, with applications to transitive orientation, interval graph recognition and consecutive ones testing. Theor. Comput. Sci. **234**(1–2), 59–84 (2000)
14. Habib, M., Paul, C.: A survey of the algorithmic aspects of modular decomposition. Comput. Sci. Rev. **4**(1), 41–59 (2010)
15. James, L.O., Stanton, R.G., Cowan, D.D.: Graph decomposition for undirected graphs. In: Proceedings of the 3rd Southeastern International Conference on Combinatorics, Graph Theory, and Computing, pp. 281–290 (1972)
16. McConnell, R.M.: A certifying algorithm for the consecutive-ones property. In: SODA, Proceedings of the Fifteenth Annual ACM-SIAM Symposium on Discrete Algorithms, pp. 768–777 (2004)

17. McConnell, R.M., de Montgolfier, F.: Algebraic operations on PQ trees and modular decomposition trees. In: Kratsch, D. (ed.) WG 2005. LNCS, vol. 3787, pp. 421–432. Springer, Heidelberg (2005). https://doi.org/10.1007/11604686_37

18. McConnell, R.M., de Montgolfier, F.: Linear-time modular decomposition of directed graphs. Discrete Appl. Math. **145**(2), 198–209 (2005)

19. Möhring, R., Radermacher, F.: Substitution decomposition for discrete structures and connections with combinatorial optimization. In: Proceedings of the Workshop on Algebraic Structures in Operations Research, pp. 257–355 (1984)

20. Möhring, R.H.: Algorithmic aspects of the substitution decomposition in optimization over relations, set systems and boolean functions. Ann. Oper. Res. **4**, 195–225 (1985)

Shortest-Path-Preserving Rounding

Herman Haverkort, David Kübel[✉], and Elmar Langetepe

Universität Bonn, Bonn, Germany
{haverkort,dkuebel,elmar.langetepe}@uni-bonn.de

Abstract. Various applications of graphs, in particular applications related to finding shortest paths, naturally get inputs with real weights on the edges. However, for algorithmic or visualization reasons, inputs with integer weights would often be preferable or even required. This raises the following question: given an undirected graph with non-negative real weights on the edges and an error threshold ε, how efficiently can we decide whether we can round all weights such that shortest paths are maintained, and the change of weight of each shortest path is less than ε? So far, only for path-shaped graphs a polynomial-time algorithm was known. In this paper we prove, by reduction from 3-SAT, that, in general, the problem is NP-hard. However, if the graph is a tree with n vertices, the problem can be solved in $O(n^2)$ time.

Keywords: Algorithms · Graph · Graph drawing · Rounding · Shortest path

1 Introduction

In this paper we study the following problem: given an undirected graph with non-negative real weights on the edges and an error threshold ε, decide efficiently whether we can replace all weights by integers such that shortest (least-cost) paths are maintained with maximum error less than ε.

Our research is motivated by applications whose inputs consist of graphs that have real weights on the edges, but prefer or require graphs with small integer weights. For example, consider a transportation network, modelled as an undirected graph, with a weight function on the edges that represents the time (or cost) it takes to travel each edge. We may also refer to the weights as lengths. If the weights are small integers, one could draw a zone map of the network such that the number of zone boundaries crossed by each shortest path corresponds to the weight of the path [5]. Given a graph with real edge weights, we would therefore like to normalize the weights such that weight 1 corresponds to the intended zone diameter of the map, and then round the normalized weights to integers such that shortest paths are maintained. The corresponding zone map would then provide a fairly accurate representation of travel costs, and would be easier to read and use than a map in which the true travel costs are written in full detail next to each edge.

© Springer Nature Switzerland AG 2019
C. J. Colbourn et al. (Eds.): IWOCA 2019, LNCS 11638, pp. 265–277, 2019.
https://doi.org/10.1007/978-3-030-25005-8_22

Other applications that could take advantage of rounded weights include algorithms to compute shortest paths: there are algorithms that are more efficient with small integer weights than with arbitrary, real weights [10]. Funke and Storandt [4] cite space efficiency, the speed of arithmetic operations, and stability as advantages of low-precision edge weights.

However, as argued and demonstrated by Funke and Storandt [4,8], naively rounding weights to the nearest integer values could lead to rounding errors accumulating in such a way, that the structure of optimal paths in the graph changes, which can be undesirable. When rounding weights naively, some paths may see their lengths doubled whereas other, arbitrarily long paths may see their lengths reduced to zero [4]. Funke and Storandt argue that randomized rounding is also likely to cause unacceptable errors in any graph that is large enough [4].

This brings us to the following problem statement. Consider an undirected graph, denoted by $G = (V, E, \omega)$, with vertex set V, edge set E, and a weight function $\omega : E \to \mathbb{R}_{\geq 0}$. A *simple* path in G is a sequence π of distinct vertices $v_1, \ldots, v_j$, where $\{v_i, v_{i+1}\} \in E$ for $1 \leq i \leq j - 1$. By $\omega(\pi)$ we denote the weight of the path π, that is, $\sum_{i=1}^{j-1} \omega(\{v_i, v_{i+1}\})$. A *shortest* path in G is a simple path $v_1, \ldots, v_j$ that has minimum weight among all paths from v_1 to v_j in G.

Definition 1 (path-oblivious/weak/strong ε-rounding). *Let $G = (V, E, \omega)$ be an undirected graph with a weight function $\omega : E \to \mathbb{R}_{\geq 0}$. We call $\tilde{\omega} : E \to \mathbb{N}_0$ a path-oblivious ε-rounding of G if the following condition holds:*

1. *For any shortest path π in G, we have $|\tilde{\omega}(\pi) - \omega(\pi)| < \varepsilon$, that is, between ω and $\tilde{\omega}$, the weight of any shortest path in G changes by strictly less than ε.*

We call $\tilde{\omega}$ a weak ε-rounding if in addition, the following condition holds:

2. *Any shortest path in $G = (V, E, \omega)$ is also a shortest path in $\tilde{G} = (V, E, \tilde{\omega})$.*

We call $\tilde{\omega}$ a strong ε-rounding if it is a weak ε-rounding and additionally:

3. *Any shortest path in $\tilde{G}$ is also a shortest path in G.*

Note that a weak ε-rounding does not imply a strong ε-rounding. Consider, for example, a triangle, where each edge is of weight 0.5 and shortest paths are unique. For $\varepsilon > 0.5$, rounding the weight of each edge to zero satisfies Condition 1 and 2, but violates Condition 3, since shortest paths are no longer unique.

By choosing ε as large as the diameter of the graph, there always exists a weak (but useless) ε-rounding. We would rather have an ε-rounding for a small value of ε such as $\varepsilon = 1$, but in that case, an ε-rounding does not always exist. For example, a star that consists of three edges of weight $1/2$ does not admit a 1-rounding: at least two of the three edges would have to be rounded in the same way, but if we would round two edges down, there would be a shortest path with rounding error -1; if we would round two edges up, there would be a shortest path with rounding error 1. Given an undirected graph $G = (V, E, \omega)$ and an error tolerance ε, our problem is therefore to decide whether G admits a path-oblivious, weak, or strong ε-rounding.

Our first goal is now to establish for what classes of graphs efficient exact algorithms for these problems may exist. In this paper, we show that all three versions of the problem are NP-hard for general graphs, but can be solved in quadratic time on trees. In fact, trees always admit a 2-rounding, and given a tree of n vertices, we can compute, in $O(n^2 \log n)$ time, the smallest ε such that the tree admits an ε'-rounding for any $\varepsilon' > \varepsilon$. The algorithm is constructive: it can easily be adapted to produce the corresponding weights $\tilde{\omega}$. We give references to related work and directions for further research in the last section of the paper.

2   Complexity of the Problem

We will show that it is NP-hard to decide, given an edge-weighted graph G and an error tolerance ε, whether G admits a path-oblivious, weak, or strong ε-rounding. To this end, we present a reduction from 3-SAT, which proves hardness for all three variants of the problem. For simplicity, we use $\varepsilon = 1$, but the proof is easily adapted to any $\varepsilon \in (7/8, 1]$.

The 3-SAT problem is the following. We are given a 3-CNF formula, that is, a boolean formula α in conjunctive normal form, where each of the m clauses consists of exactly three literals. Each literal is either one of n variables $x_1, x_2, \ldots, x_n$ or its negation. Decide whether α is satisfiable. W. l. o. g. we assume that every variable appears at most once in each clause of the 3-SAT formula.

In the following, we show how to construct a graph $G_\alpha = (V, E, \omega)$ for a given 3-CNF formula α such that G_α admits a strong 1-rounding if α is satisfiable, whereas G_α does not even admit a path-oblivious 1-rounding if α is not satisfiable. To describe G_α, we introduce subgraphs called *variable gadgets* and *clause gadgets*, as well as *clause-variable edges* and *shortcut edges*.

The idea of the construction is as follows. In Lemma 1, we will show that a variable gadget admits exactly two strong 1-rounding. We identify these two roundings with the assignments true and false of a boolean variable. Using clause-variable edges, the state of a variable gadget can be transferred to a clause gadget (Lemma 3). Locally, the clause gadget admits a 1-rounding if and only if one of the variable assignments (transferred via clause-variable edges) satisfies the clause (Lemma 2). We use shortcut edges to ensure that shortest paths in G_α that do not contribute to modelling α, are easy to analyse and unique—before and after rounding the weights (Lemma 4).

Due to space restrictions, the (full) proofs of the lemmata and the theorem in this section have been omitted, but they can be found in [6].

To design a *variable gadget*, first consider two edges attached to a triangle, where each edge is of weight 2.5, as illustrated in Fig. 1a. In a 1-rounding, the choice of the rounding for $e(v_{i,0})$, the edge incident on $v_{i,0}$, determines the rounding of the remaining edges; see Fig. 1b and c. To obtain a variable gadget for variable x_i, we proceed as follows. Assume that x_i appears in h literals $l_1, \ldots, l_h$ of α. We construct h triangles $\Delta_{i,1}, \ldots, \Delta_{i,h}$, where each $\Delta_{i,k}$ (for $k \in \{1, \ldots, h\}$) has a left vertex, a right vertex, and a *base* (bottom) vertex; we label the base vertex $v_{i,k}$. We chain up the triangles by including an edge between the right

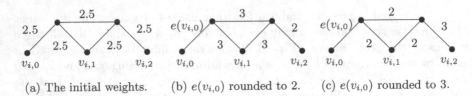

(a) The initial weights.   (b) $e(v_{i,0})$ rounded to 2.   (c) $e(v_{i,0})$ rounded to 3.

Fig. 1. A minimal variable gadget. In a 1-rounding of this gadget, any shortest path of two edges has to round one edge up and one edge down—otherwise the total rounding error on the path would be ± 1, violating Condition 1 of a 1-rounding. Thus, the top triangle edge must be rounded in the opposite way as compared to the edges $e(v_{i,0})$ and $e(v_{i,2})$, incident on $v_{i,0}$ and $v_{i,2}$, respectively. These arguments imply that $e(v_{i,0})$ and $e(v_{i,2})$ have to be rounded in the same way and *all* triangle edges are rounded in the opposite way. Note that paths containing two triangle edges have rounding error ± 1, but such paths are not shortest paths, neither before nor after rounding.

vertex of $\Delta_{i,k}$ and the left vertex of $\Delta_{i,k+1}$ for each $k \in \{1, \ldots, h-1\}$. To the left vertex of $\Delta_{i,1}$, we attach another vertex $v_{i,0}$, and to the right vertex of $\Delta_{i,h}$, we attach another vertex $v_{i,h+1}$, as shown in Fig. 2. Finally, for every $1 \le k \le h$, if $l_k = \neg x_i$, we add another vertex $\overline{v}_{i,k}$, called *inverter*, which we connect to $v_{i,k}$. All edges of the variable gadget have an initial weight of 2.5.

We call the edges of $\Delta_{i,1}, \ldots, \Delta_{i,h}$ *triangle edges*. Moreover, with $e(v_{i,0})$ we denote the unique edge of the variable gadget attached to $v_{i,0}$.

Lemma 1 (1-roundings of variable gadgets). *A variable gadget admits exactly two 1-roundings (both of which are strong 1-roundings): either all triangle edges are rounded up and all other edges down, or vice versa.*

Moreover, for any two vertices u and v of the gadget, the rounding error of the unique shortest path from u to v is either zero or equal to the rounding error on the last edge of the path (ending at v).

From Lemma 1, we obtain that in a 1-rounding, the choice of the rounding for $e(v_{i,0})$ determines the rounding of all other edges.

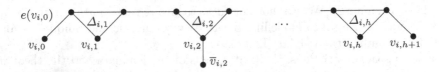

Fig. 2. The variable gadget for x_i, where x_i appears in h literals $l_1, \ldots, l_h$ of α. Here, $l_2 = \neg x_i$, so an additional vertex $\overline{v}_{i,2}$ is added and attached to $v_{i,2}$. All edges have weight 2.5, so the choice of the rounding for $e(v_{i,0})$ determines the rounding of all the other edges in a 1-rounding: triangle edges have to be rounded complementary to non-triangle edges.

To create a *clause gadget* for clause C_j, we take a cycle of nine vertices and edges, where each edge gets an initial weight of 3.6. Moreover, we attach, to every third vertex along the cycle, a new vertex, called a *knob*, with another edge of weight 2.5, called a *handle*. We denote the knobs by $c_{j,1}, c_{j,2}, c_{j,3}$, as shown in Fig. 3a. We will use the notation $e(c_{j,t})$ to denote the edge (handle) of the clause gadget that connects knob $c_{j,t}$ to the nonagon. Finally, we add a vertex which we connect to every vertex on the cycle with an edge of weight 6. Note that the weights of these edges are integer—hence, they cannot be rounded. A clause gadget has at least three strong 1-roundings; see Fig. 3b. However, there is no path-oblivious, weak, or strong 1-rounding for the clause gadget in which $e(c_{j,1})$, $e(c_{j,2})$ and $e(c_{j,3})$ are all rounded up, as the following lemma states.

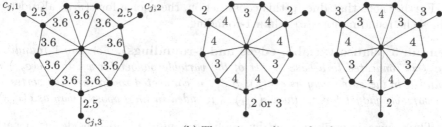

(a) The initial weight of all edges incident on the centre is 6.

(b) Three 1-roundings of a clause gadget. At most two of the edges $e(c_{j,1}), e(c_{j,2})$ and $e(c_{j,3})$ are rounded up.

Fig. 3. The clause gadget for clause C_j.

Lemma 2 (1-roundings of a clause gadget). *Consider a clause gadget for C_j, and suppose we fix, for each of its handles $e(c_{j,1}), e(c_{j,2})$ and $e(c_{j,3})$, whether its weight is rounded up or down. The clause gadget now admits a path-oblivious 1-rounding $\tilde{\omega}$ if and only if at least one of its three handles is rounded down. If there is a path-oblivious 1-rounding, there is a strong 1-rounding.*

Next, we introduce *clause-variable edges* to connect the gadgets of variables to the gadgets of clauses that contain these variables. So if variable x_i appears in clause C_j, we connect the corresponding gadgets with exactly one clause-variable edge of weight $D := 5m + 20$ according to the following rule: if the t-th literal in C_j is x_i, then we connect a base vertex of the variable gadget for x_i to $c_{j,t}$, using an edge of weight D; if the t-th literal in C_j is $\neg x_i$, then we connect an inverter of the variable gadget for x_i to $c_{j,t}$, using an edge of weight D. We do this such that exactly one clause-variable edge is connected to each inverter, and exactly one clause-variable edge is connected to each base vertex that is not attached to an inverter. By design, the variable gadgets have the right numbers of base vertices and inverters to make this possible.

Note that clause-variable edges do not invalidate Lemmas 1 and 2, that is, they still hold with respect to the shortest paths between any pair of vertices of the variable or clause gadget, respectively. This can be seen as follows. There are m clauses and each variable appears in each clause at most once. Hence, the diameter of a variable gadget is at most $(2m + 1) \cdot 2.5$. The diameter of a clause gadget is 15.8. Thus, the variable and clause gadgets all have diameter less than $D - 2$. Therefore, before rounding, no path between two vertices of the same gadget that uses a clause-variable edge can be a shortest path. Moreover, when we choose a 1-rounding for each gadget separately, the rounded weight of the shortest path between any pair of vertices within a gadget will still be less than $D - 1$, while the rounded weight of any path using a clause-variable edge will still be at least D. Therefore, also after rounding, no path between two vertices of the same gadget that uses a clause-variable edge can be a shortest path. Thus, adding the clause-variable edges does not invalidate Lemmas 1 and 2.

Further note that due to this construction, the choice for $e(v_{i,0})$ also determines the rounding for $e(c_{j,t})$ in a 1-rounding:

Lemma 3 (clause-variable edges and 1-roundings). *For any 1-rounding, if $c_{j,t}$ is connected to a base vertex of the variable gadget for x_i, then $e(c_{j,t})$ is rounded in the same way as $e(v_{i,0})$; if $c_{j,t}$ is connected to an inverter vertex of the variable gadget for x_i, then $e(c_{j,t})$ is rounded in the opposite way as $e(v_{i,0})$.*

The lemma is easily proven by applying Lemma 1 and considering a path that consists of $e(c_{j,t})$, the clause-variable edge connecting the gadgets, and an adjacent edge in the variable gadget.

Finally, to ensure that the shortest path between any pair of vertices of G_α is unique and easy to analyse, we add *shortcut edges* as follows. We include an edge $\{u, v\}$ in G_α with weight $2D$ if one of the following conditions holds:

(i) u and v belong to different variable gadgets;
(ii) u and v belong to different clause gadgets;
(iii) u belongs to a variable gadget for variable x_i and v belongs to a clause gadget for clause C_j and neither x_i nor $\neg x_i$ appears in C_j.

Lemma 4 (shortest path via shortcut edge). *Let u and v be vertices of G_α that are directly connected by a shortcut edge, and let $\tilde{\omega}$ be a 1-rounding on G_α. Then, the shortcut edge $\{u, v\}$ is the unique shortest path in G_α with respect to ω and $\tilde{\omega}$.*

Note that, just like clause-variable edges, the shortcut edges do not invalidate Lemmas 1 and 2. They do not invalidate Lemma 3 either, as its proof hinges on shortest paths of length $D + 5 < 2D - 2$.

An example for the construction is given in Fig. 4.

Theorem 1. *It is NP-hard to decide, given an edge-weighted graph G and an error tolerance ε, whether G admits (1) a path-oblivious ε-rounding; (2) a weak ε-rounding; (3) a strong ε-rounding.*

Proof Sketch: To start with, we show how to obtain a strong 1-rounding for G_α if α is satisfiable. Let ψ be an assignment of values to variables that satisfies α. So we have $\psi : \{x_1, \ldots, x_n\} \rightarrow \{0, 1\}$, where 0 denotes the logical value `false`

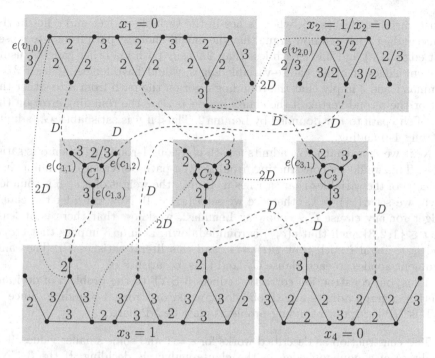

Fig. 4. A sketch of G_α for $\alpha = (x_1 \lor x_2 \lor \neg x_3) \land (x_1 \lor x_3 \lor \neg x_4) \land (\neg x_1 \lor \neg x_3 \lor x_4)$. Grey bounding boxes mark gadgets, clause-variable edges are dashed and have weight $D = 5 \cdot 3 + 20 = 35$, shortcut edges (mostly omitted) are dotted and have weight $2D$. The weights of the edges in G_α have been rounded according to the assignment of the variables, which are given in the boxes of the corresponding variable gadgets. For x_2, both assignments are given, which affects the rounded weight of several edges. The rounded weight of each of these edges is given by a/b, where a corresponds to the assignment $x_2 = 1$ and b corresponds to $x_2 = 0$. Note that for $x_1 = 0, x_3 = 1, x_4 = 0$, if we set $x_2 = 1$, we can obtain a 1-rounding for the gadget of C_1 and for G_α. If we set $x_2 = 0$, then there is no 1-rounding for the gadget of C_1, as all of its handles are rounded up.

and 1 denotes **true**. For each variable x_i, we round $e(v_{i,0})$ down if $\psi(x_i) = 1$ and up otherwise. This determines how to round the rest of each variable gadget and each clause handle according to Lemmas 1 and 3. Since each clause is satisfied, Lemma 3 now implies that at least one of the handles of each clause gadget is rounded down, so that we can complete the rounding of the clause gadgets following Lemma 2. For an example, see Fig. 4.

We should now prove that this rounding satisfies the conditions of a strong 1-rounding for the shortest paths between any pair of vertices u and v in G_α. If u and v lie in the same gadget, or if G_α contains a shortcut edge $\{u, v\}$, then the conditions are ensured by Lemmas 1, 2, or 4. Otherwise, one vertex, say u, must lie in a gadget for a variable x_i; the other vertex, v, lies in a gadget for a clause C_j that includes x_i or $\neg x_i$; and these gadgets are connected by exactly one clause-variable edge $\{s, c\}$, where s lies in the variable gadget and c lies in the clause gadget. One can now argue that the unique shortest path from u to v uses that edge $\{s, c\}$ and has length less than $2D$ (any path via other gadgets, using shortcut edges or other clause-variable edges, would have length at least $3D$). Lemmas 1 and 3 imply that the rounding error on the path from u to either the first or the second vertex of the clause handle is zero; the rounding error on the rest of the path to v is bounded by Lemma 2. Thus, if α is satisfiable, G_α admits a strong 1-rounding.

Next we show that if G_α admits a path-oblivious 1-rounding, then α is satisfiable. This we do by constructing, from a given path-oblivious 1-rounding $\tilde{\omega}$, a choice ψ of the variables that satisfies α. If in $\tilde{\omega}$, the weight of $e(v_{i,0})$ is rounded down, we set $\psi(x_i) = 1$, otherwise we set $\psi(x_i) = 0$. Now consider the clause gadget for any clause C_j. Following Lemma 2, we know that there is at least one $t \in \{1, 2, 3\}$ such that $e(c_{j,t})$ is rounded down. Lemma 3 implies that $e(c_{j,t})$ models a literal l such that $\psi(l) = 1$, and this literal satisfies C_j. The same argument applies to each clause C_j, and thus, ψ satisfies α.

Thus, our construction correctly reduces 3-SAT to the problem of deciding whether a graph admits a path-oblivious, weak, or strong 1-rounding. Since 3-SAT is NP-hard, our rounding problems must be NP-hard.   □

The construction as described works for $\varepsilon = 1$ and some smaller values. The weight w of a nonagon edge in the clause gadget is deciding. If the 3-CNF formula is satisfiable, one can obtain a maximum absolute rounding error of $\max(|10 - 3w|, |4\frac{1}{2} - w|)$. With $w = 3.6$, this is 0.9; the error is minimized to $7/8$ when we choose $w = 3\frac{5}{8}$. If we choose $\varepsilon > 1$, Lemma 1 will not hold. Thus, the construction works as long as $7/8 < \varepsilon \leq 1$.

3   A Quadratic-Time Algorithm for Trees

In this section, we will present algorithms for the case in which G is a tree. Note that in this case, there is only one simple path between any pair of vertices, so there is no difference between path-oblivious, weak, and strong ε-roundings.

Clearly, if the whole graph is a simple path with edges $e_1, \ldots, e_n$, a 1-rounding always exists, and can be computed in linear time (assuming the floor function can be computed in constant time). For example [7,8], let d_i be $\frac{1}{2} + \sum_{j=1}^{i} \omega(e_i)$; then we set $\tilde{\omega}(e_i) = \lfloor d_i \rfloor - \lfloor d_{i-1} \rfloor$. Now, for any subpath $e_a, \ldots, e_z$, we have $\sum_{i=a}^{z} \tilde{\omega}(e_i) = \lfloor d_z \rfloor - \lfloor d_{a-1} \rfloor < d_z - (d_{a-1} - 1) = 1 + \sum_{i=a}^{z} \omega(e_i)$, and $\sum_{i=a}^{z} \tilde{\omega}(e_i) = \lfloor d_z \rfloor - \lfloor d_{a-1} \rfloor > (d_z - 1) - d_{a-1} = -1 + \sum_{i=a}^{z} \omega(e_i)$; thus $\tilde{\omega}$ satisfies Condition 1 for $\varepsilon = 1$, and $\tilde{\omega}$ is a 1-rounding. Sadakane et al. [7] prove that a path of n vertices admits at most n different 1-roundings, and shows how to compute all 1-roundings in $O(n^2)$ time, and how to determine the 1-rounding with the smallest maximum absolute rounding error in the same time.

If the graph is a tree, observe that we can obtain a 2-rounding in linear time as follows. Choose any vertex of the tree as the root r. For any other vertex u, let $p(u)$ be the parent of u, and let d_u be the (unrounded) weight of the path from r to u. Now we set $\tilde{\omega}(\{p(u), u\}) = \lfloor d_u \rfloor - \lfloor d_{p(u)} \rfloor$. By the same calculation as above, for any vertex u, the absolute rounding error $|e(u, v)|$ on any path from u to an ancestor v of u is now less than one. Now, given two arbitrary vertices u and w, let v be their lowest common ancestor. The absolute rounding error on the path from u to w is at most $|e(u, v)| + |e(v, w)| < 2$.

We will now present an algorithm that decides, given a tree T and an error threshold $\varepsilon < 2$, in quadratic time, whether T admits an ε-rounding. We choose an arbitrary vertex of T as the root r. We say v is a descendant of u if u lies on the path from r to v. For any vertex u, the subtree T_u of T is the subgraph of T that is induced by all descendants of u; this vertex u is called the root of T_u; see Fig. 5. By $|T|$ we denote the number of vertices of T. By $\pi(u, v)$ we denote the path in T from u to v.

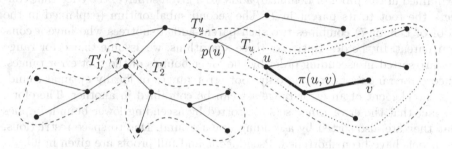

Fig. 5. A rooted tree T with root vertex r. The subtree T_u with root u can be extended to a subgraph T'_u with root $p(u)$ by adding the edge $\{u, p(u)\}$.

Definition 2 (root error range). *Let $\tilde{\omega}$ be an ε-rounding on a tree T with root r. For any $v \in T$, let $e(r, v)$ be the rounding error on $\pi(r, v)$, that is, $e(r, v) := \tilde{\omega}(\pi(r, v)) - \omega(\pi(r, v))$.*

We call the smallest interval that contains the signed rounding errors of the paths from r to all vertices of T the root error range $E(T, \tilde{\omega})$, *so*

$$E(T, \tilde{\omega}) := \left[\min_{v \in T} e(r, v), \max_{v \in T} e(r, v) \right].$$

Note that $e(r, r) = 0$, so if T is a leaf, then $E(T, \tilde{\omega}) = [0, 0]$.

We call a rounding $\tilde{\omega}$ of T locally optimal if there is no other rounding $\tilde{\omega}'$ of T such that $E(T, \tilde{\omega}')$ is smaller than and contained in $E(T, \tilde{\omega})$.

Let the error range set $\mathcal{E}(T)$ *be the set of root error ranges that can be realized by locally optimal ε-roundings of T, that is, the set $\mathcal{E}(T) := \{E(T, \tilde{\omega}) \mid \tilde{\omega}$ is a locally optimal rounding$\}$.*

Lemma 5 (error range set size). *$\mathcal{E}(T)$ has at most $2|T|$ elements.*

Proof. Observe that in any ε-rounding with $\varepsilon < 2$, the weight of any path $\pi(r, v)$ is rounded to $\lfloor \omega(\pi(r, v)) \rfloor - 1$, $\lfloor \omega(\pi(r, v)) \rfloor$, $\lceil \omega(\pi(r, v)) \rceil$, or $\lceil \omega(\pi(r, v)) \rceil + 1$.

Any root error range $E(T, \tilde{\omega})$ includes 0, since $e(r, r) = 0$. Therefore, the lower bound of any root error range $E(T, \tilde{\omega})$ is zero or negative, and it must be the rounding error on some path $\pi(r, v)$ whose rounded weight is $\lfloor \omega(\pi(r, v)) \rfloor$ or $\lfloor \omega(\pi(r, v)) \rfloor - 1$. Since v must be one of the n vertices of T, this implies that there are at most $2n$ possible values for the lower bound of any root error range.

Because $\mathcal{E}(T)$ contains only root error ranges of locally optimal ε-roundings, no two elements of $\mathcal{E}(T)$ can have the same lower bound, so the total number of elements of $\mathcal{E}(T)$ is also bounded by $2n$.   □

Our algorithm will compute the error range set for every subtree of T bottom-up. For this purpose, we need two subalgorithms. The first subalgorithm (explained in the proof of Lemma 6) adds, to a given subtree, the edge that connects the root to its parent in T. The second subalgorithm (explained in the proof of Lemma 7) combines two such parent-added subtrees who have a common parent. In the description of these algorithms, we assume that error range sets are sorted in ascending order by the lower bounds of the root error ranges. Since error range sets contain only root error ranges of locally optimal roundings, no element of an error range set can be contained in another. Therefore, the fact that the error range sets are sorted by ascending lower bound, implies that they are also sorted by ascending upper bound. Due to space restrictions, the proofs have been shortened. Pseudocode and full proofs are given in [6].

Let T'_u be the subgraph of T that consists of T_u and the edge between u and its parent $p(u)$ in T; we choose $p(u)$ as the root of T'_u, see Fig. 5.

Lemma 6 (moving up). *Given $\mathcal{E}(T_u)$, we can compute $\mathcal{E}(T'_u)$ in $O(|T_u|)$ time.*

Proof. Let f be the fractional part of the weight of $\{p(u), u\}$, that is, $f := \omega(\{p(u), u\}) - \lfloor \omega(\{p(u), u\}) \rfloor$. Any ε-rounding for T'_u must consist of an ε-rounding for T_u combined with setting $\tilde{\omega}(\{p(u), u\})$ to $\lfloor \omega(\{p(u), u\}) \rfloor + k$ for some $k \in \{-1, 0, 1, 2\}$ (because $\varepsilon < 2$, no other values for $\tilde{\omega}(\{p(u), u\})$ are allowed). For any vertex $v \in T'_u$, other than the root $p(u)$, we have $e(p(u), v) = e(p(u), u) + e(u, v) = k - f + e(u, v)$; for the root $p(u)$ we have $e(p(u), p(u)) = 0$. Thus, a choice of an ε-rounding for T_u with root error range $[a, b] \in \mathcal{E}(T_u)$, together with a choice of $k \in \{-1, 0, 1, 2\}$, results in a rounding for T'_u whose root error range is the smallest interval that includes $[a + k - f, b + k - f]$ and 0, that is, the root error range for T'_u is $[\min(a + k - f, 0), \max(0, b + k - f)]$. This rounding is an ε-rounding if and only if $-\varepsilon < a + k - f$ and $b + k - f < \varepsilon$.

Thus, the elements of $\mathcal{E}(T'_u)$ are all from the set:

$$S = \left\{ [\min(a + k - f, 0), \max(0, b + k - f)] \;\middle|\; \begin{matrix} [a, b] \in \mathcal{E}(T_u), \\ k \in \{-1, 0, 1, 2\}, \\ -\varepsilon < a + k - f, \\ b + k - f < \varepsilon \end{matrix} \right\}.$$

We can compute S in lexicographical order by first computing, for each $k \in \{-1, 0, 1, 2\}$, the set $S_k := \{[\min(a + k - f, 0), \max(0, b + k - f)] \mid [a, b] \in$

$\mathcal{E}(T_u), -\varepsilon < a + k - f, b + k - f < \varepsilon\}$ in lexicographical order from $\mathcal{E}(T_u)$, and then merging the sets S_{-1}, S_0, S_1 and S_2 into one lexicographically ordered set S.

To obtain $\mathcal{E}(T_u')$ from S, all that remains to do is to filter out the root error ranges that are not locally optimal. Given that S is ordered lexicographically, this can be done in linear time; see [6] for details. $\qquad\square$

Given two trees T_1' and T_2' that have the same root vertex r, but are otherwise disjoint as in Fig. 5, we denote by $T_1' \cup T_2'$ the union of the two trees; $T_1' \cup T_2'$ also has root r.

Lemma 7 (merging error range sets). *Given $\mathcal{E}(T_1')$ and $\mathcal{E}(T_2')$ for two trees T_1' and T_2', whose root r is the only vertex that they have in common, we can compute $\mathcal{E}(T_1' \cup T_2')$ in $O(|T_1'| + |T_2'|)$ time.*

Proof. Consider a rounding $\tilde{\omega}$ of $T_1' \cup T_2'$ that consists of an ε-rounding of T_1' with root error range $[a_1, b_1]$ and an ε-rounding of T_2' with root error range $[a_2, b_2]$. For $i \in \{1, 2\}$, let u_i and v_i be vertices in T_i' that determine the lower and upper bounds of the root error range $[a_i, b_i]$, that is: $a_i = e(r, u_i)$ and $b_i = e(r, v_i)$. The path composed of $\pi(u_1, r)$ and $\pi(r, u_2)$ is a path in $T_1' \cup T_2'$ with $e(u_1, u_2) = a_1 + a_2$; similarly, we have $e(v_1, v_2) = b_1 + b_2$. It follows that $\tilde{\omega}$ can be an ε-rounding only if $a_1 + a_2 > -\varepsilon$ and $b_1 + b_2 < \varepsilon$. These conditions are also sufficient, since any other path from a vertex $w_1 \in T_1'$ to a vertex $w_2 \in T_2'$ consists of a path with error $e(r, w_1) \in [a_1, b_1]$ and a path with error $e(r, w_2) \in [a_2, b_2]$, so the total error is within $[a_1 + a_2, b_1 + b_2]$.

Since T_1', T_2' and $T_1' \cup T_2'$ have the same root, the root error range $E(T_1' \cup T_2', \tilde{\omega})$ is the union of the root error ranges of T_1' and T_2', that is, $E(T_1' \cup T_2', \tilde{\omega}) = [\min(a_1, a_2), \max(b_1, b_2)]$. We say $\tilde{\omega}$ is of type 1 if $a_1 < a_2$, and of type 2 if $a_2 \le a_1$. To find, in linear time, a set S_1 of root error ranges for $T_1' \cup T_2'$ that includes those of all locally optimal roundings of type 1, we proceed as follows. We scan all ranges $[a_1, b_1] \in \mathcal{E}(T_1')$, while maintaining pointers to: (1) the first range $[a_2', b_2'] \in \mathcal{E}(T_2')$ that satisfies $a_2' > a_1$; (2) the first range $[a_2'', b_2''] \in \mathcal{E}(T_2')$ that satisfies $a_2'' > -\varepsilon - a_1$; and (3) the last range $[a_2''', b_2'''] \in \mathcal{E}(T_2')$ that satisfies $b_2''' < \varepsilon - b_1$. Whenever $\max(b_2', b_2'') \le b_2'''$, we include $[a_1, \max(b_1, b_2', b_2'')]$ in S_1. In a similar fashion, we find a set S_2 that includes the root error ranges of all locally optimal roundings of type 2. Finally we merge S_1 and S_2 and filter out suboptimal error ranges in linear time. For details, see [6]. $\qquad\square$

To decide whether a tree T admits an ε-rounding, we compute $\mathcal{E}(T_u)$ for all subtrees T_u of T bottom-up. Specifically, if u is a leaf, $\mathcal{E}(T_u) = [0, 0]$. If u is an internal vertex with a single child v, then $T_u = T_v'$ and we compute $\mathcal{E}(T_u) = \mathcal{E}(T_v')$ from $\mathcal{E}(T_v)$ with the algorithm of Lemma 6. If u is an internal vertex with two children v and w, we first compute $\mathcal{E}(T_v')$ and $\mathcal{E}(T_w')$ from $\mathcal{E}(T_v)$ and $\mathcal{E}(T_w)$, respectively, with the algorithm of Lemma 6, and then we compute $\mathcal{E}(T_u) = \mathcal{E}(T_v' \cup T_w')$ with the algorithm of Lemma 7. Finally, if u is an internal vertex with more than two children, we first compute $\mathcal{E}(T_v')$ from $\mathcal{E}(T_v)$ for each child v. Then we organize all children in a balanced binary merge tree M with the children of u at the leaves; for a vertex x in M, let $C(x)$ be the children of u

in the subtree of M rooted at x. With vertex x we associate the error range set $\mathcal{E}(\bigcup_{v \in C(x)} T'_v)$. We process the merge tree M bottom-up, using the algorithm of Lemma 7 for each internal vertex x of M to compute $\mathcal{E}(\bigcup_{v \in C(x)} T'_v)$ from the error range sets associated with the children of x. The error range set computed for the root of M constitutes $\mathcal{E}(T_u)$. Ultimately, we compute $\mathcal{E}(T_r)$. If and only if this error range set is non-empty, T admits an ε-rounding.

We say the *effective* height of the tree T is the height it would have when all internal vertices with more than two children were replaced by their binary merge trees. The algorithms of Lemmas 6 and 7 take time linear in the size of the subtrees that are being processed. Thus, if T has n vertices and effective height h, the above algorithm to compute $\mathcal{E}(T_r)$ runs in $O(nh)$ time. This proves:

Theorem 2. *Given an edge-weighted tree T of n vertices and an error tolerance ε, one can decide in $O(n^2)$ time whether T admits an ε-rounding.*

To find the minimal maximum rounding error, we first compute the lengths of all $O(n^2)$ simple paths in the tree in $O(n^2)$ time. We do so with a bottom-up algorithm that computes for each vertex u the lengths of all paths in T_u, and passes on the lengths of all paths in T_u that end in u to the parent of u. Each path produces up to four candidate values for the maximum rounding error, namely, for $k \in \{-1, 0, 1, 2\}$, the absolute value of (k minus the fractional part of the path length). We sort all candidate values in $O(n^2 \log n)$ time. Finally we find the smallest ε for which the decision algorithm says yes by binary search, using $O(\log n)$ calls to the decision algorithm, which take $O(n^2)$ time each.

Corollary 1. *Given an edge-weighted tree T of n vertices, we can compute a rounding of T that minimizes the maximum absolute rounding error on any simple path in the tree in $O(n^2 \log n)$ time.*

4   Conclusions and Directions for Further Work

We have shown that it is, in general, NP-hard to decide whether a path-oblivious, weak, or strong ε-rounding exists for a given graph, but the problem can be solved in polynomial time if the graph is a tree. Does this mean there is no hope of finding efficient algorithms to round weights in practical graphs other than trees? The conditions of our NP-hardness construction raise several questions:

- What is the complexity of the problem for other types of graphs, such as planar graphs or cycles? (Our hardness proof uses graphs that are highly non-planar due to the shortcut edges, even if we reduce from planar 3-SAT.)
- Is it possible to prove NP-hardness for $\varepsilon > 1$ or $\varepsilon < \frac{7}{8}$?
- Can we prove similar results for other rounding models, for example based on relative rounding errors (Funke and Storandt [4])?
- Can we establish relations to results on global roundings of hypergraphs, where the weights are not on the edges but on the vertices [1–3,9]?

References

1. Asano, T., Matsui, T., Tokuyama, T.: Optimal roundings of sequences and matrices. Nord. J. Comput. **7**(3), 241–256 (2000)
2. Asano, T., Matsui, T., Tokuyama, T.: On the complexities of the optimal rounding problems of sequences and matrices. In: Halldórsson, M.M. (ed.) SWAT 2000. LNCS, vol. 1851, pp. 476–489. Springer, Heidelberg (2000). https://doi.org/10.1007/3-540-44985-X_40
3. Asano, T., Katoh, N., Tamaki, H., Tokuyama, T.: The structure and number of global roundings of a graph. Theor. Comput. Sci. **325**(3), 425–437 (2004)
4. Funke, S., Storandt, S.: Consistent rounding of edge weights in graphs. In: Proceedings of the 9th Annual Symposium on Combinatorial Search (SOCS 2016), pp. 28–35 (2016)
5. Haverkort, H.: Embedding cues about travel time in schematic maps. In: Schematic Mapping Workshop 2014, University of Essex (2014). https://sites.google.com/site/schematicmapping/
6. Haverkort, H., Kübel, D., Langetepe, E.: Shortest-path-preserving rounding. CoRR abs/1905.08621 (2019). https://arxiv.org/abs/1905.08621
7. Sadakane, K., Takki-Chebihi, N., Tokuyama, T.: Combinatorics and algorithms for low-discrepancy roundings of a real sequence. Theor. Comput. Sci. **331**(1), 23–36 (2005)
8. Storandt, S.: Sensible edge weight rounding for realistic path planning. In: Proceedings of the 26th ACM SIGSPATIAL International Conference on Advances in Geographic Information Systems (SIGSPATIAL/GIS 2018), pp. 89–98 (2018)
9. Takki-Chebihi, N., Tokuyama, T.: Enumerating global roundings of an outerplanar graph. In: Ibaraki, T., Katoh, N., Ono, H. (eds.) ISAAC 2003. LNCS, vol. 2906, pp. 425–433. Springer, Heidelberg (2003). https://doi.org/10.1007/978-3-540-24587-2_44
10. Thorup, M.: Integer priority queues with decrease key in constant time and the single source shortest paths problem. J. Comput. Syst. Sci. **69**(3), 330–353 (2004)

Complexity and Algorithms
for Semipaired Domination in Graphs

Michael A. Henning[1], Arti Pandey[2(✉)], and Vikash Tripathi[2]

[1] Department of Mathematics and Applied Mathematics,
University of Johannesburg, Auckland Park 2006, South Africa
`mahenning@uj.ac.za`
[2] Department of Mathematics, Indian Institute of Technology Ropar,
Nangal Road, Rupnagar 140001, Punjab, India
`{arti,2017maz0005}@iitrpr.ac.in`

Abstract. For a graph $G = (V, E)$ with no isolated vertices, a set $D \subseteq V$ is called a semipaired dominating set of G if (i) D is a dominating set of G, and (ii) D can be partitioned into two element subsets such that the vertices in each two element set are at distance at most two. The minimum cardinality of a semipaired dominating set of G is called the semipaired domination number of G, and is denoted by $\gamma_{pr2}(G)$. The MINIMUM SEMIPAIRED DOMINATION problem is to find a semipaired dominating set of G of cardinality $\gamma_{pr2}(G)$. In this paper, we initiate the algorithmic study of the MINIMUM SEMIPAIRED DOMINATION problem. We show that the decision version of the MINIMUM SEMIPAIRED DOMINATION problem is NP-complete for bipartite graphs and chordal graphs. On the positive side, we present a linear-time algorithm to compute a minimum cardinality semipaired dominating set of interval graphs. We also propose a $1 + \ln(2\Delta + 2)$-approximation algorithm for the MINIMUM SEMIPAIRED DOMINATION problem, where Δ denotes the maximum degree of the graph and show that the MINIMUM SEMIPAIRED DOMINATION problem cannot be approximated within $(1 - \epsilon) \ln |V|$ for any $\epsilon > 0$ unless P = NP.

Keywords: Domination · Semipaired domination · Bipartite graphs · Chordal graphs · Interval graphs · Graph algorithm · NP-complete · Approximation algorithm

1 Introduction

A *dominating set* in a graph G is a set D of vertices of G such that every vertex in $V(G) \backslash D$ is adjacent to at least one vertex in D. The *domination number* of G, denoted by $\gamma(G)$, is the minimum cardinality of a dominating set of G. The MINIMUM DOMINATION problem is to find a dominating set of cardinality $\gamma(G)$.

Research of Michael A. Henning was supported in part by the University of Johannesburg.

C. J. Colbourn et al. (Eds.): IWOCA 2019, LNCS 11638, pp. 278–289, 2019.
https://doi.org/10.1007/978-3-030-25005-8_23

More thorough treatment of domination, can be found in the books [8,9]. A dominating set D is called a *paired dominating set* if the subgraph of G induced by D contains a perfect matching. The *paired domination number* of G, denoted by $\gamma_{pr}(G)$ is the minimum cardinality of a paired dominating set of G. The concept of paired domination was introduced by Haynes and Slater in [13].

A relaxed form of paired domination called semipaired domination was introduced by Haynes and Henning [10] and studied further in [11,12,14]. A set S of vertices in a graph G with no isolated vertices is a *semipaired dominating set*, abbreviated a semi-PD-set, of G if S is a dominating set of G and S can be partitioned into 2-element subsets such that the vertices in each 2-element set are at distance at most 2. In other words, the vertices in the dominating set S can be partitioned into 2-element subsets such that if $\{u, v\}$ is a 2-set, then the distance between u and v is either 1 or 2. We say that u and v are *semipaired*. The *semipaired domination number* of G, denoted by $\gamma_{pr2}(G)$, is the minimum cardinality of a semi-PD-set of G. Since every paired dominating set is a semi-PD-set, and every semi-PD-set is a dominating set, we have the following observation.

Observation 1 [10]. *For every isolate-free graph G, $\gamma(G) \leq \gamma_{pr2}(G) \leq \gamma_{pr}(G)$.*

By Observation 1, the semipaired domination number is squeezed between two fundamental domination parameters, namely the domination number and the paired domination number.

More formally, the minimum semipaired domination problem and its decision version are defined as follows:

MINIMUM SEMIPAIRED DOMINATION problem (MSPDP)

Instance: A graph $G = (V, E)$.
Solution: A semi-PD-set D of G.
Measure: Cardinality of the set D.

SEMIPAIRED DOMINATION DECISION problem (SPDDP)

Instance: A graph $G = (V, E)$ and a positive integer $k \leq |V|$.
Question: Does there exist a semi-PD-set D in G such that $|D| \leq k$?

In this paper, we initiate the algorithmic study of the semipaired domination problem. The main contributions of the paper are summarized below. In Sect. 2, we discuss some definitions and notations. In Sect. 3, we show that the SEMIPAIRED DOMINATION DECISION problem is NP-complete for bipartite graphs. In Sect. 4, we propose a linear-time algorithm to solve the MINIMUM SEMIPAIRED DOMINATION problem in interval graphs. In Sect. 5, we propose an approximation algorithm for the MINIMUM SEMIPAIRED DOMINATION problem in general graphs. In Sect. 6, we discuss an approximation hardness result. Finally, Sect. 7, concludes the paper.

2   Terminology and Notation

For notation and graph theory terminology, we in general follow [15]. Specifically, let $G = (V, E)$ be a graph with vertex set $V = V(G)$ and edge set $E = E(G)$, and let v be a vertex in V. The *open neighborhood* of v is the set $N_G(v) = \{u \in V \mid uv \in E\}$ and the *closed neighborhood of v* is $N_G[v] = \{v\} \cup N_G(v)$. Thus, a set D of vertices in G is a dominating set of G if $N_G(v) \cap D \neq \emptyset$ for every vertex $v \in V \backslash D$, while D is a total dominating set of G if $N_G(v) \cap D \neq \emptyset$ for every vertex $v \in V$. The *distance* between two vertices u and v in a connected graph G, denoted by $d_G(u, v)$, is the length of a shortest (u, v)-path in G. If the graph G is clear from the context, we omit it in the above expressions. We write $N(v)$, $N[v]$ and $d(u, v)$ rather than $N_G(v)$, $N_G[v]$ and $d_G(u, v)$, respectively.

For a set $S \subseteq V(G)$, the subgraph induced by S is denoted by $G[S]$. If $G[C]$, where $C \subseteq V$, is a complete subgraph of G, then C is a *clique* of G. A set $S \subseteq V$ is an *independent set* if $G[S]$ has no edge. A graph G is *chordal* if every cycle in G of length at least four has a *chord*, that is, an edge joining two non-consecutive vertices of the cycle. A vertex $v \in V(G)$ is a *simplicial* vertex of G if $N_G[v]$ is a clique of G. An ordering $\alpha = (v_1, v_2, \ldots, v_n)$ is a *perfect elimination ordering* (PEO) of vertices of G if v_i is a simplicial vertex of $G_i = G[\{v_i, v_{i+1}, \ldots, v_n\}]$ for all i, $1 \leq i \leq n$. Fulkerson and Gross [5] characterized chordal graphs, and showed that a graph G is chordal if and only if it has a PEO. A graph $G = (V, E)$ is *bipartite* if V can be partitioned into two disjoint sets X and Y such that every edge of G joins a vertex in X to a vertex in Y, and such a partition (X, Y) of $V(G)$ is called a *bipartition* of G. Further, we denote such a bipartite graph G by $G = (X, Y, E)$. A graph G is an *interval graph* if there exists a one-to-one correspondence between its vertex set and a family of closed intervals in the real line, such that two vertices are adjacent if and only if their corresponding intervals intersect. Such a family of intervals is called an *interval model* of a graph.

In the rest of the paper, all graphs considered are simple connected graphs with at least two vertices, unless otherwise mentioned specifically. We use the standard notation $[k] = \{1, \ldots, k\}$. For most of the approximation related terminologies, we refer to [1, 16].

3   NP-Completeness Results

In this section, we study the NP-completeness of the SEMIPAIRED DOMINATION DECISION problem. We show that the SEMIPAIRED DOMINATION DECISION problem is NP-complete for bipartite graphs by making polynomial reduction from the MINIMUM VERTEX COVER problem, which is already known to be NP-complete [6].

Theorem 2. *The* SEMIPAIRED DOMINATION DECISION *problem is NP-complete for bipartite graphs.*

Proof. Clearly, the SEMIPAIRED DOMINATION DECISION problem is in NP for bipartite graphs. To show the hardness, we give a polynomial reduction from the MINIMUM VERTEX COVER problem. Given a non-trivial graph $G = (V, E)$, where $V = \{v_1, v_2, \ldots, v_n\}$ and $E = \{e_1, e_2, \ldots, e_m\}$, we construct a graph $H = (V_H, E_H)$ in the following way:

Let $V_k = \{v_i^k \mid i \in [n]\}$ and $E_k = \{e_j^k \mid j \in [m]\}$ for $k \in [2]$. Also assume that $A = \{a_i \mid i \in [n]\}$, $B = \{b_i \mid i \in [n]\}$, $C = \{c_i \mid i \in [n]\}$, and $F = \{f_i \mid i \in [n]\}$.

Now define $V_H = V_1 \cup V_2 \cup E_1 \cup E_2 \cup A \cup B \cup C \cup F$, and $E_H = \{v_i^1 f_i, v_i^2 f_i, a_i b_i, b_i c_i, a_i f_i \mid i \in [n]\} \cup \{v_p^k e_i^k, v_q^k e_i^k \mid k \in [2], i \in [m], v_p, v_q \text{ are}$ endpoints of edge e_i in $G\}$. Figure 1 illustrates the construction of H from G.

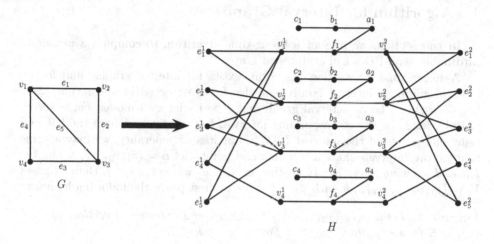

Fig. 1. An illustration of the construction of H from G in the proof of Theorem 2.

Note that the set $I_1 = V_1 \cup V_2 \cup A \cup C$ is an independent set in H. Also, the set $I_2 = E_1 \cup E_2 \cup F \cup B$ is an independent set in H. Since $V_H = I_1 \cup I_2$, the graph H is a bipartite graph. Now to complete the proof, it suffices for us to prove the following claim:

Claim. The graph G has a vertex cover of cardinality at most k if and only if the graph H has a semi-PD-set of cardinality at most $2n + 2k$.

Proof. Let $V_c = \{v_{i_1}, v_{i_2}, \ldots, v_{i_k}\}$ be a vertex cover of G of cardinality k. Then $D_p = \{v_{i_1}^1, v_{i_2}^1, \ldots, v_{i_k}^1\} \cup \{v_{i_1}^2, v_{i_2}^2, \ldots, v_{i_k}^2\} \cup B \cup F$ is a semi-PD-set of H of cardinality $2n + 2k$.

Conversely, suppose that H has a semi-PD-set D of cardinality at most $2n + 2k$. Note that $|D \cap \{a_i, b_i, c_i, f_i\}| \geq 2$ for each $i \in [n]$. Hence, without loss of generality, we may assume that $\{b_i, f_i \mid i \in [n]\} \subseteq D$, where b_i and f_i are semipaired. Hence $|D \cap (E_1 \cup E_2 \cup V_1 \cup V_2)| \leq 2k$. Let $S = (V_1 \cup E_1) \cap D$. Without loss of generality, we may also assume that $|S| \leq k$. Now, if $e_i^1 \in S$ for some

$i \in [m]$, and none of its neighbors belongs to D, then e_i^1 must be semipaired with some vertex e_j^1 where $j \in [m]\backslash\{i\}$, and also there must exist a vertex v_k^1 which is a common neighbor of e_i^1 and e_j^1. In this case, we replace the vertex e_i^1 in the set S with the vertex v_k^1 and so $S \leftarrow (S\backslash\{e_i^1\}) \cup \{v_k^1\}$ where v_k^1 and e_j^1 are semipaired. We do this for each vertex $e_i^1 \in S$ where $i \in [m]$ with none of its neighbors in the set D. For the resulting set S, $|S \cap V_1| \le k$ and every vertex e_i^1 has a neighbor in $V_1 \cap S$. The set $V_c = \{v_i \mid v_i^1 \in S\}$ is a vertex cover of G of cardinality at most k. This completes the proof of the claim.   □

Hence, the theorem is proved.   □

4   Algorithm for Interval Graphs

In this section, we present a linear-time algorithm to compute a minimum cardinality semi-PD-set of an interval graph.

A linear time recognition algorithm exists for interval graphs, and for an interval graph an interval family can also be constructed in linear time [2,7]. Let $G = (V, E)$ be an interval graph and I be its interval model. For a vertex $v_i \in V$, let I_i be the corresponding interval. Let a_i and b_i denote the left and right end points of the interval I_i. Without loss of generality, we may assume that no two intervals share a common end point. Let $\alpha = (v_1, v_2, \ldots, v_n)$ be the *left end ordering* of vertices of G, that is, $a_i < a_j$ whenever $i < j$. Define the set $V_i = \{v_1, v_2, \ldots, v_i\}$, for each $i \in [n]$. Now we first prove the following lemmas.

Lemma 1. *Let $\alpha = (v_1, v_2, \ldots, v_n)$ be the left end ordering of vertices of G. If $v_i v_j \in E$ for $i < j$, then $v_i v_k \in E$ for every $i < k < j$.*

Proof. The proof directly follows from the left end ordering of vertices of G.   □

Lemma 2. *If G is a connected interval graph, then $G[V_i]$ is also connected.*

Proof. The proof can easily be done using induction on i.   □

Let $F(v_i)$ be the least index vertex adjacent to v_i, that is, if $F(v_i) = v_p$, then $p = \min\{k \mid v_k v_i \in E\}$. In particular, we define $F(v_1) = v_1$. Let $L(v_i) = v_q$, where $q = \max\{k \mid v_k v_i \notin E \text{ and } k < i\}$. In particular, if $L(v_i)$ does not exist, we assume that $L(v_i) = v_0$ ($v_0 \notin V$). Let $G_i = G[V_i]$ and D_i denote a semi-PD-set of G_i of minimum cardinality. Recall that we only consider connected graphs with at least two vertices.

Lemma 3. *For $i \ge 2$, if $F(v_i) = v_1$, then $D_i = \{v_1, v_i\}$.*

Proof. Note that every vertex in G_i is dominated by v_1, and $d_{G_i}(v_1, v_i) = 1$. Hence, $D_i = \{v_1, v_i\}$.   □

Lemma 4. *For $i > 1$, if $F(v_i) = v_j$, $j > 1$ and $F(v_j) = v_1$, then $D_i = \{v_1, v_j\}$.*

Proof. Note that every vertex in G_i is dominated by some vertex in the set $\{v_1, v_j\}$, and $d_{G_i}(v_1, v_j) = 1$. Hence, $D_i = \{v_1, v_j\}$. □

Lemma 5. *For $r < k < j < i$, let $F(v_i) = v_j$, $F(v_j) = v_k$ $F(v_k) = v_r$. If every vertex v_l where $k < l < j$, is adjacent to at least one vertex in the set $\{v_j, v_r\}$, then the following holds:*

(a) $\{v_j, v_r\} \subseteq D_i$.
(b) v_j *is semipaired with* v_r *in* D_i.
(c) $D_i \cap \{v_{s+1}, \ldots, v_r, v_{r+1}, \ldots, v_i\} = \{v_j, v_r\}$.

Proof. The proof is omitted due to space constraints. □

Lemma 6. *For $r < k < j < i$, let $F(v_i) = v_j$, $F(v_j) = v_k$ $F(v_k) = v_r$. If every vertex v_l where $k < l < j$, is adjacent to at least one vertex in the set $\{v_j, v_r\}$, then the following holds.*

(a) $D_i = \{v_j, v_r\}$ *if* $L(v_r) = v_0$.
(b) $D_i = \{v_1, v_2, v_j, v_r\}$ *if* $L(v_r) = v_1$.
(c) $D_i = D_s \cup \{v_j, v_r\}$ *if* $L(v_r) = v_s$ *with* $s \geq 2$.

Proof.(a) Clearly $D_i = \{v_j, v_r\}$.
(b) From Lemma 5, we know that $\{v_j, v_r\} \subseteq D_i$. Also, other than v_1, all vertices are dominated by the set $\{v_j, v_r\}$. Hence, $D_i = \{v_1, v_2, v_j, v_r\}$.
(c) Clearly $D_s \cup \{v_j, v_r\}$ is a semi-PD-set of G_i. Hence $|D_i| \leq |D_s| + 2$. We also know that there exists a semi-PD-set D_i of G_i of minimum cardinality such that $D_i \cap \{v_{s+1}, v_{s+2}, \ldots, v_i\} = \{v_j, v_r\}$ (where v_j and v_r are semi-paired in D_i). Hence $D_i \backslash \{v_j, v_r\} \subseteq V(G_s)$. Also, $\{v_j, v_r\}$ dominates the set $\{v_{s+1}, v_{s+2}, \ldots, v_n\}$, implying that the set $\{v_1, v_2, \ldots, v_s\}$ is dominated by the vertices in $D_i \backslash \{v_j, v_r\}$. Hence, the set $D_i \backslash \{v_j, v_r\}$ is semi-PD-set of G_s. Therefore, $|D_s| \leq |D_i| - 2$. This proves that $|D_i| = |D_s| + 2$. Hence, $D_i = D_s \cup \{v_j, v_r\}$. □

Lemma 7. *For $r < k < j < i$, let $F(v_i) = v_j$, $F(v_j) = v_k$ $F(v_k) = v_r$, and $\{v_l \mid k < l < j\} \not\subseteq N_{G_i}[v_r] \cup N_{G_i}[v_j]$. Let $t = \max\{l \mid k < l < j$ and $v_l v_j \notin E\}$ (assume that such a t exists). Let $F(v_t) = v_b$. Then, the following holds.*

(a) $\{v_j, v_b\} \subseteq D_i$.
(b) v_j *is semipaired with* v_b *in* D_i.
(c) $D_i \cap \{v_{s+1}, \ldots, v_b, v_{b+1}, \ldots, v_i\} = \{v_j, v_b\}$.

Proof. The proof is omitted due to space constraints. □

Lemma 8. *For $r < k < j < i$, let $F(v_i) = v_j$, $F(v_j) = v_k$ $F(v_k) = v_r$, and $\{v_l \mid k < l < j\} \not\subseteq N_{G_i}[v_r] \cup N_{G_i}[v_j]$. Let $t = \max\{l \mid k < l < j$ and $v_l v_j \notin E\}$ (assume that such a t exists). Let $F(v_t) = v_b$. Then, the following holds.*

(a) $D_i = \{v_j, v_b\}$ *if* $L(v_b) = v_0$.
(b) $D_i = \{v_1, v_2, v_j, v_b\}$ *if* $L(v_b) = v_1$.

(c) $D_i = D_s \cup \{v_j, v_b\}$ if $L(v_b) = v_s$ with $s \geq 2$.

Proof. The proof is similar to the proof of Lemma 6, and hence is omitted.   □

Based on above lemmas, we present an algorithm to compute a minimum semi-PD-set of an interval graph.

Algorithm 1. SEMI-PAIRED-DOM-IG(G)

Input: An interval graph $G = (V, E)$ with a left end ordering
$\alpha = (v_1, v_2, \ldots, v_n)$ of vertices of G.
Output: A semi-PD-set D of G of minimum cardinality.
$V' = V$;
while $(V' \neq \phi)$ **do**

    Let $i = \max\{k \mid v_k \in V'\}$. **if** $(F(v_i) = v_1)$ **then**
        $D = D \cup \{v_1, v_i\}$;
        $V' = V' \setminus \{v_1, v_2, \ldots, v_i\}$;
    else if $(F(v_i) = v_j$ and $F(v_j) = v_1$ where $j > 1)$ **then**
        $D = D \cup \{v_1, v_j\}$; $V' = V' \setminus \{v_1, v_2, \ldots, v_i\}$;
    else if $(F(v_i) = v_j$ and $F(v_j) = v_k$ where $k \geq 2)$ **then**
        Let $F(v_k) = v_r$.
        if $\{v_{k+1}, v_{k+2}, \ldots, v_{j-1}\} \subseteq N_G[v_j] \cup N_G[v_r]$ **then**
            if $(L(v_r) = v_0)$ **then**
                $D = D \cup \{v_j, v_r\}$;
                $V' = V' \setminus \{v_1, v_2, \ldots, v_i\}$;
            else if $(L(v_r) = v_1)$ **then**
                $D = D \cup \{v_1, v_2, v_j, v_r\}$;
                $V' = V' \setminus \{v_1, v_2, \ldots, v_i\}$;
            else
                Let $(L(v_r) = v_s)$ where $s \geq 2$.
                $D = D \cup \{v_j, v_r\}$;
                $V' = V' \setminus \{v_{s+1}, v_{s+2}, \ldots, v_i\}$;
        else
            Let $t = \max\{l \mid k < l < j$ and $v_l \notin N_G(v_j)\}$ and $F(v_t) = v_b$.
            if $(L(v_b) = v_0)$ **then**
                $D = D \cup \{v_j, v_b\}$;
                $V' = V' \setminus \{v_1, v_2, \ldots, v_i\}$;
            else if $(L(v_b) = v_1)$ **then**
                $D = D \cup \{v_1, v_2, v_j, v_b\}$;
                $V' = V' \setminus \{v_1, v_2, \ldots, v_i\}$;
            else
                Let $(L(v_b) = v_s)$ where $s \geq 2$.
                $D = D \cup \{v_j, v_b\}$;
                $V' = V' \setminus \{v_{s+1}, v_{s+2}, \ldots, v_i\}$;

Theorem 3. *Given a left end ordering of vertices of G, the algorithm SEMI-PAIRED-DOM-IG computes a semi-PD-set of G of minimum cardinality in linear-time.*

Proof. By Lemmas 3, 4, 6 and 8, we can ensure that the algorithm SEMI-PAIRED-DOM-IG computes a semi-PD-set of G of minimum cardinality. Also, it can be easily seen that the algorithm can be implemented in $O(m + n)$ time, where $n = |V(G)|$ and $m = |E(G)|$. $\qquad\qquad\square$

5 Approximation Algorithm

In this section, we present a greedy approximation algorithm for the MINI-MUM SEMIPAIRED DOMINATION problem in graphs. We also provide an upper bound on the approximation ratio of this algorithm. The greedy algorithm is described as follows.

Algorithm 2. APPROX-SEMI-PAIRED-DOM-SET(G)

Input: A graph $G = (V, E)$ with no isolated vertex.
Output: A semi-PD-set D of G.
begin
$\quad D = \emptyset$;
$\quad i = 0;\ D_0 = \emptyset$;
$\quad$**while** $(V \setminus (D_0 \cup D_1 \cup \ldots \cup D_i) \neq \emptyset)$ **do**
$\quad\quad i = i + 1$;
$\quad\quad$ choose two distinct vertices $u, v \in V$ such that $d_G(u, v) \leq 2$ and
$\quad\quad |(N_G[u] \cup N_G[v]) \setminus (D_0 \cup D_1 \cup \ldots \cup D_{i-1})|$ is maximized;
$\quad\quad D_i = (N_G[u] \cup N_G[v]) \setminus (D_0 \cup D_1 \cup \ldots \cup D_{i-1})$;
$\quad\quad D = D \cup \{u, v\}$;
$\quad$return D;
end

Lemma 9. *The algorithm APPROX-SEMI-PAIRED-DOM-SET produces a semi-PD-set of G in polynomial time.*

Proof. Clearly, the output set D produced by the algorithm APPROX-SEMI-PAIRED-DOM-SET is a semi-PD-set of G. Also, each step of the algorithm can be computed in polynomial time. Hence, the lemma follows. $\qquad\qquad\square$

Lemma 10. *For each vertex $v \in V$, there exists exactly one set D_i which contains v.*

Proof. We note that $V = D_0 \cup D_1 \cup \cdots \cup D_{|D|/2}$. Also, if $v \in D_i$, then $v \notin D_j$ for $i < j$. Hence, the lemma follows. $\qquad\qquad\square$

By Lemma 10, there exists only one index $i \in [|D|/2]$ such that $v \in D_i$ for each $v \in V$. We now define $d_v = \frac{1}{|D_i|}$. Now we are ready to prove the main theorem of this section.

Theorem 4. *The* MINIMUM SEMIPAIRED DOMINATION *problem for a graph* G *with maximum degree* Δ *can be approximated with an approximation ratio of* $1 + \ln(2\Delta + 2)$.

Proof. The proof is omitted due to space constraints. $\square$

6   Lower Bound on Approximation Ratio

To obtain the lower bound on the approximation ratio of the MINIMUM SEMI-PAIRED DOMINATION problem, we give an approximation preserving reduction from the MINIMUM DOMINATION problem. The following approximation hardness result is already known for the MINIMUM DOMINATION problem.

Theorem 5 [3,4]. *For a graph* $G = (V, E)$, *the* MINIMUM DOMINATION *problem cannot be approximated within* $(1 - \epsilon) \ln |V|$ *for any* $\epsilon > 0$ *unless P=NP.*

Now, we are ready to prove the following theorem.

Theorem 6. *For a graph* $G = (V, E)$, *the* MINIMUM SEMIPAIRED DOMINATION *problem cannot be approximated within* $(1-\epsilon) \ln |V|$ *for any* $\epsilon > 0$ *unless P=NP.*

Proof. Let $G = (V, E)$, where $V = \{v_1, v_2, \ldots, v_n\}$ be an arbitrary instance of the MINIMUM DOMINATION problem. Now, we construct a graph $H = (V_H, E_H)$, an instance of the MINIMUM SEMIPAIRED DOMINATION problem in the following way: $V_H = \{v_i^1, v_i^2, w_i^1, w_i^2, z_i \mid i \in [n]\}$ and $E_H = \{w_i^1 v_j^1, w_i^2 v_j^2 \mid v_j \in N_G[v_i]\} \cup \{v_i^1 v_j^1, v_i^2 v_j^2, z_i z_j \mid 1 \leq i < j \leq n\} \cup \{v_i^1 z_j, v_i^2 z_j \mid i \in [n], j \in [n]\}$. Figure 2 illustrates the construction of H from G.

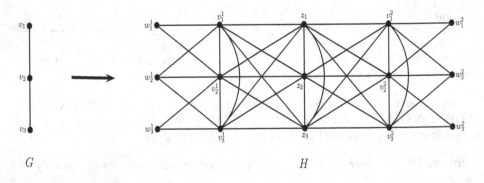

Fig. 2. An illustration of the construction of H from G in the proof of Theorem 6.

Let $V^k = \{v_i^k \mid i \in [n]\}$ and $W^k = \{w_i^k \mid i \in [n]\}$ for $k = 1, 2$. Also, assume that $Z = \{z_i \mid i \in [n]\}$. Note that $V^1 \cup Z$ is a clique in H. Also $V^2 \cup Z$ is a clique in H.

Let D^* denote a minimum dominating set of G. Then the set $D' = \{v_i^1, v_i^2 \mid v_i \in D^*\}$ is a semi-PD-set of H. Hence, if D_{sp}^* denotes a semi-PD-set of H of minimum cardinality, then $|D_{sp}^*| \leq 2|D^*|$.

Suppose that the MINIMUM SEMIPAIRED DOMINATION problem can be approximated within a ratio of α, where $\alpha = (1 - \epsilon) \ln(|V_H|)$ for some fixed $\epsilon > 0$, by some polynomial time approximation algorithm, say **Algorithm A**. Next, we propose an algorithm, which we call **APPROX-DOMINATING-SET**, to compute a dominating set of a given graph G in polynomial time.

Algorithm 3. APPROX-DOMINATING-SET(G)

Input: A graph $G = (V, E)$.
Output: A dominating set D of G.
begin
> Initialize $k = 0$;
> Construct the graph H;
> Compute a semi-PD-set D_{sp} of H using Algorithm A;
> Define $D'_{sp} = D_{sp}$; **if** $(|D'_{sp} \cap (V^1 \cup W^1)| \leq |D_{sp}|/2)$ **then**
>> k=1;
>
> **else**
>> k=2;
>
> **for** $i=1$ to n **do**
>> **if** $(N_H(w_i^k) \cap D'_{sp} == \emptyset)$ **then**
>>> $D'_{sp} = (D'_{sp} \setminus w_i^k) \cup \{v_i^k\}$;
>
> $D = \{v_i \mid v_i^k \in D'_{sp} \cap V^k\}$;
> return D;

end

Next, we show that the set D returned by Algorithm 3 is a dominating set of G. If D_{sp} is any semi-PD-set of H, then clearly either $|D_{sp} \cap (V^1 \cup W^1)| \leq |D_{sp}|/2$ or $|D_{sp} \cap (V^2 \cup W^2)| \leq |D_{sp}|/2$. Assume that $|D_{sp} \cap (V^k \cup W^k)| \leq |D_{sp}|/2$ for some $k \in [2]$. Now, to dominate a vertex $w_i^k \in W^k$, either $w_i^k \in D_{sp}$ or $v_j^k \in D_{sp}$ where $v_j^k \in N_H(w_i)$. If $N_H(w_i^k) \cap D_{sp}$ is an empty set, then we update D_{sp} by removing w_i^k and adding v_j^k for some $v_j^k \in N_H(w_i)$, and call the updated set D'_{sp}. We do this for each i from 1 to n. Note that even for the updated set D'_{sp}, we have $|D'_{sp} \cap (V^k \cup W^k)| \leq |D_{sp}|/2$. Also, in the updated set D'_{sp}, for each w_i^k, $N_H(w_i^k) \cap (D_{sp} \cap V^k)$ is non-empty. Hence $|D'_{sp} \cap V^k| \leq |D_{sp}|/2$ and $D'_{sp} \cap V^k$ dominates W^k. Therefore the set $D = \{v_i \mid v_i^k \in D'_{sp} \cap V^k\}$ is a dominating set of G. Also $|D| \leq |D_{sp}|/2$.

By above arguments, we may conclude that Algorithm 3 produces a dominating set D of the given graph G in polynomial time, and $|D| \leq |D_{sp}|/2$. Hence, $|D| \leq \frac{|D_{sp}|}{2} \leq \alpha \frac{|D_{sp}^*|}{2} \leq \alpha |D^*|$.

Also $\alpha = (1 - \epsilon) \ln(|V_H|) \approx (1 - \epsilon) \ln(|V|)$ where $|V_H| = 5|V|$. Therefore the Algorithm **APPROX-DOMINATING-SET** approximates the minimum dominating set within ratio $(1 - \epsilon) \ln(|V|)$ for some $\epsilon > 0$. By Theorem 5, if the minimum dominating set can be approximated within ratio $(1 - \epsilon) \ln(|V|)$ for some $\epsilon > 0$, then P=NP. Hence, if the MINIMUM SEMIPAIRED DOMINATION problem can be approximated within ratio $(1 - \epsilon) \ln(|V_H|)$ for some $\epsilon > 0$, then P=NP. This proves that the MINIMUM SEMIPAIRED DOMINATION problem cannot be approximated within $(1 - \epsilon) \ln(|V_H|)$ unless P=NP. $\square$

Note that the constructed graph H in Theorem 6 is also a chordal graph, as $\alpha = (w_1^1, w_2^1, \ldots, w_n^1, w_1^2, w_2^2, \ldots, w_n^2, v_1^1, v_2^1, \ldots, v_n^1, v_1^2, v_2^2, \ldots, v_n^2, z_1, z_2, \ldots, z_n)$ is a PEO of vertices of H. Hence, we have the following corollary.

Corollary 1. *For a chordal graph* $G = (V, E)$, *the* MINIMUM SEMIPAIRED DOMINATION *problem cannot be approximated within* $(1 - \epsilon) \ln |V|$ *for any* $\epsilon > 0$ *unless P=NP.*

7   Conclusion

In this paper, we initiate the algorithmic study of the MINIMUM SEMIPAIRED DOMINATION problem. We have resolved the complexity status of the problem for bipartite graphs, chordal graphs and interval graphs. We have proved that the SEMIPAIRED DOMINATION DECISION problem is NP-complete for bipartite graphs and chordal graphs. We also present a linear-time algorithm to compute a semi-PD-set of minimum cardinality for interval graphs. A $1 + \ln(2\Delta + 2)$ approximation algorithm for the MINIMUM SEMIPAIRED DOMINATION problem in general graphs is given, and we prove that it can not be approximated within any sub-logarithmic factor even for chordal graphs. It will be interesting to study better approximation algorithms for this problem for bipartite graphs and other important graph classes.

References

1. Ausiello, G., Crescenzi, P., Gambosi, G., Kann, V., Marchetti-Spaccamela, A., Protasi, M.: Complexity and Approximation. Springer, Heidelberg (1999). https://doi.org/10.1007/978-3-642-58412-1
2. Booth, K.S., Leuker, G.S.: Testing for consecutive ones property, interval graphs, and graph planarity using PQ-tree algorithms. J. Comput. Syst. Sci. **13**, 335–379 (1976)
3. Chlebík, M., Chlebíková, J.: Approximation hardness of dominating set problems in bounded degree graphs. Inf. Comput. **206**, 1264–1275 (2008)
4. Dinur, I., Steurer, D.: Analytical approach to parallel repetition. In: Proceedings of the ACM Symposium on Theory of Computing, STOC 2014, pp. 624–633. ACM, New York (2014)
5. Fulkerson, D.R., Gross, O.A.: Incidence matrices and interval graphs. Pac. J. Math. **15**, 835–855 (1965)
6. Garey, M.R., Johnson, D.S.: Computers and Interactability: A Guide to the Theory of NP-Completeness. W.H. Freeman and Co., San Francisco, New York (1979)
7. Golumbic, M.C.: Algorithmic Graph Theory and Perfect Graphs. Academic Press, New York (1980)

8. Haynes, T.W., Hedetniemi, S.T., Slater, P.J.: Fundamentals of Domination in Graphs, vol. 208. Marcel Dekker Inc., New York (1998)
9. Haynes, T.W., Hedetniemi, S.T., Slater, P.J.: Domination in Graphs: Advanced Topics, vol. 209. Marcel Dekker Inc., New York (1998)
10. Haynes, T.W., Henning, M.A.: Perfect graphs involving semitotal and semipaired domination. J. Comb. Optim. **36**, 416–433 (2018)
11. Haynes, T.W., Henning, M.A.: Semipaired domination in graphs. J. Comb. Math. Comb. Comput. **104**, 93–109 (2018)
12. Haynes, T.W., Henning, M.A.: Graphs with large semipaired domination number. Discuss. Math. Graph Theory **39**(3), 659–671 (2019). https://doi.org/10.7151/dmgt.2143
13. Haynes, T.W., Slater, P.J.: Paired domination in graphs. Networks **32**, 199–206 (1998)
14. Henning, M.A., Kaemawichanurat, P.: Semipaired domination in claw-free cubic graphs. Graphs Comb. **34**, 819–844 (2018)
15. Henning, M.A., Yeo, A.: Total Domination in Graphs. Springer, New York (2013). https://doi.org/10.1007/978-1-4614-6525-6
16. Klasing, R., Laforest, C.: Hardness results and approximation algorithms of k-tuple domination in graphs. Inf. Process. Lett. **89**, 75–83 (2004)

Computing the Rooted Triplet Distance Between Phylogenetic Networks

Jesper Jansson[1], Konstantinos Mampentzidis[2], Ramesh Rajaby[3],
and Wing-Kin Sung[3(✉)]

[1] The Hong Kong Polytechnic University, Hung Hom, Kowloon, Hong Kong
jesper.jansson@polyu.edu.hk
[2] Department of Computer Science, Aarhus University, Aarhus, Denmark
kmampent@cs.au.dk
[3] School of Computing, National University of Singapore, Singapore, Singapore
e0011356@u.nus.edu, ksung@comp.nus.edu.sg

Abstract. The *rooted triplet distance* measures the structural dissimilarity of two phylogenetic trees or networks by counting the number of rooted trees with exactly three leaf labels that occur as embedded subtrees in one, but not both of them. Suppose that $N_1 = (V_1, E_1)$ and $N_2 = (V_2, E_2)$ are rooted phylogenetic networks over a common leaf label set of size λ, that N_i has level k_i and maximum in-degree d_i for $i \in \{1, 2\}$, and that the networks' out-degrees are unbounded. Denote $n = \max(|V_1|, |V_2|)$, $m = \max(|E_1|, |E_2|)$, $k = \max(k_1, k_2)$, and $d = \max(d_1, d_2)$. Previous work has shown how to compute the rooted triplet distance between N_1 and N_2 in $O(\lambda \log \lambda)$ time in the special case $k \leq 1$. For $k > 1$, no efficient algorithms are known; a trivial approach leads to a running time of $\Omega(n^7 \lambda^3)$ and the only existing non-trivial algorithm imposes restrictions on the networks' in- and out-degrees (in particular, it does not work when non-binary nodes are allowed). In this paper, we develop two new algorithms that have no such restrictions. Their running times are $O(n^2 m + \lambda^3)$ and $O(m + k^3 d^3 \lambda + \lambda^3)$, respectively. We also provide implementations of our algorithms and evaluate their performance in practice. This is the first publicly available software for computing the rooted triplet distance between unrestricted networks of arbitrary levels.

1 Introduction

Background. Trees are commonly used in biology to represent evolutionary relationships, with the leaves corresponding to species that exist today and internal nodes to ancestor species that existed in the past. When studying the evolution of a fixed set of species, different available data and tree construction methods [7] can lead to trees that look structurally different. Quantifying this difference is essential to make better evolutionary inferences, which has led to the proposal of several tree distance measures in the literature, e.g., the Robinson-Foulds distance [17], the rooted triplet distance [5] for rooted trees, and the unrooted quartet distance [6] for unrooted trees.

© Springer Nature Switzerland AG 2019
C. J. Colbourn et al. (Eds.): IWOCA 2019, LNCS 11638, pp. 290–303, 2019.
https://doi.org/10.1007/978-3-030-25005-8_24

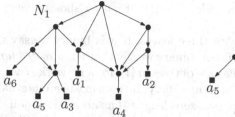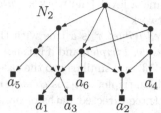

Fig. 1. N_1 is a level-2 network and N_2 is a level-3 network with $\mathcal{L}(N_1) = \mathcal{L}(N_2) = \{a_1, a_2, \ldots, a_6\}$. In this example, $D(N_1, N_2) = 33$. Some shared triplets are: $a_2|a_4|a_6$, $a_2a_4|a_6$, $a_4a_6|a_2$. Some triplets consistent with only one network are: $a_1|a_3|a_6$, $a_2a_6|a_4$.

A rooted phylogenetic network is an extension of a *rooted phylogenetic tree* (i.e., a rooted, unordered, distinctly leaf-labeled tree with no degree-1 nodes) that allows internal nodes to have more than just one parent. Such networks are designed to capture more complex evolutionary relationships when reticulation events such as horizontal gene transfer and hybridization are involved. Similarly to phylogenetic trees, it becomes useful to have distance measures for comparing phylogenetic networks. In this paper we study a natural extension, by Gambette and Huber [10], of the rooted triplet distance from the case of rooted phylogenetic trees to the case of rooted level-k phylogenetic networks.

Problem Definitions. A *rooted phylogenetic network* $N = (V, E)$ is a rooted, directed acyclic graph with one root (a node with in-degree 0), distinctly labeled leaves, and no nodes with both in-degree 1 and out-degree 1. Below, when referring to a "tree" we imply a "rooted phylogenetic tree" and when referring to a "network" we imply a "rooted phylogenetic network". For a node u in N, let $in(u)$ and $out(u)$ be the in-degree and out-degree of u. The network N can have three types of nodes. A node u is an *internal node* if $out(u) \geq 1$, a *leaf node* if $in(u) = 1$ and $out(u) = 0$, and a *reticulation node* if $out(u) \geq 1$ and $in(u) \geq 2$. By definition, N cannot have a node u with $in(u) > 1$ and $out(u) = 0$. Let $r(N)$ be the root of N and $\mathcal{L}(N)$ the set of leaves in N. A directed edge from a node u to a node v in N is denoted by $u \rightarrow v$. A path from u to v in N is denoted by $u \rightsquigarrow v$. Let the height $h(u)$ be the length (number of edges) of the longest path from u to a leaf in N. By definition, if v is a parent of u in N, we have $h(v) > h(u)$.

Let $U(N)$ be the undirected graph created by replacing every directed edge in N with an undirected edge. An undirected graph H is called *biconnected* if it has no node whose removal makes H disconnected. We call H' a *biconnected component of* $U(N)$ if H' is a maximal subgraph of $U(N)$ that is biconnected. The biconnected components of $U(N)$ are edge-disjoint but not necessarily node-disjoint. We say that N is a *level-k network*, equivalently N *has level k*, if every biconnected component of $U(N)$ contains at most k reticulation nodes. The level of a network was introduced by Choy *et al.* [4] as a parameter to measure the treelikeness of a network, with the special case of a level-0 network corresponding to a tree and a level-1 network a *galled tree* [11]. Figure 1 shows a level-2 and a level-3 network.

A *rooted triplet* τ is a tree with three leaves. If it is binary we say that τ is a *rooted resolved triplet*, and if it is non-binary we say that τ is a *rooted fan triplet*. Following [13] and similarly to the case of trees in [1], for a network N we say that the rooted fan triplet $x|y|z$ is *consistent with* N, if there exists an internal node u in N and three directed paths of non-zero length that are node-disjoint, except for u, one going from u to x, one from u to y and one from u to z. Similarly, we say that the rooted resolved triplet $xy|z$ is *consistent with* N, if N contains two internal nodes u and v such that $u \neq v$, and there are four directed paths of non-zero length that are node-disjoint, except for u and v, one going from u to v, one from v to x, one from v to y and one from u to z. See Fig. 1 for an example. From here on, by "disjoint paths" we imply "node-disjoint paths of non-zero length". Moreover, when referring to a "triplet" we imply a "rooted triplet".

Given two networks $N_1 = (V_1, E_1)$ and $N_2 = (V_2, E_2)$ built on the same leaf label set Λ of size λ, the *rooted triplet distance* $D(N_1, N_2)$, or *triplet distance* for short, is the number of triplets over Λ that are consistent with exactly one of the two input networks [10] (see also [12, Sect. 3.2] for a discussion). Let $S(N_1, N_2)$ be the total number of triplets that are consistent with both N_1 and N_2, commonly referred to as *shared triplets*. We then have:

$$D(N_1, N_2) = S(N_1, N_1) + S(N_2, N_2) - 2S(N_1, N_2) \tag{1}$$

Note that a shared triplet contributes a $+1$ to $S(N_1, N_1)$, $S(N_2, N_2)$, and $S(N_1, N_2)$, e.g., the triplet $a_2|a_4|a_6$ in Fig. 1. On the other hand, a triplet from either network that is not shared contributes a $+1$ to either $S(N_1, N_1)$ or $S(N_2, N_2)$, and a 0 to $S(N_1, N_2)$, e.g., $a_1|a_3|a_6$ from Fig. 1 contributes a $+1$ to $S(N_1, N_1)$ and a 0 to $S(N_2, N_2)$ and $S(N_1, N_2)$. Let $S_r(N_1, N_2)$ and $S_f(N_1, N_2)$ be the total number of resolved and fan triplets respectively that are consistent with both N_1 and N_2. We then have that $S(N_1, N_2) = S_r(N_1, N_2) + S_f(N_1, N_2)$.

We define the following notation that we use from here on. A network N_i is built on a leaf label set of size λ and is defined by the node set V_i and the edge set E_i. Moreover, N_i has level k_i and the maximum in-degree of every node in N_i is d_i. Two given networks N_1 and N_2 are built on the same leaf label set and $n = \max(|V_1|, |V_2|)$, $m = \max(|E_1|, |E_2|)$, $k = \max(k_1, k_2)$ and $d = \max(d_1, d_2)$.

Related Work. Table 1 lists the running times of different algorithms for computing $D(N_1, N_2)$. When $k = 0$, both N_1 and N_2 are trees. This case has been extensively studied in the literature, with the fastest algorithm in theory and practice by Brodal *et al.* [1] running in $O(\lambda \log \lambda)$ time. For $k = 1$, an $O(\lambda^{2.687})$-time algorithm based on counting 3-cycles in an auxiliary graph was given in [12], and a faster, $O(\lambda \log \lambda)$-time algorithm that transforms the input to a constant number of instances with $k = 0$ was given in [13]. All algorithms mentioned above allow nodes of arbitrary degree in the input networks. Moreover, software packages implementing the $O(\lambda \log \lambda)$-time algorithms are available.

For $k > 1$, Byrka *et al.* [2] considered the special case of networks whose roots have out-degree 2 and whose other non-leaf nodes have in-degree 2 and

Table 1. Previous and new results for computing $D(N_1, N_2)$, where N_1 and N_2 are two level-k networks built on the same leaf label set of size λ.

Year	Reference	k	Degrees	Time complexity
1980	Fortune et al. [8]	Arbitrary	Arbitrary	$\Omega(n^7\lambda^3)$
2010	Byrka et al. [2]	Arbitrary	Binary	$O(n^3 + \lambda^3)$
2010	Byrka et al. [2]	Arbitrary	Binary	$O(n + k^2n + \lambda^3)$
2017	Brodal et al. [1]	0	Arbitrary	$O(\lambda \log \lambda)$
2017	Jansson et al. [13]	1	Arbitrary	$O(\lambda \log \lambda)$
2019	New	Arbitrary	Arbitrary	$O(n^2m + \lambda^3)$
2019	New	Arbitrary	Arbitrary	$O(m + k^3d^3\lambda + \lambda^3)$

out-degree 1 or in-degree 1 and out-degree 2. Given such a network $N = (V, E)$, they defined a data structure D that can be constructed in $O(|V|^3)$ time by dynamic programming and then used to determine in $O(1)$ time if any resolved triplet $xy|z$ is consistent with N. This result was then strengthened by obtaining a new data structure D' that requires $O(|V| + k^2|V|)$ construction time, where k is the level of N. If N_1 and N_2 have arbitrary levels and follow the degree constraints of N, D can be used to compute $D(N_1, N_2)$ in $O(n^3 + \lambda^3)$ time and D' can be used to compute $D(N_1, N_2)$ in $O(n + k^2n + \lambda^3)$ time.

Contribution. The data structures D and D' of Byrka et al. [2] can only support consistency queries for resolved triplets. However, a network with nodes of arbitrary degree may contain fan triplets. Moreover, D' exploits the fact that given the degree constraints in N, all biconnected components of $U(N)$ are node-disjoint. However, even a small change in these constraints, e.g., if we allow nodes with in-degree 2 to have an out-degree 2 instead of 1, could produce a network with biconnected components that are not node-disjoint, thus making the application of D' impossible.

Without any degree constraints in N_1 and N_2 and when k_1 and k_2 are arbitrary, an algorithm for computing $D(N_1, N_2)$ that iterates over all $4\binom{\lambda}{3}$ triplets and for each triplet applies the pattern matching algorithm in [8] to determine its consistency with N_1 and N_2, has a $\Omega(n^7\lambda^3)$ running time. In this paper we give two algorithms that improve significantly upon this approach. The running time of the first algorithm is $O(n^2m+\lambda^3)$ and the second algorithm $O(m+k^3d^3\lambda+\lambda^3)$. For networks N_1 and N_2 that satisfy the degree constraint in Byrka et al. [2], we prove that our algorithms can compute $D(N_1, N_2)$ using the same time complexity as that of Byrka et al. [2]. To determine the efficiency of the two algorithms in practice, we provide an implementation as well as extensive experiments on both simulated and real datasets. We note that this is the first publicly available software that can compute the triplet distance between two unrestricted networks of arbitrary levels.

Organization of the Article. In Sect. 2 we present the first algorithm and in Sect. 3 the second algorithm. Section 4 presents an implementation of the two algorithms as well as experiments illustrating their practical performance. Due to space constraints, the proofs and most of the experimental results have been deferred to the journal version.

2   A First Approach

In this section we describe an algorithm that for two given networks N_1 and N_2 can compute $D(N_1, N_2)$ in $O(n^2 m + \lambda^3)$ time.

Overview. The algorithm consists of a preprocessing step and a triplet distance computation step. In the preprocessing step, we extend a technique introduced by Shiloach and Perl [18] in 1978 to construct suitably defined auxiliary graphs that compactly encode disjoint paths within N_1 and N_2. Two graphs, the *fan graph* and *resolved graph*, are created that enable us to check the consistency of any fan triplet and any resolved triplet, respectively, with N_1 and N_2 in $O(1)$ time. In the triplet distance computation step, we compute $D(N_1, N_2)$ by iterating over all possible $4\binom{\lambda}{3}$ triplets and using the fan and resolved graphs to check the consistency of each triplet with N_1 and N_2 efficiently.

2.1   Preprocessing

Fan Graph. For any network N_i, let the *fan graph* $N_i^f = (V_i^f, E_i^f)$ be a graph such that $V_i^f = \{s\} \cup \{(u, v, w) \mid u, v, w \in V_i, u \neq v, u \neq w, v \neq w\}$ and E_i^f includes the following edges:

1. $\{(u_1, v_1, w_1) \rightarrow (u_2, v_1, w_1) \mid u_1 \rightarrow u_2 \in E_i \wedge h(u_1) \geq \max(h(v_1), h(w_1))\}$
2. $\{(u_1, v_1, w_1) \rightarrow (u_1, v_2, w_1) \mid v_1 \rightarrow v_2 \in E_i \wedge h(v_1) \geq \max(h(u_1), h(w_1))\}$
3. $\{(u_1, v_1, w_1) \rightarrow (u_1, v_1, w_2) \mid w_1 \rightarrow w_2 \in E_i \wedge h(w_1) \geq \max(h(u_1), h(v_1))\}$
4. $\{s \rightarrow (u, v, w) \mid u \rightarrow v \in E_i \text{ and } u \rightarrow w \in E_i\}$

Note that N_i^f contains $O(|V_i|^3)$ nodes, $O(|V_i|^2 |E_i|)$ edges and also has the property described in the following lemma:

Lemma 1. *Consider a network N_i and its fan graph $N_i^f = (V_i^f, E_i^f)$. For any three different leaves x, y and z in N_i, node s can reach node (x, y, z) in N_i^f if and only if $x|y|z$ is a fan triplet in N_i.*

Corollary 1. *Let N_i be a given network and r' a dummy leaf attached to $r(N_i)$. For any two different leaves x and y in N_i that are not r', there are two paths from $r(N_i)$ to x and y that are disjoint, except for $r(N_i)$, if and only if s can reach (r', x, y) in N_i^f.*

Resolved Graph. For any network N_i, let the *resolved graph* $N_i^r = (V_i^r, E_i^r)$ be a graph such that $V_i^r = \{s\} \cup \{(u, v) \mid u, v \in V_i, u \neq v\} \cup \{(u, v, w) \mid u, v, w \in V_i, u \neq v, u \neq w, v \neq w\}$ and E_i^r includes the following edges:

1. $\{s \rightarrow (u, v) \mid u \rightarrow v \in E_i\}$

2. $\{(u_1, v_1) \rightarrow (u_2, v_1) \mid u_1 \rightarrow u_2 \in E_i, h(u_1) \geq h(v_1)\}$
3. $\{(u_1, v_1) \rightarrow (u_1, v_2) \mid v_1 \rightarrow v_2 \in E_i, h(v_1) \geq h(u_1)\}$
4. $\{(u, v) \rightarrow (u, v, w) \mid v \rightarrow w \in E_i, h(v) \geq h(u)\}$
5. $\{(u_1, v_1, w_1) \rightarrow (u_2, v_1, w_1) \mid u_1 \rightarrow u_2 \in E_i \wedge h(u_1) \geq \max(h(v_1), h(w_1))\}$
6. $\{(u_1, v_1, w_1) \rightarrow (u_1, v_2, w_1) \mid v_1 \rightarrow v_2 \in E_i \wedge h(v_1) \geq \max(h(u_1), h(w_1))\}$
7. $\{(u_1, v_1, w_1) \rightarrow (u_1, v_1, w_2) \mid w_1 \rightarrow w_2 \in E_i \wedge h(w_1) \geq \max(h(u_1), h(v_1))\}$

Note that N_i^r contains $O(|V_i|^3)$ nodes, $O(|V_i|^2|E_i|)$ edges and also has the property described in the following lemma:

Lemma 2. *Consider a network N_i and its resolved graph $N_i^r = (V_i^r, E_i^r)$. For any three different leaves x, y and z in N_i, node s can reach node (x, y, z) in N_i^r if and only if $x|yz$ is a resolved triplet in N_i.*

Corollary 2. *Let N_i be a given network and r' a dummy leaf attached to $r(N_i)$. For any two different leaves x and y in N_i that are not r', there are two paths from some internal node $z \neq r(N_i)$ in N_i, to x and y that are disjoint, except for z, if and only if s can reach (r', x, y) in N_i^r.*

Given N_i^f and N_i^r, we define the $\lambda \times \lambda \times \lambda$ *fan table* A_i^f and the $\lambda \times \lambda \times \lambda$ *resolved table* A_i^r as follows. For three different leaves x, y and z, $A_i^f[x][y][z] = 1$ if the fan triplet $x|y|z$ is consistent with N_i and $A_i^f[x][y][z] = 0$ otherwise. Similarly, $A_i^r[x][y][z] = 1$ if the resolved triplet $x|yz$ is consistent with N_i and $A_i^r[x][y][z] = 0$ otherwise. Due to Lemmas 1 and 2, both A_i^f and A_i^r can be computed by a depth first traversal (starting from s) of N_i^f and N_i^r. More precisely, $A_i^f[x][y][z] = 1$ if s can reach (x, y, z) in N_i^f and 0 otherwise. Finally, $A_i^r[x][y][z] = 1$ if s can reach (x, y, z) in N_i^r and 0 otherwise.

2.2   Triplet Distance Computation

Algorithm 1 summarizes all the procedures needed to compute the triplet distance between two given networks N_1 and N_2. For every $i \in \{1, 2\}$ the tables A_i^f and A_i^r are built in lines 2–7. These tables are then used in lines 11–12 and 16–19 to determine in $O(1)$ time if a triplet is consistent with N_1 or N_2. Procedures $S_f()$ and $S_r()$ count the number of shared fan and resolved triplets. Both procedures enumerate over all possible triplets and use the tables A_i^f and A_i^r to determine their consistency with either network. The correctness is ensured by Lemmas 1 and 2. Procedure $S()$ reports the number of shared triplets, which is the sum of the number of shared fan triplets and shared resolved triplets. The main procedure is $D()$. It uses Eq. (1) to determine $D(N_1, N_2)$.

To analyze the running time, after the preprocessing is finished, the procedures $S_f()$ and $S_r()$ require $O(\lambda^3)$ time. For the total preprocessing time, by definition, building the data structures N_i^r and N_i^f for $i \in \{1, 2\}$ in line 3, requires $O(|V_1|^2|E_1| + |V_2|^2|E_2|)$ time. Building the auxiliary arrays A_i^r and A_i^f in lines 5–7 is performed by a depth first traversal of N_i^r and N_i^f, thus requiring $O(|V_1|^2|E_1| + |V_2|^2|E_2|)$ time as well. Hence, the total time of the algorithm

---

Algorithm 1. Computing $D(N_1, N_2)$ using the data structures from Section 2.

1: **procedure** PREPROCESSING(N_1, N_2)   ▷ Building the data structures
2:     **for** $i \in \{1, 2\}$ **do**
3:         build $N_i^f = (V_i^f, E_i^f)$ and $N_i^r = (V_i^r, E_i^r)$
4:         let A_i^f, A_i^r be $\lambda \times \lambda \times \lambda$ arrays initialized with 0 entries
5:         **for** three different leaves x, y and z **do**
6:             $A_i^f[x][y][z] = 1$ if s can reach (x, y, z) in N_i^f
7:             $A_i^r[x][y][z] = 1$ if s can reach (x, y, z) in N_i^r
8:     **return** $(A_1^r, A_1^f, A_2^r, A_2^f)$

9: **procedure** $S_f(A_1^f, A_2^f)$ ▷ Finding the shared fan triplets
10:     $sharedFan = 0$
11:     **for** three different leaves x, y and z **do**
12:         **if** $A_1^f[x][y][z] = A_2^f[x][y][z] = 1$ **then** $sharedFan = sharedFan + 1$
13:     **return** $sharedFan$

14: **procedure** $S_r(A_1^r, A_2^r)$ ▷ Finding the shared resolved triplets
15:     $sharedResolved = 0$
16:     **for** three different leaves x, y and z **do**
17:         **if** $A_1^r[x][y][z] = A_2^r[x][y][z] = 1$ **then** $sharedResolved = sharedResolved + 1$
18:         **if** $A_1^r[x][z][y] = A_2^r[x][z][y] = 1$ **then** $sharedResolved = sharedResolved + 1$
19:         **if** $A_1^r[y][z][x] = A_2^r[y][z][x] = 1$ **then** $sharedResolved = sharedResolved + 1$
20:     **return** $sharedResolved$

21: **procedure** $S(A_1^r, A_1^f, A_2^r, A_2^f)$ ▷ Finding the shared triplets
22:     **return** $S_f(A_1^f, A_2^f) + S_r(A_1^r, A_2^r)$

23: **procedure** $D(N_1 = (V_1, E_1), N_2 = (V_2, E_2))$ ▷ Computing the triplet distance
24:     $(A_1^r, A_1^f, A_2^r, A_2^f) = $ PREPROCESSING(N_1, N_2)
25:     **return** $S(A_1^r, A_1^f, A_1^r, A_1^f) + S(A_2^r, A_2^f, A_2^r, A_2^f) - 2S(A_1^r, A_1^f, A_2^r, A_2^f)$

---

becomes $O(|V_1|^2|E_1| + |V_2|^2|E_2| + \lambda^3)$. By the definition of n and m from Sect. 1, the running time becomes $O(n^2 m + \lambda^3)$. Hence, we obtain the following theorem:

Theorem 1. *There exists an algorithm that computes the triplet distance between two networks N_1 and N_2 in $O(n^2 m + \lambda^3)$ time.*

Let N_1 and N_2 follow the degree constraints of Byrka et al. [2]. We then have $n = \Theta(m)$ and the bound becomes $O(n^3 + \lambda^3)$, thus matching the bound achieved by the first data structure of Byrka et al. [2].

3   A Second Approach

In this section we extend the algorithm from Sect. 2 in order to exploit the information about the level of the two input networks. More specifically, we describe an algorithm that for two given networks N_1 and N_2 can compute $D(N_1, N_2)$ in $O(m + k^3 d^3 \lambda + \lambda^3)$ time.

Overview. In the first approach, for a given network N_i we built the fan and resolved graph presented in Lemmas 1 and 2. In this second approach, for every biconnected component of $U(N_i)$ we define a network of approximately the same size as the biconnected component, which we call *contracted block network*. For this contracted block network we then build the corresponding fan and resolved graph. By carefully contracting every biconnected component of $U(N_i)$ into one node we obtain a tree, which we call *block tree*. We finally show how to combine the block tree and all the fan and resolved graphs of the contracted block networks of N_i to count triplets efficiently.

3.1 Preprocessing

Let N_i be a given network. From here on, we call a biconnected component of $U(N_i)$ a *block*. For simplicity, when we refer to a block of N_i, we imply a block of $U(N_i)$. We say that for a block B of N_i, node $r(B)$ is the root of B, if $r(B)$ has the largest height in N_i among all nodes in B. Note that because N_i has one root that can reach every node of N_i and B corresponds to a maximal subgraph of $U(N_i)$ that is biconnected, B can only contain one root. If B contains only one edge $u \rightarrow v$ such that $v \in \mathcal{L}(N_i)$, then B is called a *leaf block*, otherwise B is called a *non-leaf block*. Lemma 3 presents a property of all blocks of N_i.

Lemma 3. *All blocks of a given network N_i are edge-disjoint.*

Block Tree. From a high level perspective, we want to remove the cycles in $U(N_i)$ that are formed by the non-leaf blocks to obtain a directed tree on the leaf label set $\mathcal{L}(N_i)$. Let $T_i = (V', E')$ be a directed tree, from now on referred to as *block tree*, with the node set V' and edge set E' defined by the following steps:

- For every block B_j in N_i create a node b_j in T_i.
- Let B_1, B_2 be two blocks in N_i with $r(B_1) \neq r(B_2)$. If $r(B_2)$ is also a node in B_1 then create the edge $b_1 \rightarrow b_2$ in T_i.
- Create a root node ρ in T_i. For every block B_j that has $r(N_i)$ as a root, create the edge $\rho \rightarrow b_j$ in T_i.
- If B_j is a leaf block, rename b_j in T_i by the label of the leaf in B_j.

The set of all blocks of N_i and the node set $V' - r(T_i)$, i.e., the set of all nodes of T_i except the root, are bijective. An edge $b_1 \rightarrow b_2$ in T_i means that the corresponding blocks B_1 and B_2 in N_i do not have the same root and the root node $r(B_2)$ is a shared node between B_1 and B_2. Note that by the definition of a block, an edge connecting two nodes can define a block of its own. The following lemma presents some properties of T_i:

Lemma 4. *Let $T_i = (V', E')$ be the block tree of a given network N_i. The block tree T_i is a directed tree that has λ leaves, $|V'| = O(\lambda)$ and $|E'| = O(\lambda)$.*

Since the set of all blocks of N_i and the set $V' - r(T_i)$ are bijective, we obtain:

Corollary 3. *A network N_i contains $O(\lambda)$ blocks.*

The following lemma presents an algorithm for constructing the block tree T_i:

Lemma 5. *Let $T_i = (V', E')$ be the block tree of a given network N_i. There exists an algorithm that builds T_i in $O(|E_i|)$ time.*

Contracted Block Network. For a given network N_i, a block B in N_i and a node u in B, define L_B^u to be the set of leaves that can be reached from u without using edges in B. Let $C_B = (V', E')$ be a network, with the node set V' and edge set E' defined by the following steps:

- Let $C_B = N_i$. All operations from now on are applied on C_B.
- Remove every edge and node not in B.
- For every edge $u_1 \rightarrow u_2$ in B, if $in(u_1) = out(u_1) = in(u_2) = out(u_2) = 1$ contract the edge as follows: let $u_2 \rightarrow u_3$ be the other edge in B, then create the edge $u_1 \rightarrow u_3$, remove u_2 from B and set $L_B^{u_1} = L_B^{u_1} \cup L_B^{u_2}$.
- For every node u_1 in C_B such that $L_B^{u_1} \neq \emptyset$, we add a child leaf with label s_1 representing all leaves in $L_B^{u_1}$. We also add another child leaf s_1' as a copy leaf that will help later on to count triplets.
- Include an artificial leaf r' which is attached to the root $r(C_B)$.

Every node in C_B corresponds to a node in B and every edge between two internal nodes in C_B corresponds to a compressed path in B. We call C_B the *contracted block network* of N_i, corresponding to block B. The following lemma presents a property of C_B:

Lemma 6. *Let N_i be a network, B a block in N_i and $C_B = (V', E')$ the contracted block network of N_i that corresponds to block B. We then have that $|\mathcal{L}(C_B)| = O(k_i d_i + 1)$, $|V'| = O(k_i d_i + 1)$ and $|E'| = O(k_i d_i + 1)$.*

Constructing All Contracted Block Networks Efficiently. For a given network N_i and a block B in N_i, a leaf x in N_i is said to *associate with B* if there exists a node u in B such that $u \neq r(B)$ and $x \in L_B^u$. For any leaf x associated with some block B of N_i, let $q_B(x)$ be the node in B that has a path to x without using edges in B, i.e., $x \in L_B^{q_B(x)}$, $p_B(x)$ the leaf in C_B representing x and $p_B'(x)$ the copy leaf of $p_B(x)$.

Lemma 3 implies an algorithm for constructing every block network C_B of N_i in $O(|E_i|)$ time. As shown in the lemma below, by properly relabeling the leaves of N_i and with an additive $O(\lambda^2)$ time, it is possible to build every block network C_B so that we can afterwards compute for every leaf $l \in \mathcal{L}(N_i)$ the functions $q_B(l)$ and $p_B(l)$ in $O(1)$ time.

Lemma 7. *For a network N_i, there exists an $O(|E_i| + \lambda^2)$-time algorithm that builds all the contracted block networks C_B of N_i, such that for all blocks of N_i and leaf $l \in \mathcal{L}(N_i)$ the functions $q_B(l)$ and $p_B(l)$ are computed in $O(1)$ time.*

For the block network C_B, we denote C_B^f the fan graph of C_B and C_B^r the resolved graph of C_B. Moreover, we denote A_B^f the fan table of C_B and A_B^r the resolved table of C_B (see Sect. 2.1 for the definition of a fan graph & table and resolved graph & table). The following lemma shows the time required to build C_B^f, C_B^r, A_B^f and A_B^r for every block B of a given network N_i:

Lemma 8. *For a network N_i, building C_B^f, C_B^r, A_B^f and A_B^r for every block B of N_i requires $O(\lambda(k_i^3 d_i^3 + 1))$ time.*

3.2   Triplet Distance Computation

Let B be a block of a network N_i. We say that $x|y|z$ is a *fan triplet consistent with B*, if there exists a node u in B that has three disjoint paths in N_i to x, y and z, except for u, one going from u to x, one from u to y and one from u to z. We also say that $x|y|z$ *is rooted at u in B*. Since u is also in N_i, this means that $x|y|z$ is rooted at u in N_i as well. Similarly, we say that $xy|z$ is a *resolved triplet consistent with B*, if there exist two nodes u and v in B such that $u \neq v$, and there are four disjoint paths in N_i, except for u and v, one going from u to v, one from v to x, one from v to y and one from u to z. Moreover, we say that $xy|z$ is *rooted at u and v in B or N_i* (similarly to the fan triplet). Note that if $x|y|z$ is a fan triplet consistent with B, then it will also be consistent with N_i. Similarly, if $xy|z$ is a resolved triplet consistent with B, it will also be consistent with N_i.

Given the data structures from the preprocessing step, Lemmas 9 and 10 together show how to determine the consistency of a fan and resolved triplet with N_i in O(1) time. From a high level perspective to achieve this, for three different leaves x, y and z, we consider all the possible cases for the location of the lowest common ancestor of every pair (x, y), (x, z) and (y, z) in T_i. Since every node in T_i except $r(T_i)$ corresponds to a block in N_i, we can then use the available data structures to determine efficiently if N_i has the necessary disjoint paths that would imply the consistency of the fan triplet $x|y|z$ or resolved triplet $xy|z$ with N_i. We start by showing in Lemma 9 how to determine the consistency of a triplet with a block B of N_i. Afterwards, we show in Lemma 10 how to use Lemma 9 to determine the consistency of a triplet with N_i.

Lemma 9. *Let N_i be a given network, T_i its block tree and x, y and z three different leaves. Let w be the lowest common ancestor of x, y and z in T_i, $w \neq r(T_i)$ and B the block in N_i corresponding to w. If C_B, C_B^f, C_B^r, A_B^f and A_B^r are given, there exists an algorithm that can determine in O(1) time if the fan triplet $x|y|z$ or resolved triplet $xy|z$ is consistent with B.*

Lemma 10. *Let N_i be a given network and x, y and z three different leaves in N_i. Given T_i, C_B^f, C_B^r, A_B^f and A_B^r for every block B in N_i, there exists an algorithm that can determine in O(1) time if the fan triplet $x|y|z$ or the resolved triplet $xy|z$ is consistent with N_i.*

The final algorithm is similar to Algorithm 1, the main difference is in the preprocessing step. In this step, for every $i \in \{1, 2\}$ we start by building the block tree T_i. Then, we build a $\lambda \times \lambda$ table for T_i in order to be able later to answer lowest common ancestor queries between pairs of leaves in T_i in $O(1)$ time. Afterwards, we build all the contracted block networks of N_i. Finally, for every block B in N_i and the corresponding contracted block network C_B, we build the fan graph C_B^f and the resolved graph C_B^r, as well as the corresponding A_B^f and A_B^r tables.

From Lemma 5, building T_i for every $i \in \{1, 2\}$ requires $O(|E_1| + |E_2|)$ time. Building the two tables for answering lowest common ancestor queries requires $O(\lambda^3)$ time. From Lemma 6, building all the contracted block networks requires $O(|E_1| + |E_2| + \lambda^2)$ time. From Lemma 8, the time required to build C_B^f, C_B^r, A_B^f and A_B^r for every block B of N_1 and N_2 is $O(\lambda(k_1^3 d_1^3 + k_2^3 d_2^3 + 2))$. Hence, the total preprocessing time becomes $O(|E_1| + |E_2| + \lambda(k_1^3 d_1^3 + k_2^3 d_2^3) + \lambda^3)$.

Using the results from Lemma 10, after the preprocessing step we can determine the consistency of a triplet with N_1 or N_2 in $O(1)$ time. Since the number of triplets that need to be checked is exactly $4\binom{\lambda}{3}$, the total running time of the algorithm remains $O(|E_1| + |E_2| + \lambda(k_1^3 d_1^3 + k_2^3 d_2^3) + \lambda^3)$. By the definition of n, m, k and d from Sect. 1, the running time becomes $O(m + k^3 d^3 \lambda + \lambda^3)$. Hence, we obtain the following theorem:

Theorem 2. *There exists an algorithm that computes the triplet distance between two networks N_1 and N_2 in $O(m + k^3 d^3 \lambda + \lambda^3)$ time.*

Let N_i be a network that follows the degree constraints of Byrka et al. [2]. If for a block $B_s = (V_s, E_s)$ of N_i we define k_s to be the number of reticulation nodes in B_s, where $k_s \leq k_i$, using the same arguments as those used in the proof of Lemma 6, we get for $C_{B_s} = (V', E')$ that $|V'| = |E'| = O(k_s + 1)$. The time to build C_B^f, C_B^r, A_B^f and A_B^r for every block B of N_i then becomes $\sum_s O(k_s^3 + 1)$. Note that Lemma 8 would give a $O(\lambda k_i^3 + 1)$ time instead, because it uses λ to upper bound (from Corollary 3) the number of blocks we can have in N_i. Since $\sum_s k_s = O(|V_i|)$, the preprocessing time required by our algorithm for N_i would be $O(|V_i| + k^2|V_i|)$. Then, the time to compute $D(N_1, N_2)$ becomes $O(n + k^2 n + \lambda^3)$, thus matching the time bound required by using the second data structure of Byrka et al. [2].

4   Implementation and Experiments

This section provides an implementation of the algorithms described in Sects. 2 and 3, referred to as NTDfirst and NTDsecond respectively, as well as experiments illustrating their practical performance.

The Setup. We implemented the two algorithms in C++ and the source code is publicly available at https://github.com/kmampent/ntd. The experiments were performed on a machine with 16 GB RAM and Intel(R) Core(TM) i5-3470 CPU @ 3.20 GHz. The operating system was Ubuntu 16.04.2 LTS. The compiler used was g++ 5.4 with cmake 3.11.0.

The Input. We consider both simulated and real datasets. For the simulated datasets, we create tree-based networks [9] as follows:

1. Build a random rooted binary tree T on λ leaves in the uniform model [16] and let $N = T$. For a node w in N, let $d(w)$ be the total number of edges on the path from $r(N)$ to N.
2. Given a parameter $e \geq 0$, add e random edges in N as follows. An edge $u \to v$ is created in N if $d(u) < d(v)$. If the total number of edges that can be added happens to be y, where $y < e$, then we only add those y edges.

For the real datasets, we consider networks that have been published in the literature and are not necessarily tree-based. More precisely, we consider the 6 trees and the corresponding networks in [15, Table S4]. The trees are given in the standard Newick format, and the networks in the extended Newick format [3].

Experiments. For the simulated datasets, in Fig. 2 we illustrate the effect of e on the CPU time in seconds of the two algorithms. Every data point in the graph is the average of 20 different runs. The effect is larger on NTDsecond, as larger values for e imply fewer blocks in the given networks. We note that space is the reason behind the difference restrictions on λ, i.e., for $\lambda = 230$ the memory usage of NTDfirst approaches the limits of the available 16 GB RAM.

For the real datasets and for every $s \in \{A, B, C, D, E, F\}$, we denote T_s the tree and N_s the corresponding network, where s is a scenario in [15, Table S4], with F corresponding to scenario "E, CHAM and MELVIO resolved". For the network N_F, we use its non-tree based version from [14]. From Eq. (1) we have the following: $D(T_s, N_s) = S(T_s, T_s) + S(N_s, N_s) - 2S(T_s, N_s)$. When computing $D(T_s, N_s)$ and to have $\mathcal{L}(T_s) = \mathcal{L}(N_s)$, if a leaf x in N_s appears as several leaves $x.1, \ldots, x.i$ in T_s, we replace x in N_s with the leaves $x.1, \ldots, x.i$ that we attach under the parent of x. For the size of the leaf label sets, in the trees T_A, T_B, T_C, T_D, T_E we have 16, 20, 21, 21, 22 and 50 leaves, in every network N_s where $s \in \{A, B, C, D, E\}$ we have 8 leaves and in N_F we have 16 leaves. In Table 2 we include the experimental results. Interestingly, while the two networks N_B and N_D look structurally different, $D(N_B, N_D) = 0$. This suggests that it may be useful to extend the definition of the triplet distance to take into account the number of times that each triplet occurs in a network.

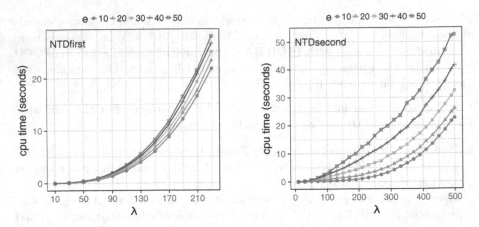

Fig. 2. Experiments on the simulated datasets: running time for different values of e.

Table 2. Experiments on the real datasets. $N_A, \ldots, N_E$ have identical leaf label sets.

s	$S(T_s, T_s)$	$S(N_s, N_s)$	$S(T_s, N_s)$	$D(T_s, N_s)$			N_A	N_B	N_C	N_D	N_E
A	560	716	443	390		N_A	0	20	19	20	10
B	1140	1870	840	1330		N_B	20	0	1	0	10
C	1330	2185	965	1585		N_C	19	1	0	1	9
D	1330	2205	964	1607		N_D	20	0	1	0	10
E	1540	1996	983	1570		N_E	10	10	9	10	0
F	19600	43710	16553	30204							

Acknowledgments. Konstantinos Mampentzidis acknowledges the support by the Danish National Research Foundation, grant DNRF84, via the Center for Massive Data Algorithmics (MADALGO).

References

1. Brodal, G.S., Mampentzidis, K.: Cache oblivious algorithms for computing the triplet distance between trees. In: Proceedings of ESA 2017, pp. 21:1–21:14 (2017)
2. Byrka, J., Gawrychowski, P., Huber, K.T., Kelk, S.: Worst-case optimal approximation algorithms for maximizing triplet consistency within phylogenetic networks. J. Discrete Algorithms **8**(1), 65–75 (2010)
3. Cardona, G., Rosselló, F., Valiente, G.: Extended Newick: it is time for a standard representation of phylogenetic networks. BMC Bioinform. **9**(1), 532 (2008)
4. Choy, C., Jansson, J., Sadakane, K., Sung, W.K.: Computing the maximum agreement of phylogenetic networks. Theor. Comput. Sci. **335**(1), 93–107 (2005)
5. Dobson, A.J.: Comparing the shapes of trees. In: Street, A.P., Wallis, W.D. (eds.) Combinatorial Mathematics III. Lecture Notes in Mathematics, vol. 452, pp. 95–100. Springer, Berlin (1975). https://doi.org/10.1007/BFb0069548
6. Estabrook, G., McMorris, F., Meacham, C.: Comparison of undirected phylogenetic trees based on subtrees of four evolutionary units. Syst. Zool. **34**(2), 193–200 (1985)
7. Felsenstein, J.: Inferring Phylogenies. Sinauer Associates, Inc., Sunderland (2004)

8. Fortune, S., Hopcroft, J., Wyllie, J.: The directed subgraph homeomorphism problem. Theor. Comput. Sci. **10**(2), 111–121 (1980)
9. Francis, A.R., Steel, M.: Which phylogenetic networks are merely trees with additional arcs? Syst. Biol. **64**(5), 768–777 (2015)
10. Gambette, P., Huber, K.T.: On encodings of phylogenetic networks of bounded level. J. Math. Biol. **65**(1), 157–180 (2012)
11. Gusfield, D., Eddhu, S., Langley, C.: Optimal, efficient reconstruction of phylogenetic networks with constrained recombination. J. Bioinform. Comput. Biol. **2**(1), 173–213 (2004)
12. Jansson, J., Lingas, A.: Computing the rooted triplet distance between galled trees by counting triangles. J. Discrete Algorithms **25**, 66–78 (2014)
13. Jansson, J., Rajaby, R., Sung, W.K.: An efficient algorithm for the rooted triplet distance between galled trees. In: Proceedings of AlCoB 2017, pp. 115–126 (2017)
14. Jetten, L., van Iersel, L.: Nonbinary tree-based phylogenetic networks. IEEE/ACM Trans. Comput. Biol. Bioinform. **1**(1), 205–217 (2018)
15. Marcussen, T., Heier, L., Brysting, A.K., Oxelman, B., Jakobsen, K.S.: From gene trees to a dated allopolyploid network: insights from the angiosperm genus viola (violaceae). Syst. Biol. **64**(1), 84–101 (2015)
16. McKenzie, A., Steel, M.: Distributions of cherries for two models of trees. Math. Biosci. **164**(1), 81–92 (2000)
17. Robinson, D., Foulds, L.: Comparison of phylogenetic trees. Math. Biosci. **53**(1), 131–147 (1981)
18. Shiloach, Y., Perl, Y.: Finding two disjoint paths between two pairs of vertices in a graph. J. ACM **25**(1), 1–9 (1978)

Parameterized Algorithms for Graph Burning Problem

Anjeneya Swami Kare[1] and I. Vinod Reddy[2(✉)]

[1] University of Hyderabad, Hyderabad, India
askcs@uohyd.ac.in
[2] Indian Institute of Technology Bhilai, Raipur, India
vinod@iitbhilai.ac.in

Abstract. The GRAPH BURNING problem is defined as follows. At time $t = 0$, no vertex of the graph is burned. At each time $t \geq 1$, we choose a vertex to burn. If a vertex is burned at time t, then at time $t + 1$ each of its unburned neighbors becomes burned. Once a vertex is burned then it remains in that state for all subsequent steps. The process stops when all vertices are burned. The burning number of a graph is the minimum number of steps needed to burn all the vertices of the graph.

Computing the burning number of a graph is NP-complete even on bipartite graphs or trees of maximum degree three. In this paper we study this problem from the parameterized complexity perspective. We show that the problem is fixed-parameter tractable (FPT) when parameterized by the distance to cluster or neighborhood diversity. We further study the complexity of the problem on restricted classes of graphs. We show that GRAPH BURNING can be solved in polynomial time on cographs and split graphs.

1 Introduction

The spread of social contagion is one of the active research areas in social network analysis. The goal is to spread a message to all users of a network as fast as possible. Graph burning is a process used to model the spread of social contagion in a graph, where the goal is to burn all the vertices in the graph as quickly as possible.

The GRAPH BURNING problem is defined as follows. At time $t = 0$, no vertex of the graph is burned. At each time $t \geq 1$, we choose a vertex to burn. If a vertex is burned at time t, then at time $t + 1$ each of its unburned neighbors becomes burned. Once a vertex is burned then it remains in that state for all subsequent steps. The process stops when all nodes are burned. The *burning number* of a graph G is the minimum number of steps needed to burn all the vertices of the graph, denoted as $b(G)$. For example, the burning number of a complete graph on n vertices is two. Computing the burning number of a graph is known to be NP-complete even for special classes of graphs, including trees of maximum degree three [1], bipartite graphs, chordal graphs, planar graphs, and binary

© Springer Nature Switzerland AG 2019
C. J. Colbourn et al. (Eds.): IWOCA 2019, LNCS 11638, pp. 304–314, 2019.
https://doi.org/10.1007/978-3-030-25005-8_25

trees [1]. Land et al. [14] gave some upper bounds on the burning number of graphs.

In this paper we study the GRAPH BURNING problem from the parameterized complexity perspective. In parameterized complexity theory the running time of an algorithm is measured as a function of input size and an additional measure called the parameter. The goal here is to design algorithms that solve the problem in time $f(k)n^{O(1)}$, where f is some computable function, k is the value of the parameter and n is the size of the problem instance. These kind of algorithms are called *fixed parameter tractable* (FPT) algorithms. In general there may be several interesting parameterizations for any given problem. However, there are two popular approaches to select a parameter for optimization problems on graphs. First, the natural parameter is the size of the solution (objective function). Second, the parameters which do not involve the objective function, which are selected based on structure of the graph. In this paper, we focus on structural graph parameters.

We give parameterized algorithms for the GRAPH BURNING problem with respect to several distance-to-triviality [11] parameters. These parameters are extensively used to design efficient algorithms for many hard graph problems. They measure how far a graph is from some class of graphs for which the problem is tractable. Our notion of distance to a graph class is the vertex deletion distance. More formally, for a class $\mathcal{F}$ of graphs we say that X is an $\mathcal{F}$-modulator of a graph G if there is a subset $X \subseteq V(G)$ such that $G \setminus X \in \mathcal{F}$. If the least size modulator to $\mathcal{F}$ is k then we say that the distance of G to the class $\mathcal{F}$ is k.

Our Contributions. We design parameterized algorithms for the GRAPH BURNING problem with respect to the distance from the following graph classes: cluster graphs (disjoint union of complete graphs). This parameter is intermediate between the vertex cover number and clique-width parameters (see Fig. 1). Studying the parameterized complexity of the problem with respect to the above mentioned distance parameters improves our understanding of the problem and closes the gap between tractability and intractability. For example, the problem is trivially FPT when parameterized by the vertex cover number and it is para-NP-hard when parameterized by the clique-width [1].

In particular we obtain the following results.

- We show that GRAPH BURNING is FPT when parameterized by the (a) distance to cluster graphs (cluster vertex deletion number) (b) neighborhood diversity.
- We show that on cographs and split graphs GRAPH BURNING can be solved in polynomial time.
- We obtain upper and lower bounds for the GRAPH BURNING problem on interval graphs in terms of the diameter of the graph.

The problem we consider in this paper is formally defined as follows.

GRAPH BURNING
Input: A graph $G = (V, E)$, and a positive integer k.
Question: Is burning number of G at most k?

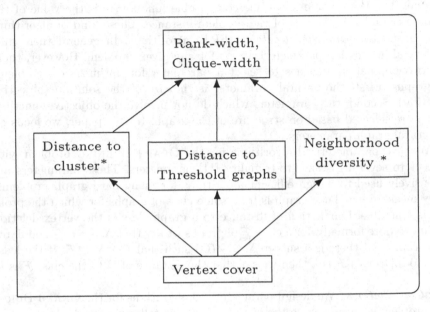

Fig. 1. A schematic showing the relation between the various graph parameters. An arrow from parameter a to b indicates that a is larger than b. Parameters marked with $*$ are studied in this paper.

2   Preliminaries

In this section, we introduce the basic notions of graph theory and parameterized complexity theory. For a basic introduction to graph theory the reader is referred to Diestel's book on graph theory [7]. All graphs we consider in this paper are undirected, connected, finite and simple. For a graph $G = (V, E)$, let $V(G)$ and $E(G)$ denote the vertex set and edge set of G respectively. We use n and m to denote the number of vertices and edges in a graph respectively. An edge in E between vertices x and y is denoted as xy for simplicity. For a subset $X \subseteq V(G)$, the graph $G[X]$ denotes the subgraph of G induced by vertices of X. Also, we abuse notation and use $G \setminus X$ to refer to the graph obtained from G after removing the vertex set X. For a vertex $v \in V(G)$, $N(v)$ denotes the set of vertices adjacent to v and $N[v] = N(v) \cup \{v\}$ is the closed neighborhood of v. The distance $d(u, v)$ between two vertices u and v in a graph G is the number of edges in a shortest path from u to v. For a vertex v in a graph G, the *eccentricity* of v

is defined as $\max\{d(u,v) \mid u \in V(G)\}$. The *radius* of G is minimum eccentricity over the set of all vertices in G. The *diameter* of G is the maximum eccentricity over the set of all vertices in G. For a given positive integer k, the k-th closed neighborhood of v is defined as $N_k[v] = \{u \in V(G) : d(u,v) \leq k\}$. A vertex is called a *universal vertex* if it is adjacent to every other vertex of the graph.

2.1 Graph Classes

We now define the graph classes which are considered in this paper. For an integer k, we denote by P_k a path on k vertices with $k-1$ edges. A graph is P_k-free if it does not contain P_k as an induced subgraph. A graph is a *split graph* if its vertices can be partitioned into a clique and an independent set. Split graphs are $(C_4, C_5, 2K_2)$-free. We denote a split graph with $G = (C, I)$ where C and I denotes the partition of G into a clique and an independent set. The class of P_4-free graphs are called *cographs*. A cluster graph is a disjoint union of complete graphs. Cluster graphs are P_3-free graphs. A graph G is called an *interval graph* if there exists a set $\{I_v \mid v \in V(G)\}$ of real intervals such that $I_u \cap I_v \neq \emptyset$ if and only if $uv \in E(G)$.

2.2 Parameterized Complexity

A parameterized problem denoted as $(I, k) \subseteq \Sigma^* \times \mathbb{N}$, where Σ is fixed alphabet and k is called the parameter. We say that the problem (I, k) is *fixed-parameter tractable* with respect to the parameter k if there exists an algorithm which solves the problem in time $f(k)|I|^{O(1)}$, where f is a computable function. A *kernel* for a parameterized problem Π is an algorithm which transforms an instance (I, k) of Π to an equivalent instance (I', k') in polynomial time such that $k' \leq g(k)$ and $|I'| \leq f(k)$ for some computable functions f and g. It is known that a parameterized problem is fixed-parameter tractable if and only of it has a kernel. For more details on parameterized complexity, we refer the reader to the texts [6,9].

Cluster graphs are P_3-free. Therefore given a graph G and integer k, the problem of deciding whether there exists a set X of vertices of size at most k whose deletion results in a cluster graph is FPT. Hence we assume that a modulator is given as part of the input.

2.3 Basics of Graph Burning Problem

In this subsection we review some results from the literature about the GRAPH BURNING problem that will be useful in our results.

Let (G, k) be an instance of GRAPH BURNING. Let $S = (x_1, x_2, \cdots, x_k)$ be a sequence of vertices, where x_i represents the vertex that we burn at time step i. We say that the sequence S is a *burning sequence* if vertex x_i is not burning at the start of time step i and G can be burned in k steps. A sequence is an *optimal* if it is the shortest among all burning sequences.

A sequence $(x_1, x_2, \cdots, x_k)$ is a burning sequence of a graph G if and only if for each pair i and j with $1 \leq i < j \leq k$, we have that $d(x_i, x_j) \geq j - i$ and

$$N_{k-1}[x_1] \cup N_{k-2}[x_2] \cup \cdots \cup N_0[x_k] = V(G)$$

Lemma 1 [1]. *Let (G, k) be an instance of* GRAPH BURNING *problem and $S \subseteq V(G)$ be an ordered subset of size k. We can test whether S is a valid burning sequence for the* GRAPH BURNING *problem in $O(n + m)$ time.*

Theorem 1 [15]. *Let G be a graph with n vertices, then $b(G) = 2$ if and only if $n \geq 2$ and G has maximum degree $n - 1$ or $n - 2$.*

Lemma 2 [2]. *If G is a graph with radius r then $b(G) \leq r + 1$.*

For a path P_n on n vertices we have that $b(P_n) = \lceil \sqrt{n} \rceil$.

3 Parameterized Algorithms

In this section, we give parameterized algorithms for GRAPH BURNING problem with respect to the parameters distance to cluster and neighborhood diversity.

3.1 Parameterization by Distance to Cluster

The cluster vertex deletion number (or distance to a cluster graph) of a graph G is the minimum number of vertices that have to be deleted from G to get a disjoint union of complete graphs. The cluster vertex deletion number is an intermediate parameter between the vertex cover number and the clique-width/rank-width [8]. In this subsection we show that GRAPH BURNING is fixed-parameter tractable when parameterized by the distance to cluster.

Lemma 3. *If $X \subseteq V(G)$ of size d such that $G \setminus X$ is a cluster graph then $b(G) \leq 3d + 3$.*

Proof. It is easy to see that the length of the longest induced path in G is at most $3d + 2$. Therefore the radius of the graph G is at most $3d + 2$. From Lemma 2, the burning number of a graph G with radius r is at most $r + 1$. Therefore $b(G) \leq 3d + 3$. □

Theorem 2. GRAPH BURNING *is fixed-parameter tractable when parameterized by the distance to cluster.*

Proof. Let G be a graph and $X \subseteq V(G)$ of size d such that $G \setminus X$ is a cluster graph. If $k \geq 3d + 3$, then from Lemma 3 the given instance is a YES instance. Therefore without loss of generality we assume that $k < 3d + 3$.

We partition the vertices of each clique C in $G \setminus X$ based on their neighborhoods in X. For every subset $Y \subseteq X$, $T_Y^C := \{x \in C \mid N(x) \cap X = Y\}$. This way we can partition vertices of a clique C into at most 2^d subsets (called types), one for each $Y \subseteq X$.

Reduction Rule 1. *Let $C \in G \setminus X$ be a clique and $Y \subseteq X$, If $|T_Y^C| > 2$ then delete $|T_Y^C| - 2$ vertices of T_Y^C.*

Lemma 4. *Reduction Rule 1 is safe.*

Proof. Let $G_1 = (V_1, E_1)$ be the graph obtained by iteratively applying the above reduction rule to T_Y^C for every $C \in G \setminus X$ and $Y \subseteq X$. Suppose that there exists a burning sequence S of size k for G. If S burns a vertex in some T_Y^C that has been deleted by the reduction rule then we can burn any one of the two non-deleted vertices from the same T_Y^C without increasing the burning number. This is possible because all vertices in T_Y^C have same neighbors and the burning number of a clique is at most two. ∎

After applying the Reduction Rule 1 on graph G, each clique C in $G_1 \setminus X$ has at most 2^{d+1} vertices (two in each T_Y^C). For each clique $C \in G_1 \setminus X$ define the type vector of C as

$$T^C = (|T_{Y_1}^C|, |T_{Y_2}^C|, \cdots, |T_{Y_{2^d}}^C|)$$

Here $Y_1, Y_2, \cdots, Y_{2^d}$ are ordered subsets of X. Now we partition the cliques in $G_1 \setminus X$ based on their type vector, i.e, two cliques C_1 and C_2 are in the same partition if $T^{C_1} = T^{C_2}$. This way we can partition the cliques in $G_1 \setminus X$ into at most 3^{2^d} many sets.

For a subset $B \subseteq \{0, 1, 2\}^{2^d}$ define

$$T(B) = \{C \mid T^C = B \text{ and } C \in G_1 \setminus X\}$$

Reduction Rule 2. *Let $B \subseteq \{0, 1, 2\}^{2^d}$. If the set $T(B)$ contains more than $3d + 3$ cliques then remove all except $3d + 3$.*

Lemma 5. *Reduction Rule 2 is safe.*

Proof. Let $G_2 = (V_2, E_2)$ be the graph obtained by iteratively applying the Reduction Rule 2 to $T(B)$ for every $B \subseteq \{0, 1, 2\}^{2^d}$. Suppose that there exists a burning sequence S of size k for G_1. If S burns a vertex in a clique in some $T(B)$ that has been deleted by the reduction rule then we can burn any vertex from a non-deleted clique from the same $T(B)$ without increasing the burning number. This is possible because of the fact that burning number of G_1 is at most $3d + 3$ and all cliques in $T(B)$ has same neighborhood in X. ∎

We remark that above reduction rules are trivially applicable in polynomial time. If the Reduction Rules 1 and 2 are not applicable to the input graph, then the size of the reduced instance is at most $d + 2(3^{2^d}(3d + 3)) = O(d3^{2^d})$. Hence, we get a kernel of size at most $O(d3^{2^d})$. ☐

3.2   Parameterization by Neighborhood Diversity

The graph parameter neighborhood diversity was introduced by Lampis [13]. This parameter is also a generalization of vertex cover in different direction (see Fig. 1). In this subsection we give a polynomial kernel for GRAPH BURNING parameterized by the neighborhood diversity.

Definition 1. *In a graph G, two vertices u and v have the same type if and only if $N(u) \setminus \{v\} = N(v) \setminus \{u\}$.*

Definition 2 (Neighborhood diversity [13]). *A graph G has neighborhood diversity d if there exists a partition of $V(G)$ into d sets $T_1, T_2, \ldots, T_d$ such that all the vertices in each set have the same type. Such a partition is called a type partition. Moreover, it can be computed in linear time [13].*

Note that all the vertices in T_i for every $i \in [d]$ have the same neighborhood in G. Moreover, each T_i either forms a clique or an independent set in G.

Theorem 3. *If neighborhood diversity of the graph is at most d, then there is a $3d$-kernel for the GRAPH BURNING problem.*

Proof. We describe the following simple reduction rules:

Reduction Rule 3. *If the partition T_i is a clique in the type partitioning of the graph, then we can remove all vertices except any two vertices of the partition T_i.*

Reduction Rule 4. *If the partition T_i is an independent set in the type partitioning of the graph, then we can remove all vertices except any three vertices of the partition T_i.*

Lemma 6. *Reduction Rules 3 and 4 are safe.*

Proof. As any two vertices in a partition have same neighborhood, burning any vertex within a partition has the same effect.

If a partition T_i is a clique, then we need to burn at most two vertices of the partition T_i which in turn burn all the vertices in the partition T_i. Hence, we can safely remove all the vertices except any two vertices of the partition T_i if it is a clique.

If a partition T_i is an independent set, then we need to burn at most three vertices of the partition T_i which in turn burn all the vertices in the partition T_i. As the graph is connected, the partition T_i which forms an independent set has a neighboring partition, If we burn a vertex in the partition T_i, which in turn burn all the vertices in the neighboring partition in the next iteration. The burned vertices in the neighboring partition in turn burn all the unburned vertices in T_i. So we need to burn at most three vertices in T_i. Hence we can safely remove all the vertices except any three vertices of the partition T_i if it is an independent set. ∎

After applying Reduction Rules 3 and 4 each partition have at most three vertices. As we have d partitions, we have a $3d$-kernel. □

4   Graph Classes

In this section, we study the complexity of the GRAPH BURNING problem on cographs, split graphs, and interval graphs.

For a fixed d, GRAPH BURNING can be solved in time $O(n^d)$ on P_d-free graphs [2]. As split graphs are P_5-free and cographs are P_4-free, using the above fact we can see that GRAPH BURNING can be solved in $O(n^5)$ and $O(n^4)$ respectively. In the next subsections, we show that the problem can be solved in linear time on these graph classes.

4.1   Cographs

We use the notion of modular decomposition [12] to solve the GRAPH BURNING problem on cographs. First, we define *modular decomposition*, which was introduced by Gallai [10]. A set $M \subseteq V(G)$ is called a *module* of G if all vertices of M have the same set of neighbors in $V(G) \setminus M$. The *trivial modules* are $V(G)$, and $\{v\}$ for all v. A *prime* graph is a graph in which all modules are trivial. The *modular decomposition* of a graph G is a rooted tree M_G that has the following properties:

1. The leaves of M_G are the vertices of G.
2. For an internal node h of M_G, let $M(h)$ be the set of vertices of G that are leaves of the subtree of M_G rooted at h. ($M(h)$ forms a module in G).
3. For each internal node h of M_G there is a graph G_h (*representative graph*) with $V(G_h) = \{h_1, h_2, \cdots, h_r\}$, where $h_1, h_2, \cdots, h_r$ are the children of h in M_G and for $1 \leq i < j \leq r$, h_i and h_j are adjacent in G_h if and only if there are vertices $u \in M(h_i)$ and $v \in M(h_j)$ that are adjacent in G.
4. G_h is either a clique, an independent set, or a prime graph and h is labeled *series* if G_h is a clique, *parallel* if G_h is an independent set, and *prime* otherwise.

A graph is a cograph if and only if its modular decomposition tree contains only parallel and series nodes [4]. The modular decomposition of a graph can be computed in linear time [5,16].

Theorem 4. GRAPH BURNING *can be solved in linear time on cographs.*

Proof. Let G be a connected cograph whose modular decomposition tree is M_G. Since G is connected the root r of tree M_G is a series node. Let the children of r be x and y and the cographs induced by leaves of the subtree rooted at x and y be G_x and G_y respectively.

If either G_x or G_y has at most two vertices then the maximum degree of the graph G is either $n-1$ or $n-2$. Hence from Theorem 1, we have $b(G) = 2$. If both G_x and G_y has at least three vertices and maximum degree of G is at most $n-3$ then $b(G) = 3$: At $t = 1, 2, 3$ burn any arbitrary three vertices u, v, w of G_x. Since every vertex of G_y is adjacent to vertex x, all the vertices of G_y are burned at $t = 2$. Similarly every vertex of G_x is adjacent to every vertex of G_y all vertices of G_x are burned at $t = 3$. Clearly the running time of the algorithm is $O(n + m)$.   □

4.2 Split Graphs

Theorem 5. GRAPH BURNING *can be solved in linear time on split graphs.*

Proof. If G has maximum degree $n - 1$ or $n - 2$ then from Theorem 1, $b(G) = 2$. Otherwise, we can burn the graph G in three steps as follows. At $t = 1$ burn one of the arbitrary clique vertex and at $t = 2, 3$ burn any arbitrary non-burned vertices of G. Clearly we can see that the running time of the above algorithm is $O(n + m)$. □

4.3 Interval Graphs

In this section, we give upper and lower bounds for the GRAPH BURNING problem on interval graphs in terms of the diameter of the graph. Let G be an interval graph and $\mathcal{I}$ be an *interval representation* of G, i.e., there is a mapping from $V(G)$ to closed intervals on the real line such that for any two vertices u and v, $uv \in E(G)$ if and only if $I_u \cap I_v \neq \emptyset$. For any interval graph, there exists an interval representation with all endpoints distinct. Such a representation is called a *distinguishing interval representation* and it can be computed starting from an arbitrary interval representation of the graph. Interval graphs can be recognized in linear time and an interval representation can be obtained in linear time [3]. Let $l(I_u)$ and $r(I_u)$ denote the left and right end points of the interval corresponding to the vertex u respectively. We call an interval $I \in \mathcal{I}$ as *rightmost interval* (resp. *leftmost interval*) if $r(J) \leq r(I)$ (resp. $l(I) \leq l(J)$) for all $J \in \mathcal{I}$.

Lemma 7. *Let G be an interval graph with diameter d and P be a path of length d in G. Then every vertex of G is either part of P or adjacent to at least one vertex of P.*

Proof. We prove the Lemma by proof by contradiction. Suppose $S \subseteq V(G) \setminus V(P)$ be the set of vertices which do not have any neighbor in P. Let $y \in S$ with the property that $r(I_y) > r(I_z)$ for all $z \in S$, that is I_y is the rightmost interval in the interval representation of vertices in S. Let I_u and I_v be the leftmost and rightmost intervals of interval representation of interval graph induced by vertices of P respectively.

If $l(I_y) < l(I_u)$, then as y is not adjacent to u we have $r(I_y) < l(I_u)$. Since I_y is the rightmost interval among the interval representation of vertices in S, there is no path from y to u in the interval graph G. This is a contradiction to the fact that G is connected. Similarly we can obtain a contradiction for the case $r(I_y) < r(I_v)$.

If $r(I_u) < l(I_y) < r(I_y) < l(I_v)$, as y is not adjacent to any vertex of P. Then there is no path from u to v in G, which is again contradiction to the fact that there is a path of length d from u to v. □

Lemma 8. *If G be an interval graph with diameter d then*

$$\lceil \sqrt{d+1} \rceil \leq b(G) \leq \lceil \sqrt{d+1} \rceil + 1$$

Proof. The first inequality follows from the fact that for a path P_n [15] we have that $b(G) = \lceil \sqrt{n} \rceil$. The second inequality follows from Lemma 7 as every vertex is either a part of diameter path or adjacent to at least one vertex of diameter path. We burn the diameter path in $\lceil \sqrt{d+1} \rceil$ steps and in the next step all the rest of unburned vertices (if any) are burned.   □

5   Conclusion

In this paper, we have studied the parameterized complexity of the GRAPH BURNING problem. We have shown that the problem is FPT parameterized by (a) distance to cluster (b) neighborhood diversity. We also studied the complexity of the problem on special classes of graphs. We have shown that the problem can be solved in polynomial time on cographs and split graphs.

The following problems remain open.

- What is the parameterized complexity of GRAPH BURNING problem when parameterized by the natural parameter (burning number)?
- While the GRAPH BURNING problem solved in polynomial time on cographs and split graphs, we do not know if the problem is FPT parameterized by distance to cographs or distance to split graphs.
- It is not known if GRAPH BURNING admit a polynomial kernel parameterized by vertex cover number?
- Finally, we do not know if GRAPH BURNING can be solved in polynomial time on interval graphs or permutation graphs.

References

1. Bessy, S., Bonato, A., Janssen, J., Rautenbach, D., Roshanbin, E.: Burning a graph is hard. Discrete Appl. Math. **232**, 73–87 (2017)
2. Bonato, A., Janssen, J., Roshanbin, E.: How to burn a graph. Internet Math. **12**(1–2), 85–100 (2016)
3. Booth, K.S., Lueker, G.S.: Testing for the consecutive ones property, interval graphs, and graph planarity using PQ-tree algorithms. J. Comput. Syst. Sci. **13**(3), 335–379 (1976)
4. Corneil, D.G., Lerchs, H., Burlingham, L.S.: Complement reducible graphs. Discrete Appl. Math. **3**(3), 163–174 (1981)
5. Cournier, A., Habib, M.: A new linear algorithm for modular decomposition. In: Tison, S. (ed.) CAAP 1994. LNCS, vol. 787, pp. 68–84. Springer, Heidelberg (1994). https://doi.org/10.1007/BFb0017474
6. Cygan, M., et al.: Parameterized Algorithms. Springer, Cham (2015). https://doi.org/10.1007/978-3-319-21275-3
7. Diestel, R.: Graph Theory. Graduate Texts in Mathematics. Springer, Heidelberg (2005)
8. Doucha, M., Kratochvíl, J.: Cluster vertex deletion: a parameterization between vertex cover and clique-width. In: Rovan, B., Sassone, V., Widmayer, P. (eds.) MFCS 2012. LNCS, vol. 7464, pp. 348–359. Springer, Heidelberg (2012). https://doi.org/10.1007/978-3-642-32589-2_32

9. Downey, R.G., Fellows, M.R.: Fundamentals of Parameterized Complexity, vol. 4. Springer, London (2013). https://doi.org/10.1007/978-1-4471-5559-1
10. Gallai, T.: Transitiv orientierbare graphen. Acta Math. Hung. 18(1), 25–66 (1967)
11. Guo, J., Hüffner, F., Niedermeier, R.: A structural view on parameterizing problems: distance from triviality. In: Downey, R., Fellows, M., Dehne, F. (eds.) IWPEC 2004. LNCS, vol. 3162, pp. 162–173. Springer, Heidelberg (2004). https://doi.org/10.1007/978-3-540-28639-4_15
12. Habib, M., Paul, C.: A survey of the algorithmic aspects of modular decomposition. Comput. Sci. Rev. 4(1), 41–59 (2010)
13. Lampis, M.: Algorithmic meta-theorems for restrictions of treewidth. Algorithmica 64(1), 19–37 (2012)
14. Land, M.R., Lu, L.: An upper bound on the burning number of graphs. In: Bonato, A., Graham, F.C., Prałat, P. (eds.) WAW 2016. LNCS, vol. 10088, pp. 1–8. Springer, Cham (2016). https://doi.org/10.1007/978-3-319-49787-7_1
15. Roshanbin, E.: Burning a graph as a model of social contagion. Ph.D. thesis, PhD thesis, Dalhousie University (2016)
16. Tedder, M., Corneil, D., Habib, M., Paul, C.: Simpler linear-time modular decomposition via recursive factorizing permutations. In: Aceto, L., Damgård, I., Goldberg, L.A., Halldórsson, M.M., Ingólfsdóttir, A., Walukiewicz, I. (eds.) ICALP 2008. LNCS, vol. 5125, pp. 634–645. Springer, Heidelberg (2008). https://doi.org/10.1007/978-3-540-70575-8_52

Extension and Its Price
for the Connected Vertex Cover Problem

Mehdi Khosravian Ghadikoalei[1], Nikolaos Melissinos[2], Jérôme Monnot[1],
and Aris Pagourtzis[2(✉)]

[1] Université Paris-Dauphine, PSL University, CNRS, LAMSADE,
75016 Paris, France
{mehdi.khosravian-ghadikolaei,jerome.monnot}@dauphine.fr
[2] School of Electrical and Computer Engineering,
National Technical University of Athens, Polytechnioupoli,
15780 Zografou, Athens, Greece
nikolaosmelissinos@mail.ntua.gr, pagour@cs.ntua.gr

Abstract. We consider *extension* variants of VERTEX COVER and INDE-
PENDENT SET, following a line of research initiated in [9]. In particular,
we study the EXT-CVC and the EXT-NSIS problems: given a graph
$G = (V, E)$ and a vertex set $U \subseteq V$, does there exist a *minimal* connected
vertex cover (respectively, a *maximal* non-separating independent set) S,
such that $U \subseteq S$ (respectively, $U \supseteq S$). We present hardness results for
both problems, for certain graph classes such as bipartite, chordal and
weakly chordal. To this end we exploit the relation of EXT-CVC to
EXT-VC, that is, to the extension variant of VERTEX COVER. We also
study the *Price of Extension (PoE)*, a measure that reflects the distance
of a vertex set U to its maximum efficiently computable subset that is
extensible to a minimal connected vertex cover, and provide negative
and positive results for PoE in general and special graphs.

Keywords: Extension problems · Connected vertex cover ·
Upper connected vertex cover · Price of extension ·
Special graph classes · Approximation algorithms · NP-completeness

1 Introduction

We consider the *extension variant* of the (MINIMUM) CONNECTED VERTEX
COVER (MIN CVC) problem and its associated *price of extension (PoE)*; we
call this variant EXTENSION CONNECTED VERTEX COVER problem (EXT-CVC
for short). Intuitively, the extension variant of a minimization problem Π is the
problem of deciding whether a partial solution U for a given instance of Π can
be extended to a minimal (w.r.t. inclusion) feasible solution for that instance;
PoE refers to the maximum size subset of U that can be extended to a minimal
feasible solution. A framework for extension problems is developed in [10] where
a number of results are given for several hereditary and antihereditary graph
problems. Particular complexity results for the extension of graph problems,

© Springer Nature Switzerland AG 2019
C. J. Colbourn et al. (Eds.): IWOCA 2019, LNCS 11638, pp. 315–326, 2019.
https://doi.org/10.1007/978-3-030-25005-8_26

such as VERTEX COVER, HITTING SET, and DOMINATING SET, are given in
[2,3,6,9,20–22]. A subset $S \subseteq V$ of a connected graph $G = (V, E)$ is a *connected
vertex cover* (*CVC* for short) if S is a *vertex cover* (i.e., each edge of G is incident
to at least a vertex of S) and the subgraph $G[S]$ induced by S is connected. The
corresponding optimization problem (MINIMUM) CONNECTED VERTEX COVER
(MIN CVC) consists in finding a CVC of minimum size [12,16,17]. Given a
(connected) vertex cover S of a graph $G = (V, E)$, an edge $e \in E$ is *private* to a
vertex $v \in S$ if v is the only vertex of S incident to e. Hence, a vertex cover S of
G is *minimal* iff each vertex $v \in S$ has a private edge. A CVC S of G is *minimal*
if for every $v \in S$, $S \setminus \{v\}$ is either not connected or not a vertex cover.

In this paper we study EXTENSION CONNECTED VERTEX COVER (EXT-
CVC): given a connected graph $G = (V, E)$ together with a subset $U \subseteq V$ of
vertices, the goal is to decide whether there exists a minimal (w.r.t. inclusion)
CVC of G containing U; note that for several instances the answer is negative.
In this latter case we are also interested in a new maximization problem where
the goal is to find the largest subset of vertices $U' \subseteq U$ which can be extended
to a minimal feasible solution. This concept is defined as the *Price of Extension
(PoE)* in [9]. For the two extreme cases $U = \emptyset$ and $U = V$, we note that for the
former the question is trivial since there always exists a minimal CVC [27], while
for the latter $(U = V)$ the problem is equivalent to finding a minimal CVC of
maximum size, (called *Upper CVC* in the paper).

1.1   Graph Definitions and Terminology

Throughout this article, we consider a simple connected undirected graph with-
out loops $G = (V, E)$ on $n = |V|$ vertices and $m = |E|$ edges. Every edge $e \in E$
is denoted as $e = uv$ for $u, v \in V$. For $X \subseteq V$, $N_G(X) = \{v \in V : vx \in E,$ for
some $x \in X\}$ and $N_G[X] = X \cup N_G(X)$ denotes *the closed neighborhood* of X.
For singleton sets $X = \{x\}$, we simply write $N_G(x)$ or $N_G[x]$, even omitting G
if it is clear from the context; for a subset $X \subset V$, $N_X(v) = N_G(v) \cap X$. The
number of neighbors of x, called *degree* of x, is denoted by $d_G(x) = |N_G(x)|$ and
the *maximum degree* of the graph G is denoted by $\Delta(G) = \max_{v \in V} d_G(v)$. A
leaf is a vertex of degree one, and V_l denotes the set of leaves in G. For $X \subseteq V$,
$G[X]$ denotes the *subgraph induced by* X, that is the subgraph only containing
X as vertices and all edges of G with both endpoints in X. A connected graph
$G = (V, E)$ is *biconnected*, if for each pair of vertices x, y there is a simple cycle
containing both x and y, or equivalently, the removal of any vertex maintains
connectivity. A *cut-set* $X \subset V$ is a subset of vertices such that the deletion
of X from G strictly increases the number of connected components. A cut-set
which is a singleton is called a *cut-vertex* and a cut-set X is *minimal* if $\forall x \in X$,
$X \setminus \{x\}$ is not a cut-set. Hence, a graph is biconnected iff it is connected and
it does not contain any cut-vertex. In this article, $V_c(G)$ denotes the *set of cut-
vertices* of a graph G; we will simply write V_c if G is clear from the context.
A graph $G = (L \cup R, E)$ is *split* (resp. *bipartite*) where the vertex set $L \cup R$ is
decomposed into a clique L and an independent set R (resp. two independent
sets). A graph is *chordal* if all its cycles of length at least four have a *chord*,

that is, an edge connecting nonconsecutive vertices of the cycle. There are many characterizations of chordal graphs. One of them, known as Dirac's theorem, asserts that a graph G is chordal iff each minimal cut-set of G is a clique. Recall that $S \subseteq V$ is a vertex cover, if for each $e = uv \in E$, $S \cap \{u, v\} \neq \emptyset$ while $S \subseteq V$ is an independent set if for each pair of vertices u, v of S, $uv \notin E$; S is a vertex cover iff $V \setminus S$ is an independent set of $G = (V, E)$. The minimum vertex cover problem (MIN VC for short) asks to find a vertex cover of minimum size and the maximum independent set problem (MAX IS for short), asks to find an independent set of maximum size for a given graph.

1.2 Problem Definitions

As mentioned above, we consider the extension variants of two optimization problems: the (MINIMUM) CONNECTED VERTEX COVER problem (MIN CVC) and the (MAXIMUM) NON SEPARATING INDEPENDENT SET problem (MAX NSIS). A non separating independent set S of a connected graph $G = (V, E)$ is a subset of vertices of G which is *independent* (i.e., any two vertices in S are non adjacent) and S is not a cut-set of G. MAX NSIS asks to find a non separating independent set of maximum size. MIN CVC and MAX NSIS have been studied in [12, 15, 16, 26, 30] where it is proved that the problems are polynomially solvable in graphs of maximum degree 3, while in graphs of maximum degree 4 they are NP-hard.

EXT-CVC

Input: A connected graph $G = (V, E)$ and a *presolution* (also called set of *forced* vertices) $U \subseteq V$.

Question: Does G have a minimal connnected vertex cover S with $U \subseteq S$?

Dealing with EXT-NSIS, the goal to decide the existence of a maximal NSIS excluding vertices from $V \setminus U$.

EXT-NSIS

Input: A connected graph $G = (V, E)$ and a *frontier* subset $U \subseteq V$.

Question: Does G have a maximal NSIS S with $S \subseteq U$?

Considering the possibility that some set U might not be extensible to any minimal solution, one might ask how far is U from an extensible set. This concept, introduced in [9], is called *Price of Extension (PoE)*. This notion is defined in an attempt to understand what effect the additional presolution constraint has on the possibility of finding minimal solutions. A similar approach has already been used in the past under the name *the Price of Connectivity* in [7] for the context of connectivity because it is a crucial issue in networking applications. This notion has been introduced in [7] for MIN VC and is defined as the maximum ratio between the connected vertex cover number and the vertex cover number. In our context, the goal of PoE is to quantify how close efficiently computable extensible subsets of the given presolution U are to U or to the largest possible extensible subsets of U. To formalize this, we define two optimization problems as follows:

MAX EXT-CVC
Input: A connected graph $G = (V, E)$ and a set of vertices $U \subseteq V$.
Feasible Solution: Minimal connected vertex cover S of G.
Goal: Maximize $|S \cap U|$.

MIN EXT-NSIS
Input: A connected graph $G = (V, E)$ and a set of vertices $U \subseteq V$.
Feasible Solution: Maximal non separating independent set S of G.
Goal: Minimize $|S \cup U|$.

For $\Pi \in \{$MAX EXT-CVC, MIN EXT-NSIS$\}$, we denote by $opt_\Pi(G, U)$ the value of an optimal solution. Since for both of them $opt_\Pi(G, U) = |U|$ iff (G, U) is a *yes*-instance of the extension variant, we deduce that MAX EXT-CVC and MIN EXT-NSIS are NP-hard since EXT-CVC and EXT-NSIS are NP-complete. Actually, for any class of graphs $\mathcal{G}$, MAX EXT-CVC is NP-hard in $\mathcal{G}$ iff MIN EXT-NSIS is NP-hard in $\mathcal{G}$ since for any graph $G \in \mathcal{G}$ it can be shown that:

$$opt_{\text{MAX EXT-CVC}}(G, U) + opt_{\text{MIN EXT-NSIS}}(G, V \setminus U) = |V|. \tag{1}$$

The price of extension PoE is defined exactly as the ratio of approximation, i.e., the best possible lower (resp. upper) bound on $\frac{apx}{opt}$ that can be achieved in polynomial time. In particular, we say that MAX EXT-CVC (resp. MIN EXT-NSIS) admits a polynomial ρ-PoE if for every instance (G, U), we can efficiently compute a solution S of G which satisfies $|S \cap U|/opt_{\text{MAX EXT-CVC}}(G, U) \geqslant \rho$ (resp., $|S \cup U|/opt_{\text{MIN EXT-NSIS}}(G, U) \leqslant \rho$).

Considering MAX EXT-CVC on $G = (V, E)$ in the particular case $U = V$, we obtain a new problem called UPPER CONNECTED VERTEX COVER (UPPER CVC) where the goal is to find the largest minimal CVC. To our best knowledge, this problem has never been studied, although UPPER VC has been extensively studied [5, 14, 25].

UPPER CVC
Input: A connected graph $G = (V, E)$.
Feasible Solution: Minimal connected vertex cover $S \subseteq V$.
Goal: Maximize $|S|$.

1.3   Related Work

Garey and Johnson proved that (minimum) CVC is NP-hard in planar graphs of maximum degree 4 [16]. Moreover, it is shown in [28, 30] that the problem is polynomially solvable for graphs of maximum degree 3, while NP-hardness proofs for bipartite and for bi-connected planar graphs of maximum degree 4, are presented in [12, 15, 26]. The approximability of MIN CVC has been considered in some more recent studies. The NP-hardness of approximating MIN CVC within $10\sqrt{5} - 21$ is proven in [15] while a 2-approximation algorithm is presented in [27]. Moreover, in [12] the problem is proven APX-complete in bipartite graphs of maximum degree 4. They also propose a $\frac{5}{3}$-approximation algorithm for MIN CVC for any class of graphs where MIN VC is polynomial-time solvable. Parameterized complexity for MIN CVC and MAX NSIS have been studied in [23, 24] while the enumeration of minimal connected vertex covers is investigated in [18] where it is shown that the number of minimal connected vertex covers of a graph of n vertices is at most 1.8668^n, and these sets can be enumerated in time $O(1.8668^n)$. For chordal graphs (even for chordality at most 5), the authors are able to give a better upper bound. The question to better understand the close relation between enumerations and extension problems is relevant because in this article we prove that EXT-CVC and MAX EXT-CVC are polynomial-time solvable in chordal graphs. Finally, one can find problems that are quite related to MIN CVC in [8].

Maximum minimal optimization variants have been studied for many classical graph problems in recent years, for example, in [5], Boria et al. have studied the MAXIMUM MINIMAL VERTEX COVER PROBLEM (UPPER VC in short) from the approximability and parameterized complexity point of views. The MINIMUM MAXIMAL INDEPENDENT SET problem, also called MINIMUM INDEPENDENT DOMINATING SET (MIN ISDS) asks, given a graph $G = (V, E)$, for a subset $S \subseteq V$ of minimum size that is simultaneously independent and dominating. From the NP-hardness and exact solvability point of views, MIN IDS is equivalent to UPPER VC [25], but they seem to behave differently in terms of approximability and parameterized complexity [1]. Although MIN IDS is polynomially solvable in strongly chordal graphs [13], it is hard to approximate within $n^{(1-\epsilon)}$, for any $\epsilon > 0$, in certain graph classes [11, 13]. Regarding parameterized complexity, Fernau [14] presents an FPT-algorithm for UPPER VC with running time $\mathcal{O}^*(2^k)$, where k is the size of an optimum solution, while it is proved that MIN IDS with respect to the standard parameter is $W[2]$-hard. Moreover, Boria et al. [5] provide a tight approximation result for UPPER VC in general graphs: they present a $n^{\frac{1}{2}}$ approximation algorithm together with a proof that UPPER VC is NP-hard to approximate within $n^{\frac{1}{2}-\epsilon}$, for any $\epsilon > 0$. Furthermore, they present a parameterized algorithm with running time (1.5397^k) where k is the standard parameter, by modifying the algorithm of [14]; they also show that weighted versions of UPPER VC and MIN IDS are in FPT with respect to the treewidth.

Regarding the extension variant of DOMINATING SET, namely EXT-DS, it is proven in [21, 22] that it is NP-complete, even in special graph classes like split graphs, chordal graphs, and line graphs. Moreover, a linear time algorithm for split graphs is given in [20] when X, Y is a partition of the clique part. In [9], it is proved that EXT-VC is NP-complete in cubic graphs and in planar graphs of maximum degree 3, while it is polynomially decidable in chordal and circular-arc graphs.

1.4 Summary of Results and Organization

The rest of the paper is organized as follows. In Sect. 2, after showing the relation between EXT-VC and EXT-CVC, we provide additional hardness results for

EXT-CVC in bipartite graphs and weakly triangulated graphs, the latter leading to hardness results for UPPER VC and UPPER CVC. We then focus on bounds for PoE in Sect. 3, providing inapproximability results for MAX EXT-CVC in general and bipartite graphs. In Sect. 4 we discuss the (in)approximability of a special case of MAX EXT-CVC, namely UPPER CVC. Note that all results given in the paper for EXT-CVC are valid for EXT-NSIS as well. Due to lack of space the proofs of statements marked with (*) are deferred to the full version of the paper.

2 Solvability and Hardness of Extension Problems

Let us begin by some simple observations: (G, U) with $G = (V, E)$ and $U \subseteq V$ is a *yes-instance* of EXT-CVC iff $(G, V \setminus U)$ is a *yes-instance* of EXT-NSIS. Hence, all complexity results given in this section for EXT-CVC are valid for EXT-NSIS as well. A leaf $(v \in V_l)$ never belongs to a minimal connected vertex cover S (apart from the extreme case where G consists of a single edge), while any cut-vertex $v \in V_c$ necessarily belongs to S. This implies that for trees, we have a simple characterization of *yes-instances* for $n \geqslant 3$: (T, U), where $T = (V, E)$ is a tree, is a *yes-instance* of EXT-CVC iff U is a subset of cut-set V_c, or equivalently $U \subseteq V_c = V \setminus V_l$. For an edge or a cycle $C_n = (V, E)$, (C_n, U) is a *yes-instance* iff $U \neq V$; since a path $P_n = (V, E)$ is a special tree the case of graphs of maximum degree 2 is settled. Dealing with split graphs, a similar but more complicated characterization can be given. In the next subsection we will deduce more general results for EXT-CVC by showing and exploiting relations to EXT-VC.

2.1 Relation Between Ext-VC and Ext-CVC

The following two properties allow to make use of known results for EXT-VC to obtain results for EXT-CVC.

Proposition 1. (*) EXT-CVC *is polynomially reducible to* EXT-VC *in chordal graphs.*

Proposition 2. EXT-CVC *is* NP-*complete in graphs of maximum degree* $\Delta + 1$ *if* EXT-VC *is* NP-*complete in graphs of maximum degree* Δ, *and this holds even for bipartite graphs.*

Proof. Given an instance (G, U) of EXT-VC, where $G = (V, E)$ with $V = \{v_1, \ldots, v_n\}$ and $U \subseteq V$, we build an instance $(G' = (V', E'), U')$ of EXT-CVC by adding a component $H = (V_H, E_H)$ to the original graph G. The construction of H is depicted to the right where $V_H = \{v_i', v_i'' : 1 \leqslant i \leqslant n\}$ is the vertex set. The new instance of EXT-CVC is given by (G', U') and consists of connecting the component H to G by linking $v_i v_i'$ for each $1 \leqslant i \leqslant n$ and by setting $U' = U$.

Clearly G' is of maximum degree $\Delta + 1$ if G is of maximum degree Δ. Moreover, it is not difficult to see that (G, U) is a yes-instance of EXT-VC iff (G', U') is a yes-instance of EXT-CVC. To maintain bipartiteness, we apply an appropriate subdivision of H. □

Using polynomial time decidability of EXT-VC in chordal graphs, parameterized complexity results (considering that the reduction increases the size of the instances only linearly), and NP-completeness in cubic bipartite graphs [9], we deduce:

Corollary 3. EXT-CVC *is polynomial-time decidable in chordal graphs and* NP-*complete in bipartite graphs of maximum degree 4.* EXT-CVC *parameterized with* $|U|$ *is* W[1]-*complete, and there is no* $2^{o(n+m)}$-*algorithm for n-vertex, m-edge bipartite graphs of maximum degree 4, unless ETH fails.*

2.2 Additional Hardness Results

We first strengthen the hardness result of Corollary 3 to bipartite graphs of maximum degree 3. This result could appear surprising since the optimization problem MIN CVC is polynomial-time solvable in graphs of maximum degree 3.

Theorem 4. EXT-CVC *is* NP-*complete in bipartite graphs of maximum degree 3 even if* U *is an independent set.*

Proof. We reduce from 2-BALANCED 3-SAT, denoted $(3, B2)$-SAT, where an instance $I = (C, X)$ is given by a set C of CNF clauses over a set of Boolean variables X such that each clause has exactly 3 literals and each variable appears exactly 4 times, twice negative and twice positive. Deciding whether an instance of $(3, B2)$-SAT is satisfiable is NP-complete by [4].

Consider an instance $(3, B2)$-SAT with clauses $C = \{c_1, \ldots, c_m\}$ and variables $X = \{x_1, \ldots, x_n\}$. We build a bipartite graph $G = (V, E)$ together with a set of forced vertices U as follows:

- For each clause $c = \ell_1 \vee \ell_2 \vee \ell_3$ where ℓ_1, ℓ_2, ℓ_3 are literals, introduce a subgraph $H(c) = (V_c, E_c)$ with 6 vertices and 6 edges. V_c contains three specified literal vertices $\ell_c^1, \ell_c^2, \ell_c^3$. The set of forced vertices in $H(c)$, denoted by U_c is given by $U_c = \{\ell_c^1, \ell_c^2, \ell_c^3\}$. The gadget $H(c)$ is illustrated in the left part of Fig. 1.
- For each variable x introduce 21 new vertices which induce the subgraph $H(x) = (V_x, E_x)$ illustrated in Fig. 1. The vertex set V_x contains four special vertices $t_x^{c_1}, t_x^{c_2}, f_x^{c_3}$ and $f_x^{c_4}$, where it is implicitly assumed (w.l.o.g.) that variable x appears positively in clauses c_1, c_2 and negatively in clauses c_3, c_4. The independent set $U_x = \{1_x, 3_x, 5_x, 6_x, 8_x, 10_x, 12_x\}$ is in U (i.e., forced to be in each feasible solution). The subgraph $H_x - U_x$ induced by $V_x \setminus U_x$ consists of an induced matching of size 5 and of 4 isolated vertices.
- We connect each gadget $H(x_i)$ to $H(x_{i+1})$ by linking vertex 12_{x_i} to vertex $6_{x_{i+1}}$ using an intermediate vertex $r_{i,i+1}$ for all $1 \leqslant i \leqslant n - 1$. We also add a pendant edge incident to each $r_{i,i+1}$ with leaf $r'_{i,i+1}$; an illustration of this connection is depicted on the right of Fig. 1.

- We interconnect $H(x)$ and $H(c)$ where x is a variable occurring in literal ℓ_i of clause c by adding edge $\ell_c^i t_x^c$ (resp., $\ell_c^i f_x^c$), where t_x^c (resp., f_x^c) is in $H(x)$ and ℓ_c^i is in $H(c)$, if x appears positively (resp., negatively) in clause c. These edges are called *crossing edges*.

Let $U = (\bigcup_{c \in C} U_c) \cup (\bigcup_{x \in X} U_x)$. This construction takes polynomial time and G is a bipartite graph of maximum degree 3.

Claim. (*) $I = (C, X)$ is satisfiable iff G admits a minimal CVC containing U.

The proof of the claim, deferred to the full version of the paper, completes the proof of the theorem.   □

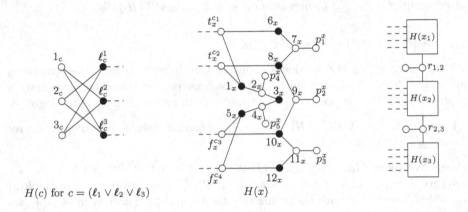

$H(c)$ for $c = (\ell_1 \vee \ell_2 \vee \ell_3)$ $H(x)$

Fig. 1. Clause gadget $H(c)$ and Variable gadget $H(x)$ for EXT-CVC are shown on the left and in the middle of the figure respectively. Forced vertices (in U) are marked in Black. On the right, the way of connecting variable gadgets is depicted. Crossing edges between $H(c)$ and $H(x)$ are marked with dashed lines.

Now, we will prove that the polynomial-time decidability of EXT-CVC in chordal graphs given in Corollary 3 cannot be extended to the slightly larger class of *weakly chordal* (also called *weakly triangulated*[1]) graphs which are contained in the class of *4-chordal* graphs. For any integer $k \geqslant 3$, a graph is called *k-chordal* if it has no induced cycle of length greater than k. Thus, chordal graphs are precisely the 3-chordal graphs. The problem of determining whether a graph is k-chordal is known to be co-NP-complete when k is a part of the instance [29].

Theorem 5. (∗) EXT-CVC *is* NP-*complete in weakly triangulated graphs.*

---

[1] This class is introduced in [19], as the class of graphs $G = (V, E)$ with no chordless cycle of five or more vertices in G or in its complement $\overline{G} = (V, \overline{E})$.

3   Bounds on the Price of Extension of Max Ext-CVC

Using Propositions 1 and 2, we can derive negative and positive approximation results for MAX EXT-CVC.

First, let us observe MIN EXT-NSIS does not admit $O(n^{1-\varepsilon})$-PoE even in the simplest case $U = \emptyset$ because there is a simple reduction from MIN ISDS (also known as *minimum maximal independent set* [11,13]) to MIN EXT-NSIS when $U = \emptyset$ by adding to the original graph $G = (V, E)$ two new vertices ℓ_0, ℓ_1 and edges $\ell_0\ell_1$ together with $\ell_1 v$ for $v \in V$ (so, ℓ_1 is an universal vertex); ℓ_1 never belongs to a NSIS (or equivalently ℓ_0 is a part of all maximal NSIS) because otherwise ℓ_0 will become isolated. For general graphs, the price of extension associated to MAX EXT-CVC is one of the hardest problems to approximate.

Theorem 6. (∗) *For any constant $\varepsilon > 0$ and any $\rho \in \Omega\left(\frac{1}{\Delta^{1-\varepsilon}}\right)$ and $\rho \in \Omega\left(\frac{1}{n^{1-\varepsilon}}\right)$, MAX EXT-CVC does not admit a polynomial ρ-PoE for general graphs of n vertices and maximum degree Δ, unless $\mathsf{P} = \mathsf{NP}$.*

Although Proposition 2 preserves bipartiteness, we cannot immediately conclude the same kind of results since in [9] it is proved that MAX EXT-VC admits a polynomial $\frac{1}{2}$-PoE for bipartite graphs.

Theorem 7. (∗) *For any constant $\varepsilon > 0$ and any $\rho \in \Omega\left(\frac{1}{n^{1/2-\varepsilon}}\right)$, MAX EXT-CVC does not admit a polynomial ρ-PoE for bipartite graphs of n vertices, unless $\mathsf{P} = \mathsf{NP}$.*

We next present a positive result, showing that the price of extension is equal to 1 in chordal graphs.

Proposition 8. (∗) *MAX EXT-CVC is polynomial-time solvable in chordal graphs.*

4   Approximability of UPPER CVC

UPPER CVC is a special case of MAX EXT-CVC where $U = V$. Regarding the approximability of UPPER CVC, we first show that an adaptation of Theorem 7 allows us to derive:

Corollary 9. (∗) *For any constant $\varepsilon > 0$, unless $\mathsf{NP} = \mathsf{P}$, UPPER CVC is not $\Omega\left(\frac{1}{n^{1/3-\varepsilon}}\right)$-approximable in polynomial time for bipartite graphs on n vertices.*

On the positive side, we show that any minimal CVC is a $\frac{2}{\Delta(G)}$ approximation for UPPER CVC. To do this, we first give a structural property that holds for any minimal connected vertex cover. For a given connected graph $G = (V, E)$ let $S^\star$ be an optimal solution of UPPER CVC and S be a minimal connected vertex cover of G. Denote by $A^\star = S^\star \setminus S$ and $A = S \setminus S^\star$ the *proper* parts of $S^\star$ and S respectively, while $B = S \cap S^\star$ is the *common* part. Finally, $R = V \setminus (S^\star \cup S)$ denotes the *rest* of vertices. Also, for $X = A^\star$ or $X = A$, we set $X_c = \{v \in X : N_G(v) \subseteq B\}$ which is exactly the vertices of X not having a neighbor in $(S \cup S^\star) \setminus X$. Actually, $(S \cup S^\star) \setminus X$ is either S or $S^\star$.

Lemma 10. (∗) *The following properties hold:*

(i) *For $X = A^\star$ or $X = A$, $X \cup R$ is an independent set of G, $G[X \cup B]$ is connected and X_c is a subset of cut-set of $G[X \cup B]$.*

(ii) *Set B is a dominating set of G.*

The following theorem describes an interesting graph theoretic property. It relates the size of an arbitrary minimal connected vertex cover of a (connected) graph to the size of the largest minimal connected vertex cover.

Theorem 11. *Any minimal CVC of a connected graph G is a $\dfrac{2}{\Delta(G)}$-approximation for UPPER CVC.*

Proof. Let $G = (V, E)$ be a connected graph. Let S and $S^\star$ be a minimal CVC and an optimal one for UPPER CVC, respectively, and w.l.o.g., assume $|S| < |S^\star|$. We prove the following inequalities:

$$|A^\star| \leqslant (\Delta(G) - 1)|B| \quad \text{and} \quad |A^\star| \leqslant (\Delta(G) - 1)|A| \qquad (2)$$

Let us prove the first part $|A^\star| \leqslant (\Delta(G) - 1)|B|$ of inequality (2). Consider $v_1 \in B$ maximizing its number of neighbors in $A^\star$, i.e. $v_1 = \arg\max\{|N_{A^\star}(v)|: v \in B\}$. Since S is a minimal CVC with $|S| < |S^\star|$, we have $\Delta(G) \geqslant |N_{A^\star}(v_1)| + 1$ from (i) of Lemma 10 (otherwise $B = \{v_1\}$ with $d_G(v_1) = \Delta(G)$). In addition, from (ii) of Lemma 10 we have $N_{A^\star}(B) = A^\star$ and then $\sum_{v \in B} |N_{A^\star}(v)| \geqslant |N_{A^\star}(B)| = |A^\star|$. Putting together these inequalities we get $|A^\star| \leqslant |B|(\Delta(G) - 1)$.

Let us prove the second part $|A^\star| \leqslant (\Delta(G) - 1)|A|$ of inequality (2) using the following Claim:

Claim. (∗) There are at least $|A_c^\star| + |A|$ edges between A and B in $G[S]$.

Each vertex in $A^\star \setminus A_c^\star$ has by definition at least one neighbor in A, so we deduce: $\sum_{v \in A} |N(v)| \geqslant |A^\star \setminus A_c^\star| + |A| + |A_c^\star| = |A| + |A^\star|$. Now, by setting $a_1 = \arg\max\{|N_G(v)|: v \in A\}$, we obviously get $|A||N(a_1)| \geqslant \sum_{v \in A} |N(v)|$. Putting together these inequalities, we obtain: $|A|\Delta(G) \geqslant |A||N(a_1)| \geqslant |A^\star| + |A|$ which leads to $|A^\star| \leqslant (\Delta(G) - 1)|A|$. The inequality $|S| \geqslant \dfrac{2}{\Delta(G)}$ follows by considering the two cases $|A| \geqslant |B|$ and $|A| < |B|$. $\qquad \square$

A tight example of Theorem 11 for any $\Delta(G) \geqslant 3$ is illustrated to the right. The optimal solution for UPPER CVC contains $\Delta(G)$ vertices $\{a\} \cup \{v_1, \ldots, v_{\Delta(G)-1}\}$ while $\{a, b\}$ is a minimal connected vertex cover of size 2.

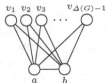

References

1. Ausiello, G., Crescenzi, P., Gambosi, G., Kann, V., Marchetti-Spaccamela, A., Protasi, M.: Complexity and Approximation: Combinatorial Optimization Problems and Their Approximability Properties. Springer, Heidelberg (2012). https://doi.org/10.1007/978-3-642-58412-1
2. Bazgan, C., Brankovic, L., Casel, K., Fernau, H.: On the complexity landscape of the domination chain. In: Govindarajan, S., Maheshwari, A. (eds.) CALDAM 2016. LNCS, vol. 9602, pp. 61–72. Springer, Cham (2016). https://doi.org/10.1007/978-3-319-29221-2_6
3. Bazgan, C., et al.: The many facets of upper domination. Theor. Comput. Sci. **717**, 2–25 (2018)
4. Berman, P., Karpinski, M., Scott, A.D.: Approximation hardness of short symmetric instances of MAX-3SAT. Electronic Colloquium on Computational Complexity (ECCC) (049) (2003)
5. Boria, N., Croce, F.D., Paschos, V.T.: On the max min vertex cover problem. Discrete Appl. Math. **196**, 62–71 (2015)
6. Boros, E., Gurvich, V., Hammer, P.L.: Dual subimplicants of positive Boolean functions. Optim. Methods Softw. **10**(2), 147–156 (1998)
7. Cardinal, J., Levy, E.: Connected vertex covers in dense graphs. Theor. Comput. Sci. **411**(26–28), 2581–2590 (2010)
8. Casel, K., et al.: Complexity of independency and cliquy trees. Discrete Appl. Math. (2019, in press)
9. Casel, K., Fernau, H., Ghadikolaei, M.K., Monnot, J., Sikora, F.: Extension of vertex cover and independent set in some classes of graphs and generalizations. In: CIAC 2019. LNCS (2018, accepted). See also CoRR, abs/1810.04629
10. Casel, K., Fernau, H., Ghadikolaei, M.K., Monnot, J., Sikora, F.: On the complexity of solution extension of optimization problems. CoRR, abs/1810.04553 (2018)
11. Damian-Iordache, M., Pemmaraju, S.V.: Hardness of approximating independent domination in circle graphs. ISAAC 1999. LNCS, vol. 1741, pp. 56–69. Springer, Heidelberg (1999). https://doi.org/10.1007/3-540-46632-0_7
12. Escoffier, B., Gourvès, L., Monnot, J.: Complexity and approximation results for the connected vertex cover problem in graphs and hypergraphs. J. Discrete Algorithms **8**(1), 36–49 (2010)
13. Farber, M.: Domination, independent domination, and duality in strongly chordal graphs. Discrete Appl. Math. **7**(2), 115–130 (1984)
14. Fernau, H.: Parameterized algorithmics: A graph-theoretic approach. Habilitationsschrift, Universität Tübingen, Germany (2005)
15. Fernau, H., Manlove, D.F.: Vertex and edge covers with clustering properties: complexity and algorithms. J. Discrete Algorithms **7**(2), 149–167 (2009)
16. Garey, M.R., Johnson, D.S.: The rectilinear steiner tree problem is NP-complete. SIAM J. Appl. Math. **32**(4), 826–834 (1977)
17. Garey, M.R., Johnson, D.S.: Computers and Intractability: A Guide to the Theory of NP-Completeness. W. H. Freeman & Co., New York (1979)
18. Golovach, P.A., Heggernes, P., Kratsch, D.: Enumeration and maximum number of minimal connected vertex covers in graphs. Eur. J. Comb. **68**, 132–147 (2018). Combinatorial Algorithms, Dedicated to the Memory of Mirka Miller
19. Hayward, R.B.: Weakly triangulated graphs. J. Comb. Theory Ser. B **39**(3), 200–208 (1985)

20. Kanté, M.M., Limouzy, V., Mary, A., Nourine, L.: On the enumeration of minimal dominating sets and related notions. SIAM J. Discrete Math. **28**(4), 1916–1929 (2014)

21. Kanté, M.M., Limouzy, V., Mary, A., Nourine, L., Uno, T.: Polynomial delay algorithm for listing minimal edge dominating sets in graphs. In: Dehne, F., Sack, J.-R., Stege, U. (eds.) WADS 2015. LNCS, vol. 9214, pp. 446–457. Springer, Cham (2015). https://doi.org/10.1007/978-3-319-21840-3_37

22. Kanté, M.M., Limouzy, V., Mary, A., Nourine, L., Uno, T.: A polynomial delay algorithm for enumerating minimal dominating sets in chordal graphs. In: Mayr, E.W. (ed.) WG 2015. LNCS, vol. 9224, pp. 138–153. Springer, Heidelberg (2016). https://doi.org/10.1007/978-3-662-53174-7_11

23. Kowalik, L., Mucha, M.: A 9k kernel for nonseparating independent set in planar graphs. Theor. Comput. Sci. **516**, 86–95 (2014)

24. Krithika, R., Majumdar, D., Raman, V.: Revisiting connected vertex cover: FPT algorithms and lossy kernels. Theory Comput. Syst. **62**(8), 1690–1714 (2018)

25. Manlove, D.F.: On the algorithmic complexity of twelve covering and independence parameters of graphs. Discrete Appl. Math. **91**(1–3), 155–175 (1999)

26. Priyadarsini, P.K., Hemalatha, T.: Connected vertex cover in 2-connected planar graph with maximum degree 4 is NP-complete. Int. J. Math. Phys. Eng. Sci. **2**(1), 51–54 (2008)

27. Savage, C.: Depth-first search and the vertex cover problem. Inf. Process. Lett. **14**(5), 233–235 (1982)

28. Speckenmeyer, E.: On feedback vertex sets and nonseparating independent sets in cubic graphs. J. Graph Theory **12**(3), 405–412 (1988)

29. Uehara, R.: Tractable and intractable problems on generalized chordal graphs. Technical report, Technical Report COMP98-83, IEICE (1999)

30. Ueno, S., Kajitani, Y., Gotoh, S.: On the nonseparating independent set problem and feedback set problem for graphs with no vertex degree exceeding three. Discrete Math. **72**(1–3), 355–360 (1988)

An Improved Fixed-Parameter Algorithm for Max-Cut Parameterized by Crossing Number

Yasuaki Kobayashi[1]($\boxtimes$), Yusuke Kobayashi[2], Shuichi Miyazaki[3], and Suguru Tamaki[4]

[1] Graduate School of Informatics, Kyoto University, Kyoto, Japan
kobayashi@iip.ist.i.kyoto-u.ac.jp
[2] Research Institute for Mathematical Sciences, Kyoto University, Kyoto, Japan
[3] Academic Center for Computing and Media Studies, Kyoto University, Kyoto, Japan
[4] School of Social Information Science, University of Hyogo, Kobe, Japan

Abstract. The Max-Cut problem is known to be NP-hard on general graphs, while it can be solved in polynomial time on planar graphs. In this paper, we present a fixed-parameter tractable algorithm for the problem on "almost" planar graphs: Given an n-vertex graph and its drawing with k crossings, our algorithm runs in time $O(2^k(n+k)^{3/2}\log(n+k))$. Previously, Dahn, Kriege and Mutzel (IWOCA 2018) obtained an algorithm that, given an n-vertex graph and its 1-planar drawing with k crossings, runs in time $O(3^k n^{3/2}\log n)$. Our result simultaneously improves the running time and removes the 1-planarity restriction.

Keywords: Crossing number · Fixed-parameter tractability · Max-Cut

1 Introduction

The Max-Cut problem is one of the most basic graph problems in theoretical computer science. In this problem, we are given an edge-weighted graph, and asked to partition the vertex set into two parts so that the total weight of edges having endpoints in different parts is maximized. This is one of the 21 problems shown to be NP-hard by Karp's seminal work [12]. To overcome this intractability, numerous researches have been done from the viewpoints of approximation algorithms [8,11,13,23], exponential-time exact algorithms [7,24], and fixed-parameter algorithms [3,16,17,21]. There are several graph classes for which the Max-Cut problem admits polynomial time algorithms [1,9]. Among others, one of the most remarkable tractable classes is the class of planar graphs.

This work is partially supported by JSPS KAKENHI Grant Numbers JP16H02782, JP16K00017, JP16K16010, JP17H01788, JP18H04090, JP18H05291, JP18K11164, and JST CREST JPMJCR1401.

Orlova and Dorfman [18] and Hadlock [10] developed polynomial time algorithms for the unweighted Max-Cut problem on planar graphs, which are subsequently extended to the weighted case by Shih et al. [22] and Liers and Pardella [14].

Dahn et al. [4] recently presented a fixed-parameter algorithm for 1-planar graphs. A graph is called *1-planar* if it can be embedded into the plane so that each edge crosses at most once. Their algorithm runs in time $O(3^k n^{3/2} \log n)$, where n is the number of vertices and k is the number of crossings of a given 1-planar drawing. Their algorithm is a typical branching algorithm: at each branch, it removes a crossing by yielding three sub-instances. After removing all of the k crossings, we have at most 3^k Max-Cut instances on planar graphs. Each of these problems can be solved optimally in $O(n^{3/2} \log n)$ time by reducing to the maximum weight perfect matching problem with small separators [15], thus giving the above mentioned time complexity.

Our Contributions. To the best of the authors' knowledge, it is not known whether the Max-Cut problem on 1-planar graphs is solvable in polynomial time. In this paper, we show that it is NP-hard even for unweighted graphs.

Theorem 1. *The Max-Cut problem on unweighted 1-planar graphs is NP-hard even when a 1-planar drawing is given as input.*

Next, we give an improved fixed-parameter algorithm, which is the main contribution of this paper:

Theorem 2. *Given a graph G and its drawing with k crossings, the Max-Cut problem can be solved in $O(2^k (n + k)^{3/2} \log(n + k))$ time.*

Note that our algorithm not only improves the running time of Dahn et al.'s algorithm [4], but also removes the 1-planarity restriction. An overview of our algorithm is as follows: Using a polynomial-time reduction in the proof of Theorem 1, we first reduce the Max-Cut problem on general graphs (with a given drawing in the plane) to that on 1-planar graphs with a 1-planar drawing, without changing the number of crossings. We then give a faster fixed-parameter algorithm than Dahn et al. [4]'s for Max-Cut on 1-planar graphs.

The main idea for improving the running time is as follows: Similarly to Dahn et al. [4], we use a branching algorithm, but it yields not three but only two sub-instances. Main drawbacks of this advantage are that these two sub-instances are not necessarily on planar graphs, and not necessarily ordinary Max-Cut instances but with some condition, which we call the "constrained Max-Cut" problem. To solve this problem, we modify the reduction of [14] and reduce the constrained Max-Cut problem to the maximum weight b-factor problem, which is known to be solvable in polynomial time in general [5]. We investigate the time complexity of the algorithm in [5], and show that it runs in $O((n + k)^{3/2} \log(n + k))$-time in our case, which proves the running time claimed in Theorem 2.

Independent Work. Chimani et al. [2] independently and simultaneously achieved the same improvement by giving an $O(2^k (n + k)^{3/2} \log(n + k))$ time algorithm for the Max-Cut problem on embedded graphs with k crossings. They used a different branching strategy, which yields 2^k instances of the Max-Cut problem on planar graphs.

Related Work. The Max-Cut problem is one of the best studied problems in several areas of theoretical computer science. This problem is known to be NP-hard even for co-bipartite graphs [1], comparability graphs [20], cubic graphs [25], and split graphs [1]. From the approximation point of view, the best known approximation factor is 0.878 [8], which is tight under the Unique Games Conjecture [13]. In the parameterized complexity setting, there are several possible parameterizations for the Max-Cut problem. Let k be a parameter and let G be an unweighted graph with n vertices and m edges. The problems of deciding if G has a cut of size at least k [21], at least $m - k$ [19], at least $m/2 + k$ [17], $m/2+(n-1)/4+k$ [3], or at least $n-1+k$ [16] are all fixed-parameter tractable. For sparse graphs, there are efficient algorithms for the Max-Cut problem. It is well known that the Max-Cut problem can be solved in $O(2^t tn)$ time [1], where t is the treewidth of the input graph. The Max-Cut problem can be solved in polynomial time on planar graphs [10,14,18,22], which has been extended to bounded genus graphs by Galluccio et al. [6].

2   Preliminaries

In this paper, an edge $\{u, v\}$ is simply denoted by uv, and a cycle $\{v_0, v_1, \ldots, v_k\}$ with edges $\{v_i, v_{(i+1) \mod k}\}$ for $0 \le i \le k$ is denoted by $v_0v_1 \ldots v_k$.

A graph is *planar* if it can be drawn into the plane without any edge crossing. A *crossing* in the drawing is a non-empty intersection of edges distinct from their endpoints. If we fix a plane embedding of a planar graph G, then the edges of G separate the plane into connected regions, which we call *faces*.

Consider a drawing not necessarily being planar, where no three edges intersect at the same point. We say that a drawing is *1-planar* if every edge is involved in at most one crossing. A graph is *1-planar* if it admits a 1-planar drawing. Note that not all graphs are 1-planar: for example, the complete graph with seven vertices does not admit any 1-planar drawing.

Let $G = (V, E, w)$ be an edge weighted graph with $w : E \to \mathbb{R}$. A *cut* of G is a pair of vertex sets $(S, V \setminus S)$ with $S \subseteq V$. We denote by $E(S, V \setminus S)$ the set of edges having one endpoint in S and the other endpoint in $V \setminus S$. We call an edge in $E(S, V \setminus S)$ a *cut edge*. The *size* of a cut $(S, V \setminus S)$ is the sum of the weights of cut edges, i.e., $\sum_{e \in E(S, V \setminus S)} w(e)$. The Max-Cut problem asks to find a maximum size of a cut, denoted $mc(G)$, of an input graph G. We assume without loss of generality that G has no degree-one vertices since such a vertex can be trivially accommodated to either side of the bipartition so that its incident edge contributes to the solution. Therefore, we can first work with removing all degree-one vertices, and after obtaining a solution, we can put them back optimally.

An instance of the *constrained Max-Cut* problem consists of an edge weighted graph $G = (V, E, w)$, together with a set C of pairs of vertices of V. A feasible solution, called a *constrained cut*, is a cut in which all the pairs in C are separated.

The *size* of a constrained cut is defined similarly as that of a cut. The problem asks to find a maximum size of a constrained cut of G, denoted $\mathrm{cmc}(G)$.

Let $G = (V, E, w)$ be an edge weighted graph with $w : E \to \mathbb{R}$ and let $b : V \to \mathbb{N}$. A *b-factor* of G is a subgraph $H = (V, F)$ such that $F \subseteq E$ and every vertex v has degree exactly $b(v)$ in H. The *cost* of a b-factor $H = (V, F)$ is the sum of the weights of edges in F, i.e., $\sum_{e \in F} w(e)$. The maximum weighted b-factor problem asks to compute a maximum cost of a b-factor of G, denoted by $\mathrm{mb}(G)$. This problem can be seen as a generalization of the maximum weight perfect matching problem and is known to be solvable in polynomial time [5].

3   NP-Hardness on 1-Planar Graphs

In this section, we prove Theorem 1, i.e., NP-hardness of the unweighted Max-Cut problem on 1-planar graphs. The reduction is performed from the unweighted Max-Cut problem on general graphs. Since we use this reduction in the next section for weighted graphs, we exhibit the reduction for the weighted case. When considering unweighted case, we may simply let $w(e) = 1$ for all e.

Proof (of Theorem 1). Fix an edge weighted graph $G = (V, E, w)$. For an edge $e = uv$ of G, define a path P_e consisting of three edges uu', $u'v'$, and $v'v$, each of weight $w(e)$, where u' and v' are newly introduced vertices (see Fig. 1).

Fig. 1. Replacing an edge e with a path P_e.

Let G' be the graph obtained from G by replacing e by P_e. The following lemma is crucial for our reduction.

Lemma 1. $\mathrm{mc}(G') = \mathrm{mc}(G) + \max(0, 2w(e))$.

Proof. Suppose first that $w(e) \geq 0$. Consider a maximum size cut $(S, V \setminus S)$ of G. If u and v are in the same side of the partition, we extend the cut to obtain a cut of G' by putting u' and v' into the other side. Otherwise, we put u' to v's side and v' to u's side. In both cases, the cut size increases by exactly $2w(e)$, so $\mathrm{mc}(G') \geq \mathrm{mc}(G) + 2w(e)$. Conversely, let $(S', V' \setminus S')$ be a maximum size cut of G', where $V' = V \cup \{u', v'\}$. If u and v are in the same side, at least one of u' and v' must be in the other side, as otherwise, we can increase the size of the cut by moving u' or v' to the other side, contradicting the maximality of $(S', V' \setminus S')$. This implies that exactly two edges of the path P_e contribute to the cut $(S', V' \setminus S')$. Similarly, if u and v are in the different side, we can see that every edge of P_e contribute to $(S', V' \setminus S')$. Thus, the cut $(S' \setminus \{u', v'\}, (V' \setminus S') \setminus \{u', v'\})$ of G is of size $\mathrm{mc}(G') - 2w(e)$ and hence we have $\mathrm{mc}(G) \geq \mathrm{mc}(G') - 2w(e)$. From the above two inequalities, we have that $\mathrm{mc}(G') = \mathrm{mc}(G) + 2w(e)$.

Suppose otherwise that $w(e) < 0$. Similarly to the first case, we extend a maximum cut $(S, V \setminus S)$ of G to a cut of G'. This time, we can do so without changing the cut size, implying that $\mathrm{mc}(G') \geq \mathrm{mc}(G)$: If u and v are in the same side, we put both u' and v' into the same side as u and v. Otherwise, we put u' in u's side and v' in v's side. For the converse direction, let $(S', V' \setminus S')$ be a maximum size cut of G'. If u and v are in the same (resp. different) side of the cut, the maximality of the cut implies that no (resp. exactly one) edge of P_e contributes to the cut. Hence the cut $(S' \setminus \{u', v'\}, (V' \setminus S') \setminus \{u', v'\})$ of G has size $\mathrm{mc}(G')$, so that $\mathrm{mc}(G) \geq \mathrm{mc}(G')$. Therefore $\mathrm{mc}(G') = \mathrm{mc}(G)$, completing the proof. □

Now, suppose we are given a Max-Cut instance G with its arbitrary drawing. We will consider here a crossing as a pair of intersecting edges. We say that two crossings are *conflicting* if they share an edge, and the shared edge is called a *conflicting edge*. With this definition, a drawing is 1-planar if and only if it has no conflicting crossings.

Suppose that the drawing has two conflicting crossings $\{e, e'\}$ and $\{e, e''\}$ with the conflicting edge e (see Fig. 2). Replace e by a path P_e defined above and locally redraw the graph as in Fig. 2. Then, this conflict is eliminated, and by Lemma 1, the optimal value increases by exactly $2w(e)$. Note that this operation also works for eliminating two conflicting crossings caused by the same pair of edges.

Fig. 2. Eliminating a conflicting crossing.

We repeat this elimination process as long as the drawing has conflicting crossings, eventually obtaining a 1-planar graph. From a maximum cut of the obtained graph, we can obtain a maximum cut of the original graph G by simply replacing each path P_e by the original edge e. The reduction described above is obviously done in polynomial time. Since the Max-Cut problem on general graphs is NP-hard, this reduction implies Theorem 1 and hence the proof is completed. □

Note that the above reduction is in fact parameter-preserving in a strict sense, that is, the original drawing has k crossings if and only if the reduced 1-planar drawing has k crossings.

4   An Improved Algorithm

In this section, we prove Theorem 2. As mentioned in Sect. 1, we first reduce a general graph to a 1-planar graph using the polynomial-time reduction given in the previous section. Recall that this transformation does not increase the number of crossings. Hence, to prove Theorem 2, it suffices to provide an $O(2^k(n+k)^{3/2}n)$ time algorithm for 1-planar graphs and its 1-planar drawing with at most k crossings.

4.1   Algorithm

Our algorithm consists of the following three phases.

Preprocessing. We first apply the following preprocessing to a given graph G: For each crossing $\{ac, bd\}$, we apply the replacement in Fig. 1 twice, once to ac and once to bd (and take the cost change in Lemma 1 into account) (see Fig. 3). As a result of this, for each crossing, all the new vertices, a', b', c' and d', concerned with this crossing have degree two and there is no edge among these four vertices. This preprocessing is needed for the subsequent phases.

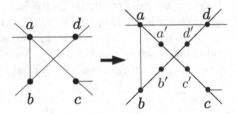

Fig. 3. Preprocessing

Branching. As Dahn et al. [4]'s algorithm, our algorithm branches at each crossing and yields sub-instances. Consider a crossing $\{ac, bd\}$. Obviously, any optimal solution lies in one of the following two cases (1) $|S \cap \{a, b\}| \neq 1$ and (2) $|S \cap \{a, b\}| = 1$. To handle case (1), we construct a sub-instance by contracting the pair $\{a, b\}$ into a single vertex. For case (2), we add four edges ab, bc, cd, and da of weight zero (see Fig. 4). Thanks to the preprocessing phase, adding these four edges does not create a new crossing. Note that these edges do not affect the size of any cut. These edges are necessary only for simplicity of the correctness proof. We then add the constraint that a and b must be separated. Therefore, the subproblem in this branch is the *constrained* Max-Cut problem. Note also that in this branch, we do not remove the crossing. We call the inner region surrounded by the cycle $abcd$ a *pseudo-face*, and the edge ab (that must be a cut edge) a *constrained edge*. Note that the better of the optimal solutions of the two sub-instances coincides the optimal solution of the original problem. After k branchings, we obtain 2^k constrained Max-Cut instances.

Solving the Constrained Max-Cut Problem. In this last phase, we solve 2^k constrained Max-Cut instances obtained above, and output the best solution among them. To solve each problem, we reduce it to the maximum weighted

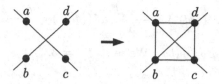

Fig. 4. Adding four edges

b-factor problem (see Sect. 2 for the definition), which is shown in Sect. 4.3 to be solvable in $O((n + k)^{3/2} \log(n + k))$ time in our case. Hence the whole running time of our algorithm is $O(2^k(n + k)^{3/2} \log(n + k))$.

Let G be a graph (with a drawing) obtained by the above branching algorithm. If there is a face with more than three edges, we triangulated it by adding zero-weight edges without affecting the cut size of a solution. By doing this repeatedly, we can assume that every face of G (except for pseudo-faces) has exactly three edges. Recall that, in the preprocessing phase, we subdivided each crossing edge twice. Due to this property, no two pseudo-faces share an edge.

Let $G = (V, E, w)$ be an instance of the constrained Max-Cut problem. We reduce it to an instance $(G^* = (V^*, E^*, w^*), b)$ of the maximum weighted b-factor problem. The reduction is basically constructing a dual graph. For each face f of G, we associate a vertex f^* of G^*. Recall that f is surrounded by three edges, say xy, yz and zx. Corresponding to these edges, f^* has the three edges $(xy)^*, (yz)^*$ and $(zx)^*$ incident to vertices corresponding to the three neighborhood faces or pseudo-faces (see Fig. 5). The weight of these edges are defined as $w^*((xy)^*) = w(xy)$, $w^*((yz)^*) = w(yz)$, and $w^*((zx)^*) = w(zx)$. We also add a self-loop l with $w^*(l) = 0$ to f^*. Note that putting this self-loop to a b-factor contributes to the degree of f^* by 2. Finally, we set $b(f^*) = 2$. (In case some edge surrounding f is shared with a pseudo-face, we do some exceptional handling, which will be explained later.)

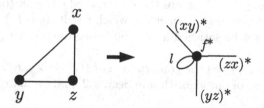

Fig. 5. Reduction for a face

For each pseudo-face f of G, we associate a vertex f^*. Corresponding to the three edges bc, cd, and da, we add the edges $(bc)^*, (cd)^*$ and $(da)^*$ to E^*. (Note that we do not add an edge corresponding to ab.) We also add a self-loop l to f^* (see Fig. 6). The weight of these edges are defined as follows:

$$w^*((bc)^*) = \frac{\beta - 2\alpha}{3}, w^*((cd)^*) = \frac{\alpha + \beta}{3}, w^*((da)^*) = \frac{\alpha - 2\beta}{3}, w^*(l) = \frac{2\alpha + 2\beta}{3},$$

where $\alpha = w(ac)$ and $\beta = w(bd)$. We set $b(f^*) = 3$.

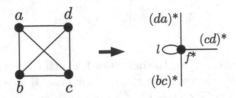

Fig. 6. Reduction for a pseudo-face

Now we explain an exception mentioned above. Consider a (normal) face f consisting of three vertices x, y, and z, and let f^* be the vertex of G^* corresponding to f. Suppose that some edge, say xy, is shared with a pseudo-face g, whose corresponding vertex in G^* is g^*. In this case, the edge $(xy)^*$ in E^*, connecting f^* and g^*, is defined according to the translation rule for the pseudo-face g. Specifically, if the edge xy is identical to the edge bc in Fig. 6, then the weight $w^*((xy)^*)$ is not $w(xy)$ but $\frac{\beta - 2\alpha}{3}$. When xy is identical to cd or da in Fig. 6, $w^*((xy)^*)$ is defined in the similar manner.

In case xy is identical to the constrained edge ab in Fig. 6, the rule is a bit complicated: First, we do not add an edge $(xy)^*$ to E^* (which matches the absence of $(ab)^*$ in Fig. 6). Next, we subtract one from $b(f^*)$; hence in this case, we have $b(f^*) = 1$ instead of the normal case of $b(f^*) = 2$. As one can see later, the absence of $(xy)^*$ and subtraction of $b(f^*)$ implicitly mean that $(xy)^*$ is already selected as a part of a b-factor, which corresponds to the constraint that a and b must be separated in any constrained cut of G. Here we stress that this subtraction is accumulated for boundary edges of f. For example, if all the three edges xy, yz, and zx are the constrained edges of (different) pseudo-faces, then we subtract three from $b(f^*)$, which results in $b(f^*) = -1$ (of course, this condition cannot be satisfied at all and hence the resulting instance has no feasible b-factor).

Now we have completed the construction of $(G^* = (V^*, E^*, w^*), b)$. We then show the correctness of our algorithm in Sect. 4.2, and evaluate its running time in Sect. 4.3.

4.2   Correctness of the Algorithm

To show the correctness, it suffices to show that the reduction in the final phase preserves the optimal solutions. Recall that $\text{cmc}(G)$ is the size of a maximum constrained cut of G, and $\text{mb}(G^*)$ is the cost of the maximum b-factor of G^*.

Lemma 2. $\mathrm{cmc}(G) = \mathrm{mb}(G^*)$.

Proof. We first show $\mathrm{cmc}(G) \leq \mathrm{mb}(G^*)$. Let S be a maximum constrained cut of G with cut size $\mathrm{cmc}(G)$. We construct a b-factor H of G^* with cost $\mathrm{cmc}(G)$. Informally speaking, this construction is performed basically by choosing dual edges of cut edges.

Formally, consider a face f surrounded by edges xy, yz and zx. First suppose that none of these edges are constrained edges. Then by construction of G^*, the degree of f^* is 5 (including the effect of the self-loop) and $b(f^*) = 2$. It is easy to see that zero or two edges among xy, yz and zx are cut edges of S. In the former case, we add only the self-loop l to H. In the latter case, if the two cut edges are e and e', then we add corresponding two edges $(e)^*$ and $(e')^*$ to H. Note that in either case, the constraint $b(f^*) = 2$ is satisfied.

Next, suppose that one edge, say xy, is a constrained edge. In this case, the degree of f^* is 4 and $b(f^*) = 1$. Since xy is a constrained edge, we know that xy is a cut edge of S. Hence exactly one of yz and zx is a cut edge. If yz is a cut edge, we add $(yz)^*$ to H; otherwise, we add $(zx)^*$ to H.

If two edges, say xy and yz, are constrained edges, we have that $b(f^*) = 0$. In this case, we do not select edges incident to f^*.

Finally, suppose that all the three edges are constrained edges. Clearly it is impossible to make all of them cut edges. Thus G admits no constrained cut, which contradicts the assumption that S is a constrained cut.

Next, we move to pseudo-faces. For each pseudo-face f with a cycle $abcd$ where ab is a constrained edge, we know that a and b are separated in S. There are four possible cases, depicted in Fig. 7, where vertices in the same side are labeled with the same color, and bold edges represent cut edges. Corresponding to each case of Fig. 7, we select edges in G^* as shown in Fig. 8 and add them to H. Note that in all four cases, the constraint $b(f^*) = 3$ is satisfied.

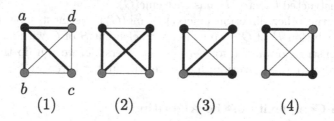

$$(1) \qquad (2) \qquad (3) \qquad (4)$$

Fig. 7. Feasible cuts of G in which a and b are separated

We have constructed a subgraph H of G^* and shown that for any vertex of G^*, the degree constraint b is satisfied in H. Hence H is in fact a b-factor.

It remains to show that the cost of H is $\mathrm{cmc}(G)$. The edges of G are classified into the following two types; (1) edges on the boundary of two (normal) faces and (2) edges constituting pseudo-faces (i.e., those corresponding to one of six edges of K_4 in Fig. 6 left). For a type (1) edge e, e is a cut edge if and only if its

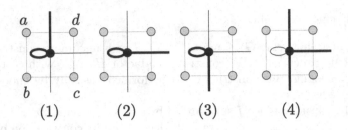

a d
b c

(1)   (2)   (3)   (4)

Fig. 8. Solution for G^* constructed from each cut of Fig. 7

dual $(e)^*$ is selected in H, and $w(e) = w^*((e^*))$. For type (2) edges, we consider six edges corresponding to one pseudo-face simultaneously. Note that there are four feasible cut patterns given in Fig. 7, and we determine edges of H according to Fig. 8. We examine that the total weight of cut edges and that of selected edges coincide in every case:

Case (1): The weight of cut edges is $w(da) + w(ac) + w(ab) = w(ac)$. The weight of selected edges is $w^*((da)^*) + w^*(l) = \frac{\alpha - 2\beta}{3} + \frac{2\alpha + 2\beta}{3} = \alpha = w(ac)$.

Case (2): The weight of cut edges is $w(ab) + w(ac) + w(bd) + w(cd) = w(ac) + w(bd)$. The weight of selected edges is $w^*((cd)^*) + w^*(l) = \frac{\alpha + \beta}{3} + \frac{2\alpha + 2\beta}{3} = \alpha + \beta = w(ac) + w(bd)$.

Case (3): The weight of cut edges is $w(ab) + w(bd) + w(bc) = w(bd)$. The weight of selected edges is $w^*((bc)^*) + w^*(l) = \frac{\beta - 2\alpha}{3} + \frac{2\alpha + 2\beta}{3} = \beta = w(bd)$.

Case (4): The weight of cut edges is $w(ab) + w(bc) + w(cd) + w(da) = 0$. The weight of selected edges is $w^*((bc)^*) + w^*((cd)^*) + w^*((da)^*) = \frac{\beta - 2\alpha}{3} + \frac{\alpha + \beta}{3} + \frac{\alpha - 2\beta}{3} = 0$.

Summing the above equalities over the whole graphs G and G^*, we can conclude that the constructed b-factor H has cost $\mathrm{cmc}(G)$.

To show the other direction $\mathrm{cmc}(G) \geq \mathrm{mb}(G^*)$, we must show that, from an optimal solution H of G^*, we can construct a cut S of G with size $\mathrm{mb}(G^*)$. Since the construction in the former case is reversible, we can do the opposite argument to prove this direction; hence we will omit it here. □

4.3   Time Complexity of the Algorithm

As mentioned previously, to achieve the claimed running time, it suffices to show that a maximum weight b-factor of G^* can be computed in $O((n+k)^{3/2} \log(n+k))$ time. The polynomial time algorithm presented by Gabow [5] first reduces the maximum weighted b-factor problem to the maximum weight perfect matching problem, and then solves the latter problem using a polynomial time algorithm. We follow this line but make a careful analysis to show the above mentioned running time.

For any vertex v of G^*, we can assume without loss of generality that $b(v) \leq d(v)$, where $d(v)$ is the degree of v, as otherwise G^* obviously does not have a

b-factor. Gabow's reduction [5] replaces each vertex v with a complete bipartite graph $K_{d(v),d(v)-b(v)}$ as shown in Fig. 9, where newly introduced edges have weight zero. Since the maximum degree of G^* is at most five, each vertex is replaced by a constant sized gadget. Also, since G^* is a planar graph with $O(n + k)$ vertices, the created graph has $O(n + k)$ vertices and admits a balanced separator of size $O(\sqrt{n + k})$. It is easy to see that this reduction can be done in $O(n + k)$ time, and G^* has a b-factor if and only if the created graph has a perfect matching of the same weight.

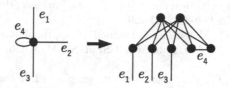

Fig. 9. Gabow's reduction

Lipton and Tarjan [15] present an algorithm for the maximum weight perfect matching problem that runs in time $O(n^{3/2} \log n)$ for an n-vertex graph having a balanced separator of size $O(\sqrt{n})$. Thus, by using it, we obtain the claimed running time of $O((n + k)^{3/2} \log(n + k))$.

Acknowledgements. The authors deeply thank anonymous referees for giving us valuable comments. In particular, one of the referees pointed out a flaw in an early version of Lemma 1, which has been fixed in the current paper.

References

1. Bodlaender, H.L., Jansen, K.: On the complexity of the maximum cut problem. Nordic J. Comput. **7**(1), 14–31 (2000)
2. Chimani, M., Dahn, C., Juhnke-Kubitzke, M., Kriegem, N.M., Mutzel, P., Nover, A.: Maximum Cut Parameterized by Crossing Number. arXiv:1903.06061 (2019)
3. Crowston, R., Jones, M., Mnich, M.: Max-cut parameterized above the Edwards-Erdős Bound. Algorithmica **72**(3), 734–757 (2015)
4. Dahn, C., Kriege, N.M., Mutzel, P.: A fixed-parameter algorithm for the max-cut problem on embedded 1-planar graphs. In: Iliopoulos, C., Leong, H.W., Sung, W.-K. (eds.) IWOCA 2018. LNCS, vol. 10979, pp. 141–152. Springer, Cham (2018). https://doi.org/10.1007/978-3-319-94667-2_12
5. Gabow, H.N.: An efficient reduction technique for degree-constrained subgraph and bidirected network flow problems. In: Proceedings of STOC 1983, pp. 448–456 (1983)
6. Galluccio, A., Loebl, M., Vondrák, J.: Optimization via enumeration: a new algorithm for the max cut problem. Math. Program. **90**(2), 273–290 (2001)
7. Gaspers, S., Sorkin, G.B.: Separate, measure and conquer: faster polynomial-space algorithms for max 2-CSP and counting dominating sets. ACM Trans. Algorithms **13**(4), 44:1–44:36 (2017)

8. Goemans, M.X., Williamson, D.P.: Improved approximation algorithms for maximum cut and satisfiability problem using semidefinite programming. J. ACM **42**(6), 1115–1145 (1995)
9. Guruswami, V.: Maximum cut on line and total graphs. Discrete Appl. Math. **92**, 217–221 (1999)
10. Hadlock, F.: Finding a maximum cut of a planar graph in polynomial time. SIAM J. Comput. **4**(3), 221–255 (1975)
11. Håstad, J.: Some optimal inapproximability results. J. ACM **48**(4), 798–859 (2001)
12. Karp, R.M.: Reducibility among combinatorial problems. In: Miller, R.E., Thatcher, J.W., Bohlinger, J.D. (eds.) Complexity of Computer Computations. The IBM Research Symposia Series, pp. 85–103. Springer, Boston (1972). https://doi.org/10.1007/978-1-4684-2001-2_9
13. Khot, S., Kindler, G., Mossel, E., O'Donnell, R.: Optimal inapproximability results for MAX-CUT and other 2-variable CSPs? SIAM J. Comput. **37**(1), 319–357 (2007)
14. Liers, F., Pardella, G.: Partitioning planar graphs: a fast combinatorial approach for max-cut. Comput. Optim. Appl. **51**(1), 323–344 (2012)
15. Lipton, R.J., Tarjan, R.E.: Applications of a planar separator theorem. SIAM J. Comput. **9**(3), 615–627 (1980)
16. Madathil, J., Saurabh, S., Zehavi, M.: MAX-CUT ABOVE SPANNING TREE is fixed-parameter tractable. In: Fomin, F.V., Podolskii, V.V. (eds.) CSR 2018. LNCS, vol. 10846, pp. 244–256. Springer, Cham (2018). https://doi.org/10.1007/978-3-319-90530-3_21
17. Mahajan, M., Raman, V.: Parameterizing above guaranteed values: MaxSat and MaxCut. J. Algorithms **31**(2), 335–354 (1999)
18. Orlova, G.I.: Dorfman: finding the maximal cut in a graph. Eng. Cybern. **10**(3), 502–506 (1972)
19. Pilipczuk, M., Pilipczuk, M., Wrochna, M.: Edge bipartization faster then 2^k. In: Proceedings of IPEC 2016. LIPIcs. vol. 62, pp. 26:1–26:13 (2016)
20. Pocai, R.V.: The complexity of SIMPLE MAX-CUT on comparability graphs. Electron. Notes Discrete Math. **55**, 161–164 (2016)
21. Raman, V., Saurabh, S.: Improved fixed parameter tractable algorithms for two "edge" problems: MAXCUT and MAXDAG. Inf. Process. Lett. **104**(2), 65–72 (2007)
22. Shih, W.-K., Wu, S., Kuo, Y.S.: Unifying maximum cut and minimum cut of a planar graph. IEEE Trans. Comput. **39**(5), 694–697 (1990)
23. Trevisan, L., Sorkin, G.B., Sudan, M., Williamson, D.P.: Gadgets, approximation, and linear programming. SIAM J. Comput. **29**(6), 2074–2097 (2000)
24. Williams, R.: A new algorithm for optimal 2-constraint satisfaction and its implications. Theor. Comput. Sci. **348**(2–3), 357–365 (2005)
25. Yannakakis, M.: Node- and edge-deletion NP-complete problems. In: Proceedings of STOC 1978, pp. 253–264 (1978)

An Efficient Algorithm for Enumerating Chordal Bipartite Induced Subgraphs in Sparse Graphs

Kazuhiro Kurita[1($\boxtimes$)], Kunihiro Wasa[2], Takeaki Uno[2], and Hiroki Arimura[1]

[1] IST, Hokkaido University, Sapporo, Japan
{k-kurita,arim}@ist.hokudai.ac.jp
[2] National Institute of Informatics, Tokyo, Japan
{wasa,uno}@nii.ac.jp

Abstract. A chordal bipartite graph is a bipartite graph without induced cycles with length six or more. As the main result of our paper, we propose an enumeration algorithm ECB-IS which enumerates all chordal bipartite induced subgraphs in $O(kt\Delta^2)$ time per solution on average, where k is the degeneracy of G, t is the maximum size of $K_{t,t}$ as an induced subgraph of G, and Δ is the maximum degree of G. To achieve the above time complexity, we introduce a new characterization of chordal bipartite graphs, called CBEO. This characterization is based on the relation between a β-acyclic hypergraph and its incidence graph. As a corollary, ECB-IS achieves constant amortized time enumeration for bounded degree graphs.

1 Introduction

A graph G is *chordal* if there are no induced cycles with length four or more. It is known that chordal graphs have several good properties, e.g. a perfect elimination ordering. By using this ordering, efficient algorithms have been developed to enumerate subgraphs and supergraphs with chordality [12,13].

Chordality of a bipartite graph has also been well studied. A chordal bipartite graph is a bipartite graph without any induced cycles with length six or more. Chordal bipartite graphs have several equivalent characterizations, and are closely related to strongly chordal graphs and β-acyclic hypergraphs [3,8,9,16,18]. In particular, a chordal bipartite graph also has a vertex elimination ordering, called *weak elimination ordering* (WEO) [18]. In this paper, we introduce a new vertex elimination ordering, called a *chordal bipartite elimination ordering* (CBEO, in short). CBEO is defined by the following operation: Recursively remove a weak-simplicial vbertex [18]. We show that a graph G is chordal bipartite if and only if G has CBEO. Interestingly, CBEO is a relaxed version of a vertex ordering proposed by Uehara [18]. CBEO plays the key role in our proposed enumeration algorithm. To show this characterization, we use the following relation between β-acyclic hypergraphs and chordal bipartite graphs:

© Springer Nature Switzerland AG 2019
C. J. Colbourn et al. (Eds.): IWOCA 2019, LNCS 11638, pp. 339–351, 2019.
https://doi.org/10.1007/978-3-030-25005-8_28

A hypergraph is β-acyclic if and only if its bipartite incidence graph is chordal bipartite [1].

A subgraph enumeration problem is defined as follows: Output all subgraphs satisfying a given constraint [10]. The efficiency of enumeration algorithms is often measured by the size of input and the number of outputs, called *output-sensitive analysis*. We say that an enumeration algorithm runs in *amortized polynomial time* if the total running time is $O(M \cdot poly(N))$ time, where M is the number of solutions, N is the input size, and $poly(\cdot)$ is a polynomial function. The view point of output-sensitive analysis, many efficient enumeration algorithms for sparse input graphs have been developed so far [5,7,11,14,19]. Especially, the *degeneracy* [15] has been payed much attention for constructing efficient enumeration algorithms. In this paper, we especially focus on the degeneracy and the size of maximum biclique $K_{t,t}$ as a sparsity measure of graphs.

Main Result: We propose an algorithm ECB-IS for chordal bipartite induced subgraphs. So far, several efficient algorithms have been developed for enumerating subgraphs or supergraphs satisfying chordality [6,13,13,20]. However, no algorithm is known for chordal bipartite induced subgraphs of sparse graphs. ECB-IS is based on the *reverse search* [2]. Roughly speaking, an enumeration algorithm based on the reverse search outputs solutions by traversing on a *family tree* spanning all solutions (See Sect. 4 for the detail). The family tree is defined by the parent-child relation among solutions. To construct an efficient enumeration algorithm, we have to give a *good* family tree, that is, the *good* parent for each solution. CBEO plays the key role for defining such parents. Moreover, from the perspective of the time complexity, a degeneracy ordering [17] and a local structure of a weak-simplicial vertex are the key points. To improve a Δ factor of ECB-IS, the above two points are important. As the main result of this paper, we propose ECB-IS which outputs all chordal bipartite induced subgraphs in a given graph G in amortized $O(kt\Delta^2)$ time, where k is the degeneracy of G, t is the maximum size of $K_{t,t}$ as an induced subgraph, and Δ is the maximum degree of G. Note that k is at most Δ and t is at most k. Moreover, the space usage of ECB-IS is $O(n + m)$.

2   Preliminaries

Let $G = (V, E)$ be a simple graph, that is, there are no self loops and multiple edges. The two vertices $u, v \in V$ are *adjacent* if there is an edge $\{u, v\} \in E$. The sequence of distinct vertices $\pi = (v_1, \ldots, v_k)$ is a *path* if v_i and v_{i+1} are adjacent for each $1 \le i \le k - 1$. If $v_1 = v_k$ and $k \ge 3$ hold in a path $C = (v_1, \ldots, v_k)$, then we call C a *cycle*. The distance $dist(u, v)$ between u and v is the length of a shortest path between u and v. We call a graph $H = (U, F)$ a *subgraph* of $G = (V, E)$ if $U \subseteq V$ and $F \subseteq E$ hold. A subgraph $H = (U, F)$ is an *induced subgraph* of G if $F = \{\{u, v\} \in E \mid u, v \in U\}$, and we denote an induced subgraph as $G[U]$. A vertex u is a *neighbor of* v if u and v are adjacent. The *neighborhood of* v is the set of vertices $\{u \in V \mid \{u, v\} \in E\}$ and is denoted by $N_G(v)$. If there is no confusion, we denote $N(v)$ as $N_G(v)$. We denote $N(v) \cap X$ as $N_X(v)$, where

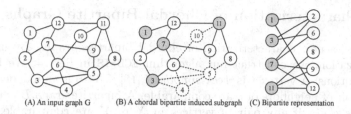

(A) An input graph G   (B) A chordal bipartite induced subgraph   (C) Bipartite representation

Fig. 1. (A) shows an input graph G and (B) shows one of the solutions $B = (X, Y, E)$, where $X = \{1, 3, 7, 11\}$ and $Y = \{2, 6, 8, 9, 12\}$. (C) shows the graph B drawn by dividing X and Y.

X is a subset of V. The pair of vertices u and v are *twin* if $N(v) = N(u)$. The set of vertices $N[v] = N(v) \cup \{v\}$ is called the *closed neighborhood*. We define the *neighborhood with distance k* and the *neighborhood with distance at most k* as $N^k(v) = \{u \in V \mid dist(u, v) = k\}$ and $N^{k:\ell}(v) = \bigcup_{k \leq i \leq \ell} N(v)^i$, respectively. $\mathcal{N}(v)$ is the *neighborhood set of the neighbors* defined as $\{\bar{N}(u) \mid u \in N(v)\}$. The *degree of v* $d(v)$ is the size of $N(v)$. The degree of a graph G is the maximum number of $d(v)$ in V. Let U be a subset of V. For vertices $u, v \in V$, u and v are *comparable* if $N(v) \subseteq N(u)$ or $N(v) \supseteq N(u)$ hold. Otherwise, u and v are *incomparable*. We call a bipartite graph $B = (X, Y, E)$ *chordal bipartite* if there are no induced cycles with length six or more. A bipartite graph B is *biclique* if any pair of vertices $x \in X$ and $y \in Y$ are adjacent. We denote a biclique as $K_{a,b}$ if $|X| = a$ and $|Y| = b$. In this paper, we consider only the case $a = b$ and say that the size of a biclique $K_{t,t}$ is t.

Let $\mathcal{H} = (V, \mathcal{E})$ be a *hypergraph*, where V is a set of vertices and $\mathcal{E}$ is a set of subsets of V. We call an element of $\mathcal{E}$ a *hyperedge*. For a vertex v, let $\mathcal{H}(v)$ be the set of edges $\{e \in \mathcal{H} \mid v \in e\}$. A sequence of edges $C = (e_1, \ldots, e_k)$ is a *berge cycle* if there exist k distinct vertices $v_1, \ldots, v_k$ such that $v_k \in e_1 \cap e_k$ and $v_i \in e_i \cap e_{i+1}$ for each $1 \leq i < k$. A berge cycle $C = (e_1, \ldots, e_k)$ is a *pure cycle* if $k \geq 3$ and $e_i \cap e_j \neq \emptyset$ hold for any distinct i and j, where i and j satisfy one of the following three conditions: (I) $|i - j| = 1$, (II) $i = 1$ and $j = k$, or (III) $i = k$ and $j = 1$. A cycle $C = (e_1, \ldots, e_k)$ is a *β-cycle* if the sequence of $(e'_1, \ldots, e'_k)$ is a pure cycle, where $e'_i = e_i \setminus \bigcap_{1 \leq j \leq k} e_j$. We call a hypergraph $\mathcal{H}$ *β-acyclic* if $\mathcal{H}$ has no β-cycles. We call a vertex v a *β-leaf* if $e \subseteq f$ or $e \supseteq f$ hold for any pair of edges $e, f \in \mathcal{H}(v)$. A bipartite graph $\mathcal{I}(\mathcal{H}) = (X, Y, E)$ is a *incidence graph* of a hypergraph $\mathcal{H} = (V, \mathcal{E})$ if $X = V$, $Y = \mathcal{E}$, and E contains an edge $\{v, e\}$ if $v \in e$, where $v \in V$ and $e \in \mathcal{E}$. Let $\mathcal{V}$ be a set of vertex subsets. $\mathcal{V}$ is *totally ordered* if for any pair $X, Y \in \mathcal{V}$ of vertex subsets, either $X \subseteq Y$ or $X \supseteq Y$. we assume that $\mathcal{H}$ has more than one vertex.

Finally, we define our problem, chordal bipartite induced subgraph enumeration problem. In Fig. 1, we show an input graph G and one of the solutions.

Problem 1 (Chordal bipartite induced subgraph enumeration problem). Output all chordal bipartite induced subgraphs in an input graph without duplication.

3   A Characterization of Chordal Bipartite Graphs

We propose a new characterization of chordal bipartite graphs. By using this characterization, we construct our algorithm ECB-IS in Sect. 4. We first give some definitions. A vertex v is *weak-simplicial* [18] if $N(v)$ is an independent set and any pair of neighbors of v are comparable. A bipartite graph $B = (X, Y, E)$ is *bipartite chain* if any pair of vertices in X or Y are comparable, that is, $N(u) \subseteq N(v)$ or $N(u) \supseteq N(v)$ holds for any $u, v \in X$ or $u, v \in Y$. To show our result, we use the following two theorems.

Theorem 1 (Theorem 1 of [1]). *Let $\mathcal{H}$ be a hypergraph. Then $\mathcal{I}(\mathcal{H})$ is chordal bipartite if and only if $\mathcal{H}$ is β-acyclic.*

Theorem 2 (Theorem 3.9 of [4]). *A β-acyclic hypergraph $\mathcal{H} = (\mathcal{V}, \mathcal{E})$ with at least two vertices has two distinct β-leaves that are not neighbors in $\mathcal{H}' = (\mathcal{V}, \mathcal{E} \setminus \{\mathcal{V}\})$.*

Brault [4] gives a vertex elimination ordering $(v_1, \ldots, v_n)$ for a hypergraph $\mathcal{H}$, called a β-*elimination ordering*. The definition of the ordering is as follows: For any $1 \leq i \leq n$, v_i is a β-leaf in $\mathcal{H}[V_{i:n}]$, where $V_{i:j} = \{v_k \in V \mid i \leq k \leq j\}$. He also showed that $\mathcal{H}$ is β-acyclic if and only if there is a β-elimination ordering of $\mathcal{H}$ [4]. Similarly, in this paper, for any graph G, we define a new vertex elimination ordering $(v_1, \ldots, v_n)$ for G, called CBEO, as follows: For any $1 \leq i \leq n$, v_i is a weak-simplicial in $G[V_{i:n}]$. In the remainder of this section, we show that a graph is chordal bipartite if and only if G has CBEO. Lemma 1 shows that a β-leaf of a hypergraph is weak-simplicial in its incidence graph.

Lemma 1. *Let $\mathcal{H} = (\mathcal{V}, \mathcal{E})$ be a hypergraph, v be a vertex in $\mathcal{V}$, and v' be a corresponding vertex of v in X of $\mathcal{I}(\mathcal{H})$. Then v is a β-leaf in $\mathcal{H}$ if and only if v' is a weak-simplicial vertex in $\mathcal{I}(\mathcal{H})$.*

Proof. We assume that v is a β-leaf in $\mathcal{H}$. Let v' be the vertex corresponding to v in $\mathcal{I}(\mathcal{H})$. From the definition of a β-leaf, $\mathcal{N}(v)$ is also totally ordered in $\mathcal{I}(\mathcal{H})$. Thus, v is a weak-simplicial vertex in $\mathcal{I}(\mathcal{H})$. We next assume that v' is weak-simplicial in X of $\mathcal{I}(\mathcal{H})$. From the definition, $\mathcal{N}(v)$ is totally ordered. Thus, $\mathcal{H}(v)$ is totally ordered. Therefore, v is a β-leaf in $\mathcal{H}$ and the statement holds. □

From Lemma 1, a β-leaf v of $\mathcal{H}$ corresponds to a weak-simplicial vertex of the incidence graph $\mathcal{I}(\mathcal{H})$. We next show that a chordal bipartite graph has at least one weak-simplicial vertex from Theorems 1, 2, and Lemma 1.

Lemma 2. *Let $B = (X, Y, E)$ be a chordal bipartite graph with at least two vertices. If there is no vertex v in B such that $N(v) = X$ or $N(v) = Y$, then B has at least two weak-simplicial vertices which are not adjacent.*

Proof. From Theorems 1, 2, and Lemma 1, if B is chordal bipartite and has no twins, then G has at two weak-simplicial vertices which are not adjacent. We now assume B has twins. We construct B' as follows: Let $\mathcal{T}$ be a set of twins

---

Algorithm 1. Enumeration algorithm for all chordal bipartite induced subgraphs

---

1 **Procedure** ECB-IS(G)   // $G = (V, E)$: **an input graph**
2 | RecECB-IS($(\emptyset, \emptyset, V, G)$);
3 **Procedure** RecECB-IS($X, WS(X), AWS(X), G$)
4 | Output X and Compute $C(X)$ from $AWS(X)$;
5 | **for** $v \in C(X)$ **do**
6 | | $Y \leftarrow X \cup \{v\}$;
7 | | **if** $\mathcal{P}(Y) = X$ **then** RecECB-IS($Y, WS(Y), AWS(Y), G$) ;

---

in B. For each twins $T \in \mathcal{T}$, remove all vertices in T except one twin of T. Note that B' is still chordal bipartite since vertex deletion does not destroy chordality. Since B' has no twins, B' has at least two weak-simplicial vertices u and v which are not adjacent. Since the set inclusion relation between B and B' is same, u and v also weak-simplicial in B. Hence, the statement holds. □

Theorem 3. *Let B be a bipartite graph. B is chordal bipartite if and only if B has CBEO.*

Proof. From Lemma 2, the only if part holds. We consider the contrapositive of the if part. Suppose that B is not chordal bipartite. Then B has an induced cycle C with length six or more. Since a vertex in C is not weak-simplicial, we cannot eliminate all vertices from B and the statement holds. □

We next show that a vertex v is weak-simplicial in a bipartite graph B if and only if $B[N^{1:2}[v]]$ is bipartite chain. The following lemma is used to improve the time complexity of ECB-IS in Sect. 5.

Lemma 3. *Let $B = (X, Y, E)$ be a chordal bipartite graph and v be a vertex in B. Then v is weak-simplicial if and only if an induced subgraph $B[N^{1:2}[v]]$ is bipartite chain.*

Proof. We assume that $B[N^{1:2}[v]]$ is bipartite chain. From the definition, any pair of vertices in $N(v)$ are comparable. Hence, v is weak-simplicial.

We next prove the other direction. We assume that v is weak-simplicial. Let x and y be vertices in $N^2(v)$. If x and y are incomparable in $B[N^{1:2}[v]]$, then there are two vertices $z \in N(x) \setminus N(y)$ and $z' \in N(y) \setminus N(x)$. Note that z and z' are neighbors of v. This contradicts that any pair of vertices in $N(v)$ are comparable. Hence, $x, y \in N^2(v)$ are comparable and $B[N^{1:2}[v]]$ is bipartite chain. □

4 Enumeration of Chordal Bipartite Induced Subgraphs

In this section, we propose an enumeration algorithm ECB-IS which is based on reverse search [2]. ECB-IS enumerates all solutions by traversing on a tree structure $\mathcal{F}(G) = (\mathcal{S}(G), \mathcal{E}(G))$, called a *family tree*, where $\mathcal{S}(G)$ is a set of

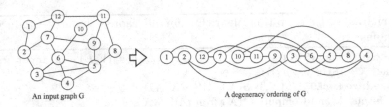

An input graph G   A degeneracy ordering of G

Fig. 2. It is a degeneracy ordering of G. The degeneracy of G is three. In this ordering, $|N_{1:v}^{1:2}(v)|$ is at most $k\Delta$ for any vertex v.

solutions in an input graph G and $\mathcal{E}(G) \subseteq \mathcal{S}(G) \times \mathcal{S}(G)$. Note that $\mathcal{F}(G)$ is directed. More precisely, $\mathcal{E}(G)$ is defined by the parent-child relationship among solutions based on Theorem 3. Let X be a vertex subset that induces a solution. We denote the set of weak-simplicial vertices in $G[X]$ as $WS(X)$. In what follows, we number the vertex indices from 1 to n and compare the vertices with their indices, where n is the number of vertices in G. The *parent* of X is defined as $\mathcal{P}(X) = X \setminus \{\arg\max WS(X)\}$. X is a *child* of Y if $\mathcal{P}(X) = Y$. Let $ch(X)$ be the set of children of X. We define the *parent vertex* $pv(X)$ as $\arg\max WS(X)$ which induces the parent. For any pair of solutions X and Y, $(X, Y) \in \mathcal{E}(G)$ if $Y = \mathcal{P}(X)$. From Theorem 3, any solution can reach the empty set by recursively removing the parent vertex from the solution. Hence, the following lemma holds.

Lemma 4. *The family tree forms a tree.*

Next, we show that ECB-IS enumerates all solutions. For any vertex subset $X \subseteq V$, we denote $X_{1:i} = X \cap V_{1:i}$. An *addible weak-simplicial vertex set* is $AWS(X) = \{v \in V \setminus X \mid v \in WS(X \cup \{v\})\}$, that is, any vertex v in $AWS(X)$ generates a solution $X \cup \{v\}$. We define a *candidate set* $C(X)$ as follows: $C(X) = AWS(X)_{v:n} \cup (AWS(X) \cap N^{1:2}(v))$, where $v = pv(X)$. Note that $C(X)$ is a subset of $AWS(X)$. The following lemma shows that we can enumerate all children if we have $C(X)$.

Lemma 5. *Let X and Y be distinct solutions. If Y is a child of X, then $pv(Y) \in C(X)$.*

Proof. Suppose that Y is a child of X. Let $v = pv(Y)$ and $u = pv(X)$. Note that v belongs to $AWS(X)$. If $u < v$, then $v \in AWS_{u:n}(X)$ and thus $v \in C(X)$. Otherwise, u is not contained in $WS(Y)$ since v has the maximum index in $WS(Y)$. From the definition of a weak-simplicial vertex, there are two vertices in $N_Y(u)$ which are incomparable in $G[Y]$. Since $\mathcal{N}(u)$ is totally ordered in $G[X]$, v must be in $N_Y^{1:2}(u)$. Hence, the statement holds. $\square$

In what follows, we say a vertex $v \in C(X)$ *generates a child* if $X \cup \{v\}$ is a child of X. From Lemmas 4 and 5, ECB-IS enumerates all solution by the DFS traversing on $\mathcal{F}(G)$.

Theorem 4. ECB-IS *enumerates all solutions without duplication.*

5   Time Complexity Analysis

Our proposed algorithm ECB-IS has two bottlenecks. (1) Some vertices in $C(X)$ do not generate a child and (2) the maintenance of $WS(X)$ and $AWS(X)$ consumes time. A trivial bound of the number of redundant vertices in $C(X)$ is $O(\Delta^2)$ since only vertices in $(AWS(X) \cap N^{1:2}(pv(X))$ may not generate a child, where Δ is the maximum degree of an input graph. To reduce the number of such redundant vertices, we use a *degeneracy ordering*. A graph G is *k-degenerate* if any induced subgraph of G has a vertex with degree k or less [15]. The *degeneracy* of a graph is the smallest such number k. Matula and Beck [17] show that a k-degenerate graph G has a following vertex ordering, called a *degeneracy ordering*: For each vertex v, the number of neighbors smaller than v is at most k. See Fig. 2. They also show that a k-degenerate ordering of G can be obtained in linear time. Note that there can be several degeneracy orderings for a graph. In what follows, we fix the reverse ordering of a degeneracy ordering and $WS(X)$ and $AWS(X)$ are sorted in this ordering. We first show that the number of redundant vertices is at most $2k\Delta$.

Algorithm 2. Update algorithms for WS and AWS

1 **Procedure** UpdateWS$(X, v, WS(X), G)$
2 $\quad$ $WS \leftarrow WS(X)$;
3 $\quad$ **for** $u \in N_X(v)$ **do**
4 $\quad\quad$ **if** $u \in WS \wedge$ *there is a vertex* $w \in N_X(u)$ *is incomparable to* u. **then** $WS \leftarrow WS \setminus \{u\}$;
5 $\quad\quad$ **for** $w \in N_X(u) \cap WS$ **do**
6 $\quad\quad\quad$ **if** w *and* v *are incomparable* **then** $WS \leftarrow WS \setminus \{w\}$;
7 $\quad$ **return** WS;
8 **Procedure** UpdateAWS$(X, v, AWS(X), G)$
9 $\quad$ $AWS \leftarrow AWS(X)$;
10 $\quad$ **for** $u \in N(v)$ **do**
11 $\quad\quad$ **if** $u \in AWS$ **then**
12 $\quad\quad\quad$ **if** *There is a vertex* $w \in N_X(u)$ *which is incomparable to* u. **then** $AWS \leftarrow AWS \setminus \{u\}$;
13 $\quad\quad$ **else if** $u \in X$ **then**
14 $\quad\quad\quad$ **for** $w \in N(u) \cap AWS$ **do**
15 $\quad\quad\quad\quad$ **if** w *and* v *are incomparable.* **then** $AWS \leftarrow AWS \setminus \{w\}$;
16 $\quad$ **return** AWS;

Lemma 6. *Let X be a solution. The number of vertices in $C(X)$ which do not generate a child is at most $2k\Delta$.*

Proof. Let v be a vertex in $C(X)$ and p be a vertex $pv(X)$. If $p < v$, then v generates a child. We assume that $v < p$. Since v is in $C(X)$, $v \in N^{1:2}(p) \cap AWX_{1:p}(X)$. We estimate the size of $N^{1:2}(p) \cap AWX_{1:p}(X)$. We consider a vertex

$u \in N_{1:v}(v)$. $\sum_{u \in N_{1:v}(v)} |N(u)|$ is at most $k\Delta$ since $|N_{1:v}(v)|$ is at most k. We next consider a vertex $u \in N_{v:n}(v)$. Since u is larger than v, a vertex in $N_{u:n}(u)$ is larger than v. Hence, we consider vertices $N_{1:u}(u)$. For each u, $|N_{1:u}(u)|$ is at most k. Hence, $\sum_{u \in N_{v:n}(v)} |N(u)_{1:u}|$ is at most $k\Delta$ and the statement holds.　$\square$

We next show how to compute $C(Y)$ from $C(X)$, where X is a solution and Y is a child of X. From the definition of $C(X)$, we can compute $C(Y)$ in $O(|C(Y)| + k\Delta)$ time if we have $AWS(Y)$ and $pv(Y)$. Moreover, if we have $WS(X \cup \{v\})$, then we can determine whether $X \cup \{v\}$ is a child of X or not in constant time since $WS(X \cup \{v\})$ is sorted. Hence, to obtain the children of X, computing $AWS(X)$ and $WS(X)$ dominates the computation time of each iteration. Here, we define two vertex sets as follows:

$$Del_W(X, v) = \{u \in N^{1:2}(v) \cap WS(X) \mid u \notin WS(X \cup \{v\})\} \text{ and}$$

$$Del_A(X, v) = \{u \in N^{1:2}(v) \cap AWS(X) \mid u \notin WS(X \cup \{u, v\})\}.$$

These vertex sets are the sets of vertices that are removed from $WS(X)$ and $AWS(X)$ after adding v to X, respectively. In the following lemmas, we show that $WS(X)$ and $AWS(X)$ can be updated if we have $Del_W(X, v)$ and $Del_A(X, v)$.

Lemma 7. *Let X be a solution, Y be a child of X, and $v = pv(Y)$. Then $WS(Y) = (WS(X) \setminus Del_W(X, v)) \cup \{v\}$.*

Proof. Let u be a vertex in $WS(Y)$. We prove u is contained in $(WS(X) \setminus Del_W(X, v)) \cup \{v\}$. If $u = v$ holds, then $u \in WS(Y)$ since $v \in pv(Y)$. We assume that $u \neq v$. Since u is weak-simplicial in Y, $u \in WS(X)$. If $u \in Del_W(X, v)$, then u is not weak-simplicial in Y. This contradicts the assumption. Hence, $u \notin Del_W(X, v)$. We prove the other direction. Let u be a vertex in $(WS(X) \setminus Del_W(X, v)) \cup \{v\}$. We assume that $u \neq v$. If $u \notin WS(Y)$, then $u \in Del_W(X, v)$. This is a contradiction and thus the statement holds.　$\square$

Lemma 8. *Let X be a solution, Y be a child of X, and $v = pv(Y)$. Then $AWS(Y) = AWS(X) \setminus Del_A(X, v)$.*

Proof. Let u be a vertex in $AWS(Y)$. We prove u is contained in $AWS(X) \setminus Del_A(X, v)$. From the definition of $AWS(X)$, $u \in AWS(X)$ holds. If $u \in Del_A(X, v)$, then u is not weak-simplicial in $Y \cup \{u\}$. Since $u \in AWS(Y)$, $u \notin Del_A(X, v)$. We prove the other direction. Let u be a vertex in $AWS(X) \setminus Del_A(X, v)$. From the definition of $AWS(X)$ and $Del_A(X, v)$, u is weak-simplicial in $X \cup \{v, u\}$. Hence, $u \in AWS(Y)$ and the statement holds.　$\square$

Note that by just removing redundant vertices, $WS(Y)$ and $AWS(X)$ can be easily sorted if $WS(X)$ and $AWS(X)$ were already sorted. We next consider how to compute $Del_W(X, v)$ and $Del_A(X, v)$. We first show a characterization of a vertex in $Del_W(X, v)$ and $Del_A(X, v)$. In the following lemmas, let X be a solution, v be a vertex in $AWS(X)$, and Y be a solution $X \cup \{v\}$.

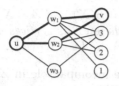
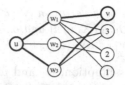

(A)When a vertex v is added,
u is a weak-simplicial

(B)When a vertex v is added,
u is not weak-simplicial

Fig. 3. Let X be a set of vertices $\{u, w_1, w_2, w_3, 1, 2, 3\}$. In case (A), a vertex u is still weak-simplicial. In case (B), however, u is not weak-simplicial since w_1 and w_3 are incomparable.

Lemma 9. *Let u be a vertex in $N_Y(v) \cap WS(X)$. Then u is contained in $Del_W(X, v)$ if and only if $N_X(u)$ contains w which is incomparable to v.*

Proof. If u has a neighbor w in X which is incomparable to v, then from the definition, u is not weak-simplicial.

Let u be a vertex in $Del_W(X, v)$. There are vertices w_1 and w_2 in $N_Y(u)$ which are incomparable. If w_1 or w_2 is equal to v, then the statement holds. Hence, we assume that both w_1 and w_2 are not equal to v. Since $G[Y]$ is bipartite, w_1 and w_2 are not adjacent to v. Hence, $N_X(w_1) = N_Y(w_1)$ and $N_X(w_2) = N_Y(w_2)$ hold. This contradicts that X is a solution and the statement holds. $\square$

Lemma 10. *Let u be a vertex in $N_Y^2(v) \cap WS(X)$. Then u is contained in $Del_W(X, v)$ if and only if there exist vertices $w_1, w_2 \in N_X(u)$ which satisfy $N_X(w_1) \subset N_X(w_2)$, $v \in N_Y(w_1)$, and $v \notin N_Y(w_2)$.*

Proof. The if part is easily shown by the assumption of the incomparability of w_1 and w_2. We next prove the other direction. We assume that $u \in Del_W(X, v)$. Hence, there are neighbors w_1 and w_2 of u such that they are incomparable in Y. Without loss of generality, $N_X(w_1) \subset N_X(w_2)$ holds since u is in $WS(X)$. If $N_X(w_1) = N_X(w_2)$, then w_1 and w_2 are comparable in Y. Since w_1 and w_2 are incomparable in Y, w_1 is adjacent to v and w_2 is not adjacent to v (Fig. 3). Thus, the statement holds. $\square$

Lemma 11. *Let u be a vertex in $N(v) \cap AWS(X)$ and $Z = X \cup \{u, v\}$. Then $u \in Del_A(X, v)$ if and only if u has a neighbor $w \in Z$ that is incomparable to v.*

Proof. The if part is trivial from the definition of weak-simplicial. We prove the only if part. We assume that $u \in Del_A(X, v)$ holds. Since $u \in Del_A(X, v)$, u has neighbors w_1 and w_2 which are incomparable in Z. If w_1 or w_2 is equal to v, then u has a neighbor w which is incomparable to v and the statement holds. We next assume that w_1 and w_2 are distinct from v. v is not adjacent to w_1, w_2, or both of them since $G[Y]$ is bipartite. Hence, w_1 and w_2 are comparable in Z since w_1 and w_2 are comparable in $G[X]$. This contradicts that w_1 and w_2 are incomparable in Z and the statement holds. $\square$

Lemma 12. *Let u be a vertex in $N^2(v) \cap AWS(X)$ and $Z = X \cup \{u, v\}$. Then u is contained in $Del_A(X, v)$ if and only if there exists vertices $w_1, w_2 \in N_Z(u)$ which satisfy $N_Z(w_1) \subset N_Z(w_2)$, $v \in N_Z(w_1)$, and $v \notin N_Z(w_2)$.*

Proof. From the assumption, w_1 and w_2 are incomparable in Z. Hence, the if part holds. We prove the other direction. We assume that $u \in Del_A(X, v)$. Hence, u has vertices w_1' and w_2' which are incomparable in Z. Without loss of generality, $N_{X \cup \{u\}}(w_1') \subset N_{X \cup \{u\}}(w_2')$ holds. Since w_1' and w_2' are incomparable in $G[Z]$, w_1' is adjacent to v. Thus, the statement holds.   □

Next, we consider the time complexity of computing $Del_W(X, v)$ and $Del_A(X, v)$. For analysing these computing time more precisely, we give two upper bounds with respect to the number of edges and the size of $N^2(v)$ in bipartite chain graphs. Note that t is the maximum size of a biclique $K_{t,t}$ that appears in B as an induced subgraph.

Lemma 13. *Let B be a bipartite chain graph and v be a vertex in B. Then the size of $N^2(v)$ is at most Δ.*

Proof. Let u be a vertex in $N(v)$ which satisfies $N(w) \subseteq N(u)$ for any $w \in N(v)$. Since $N^2(v) = \bigcup_{w \in N(v)} N(w)$, $N^2(v) = N(u)$. Hence, the statement holds.   □

Lemma 14. *Let $B = (X, Y, E)$ be a bipartite chain graph. Then the number of edges in B is $O(t\Delta)$.*

Proof. Let w be a vertex in X which satisfies for $N(u) \supseteq N(w)$ for any $u \in X$. If $d(w) \leq t$, then the statement holds from Lemma 13. We assume that $d(w) > t$. We consider the number of edges in B. Let $(u_1, \ldots, u_{d(w)})$ be a sequence of vertices in $N(w)$ such that $N(u_i) \subseteq N(u_{i+1})$ for $1 \leq i < d(w)$. For each $d(w) - t + 1 \leq i \leq d(w)$, the sum of $|N(u_i)|$ is at most $O(t\Delta)$. We next consider the case for $1 \leq i \leq d(w) - t$. Since $N(u_i)$ is a subset of $N(u_j)$ for any $i < j$, $|N(u_i)|$ is at most t. If $|N(u_i)|$ is greater than t, then B has a biclique $K_{t+1,t+1}$. Hence, the number of edges in B is $O(t\Delta)$ and the statement holds.   □

Lemma 15. *Let v be a vertex in $C(X)$. Then we can compute $Del_W(X, v)$ in $O(t\Delta)$ time.*

Proof. Let Y be a solution $X \cup \{v\}$. We first compute vertices in $WS(X) \cap N_X^2(v)$ that remain in $WS(Y)$. From Lemma 7 and Lemma 10, $w \notin WS(Y)$ if and only if there exists vertices $w_1, w_2 \in N(u)$ which satisfy $N_X(w_1) \subset N_X(w_2)$, $v \in N_Y(w_1)$, and $v \notin N_Y(w_2)$. By scanning vertices with distance two from v, this can be done in linear time in the size of $\sum_{u \in N_X(v)} |N_X(u)|$. Since $G[N^{1:2}(v)]$ is bipartite chain from Lemma 3, $\sum_{u \in N_X(v)} |N_X(u)|$ is at most $O(t\Delta)$ from Lemma 14. Moreover, it can be determined whether $w \in N^2(v)$ and v are comparable or not in this scan operation.

We next compute vertices in $WS(X) \cap N_X(v)$ that remain in $WS(Y)$. From Lemma 9, u is contained in $Del_W(X, v)$ if and only if u has a neighbor w which is incomparable to v. Since $G[Y]$ is bipartite, $N_X(u)$ is contained in $N_Y^2(v)$. In the previous scan operation, we already know whether w and v are comparable or not. Hence, we can compute whether $w \in WS(Y)$ or not in $O(|N(u)|)$ time. Since $O(\sum_{u \in N_Y(v)} |N(u)|) = O(t\Delta)$ holds from Lemma 14, we can find $N_X(v) \cap Del_W(X, v)$ in $O(t\Delta)$ total time. Hence, the statement holds.   □

Lemma 16. *Let v be a vertex in $C(X)$. Then we can compute $Del_A(X, v)$ in $O(\Delta^2)$ time.*

Proof. Since v is contained in $Del_A(X, v)$, $G[X \cup \{v\}]$ is a chordal bipartite induced subgraph. Here, let Y be a set of vertices $X \cup \{v\}$. In the same fashion as Lemma 15, we can decide $u \in AWS(X \cup \{v\} \cap N_Y^2(v))$ in $O(\Delta^2)$ total time. By applying the above procedure for all vertices distance two from v, we can obtain all vertices in $Del_A(X, v)$ in $O(\Delta^2)$ time since the number of edges can be bounded in $O(\Delta^2)$.   □

From Lemmas 15 and 16, we can compute $Del_W(X, v)$ and $Del_A(X, v)$ in $O(t\Delta)$ and $O(\Delta^2)$ time for each $v \in C(X)$, respectively.

Hence, we can enumerate all children in $O(|C(X)| t\Delta + |ch(X)| \Delta^2)$ time from Lemmas 5, 7 and 8. In the following theorem, we show the amortized time complexity and the space usage of ECB-IS.

Theorem 5. *ECB-IS enumerates all solutions in amortized $O(kt\Delta^2)$ time by using $O(n + m)$ space, where k is the degeneracy of an input graph G, t is the maximum size of $K_{t,t}$ that appears in G as an induced subgraph, n is the number of vertices of G, and m is the number of edges of G.*

Proof. ECB-IS uses $AWS(X)$ and $WS(X)$ as data structures. Each data structure demands linear space and the total space usage of ECB-IS is $O(n+m)$ space. Hence, the total space usage of ECB-IS is linear in the size of the input. We next consider the amortized time complexity of ECB-IS. From Lemma 4, ECB-IS enumerates all solutions. From Lemmas 15 and 16, ECB-IS computes all children and updates all data structures in $O(|C(X)| t\Delta + |ch(X)| \Delta^2)$ time. From Lemma 6, $|C(X)|$ is at most $|ch(X)| + k\Delta$. Hence, we need $O((|ch(X)| + k\Delta)t\Delta + |ch(X)| \Delta^2)$ time to generate all children. Note that this computation time is bounded by $O((|ch(X)| + kt)\Delta^2)$. We consider the total time for enumerating all solutions. Since each iteration X needs $O((|ch(X)| + kt)\Delta^2)$ time, the total time is $O(\sum_{X \in \mathcal{S}}(|ch(X)| + kt)\Delta^2)$ time, where $\mathcal{S}$ is the set of solutions. Since $O(\sum_{X \in \mathcal{S}} |ch(X)| \Delta^2)$ is bounded by $O(|\mathcal{S}| \Delta^2)$, the total time is $O(|\mathcal{S}| kt\Delta^2)$ time. Therefore, ECB-IS enumerates all solutions in amortized $O(kt\Delta^2)$ time.   □

6   Conclusion

In this paper, we propose a new vertex ordering CBEO by relaxing a vertex ordering proposed by Uehara [18]. A bipartite graph B is chordal bipartite if and only if B has CBEO, that is, this vertex ordering characterizes chordal bipartite graphs. This ordering comes from hypergraph acyclicity and the relation between

β-acyclic hypergraphs and chordal bipartite graphs. In addition, we also show that a vertex v is weak-simplicial if and only if $G[N^{1:2}[v]]$ is bipartite chain. By using these facts, we propose an amortized $O(kt\Delta^2)$ time algorithm ECB-IS.

As future work, the following two enumeration problems are interesting: Enumeration of bipartite induced subgraph for dense graphs and enumeration of chordal bipartite subgraph. For dense graphs, ECB-IS does not achieve an amortized linear time enumeration. If an input graph is biclique, then the time complexity of ECB-IS is $O(nm)$ time, where n and m are the number of vertices and the number of edges in an input graph, respectively. Hence, it is still open whether there is an amortized linear time enumeration algorithm for chordal bipartite induced subgraph enumeration problem or not. In addition, in the chordal bipartite subgraph enumeration problem, we cannot use CBEO for the parent-child relation. Therefore, we need to consider another parent-child relation.

References

1. Ausiello, G., D'Atri, A., Moscarini, M.: Chordality properties on graphs and minimal conceptual connections in semantic data models. J. Comput. Syst. Sci. **33**(2), 179–202 (1986)
2. Avis, D., Fukuda, K.: Reverse search for enumeration. Discrete Appl. Math. **65**(1), 21–46 (1996)
3. Brandstädt, A., Spinrad, J.P., et al.: Graph Classes: A Survey, vol. 3. Siam, Philadelphia (1999)
4. Brault-Baron, J.: Hypergraph acyclicity revisited. ACM Comput. Surv. **49**(3), 54 (2016)
5. Conte, A., Grossi, R., Marino, A., Versari, L.: Sublinear-space bounded-delay enumeration for massive network analytics: maximal cliques. In: Proceedings of ICALP 2016. LIPIcs, vol. 55, pp. 148:1–148:15 (2016)
6. Daigo, T., Hirata, K.: On generating all maximal acyclic subhypergraphs with polynomial delay. In: Nielsen, M., Kučera, A., Miltersen, P.B., Palamidessi, C., Tůma, P., Valencia, F. (eds.) SOFSEM 2009. LNCS, vol. 5404, pp. 181–192. Springer, Heidelberg (2009). https://doi.org/10.1007/978-3-540-95891-8_19
7. Eppstein, D., Löffler, M., Strash, D.: Listing all maximal cliques in large sparse real-world graphs. J. Exp. Algorithmics **18**, 3-1 (2013)
8. Farber, M.: Characterizations of strongly chordal graphs. Discrete Math. **43**(2–3), 173–189 (1983)
9. Huang, J.: Representation characterizations of chordal bipartite graphs. J. Comb. Theory Ser. B **96**(5), 673–683 (2006)
10. Johnson, D.S., Yannakakis, M., Papadimitriou, C.H.: On generating all maximal independent sets. Inf. Process. Lett. **27**(3), 119–123 (1988)
11. Kanté, M.M., Limouzy, V., Mary, A., Nourine, L.: Enumeration of minimal dominating sets and variants. In: Owe, O., Steffen, M., Telle, J.A. (eds.) FCT 2011. LNCS, vol. 6914, pp. 298–309. Springer, Heidelberg (2011). https://doi.org/10.1007/978-3-642-22953-4_26
12. Kiyomi, M., Kijima, S., Uno, T.: Listing chordal graphs and interval graphs. In: Fomin, F.V. (ed.) WG 2006. LNCS, vol. 4271, pp. 68–77. Springer, Heidelberg (2006). https://doi.org/10.1007/11917496_7

13. Kiyomi, M., Uno, T.: Generating chordal graphs included in given graphs. IEICE Trans. Inf. Syst. **89**(2), 763–770 (2006)
14. Kurita, K., Wasa, K., Arimura, H., Uno, T.: Efficient enumeration of dominating sets for sparse graphs. In: Proceedings of ISAAC 2018, pp. 8:1–8:13 (2018)
15. Lick, D.R., White, A.T.: k-degenerate graphs. Can. J. Math. **22**, 1082–1096 (1970)
16. Lubiw, A.: Doubly lexical orderings of matrices. SIAM J. Comput. **16**(5), 854–879 (1987)
17. Matula, D.W., Beck, L.L.: Smallest-last ordering and clustering and graph coloring algorithms. J. ACM **30**(3), 417–427 (1983)
18. Uehara, R.: Linear time algorithms on chordal bipartite and strongly chordal graphs. In: Widmayer, P., Eidenbenz, S., Triguero, F., Morales, R., Conejo, R., Hennessy, M. (eds.) ICALP 2002. LNCS, vol. 2380, pp. 993–1004. Springer, Heidelberg (2002). https://doi.org/10.1007/3-540-45465-9_85
19. Wasa, K., Uno, T.: Efficient enumeration of bipartite subgraphs in graphs. In: Wang, L., Zhu, D. (eds.) COCOON 2018. LNCS, vol. 10976, pp. 454–466. Springer, Cham (2018). https://doi.org/10.1007/978-3-319-94776-1_38
20. Wasa, K., Uno, T., Hirata, K., Arimura, H.: Polynomial delay and space discovery of connected and acyclic sub-hypergraphs in a hypergraph. In: Fürnkranz, J., Hüllermeier, E., Higuchi, T. (eds.) DS 2013. LNCS (LNAI), vol. 8140, pp. 308–323. Springer, Heidelberg (2013). https://doi.org/10.1007/978-3-642-40897-7_21

Complexity of Fall Coloring for Restricted Graph Classes

Juho Lauri[1](✉) and Christodoulos Mitillos[2]

[1] Nokia Bell Labs, Dublin, Ireland
juho.lauri@gmail.com
[2] Illinois Institute of Technology, Chicago, USA
cmitillo@iit.edu

Abstract. We strengthen a result by Laskar and Lyle (Discrete Appl. Math. (2009), 330–338) by proving that it is NP-complete to decide whether a bipartite planar graph can be partitioned into three independent dominating sets. In contrast, we show that this is always possible for every maximal outerplanar graph with at least three vertices. Moreover, we extend their previous result by proving that deciding whether a bipartite graph can be partitioned into k independent dominating sets is NP-complete for every $k \geq 3$. We also strengthen a result by Henning et al. (Discrete Math. (2009), 6451–6458) by showing that it is NP-complete to determine if a graph has two disjoint independent dominating sets, even when the problem is restricted to triangle-free planar graphs. Finally, for every $k \geq 3$, we show that there is some constant t depending only on k such that deciding whether a k-regular graph can be partitioned into t independent dominating sets is NP-complete. We conclude by deriving moderately exponential-time algorithms for the problem.

Keywords: Fall coloring · Independent domination · Computational complexity

1 Introduction

Domination and independence are two of the most fundamental and heavily-studied concepts in graph theory. In particular, a partition of the vertices of a graph into independent sets is known as *graph coloring*—a central problem with several practical applications in e.g., scheduling [18], timetabling, and seat planning [16]. In addition, independence and domination are central to various problems in telecommunications, such as adaptive clustering in distributed wireless networks and various channel assignment type problems such as code assignment, frequency assignment, and time-slot assignment. For an overview, see [20, Chap. 30].

Let $G = (V, E)$ be a graph and let $S \subseteq V$ be a subset of its vertices. Here, S is an *independent set* if the vertices in S are pairwise non-adjacent. We say that

© Springer Nature Switzerland AG 2019
C. J. Colbourn et al. (Eds.): IWOCA 2019, LNCS 11638, pp. 352–364, 2019.
https://doi.org/10.1007/978-3-030-25005-8_29

S is a *dominating set* when every vertex of V either is in S or is adjacent to a vertex in S. Combining these properties, Dunbar et al. [8] studied the problem of partitioning the vertex set of a graph into sets that are both independent and dominating. The authors viewed this problem as a kind of a graph coloring defined as follows. Let $\Pi = \{V_1, V_2, \ldots, V_k\}$ be a partition of V. We say that a vertex $v \in V_i$ for $i \in \{1, 2 \ldots, k\}$ is *colorful* if v is adjacent to at least one vertex in each color class V_j for $i \neq j$. Π is a *fall k-coloring* if each V_i is independent and every vertex $v \in V$ is colorful. Informally, in a fall coloring, each vertex has in its immediate neighborhood each of the colors except for its own. For an illustration of the concept, see Fig. 1. For possible applications of fall k-coloring, including transceiver frequency allocation and timetabling, see [19, Sect. 4.2].

The maximum k for which a graph G has a fall k-coloring is known as the *fall achromatic number*, denoted by $\psi_{fall}(G)$. Clearly, we have $\psi_{fall}(G) \leq \delta(G) + 1$, where $\delta(G)$ is the minimum degree of G. Similarly, the minimum k for which a graph G has a fall k-coloring is known as the *fall chromatic number*, denoted by $\chi_{fall}(G)$. Here, it holds that $\chi(G) \leq \chi_{fall}(G)$, where $\chi(G)$ is the chromatic number of G (see [8]). The *fall set* of a graph G, denoted by $\mathrm{Fall}(G)$, is the set of integers k such that G admits a fall k-coloring. In general, $\mathrm{Fall}(G)$ is not guaranteed to be non-empty, but it is finite for finite graphs. For example, we have that $\mathrm{Fall}(C_6) = \{\chi_{fall}(C_6), \psi_{fall}(C_6)\} = \{2, 3\}$ with a fall 3-coloring shown in Fig. 1. To obtain a fall 2-coloring for C_6, it suffices to observe that any 2-coloring of a connected bipartite graph is a fall 2-coloring.

In this work, our focus is on the computational complexity of fall coloring. In this context, it was shown by Heggernes and Telle [12] that for every $k \geq 3$, it is NP-complete to decide whether a given graph G has $k \in \mathrm{Fall}(G)$. Laskar and Lyle [14] improved on this in the case of $k = 3$ by showing that it is NP-complete to decide whether a given bipartite graph H has $3 \in \mathrm{Fall}(H)$. On a positive side, it was shown by Telle and Proskurowski [22] that deciding whether $k \in \mathrm{Fall}(G)$ can be done in polynomial time when G has bounded cliquewidth (or treewidth). This can also be derived from the fact that the property of being fall k-colorable can be expressed in monadic second order logic (for details, see [7, Sect. 7.4]). For chordal graphs G, it is known that the fall set is either empty or contains exactly $\delta(G) + 1$. To the best of our knowledge, the complexity of deciding this case is open. For subclasses of chordal graphs, the fall sets of threshold and split graphs can be characterized in polynomial time [19]. Despite independence and domination being central concepts in graph theory, we are unaware of any further hardness results for fall coloring (see also e.g., [11, Sect. 7]).

We extend and strengthen previous hardness results for fall coloring, and provide new results as follows:

- In Sect. 3, we extend the result of Laskar and Lyle [14] by proving that for $k \geq 3$, it is NP-complete to decide whether a bipartite graph is fall k-colorable. Further, for the case of $k = 3$, we strengthen their result considerably by showing it is NP-complete to decide whether a bipartite *planar* graph is fall 3-colorable.

If we do not insist on a partition, we prove that deciding whether a triangle-free planar graph contains two disjoint independent dominating sets is NP-complete, strengthening the result of Henning et al. [13] who only showed it for general graphs.

- In Sect. 4, we turn our attention to regular graphs. While fall coloring 2-regular graphs is easy, we prove that for every $k \geq 3$, there is some t—dependant only on k—such that it is NP-complete to decide whether a k-regular graph G is fall t-colorable (see the section for precise statements).
- In Sect. 5, we conclude by detailing some further algorithmic consequences of our hardness results presented in Sect. 3. In addition, we derive moderately exponential-time algorithms for fall coloring.

2   Preliminaries

For a positive integer n, we write $[n] = \{1, 2, \ldots, n\}$. All graphs we consider in this work are undirected and finite.

Graph Theory. Let $G = (V, E)$ be a graph. For any $s \geq 1$, we denote by G^s the *sth power* of G, which is G with edges added between every two vertices at a distance no more than s. In particular, G^2 is called the *square* of G. By $G^{\frac{1}{s}}$, we mean G with each of its edges subdivided $s - 1$ times.

A *k-coloring* of a graph G is a function $c : V \to [k]$. A *coloring* is a k-coloring for some $k \leq |V|$. We say that a coloring c is *proper* if $c(u) \neq c(v)$ for every edge $uv \in E$. In particular, if G admits a proper k-coloring, we say that G is *k-colorable*. The *chromatic number* of G, denoted by $\chi(G)$, is the smallest k such that G is k-colorable.

Computational Problems. The problem of deciding whether a given graph G has $\chi(G) \leq k$ is NP-complete for every $k \geq 3$ (see e.g., [10]). We refer to this computational problem as k-COLORING. In a closely related problem known as EDGE k-COLORING, the task is to decide whether the *edges* of the input graph can be assigned k colors such that every two adjacent edges receive a distinct color. Similarly, for every $k \geq 3$, this problem is also NP-complete even when the input graph is k-regular as shown by Leven and Galil [15]. Our focus is on the following problem and its computational complexity.

FALL k-COLORING
Instance: A graph $G = (V, E)$.
Question: Can V be partitioned into k independent dominating sets, i.e., is $k \in \text{Fall}(G)$?

---

3   Hardness Results for Planar and Bipartite Graphs

In this section, we prove that deciding whether a bipartite planar graph can be fall 3-colored is NP-complete. Moreover, we show that for every $k \geq 3$, it is NP-complete to decide whether a bipartite graph can be fall k-colored.

We begin with the following construction that will be useful to us throughout the section.

Lemma 1. 3-COLORING *reduces in polynomial-time to* FALL 3-COLORING.

Proof. Let G be an instance of 3-COLORING. In polynomial time, we will create the following instance G' of FALL 3-COLORING, such that G is 3-colorable if and only if G' is fall 3-colorable.

The graph $G' = (V', E')$ is obtained from G by subdividing each edge, and by identifying each vertex in V with a copy of C_6. Formally, we let

$$V' = V \cup \{x_{uv} \mid uv \in E\} \cup \{w_{vi} \mid v \in V, i \in [4]\}, \text{ and}$$

$$E' = \{ux_{uv}, vx_{uv} \mid uv \in E\} \cup \{vw_{v1}, vw_{v5}, w_{vi}w_{vi+1} \mid v \in V, i \in [4]\}.$$

This finishes the construction of G'.

Let $c : V \to [3]$ be a proper vertex-coloring of G, and let us construct a fall 3-coloring $c' : V' \to [3]$ as follows. We retain the coloring of the vertices in V, that is, $c'(v) = c(v)$ for every $v \in V$. Then, as the degree of each x_{uv} is two, it holds that in any valid fall 3-coloring of G', the colors from $[3]$ must be bijectively mapped to the closed neighborhood $\{u, v, x_{uv}\}$ of v_{uv}. Thus, we set $c'(x_{uv}) = f$, where f is the unique color in $[3]$ neither $c(u)$ nor $c(v)$. Finally, consider an arbitrary vertex $v \in V$. Without loss of generality, suppose $c(v) = 1$. We will then finish the vertex-coloring c' as follows (see Fig. 1, where $C_6 \simeq F_3$):

$$c'(w_{v3}) = 1, \ c'(w_{v1}) = 2, \ c'(w_{v4}) = 2, \ c'(w_{v2}) = 3, \ c'(w_{v5}) = 3.$$

It is straightforward to verify that c' is indeed a fall 3-coloring of G'.

For the other direction, let c' be a fall 3-coloring of G'. Again, because the degree of each x_{uv} is two, it holds that $c'(u) \neq c'(v)$. Therefore, c' restricted to G is a proper 3-coloring for G. This concludes the proof. □

Combining the previous lemma with the well-known fact that deciding whether a planar graph of maximum degree 4 can be properly 3-colored is NP-complete [10], we obtain the following.

Corollary 2. *It is* NP-*complete to decide whether a bipartite planar graph G of maximum degree 6 is fall 3-colorable.*

Proof. It suffices to observe that the construction of Lemma 1 does not break planarity (i.e., if G is planar, so is G') and that after subdividing the edges of G the resulting graph G' is bipartite. Finally, a vertex v of degree $\Delta \leq 4$ in G has degree $\Delta + 2 \leq 6$ in G' after v is identified with a copy of C_6, whereas the new vertices (in copies of C_6 or from subdividing) have degree 2. □

In order to show that fall k-coloring is hard for every $k \geq 3$ for the class of bipartite graphs, we make use of the following construction. As a reminder, $G \times H$ is the *categorical product* of graphs G and H with $V(G \times H) = V(G) \times V(H)$ and $(u_1, v_1)(u_2, v_2) \in E(G \times H)$ when $u_1u_2 \in E(G)$ and $v_1v_2 \in E(H)$.

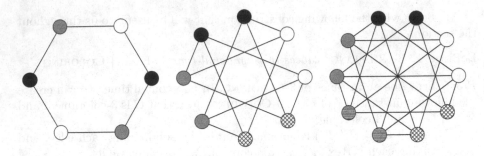

Fig. 1. The graphs $F_k = K_2 \times K_k$ for $3 \leq k \leq 5$ each with a fall k-coloring shown.

Proposition 3. *For every $k \geq 3$, the graph $F_k = K_2 \times K_k$ is bipartite and uniquely fall k-colorable.*

Proof. It is well-known that $G \times H$ is bipartite if either G or H is bipartite. Thus, as K_2 is bipartite, so is F_k.

It follows from Dunbar et al. [8, Theorem 6] that if s and k are distinct positive integers both greater than one, then $K_s \times K_k$ has a fall k-coloring. In our case, $s = 2$ and $k \geq 3$, so F_k admits a fall k-coloring. The fact that F_k has a unique fall k-coloring follows from [19, Theorem 15]. □

We are then ready to proceed with the reduction, following the idea of Lemma 1.

Lemma 4. *For every $k \geq 4$, it is NP-complete to decide whether a bipartite graph G is fall k-colorable.*

Proof. We show this by extending the method in Lemma 1. Given a graph $G = (V, E)$, we construct in polynomial time a bipartite graph $G' = (V', E')$, so that G' is fall k-colorable if and only if G is k-colorable. Then, since it is NP-complete to decide whether G is k-colorable, the result will follow.

As before, we begin by subdividing every edge of G once, and identifying each vertex of V with a copy of F_k. Then, for each vertex x_{uv} created by subdividing some edge uv of G, we create $k - 3$ disjoint copies of F_k, and arbitrarily select one vertex in each such copy to make adjacent to x_{uv}. Note that when $k = 3$, this simplifies to the construction in Lemma 1.

First, we observe that the resulting graph G' is bipartite. It consists of one copy of $G^{\frac{1}{2}}$ and multiple disjoint copies of $F_k = K_2 \times K_k$, connected to $G^{\frac{1}{2}}$ by either cut-vertices or cut-edges. Since $G^{\frac{1}{2}}$ and F_k are both bipartite, G' is bipartite as well.

Let c be a proper k-coloring of G. We extend it to a fall k-coloring c' of G' as follows. For every edge $uv \in E$, the vertex x_{uv} is colored arbitrarily with some color distinct from both $c(u)$ and $c(v)$. Then, its remaining $k - 3$ neighbors are each given a different color, so that x_{uv} is colorful. Now every copy of F_k in the graph has exactly one colored vertex; since F_k has a unique fall k-coloring (up to isomorphism) by Proposition 3, we use this to complete c'.

For the other direction, let c' be a fall k-coloring of G'. Then, since each x_{uv} has $k-1$ neighbours and is colorful, $c'(u) \neq c'(v)$. Restricting c' to V, we obtain a proper k-coloring of G. $\qquad\square$

Theorem 5. *For every $k \geq 3$, it is* NP-*complete to decide whether a bipartite graph G is fall k-colorable.*

Proof. The proof follows by combining Lemmas 1 and 4. $\qquad\square$

We also observe the following slightly stronger corollary.

Corollary 6. *For every $k \geq 3$, it is* NP-*complete to decide whether a bipartite graph G of maximum degree $3(k-1)$ is fall k-colorable.*

Proof. We use Lemmas 1 and 4 with the fact that deciding whether a graph G has $\chi(G) \leq k$ is NP-complete for every $k \geq 3$ even when G has maximum degree $\Delta = 2k - 2$ (see [17, Theorem 3]). Now, $F_k = K_2 \times K_k$ is $(k-1)$-regular, so a vertex of degree Δ in G has degree $\Delta + k - 1 = 3k - 3$ in G'. At the same time, the new vertices (from subdividing, or copies of F_k) have degree at most $k < 3k - 3$. The claim follows. $\qquad\square$

After Corollary 2, it is natural to wonder what are the weakest additional constraints to place on the structure of a planar graph so that say fall 3-coloring is solvable in polynomial time. In the following, we show that maximal outerplanar graphs with at least three vertices admit a fall 3-coloring, and in fact no other fall colorings. We begin with the following two propositions; for short proofs of both we refer the reader to [19].

Proposition 7. *Let G be a chordal graph. Then either $\mathrm{Fall}(G) = \emptyset$ or $\mathrm{Fall}(G) = \{\delta(G) + 1\}$.*

Proposition 8. *If G is a uniquely k-colorable graph, then G is fall k-colorable.*

These results will be combined with the following theorem.

Theorem 9 (Chartrand and Geller [5]). *An outerplanar graph G with at least three vertices is uniquely 3-colorable if and only if it is maximal outerplanar.*

The claimed result is now obtained as follows.

Theorem 10. *Let G be a maximal outerplanar graph with at least three vertices. Then $\mathrm{Fall}(G) = \{3\}$.*

Proof. As every maximal outerplanar graph G is chordal, it follows by Proposition 7 that $\mathrm{Fall}(G) = \emptyset$ or $\mathrm{Fall}(G) = \{\delta(G) + 1\}$. It is well-known that every maximal outerplanar graph has at least two vertices of degree two. By combining Theorem 9 with Proposition 8, we have that $\mathrm{Fall}(G) = \{\delta(G) + 1\} = \{3\}$. $\qquad\square$

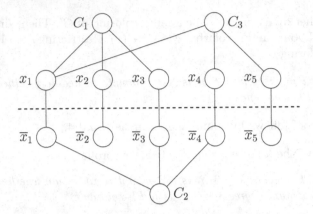

Fig. 2. An instance $\varphi = (x_1 \vee x_2 \vee x_3) \wedge (\overline{x}_1 \vee \overline{x}_3 \vee \overline{x}_4) \wedge (x_1 \vee x_4 \vee x_5)$ of PLANAR MONOTONE 3-SAT, which always admits a planar drawing $G(\varphi)$. Conceptually, the dashed horizontal line separates the upper part containing all positive literals and clauses from the lower part containing all negative literals and clauses.

Also, in the light of Corollary 2, it should be recalled that any proper 2-coloring of a connected bipartite graph is a fall 2-coloring. Thus, the statement of Corollary 2 would not hold for the case of $k = 2$ colors (unless P = NP). However, what if we do not insist on a partition of the vertices but are merely interested in the existence of two disjoint independent dominating sets? As we will show, this problem is NP-complete for planar graphs; in fact even those that are triangle-free. This result is a considerable strengthening of an earlier result of Henning et al. [13], who showed it only for general graphs.

In the PLANAR MONOTONE 3-SAT problem, we are given a 3-SAT formula φ with m clauses over n variables $x_1, x_2, \ldots x_n$, where each clause $c_1, c_2, \ldots, c_m$ comprises either three positive literals or three negative literals. We call such clauses *positive* and *negative*, respectively. Moreover, the associated graph $G(\varphi)$ has a 2-clique (i.e., an edge) $\{x_i, \overline{x}_i\}$ for each variable x_i, a vertex for each c_j, and an edge between a literal contained in a clause and the corresponding clause. In particular, $G(\varphi)$ admits a planar drawing such that every 2-clique sits on a horizontal line with the line intersecting their edges. In addition, every positive clause is placed above the line, while every negative clause is placed below the line (see Fig. 2). The fact that PLANAR MONOTONE 3-SAT is NP-complete and that $G(\varphi)$ admits the claimed planar drawing follows from de Berg and Khosravi [1].

Theorem 11. *It is NP-complete to decide whether a given triangle-free planar graph has two disjoint independent dominating sets.*

Proof. The proof is by a polynomial-time reduction from PLANAR MONOTONE 3-SAT, whose input is a monotone 3-SAT instance φ with a set of m clauses $\mathcal{C} = \{C_1, C_2, \ldots, C_m\}$ over the n variables $\mathcal{X} = \{x_1, x_2, \ldots, x_n\}$. Since our goal

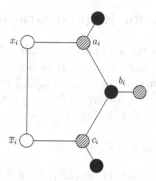

Fig. 3. The variable gadget X_i. If I_1 and I_2 are two disjoint independent dominating sets, then a_i and c_i must both be in either I_1 or I_2. Furthermore, exactly one of x_i and $\overline{x}_i$ can be in $I_1 \cup I_2$.

is to construct a graph G' that is both triangle-free and planar, it is convenient to start from a planar drawing of $G(\varphi)$ as described, and proceed as follows.

For each variable x_i, we extend its corresponding 2-clique in $G(\varphi)$ by replacing it with the following variable gadget X_i (see Fig. 3). Here, X_i is a 5-cycle on the vertices x_i, a_i, b_i, c_i, and $\overline{x}_i$ (in clockwise order) with a pendant vertex attached to each of a_i, b_i, and c_i. Otherwise, we retain the structure of $G(\varphi)$ finishing our construction of G'. Clearly, as $G(\varphi)$ is planar and triangle-free, so is G'. We will then prove that φ is satisfiable if and only if G' contains two disjoint independent dominating sets.

Let φ be satisfiable under the truth assignment $\tau = \{0, 1\}^n$. We construct two disjoint independent dominating sets I_1 and I_2 as follows. For each $i \in [n]$, if τ sets x_i to 1, put x_i to I_1. Otherwise, if τ sets x_i to 0, put $\overline{x}_i$ to I_1. Put every b_i to I_1, and every a_i and c_i to I_2. The pendant vertices of a_i and c_i are put to I_1, while the pendant vertices of b_i are put to I_2. For each $j \in [m]$, put C_j in I_2. Observe that both I_1 and I_2 are independent. Moreover, every vertex of X_i is dominated by a vertex in I_1, and also by a vertex in I_2. Every vertex C_j is dominated by a vertex in I_2, and since τ is a satisfying assignment, C_j must also be adjacent to a vertex in I_1. We conclude that I_1 and I_2 are disjoint independent dominating sets of G'.

Conversely, suppose that I_1 and I_2 are two disjoint independent dominating sets of G'. Clearly, each clause C_j for $j \in [m]$ must be dominated by at least one x_i (or $\overline{x}_i$ in the case of a negative clause). For each $i \in [n]$, observe that a_i and c_i must both be in I_1 or I_2 (for otherwise the pendant of b_i could not be dominated by both a vertex of I_1 and a vertex of I_2). It follows that at most one of x_i and $\overline{x}_i$ can be in $I_1 \cup I_2$. Thus, the vertices in $I_1 \cup I_2$ corresponding to variable vertices encode a satisfying assignment τ for φ. Finally, notice that neither x_i or $\overline{x}_i$ are in $I_1 \cup I_2$, the truth value of the corresponding variable does not affect the satisfiability of φ, and can thus be set arbitrarily in τ.   □

4   Hardness Results for Regular Graphs

In this section, we consider the complexity of fall coloring regular graphs. For connected 2-regular graphs (i.e., cycles), it is not difficult to verify that $2 \in$ Fall(C_n) if and only if $2 \mid n$ and that $3 \in$ Fall(C_n) if and only if $3 \mid n$ with no other integer being in Fall(C_n), for any n (see e.g., [8,19]). However, as we will show next, the problem of fall coloring 3-regular graphs is considerably more difficult.

We begin by recalling the following result.

Theorem 12 (Heggernes and Telle [12]**).** *It is* NP-*complete to decide if the square of a cubic graph is 4-chromatic.*

In addition, we make use of the following fact.

Theorem 13 ([19]**).** *A k-regular graph G is fall $(k+1)$-colorable if and only if G^2 is $(k+1)$-chromatic.*

By combining the two previous theorems, we arrive at the following.

Theorem 14. *It is* NP-*complete to decide whether a 3-regular graph G is fall 4-colorable.*

The previous result suggests that there may be similar intractable fall-colorability problems for regular graphs of higher degree. With this in mind, we use different constructions for regular graphs, to show that fall coloring k-regular graphs for $k > 3$ is NP-complete as well.

Theorem 15. *For every $k \geq 3$, it is* NP-*complete to decide whether a $(2k-2)$-regular graph G is fall k-colorable.*

Proof. The proof is by a polynomial-time reduction from EDGE k-COLORING, where we assume that $k \geq 3$ and that the input graph G is k-regular. Let $G' = L(G)$, that is, G' is the line graph of G. Because G is k-regular, it is straightforward to verify that G' is $(2k-2)$-regular. We then prove that G admits a proper edge k-coloring if and only if G' admits a fall k-coloring.

Let h be a proper edge k-coloring of G. We construct a vertex-coloring c' of G' as follows. Let $c'(x_{uv}) = h(uv)$, where $uv \in E(G)$ and x_{uv} is the vertex of G' corresponding to the edge uv. By construction, x_{uv} for every $uv \in E(G)$ is adjacent to precisely the vertices corresponding to the edges adjacent to u and v. Since h is a proper edge-coloring, h has colored these edges differently from $h(uv) = c'(x_{uv})$. Furthermore, all the k edges incident to the same vertex will receive k different colors. As such, the corresponding vertices in G' will all be colorful. We conclude that c' is a fall k-coloring for G'.

In the other direction, let c' be a fall k-coloring of G'. Consider any x_{uv} of G'. Because c' is a fall k-coloring, each neighbor of x_{uv} has received a distinct color. Again, x_{uv} is adjacent to precisely the vertices that correspond to edges adjacent to u and v in G. Thus, we obtain immediately a proper edge k-coloring h from c', concluding the proof.   □

We can get a similar result for regular graphs with vertices of odd degree by using the *Cartesian product* of graphs G and H, denoted as $G\square H$. As a reminder, $V(G\square H) = V(G) \times V(H)$ and $(u_1, v_1)(u_2, v_2) \in E(G\square H)$ when either $u_1 u_2 \in E(G)$ and $v_1 = v_2$ or $u_1 = u_2$ and $v_1 v_2 \in E(H)$.

Theorem 16. *For every $k \geq 3$, it is NP-complete to decide whether a $(2k-1)$-regular graph G is fall k-colorable.*

Proof. It suffices to modify the graph G' from the proof of Theorem 15 to obtain a graph with mostly the same structure; in particular, a graph which can be fall k-colored exactly when G' can, but whose vertices have common degree on more than those of G'. One such construction is $G'' = G'\square K_2$. It is easy to see that $\mathrm{Fall}(G'') = \mathrm{Fall}(G')$ (see [19]), so G'' has exactly the properties we require. $\square$

5 Further Algorithmic Consequences

In this section, we give further algorithmic consequences of our hardness results.

A popular measure—especially from an algorithmic viewpoint—for the "tree-likeness" of a graph is captured by the notion of treewidth. Here, a *tree decomposition* of G is a pair $(T, \{X_i : i \in I\})$ where $X_i \subseteq V$, $i \in I$, and T is a tree with elements of I as nodes such that:

1. for each edge $uv \in E$, there is an $i \in I$ such that $\{u, v\} \subseteq X_i$, and
2. for each vertex $v \in V$, $T[\{i \in I \mid v \in X_i\}]$ is a tree with at least one node.

The *width* of a tree decomposition is $\max_{i \in I} |X_i| - 1$. The *treewidth* of G, denoted by $\mathrm{tw}(G)$, is the minimum width taken over all tree decompositions of G.

The following result is easy to observe, but we include its proof for completeness.

Theorem 17. *Let G be a graph of bounded treewidth. The fall set $\mathrm{Fall}(G)$ can be determined in polynomial time.*

Proof. It is well-known that every graph of treewidth at most p has a vertex of degree at most p. It follows that the largest integer in $\mathrm{Fall}(G)$ is $\psi_{fall}(G) \leq p+1$. Thus, it suffices to test whether $i \in \mathrm{Fall}(G)$ for $i \in \{1, 2, \ldots, p+1\}$. Furthermore, the fall i-colorability of G can be tested in polynomial time by the result of Telle and Proskurowski [22]. (Alternatively, this can be seen by observing that fall i-colorability can be characterized in monadic second order logic, and then applying the result of Courcelle [6]). The claim follows. $\square$

At this point, it will be useful to recall that a *parameterized problem* I is a pair (x, k), where x is drawn from a fixed, finite alphabet and k is an integer called the *parameter*. Then, a *kernel* for (x, k) is a polynomial-time algorithm that returns an instance (x', k') of I such that (x, k) is a YES-instance if and only if (x', k') is a YES-instance, and $|x'| \leq g(k)$, for some computable function $g : \mathbb{N} \to \mathbb{N}$. If $g(k)$ is a polynomial (exponential) function of k, we say that I admits a polynomial (exponential) kernel (for more, see Cygan et al. [7]).

A consequence of Theorem 17 is that for every $k \geq 1$, FALL k-COLORING admits an exponential kernel. Here, we will observe that Lemma 1 actually proves that this is the best possible, i.e., that there is no polynomial kernel under reasonable complexity-theoretic assumptions.

First, the gadget C_6 each vertex of G is identified with in Lemma 1 has treewidth two. Second, this identification increases the treewidth of G by only an additive constant. To make use of these facts, we recall that Bodlaender et al. [3] proved that 3-COLORING does not admit a polynomial kernel parameterized by treewidth unless NP $\subseteq$ coNP/poly. At this point, it is clear that the proof of Lemma 1 is actually a parameter-preserving transformation (see [7, Theorem 15.15] or [4, Sect. 3]) guaranteeing $\text{tw}(G') \leq \text{tw}(G) + 2$. We obtain the following.

Theorem 18. FALL 3-COLORING *parameterized by treewidth does not admit a polynomial kernel unless* NP $\subseteq$ coNP/poly.

A further consequence of Lemma 1 is that fall k-coloring is difficult algorithmically, even when the number of colors is small and the graph is planar. To make this more precise, we recall the well-known *exponential time hypothesis* (ETH), which is a conjecture stating that there is a constant $c > 0$ such that 3-SAT cannot be solved in time $O(2^{cn})$, where n is the number of variables.

Corollary 19. FALL 3-COLORING *for planar graphs cannot be solved in time* $2^{o(\sqrt{n})}$ *unless ETH fails, where n is the number of vertices. However, the problem admits an algorithm running in time $2^{O(\sqrt{n})}$ for planar graphs.*

Proof. It suffices to observe that the graph G' obtained in the proof of Lemma 1 has size linear in the size of the input graph G. The claimed lower bound then follows by a known chain of reductions originating from 3-SAT (see e.g., [7, Theorem 14.3]).

The claimed upper bound follows from combining the single-exponential dynamic programming algorithm on a tree decomposition of van Rooij et al. [21] with the fact that an n-vertex planar graph has treewidth $O(\sqrt{n})$ (for a proof, see [9, Theorem 3.17]).   □

Finally, observe that the naive exponential-time algorithm for deciding whether $k \in \text{Fall}(G)$ enumerates all possible k-colorings of $V(G)$ and thus requires $k^n n^{O(1)}$ time. A much faster exponential-time algorithm is obtained as follows.

Theorem 20. FALL k-COLORING *can be solved in $3^n n^{O(1)}$ time and polynomial space. In exponential space, the time can be improved to $2^n n^{O(1)}$.*

Proof. The claimed algorithms are obtained by reducing the problem to SET PARTITION, in which we are given a universe $U = [n]$, a set family $\mathcal{F} \subseteq 2^U$, and an integer k. The goal is to decide whether U admits a partition into k members.

We enumerate all the 2^n vertex subsets of the n-vertex input graph G and add precisely those to $\mathcal{F}$ that form an independent dominating set, a property decidable in polynomial time. To finish the proof, we apply the result of

Björklund et al. [2, Thms. 2 and 5] stating that SET PARTITION can be solved in $2^n n^{O(1)}$ time. Further, if membership in $\mathcal{F}$ can be decided in $n^{O(1)}$ time, then SET PARTITION can be solved in $3^n n^{O(1)}$ time and $n^{O(1)}$ space. $\square$

6 Conclusions

We further studied the problem of partitioning a graph into independent dominating sets, also known as fall coloring. Despite the centrality of the concepts involved, independence and domination, a complete understanding of the complexity fall coloring is lacking. Towards this end, our work gives new results and strengthens previously known hardness results on structured graph classes, including various planar graphs, bipartite graphs, and regular graphs.

An interesting direction for future work is finding combinatorial algorithms for fall coloring classes of bounded treewidth (or in fact, bounded cliquewidth). Indeed, the algorithms following from the proof of Theorem 17 are not practical. For concreteness, one could consider outerplanar graphs or cographs.

References

1. de Berg, M., Khosravi, A.: Optimal binary space partitions for segments in the plane. Int. J. Comput. Geom. Appl. **22**(03), 187–205 (2012)
2. Björklund, A., Husfeldt, T., Koivisto, M.: Set partitioning via inclusion-exclusion. SIAM J. Comput. **39**(2), 546–563 (2009)
3. Bodlaender, H.L., Downey, R.G., Fellows, M.R., Hermelin, D.: On problems without polynomial kernels. J. Comput. Syst. Sci. **75**(8), 423–434 (2009)
4. Bodlaender, H.L., Thomassé, S., Yeo, A.: Kernel bounds for disjoint cycles and disjoint paths. Theor. Comput. Sci. **412**(35), 4570–4578 (2011)
5. Chartrand, G., Geller, D.P.: On uniquely colorable planar graphs. J. Comb. Theory **6**(3), 271–278 (1969)
6. Courcelle, B.: The monadic second-order logic of graphs. I. Recognizable sets of finite graphs. Inf. Comput. **85**(1), 12–75 (1990)
7. Cygan, M., et al.: Parameterized Algorithms. Springer, Cham (2015). https://doi.org/10.1007/978-3-319-21275-3
8. Dunbar, J.E., et al.: Fall colorings of graphs. J. Comb. Math. Comb. Comput. **33**, 257–274 (2000)
9. Fomin, F.V., Thilikos, D.M.: New upper bounds on the decomposability of planar graphs. J. Graph Theory **51**(1), 53–81 (2006)
10. Garey, M., Johnson, D., Stockmeyer, L.: Some simplified NP-complete graph problems. Theor. Comput. Sci. **1**(3), 237–267 (1976)
11. Goddard, W., Henning, M.A.: Independent domination in graphs: a survey and recent results. Discrete Math. **313**(7), 839–854 (2013)
12. Heggernes, P., Telle, J.A.: Partitioning graphs into generalized dominating sets. Nord. J. Comput. **5**(2), 128–142 (1998)
13. Henning, M.A., Löwenstein, C., Rautenbach, D.: Remarks about disjoint dominating sets. Discrete Math. **309**(23), 6451–6458 (2009)
14. Laskar, R., Lyle, J.: Fall colouring of bipartite graphs and cartesian products of graphs. Discrete Appl. Math. **157**(2), 330–338 (2009)

15. Leven, D., Galil, Z.: NP-completeness of finding the chromatic index of regular graphs. J. Algorithms **4**(1), 35–44 (1983)
16. Lewis, R.: A Guide to Graph Colouring. Springer, Cham (2015). https://doi.org/10.1007/978-3-319-25730-3
17. Maffray, F., Preissmann, M.: On the NP-completeness of the k-colorability problem for triangle-free graphs. Discrete Math. **162**(1–3), 313–317 (1996)
18. Marx, D.: Graph colouring problems and their applications in scheduling. Electr. Eng. **48**(1–2), 11–16 (2004)
19. Mitillos, C.: Topics in graph fall-coloring. Ph.D. thesis, Illinois Institute of Technology (2016)
20. Resende, M.G., Pardalos, P.M.: Handbook of Optimization in Telecommunications. Springer, Boston (2008). https://doi.org/10.1007/978-0-387-30165-5
21. van Rooij, J.M.M., Bodlaender, H.L., Rossmanith, P.: Dynamic programming on tree decompositions using generalised fast subset convolution. In: Fiat, A., Sanders, P. (eds.) ESA 2009. LNCS, vol. 5757, pp. 566–577. Springer, Heidelberg (2009). https://doi.org/10.1007/978-3-642-04128-0_51
22. Telle, J.A., Proskurowski, A.: Algorithms for vertex partitioning problems on partial k-trees. SIAM J. Discrete Math. **10**, 529–550 (1997)

Succinct Representation of Linear Extensions via MDDs and Its Application to Scheduling Under Precedence Constraints

Fumito Miyake[1], Eiji Takimoto[1], and Kohei Hatano[1,2]

[1] Department of Informatics, Kyushu University, Fukuoka, Japan
{fumito.miyake,eiji,hatano}@inf.kyushu-u.ac.jp
[2] RIKEN AIP, Tokyo, Japan

Abstract. We consider a single machine scheduling problem to minimize total flow time under precedence constraints, which is NP-hard. Matsumoto et al. proposed an exact algorithm that consists of two phases: first construct a Multi-valued Decision Diagram (MDD) to represent feasible permutations of jobs, and then find the shortest path in the MDD which corresponds to the optimal solution. Although their algorithm performs significantly better than standard IP solvers for problems with dense constraints, the performance rapidly diminishes when the number of constraints decreases, which is due to the exponential growth of MDDs. In this paper, we introduce an equivalence relation among feasible permutations and show that it suffices to construct an MDD that maintains only one representative for each equivalence class. Experimental results show that our algorithm outperforms Matsumoto et al.'s algorithm for problems with sparse constraints, while keeping good performance for dense constraints. Moreover, we show that Matsumoto et al.'s algorithm can be extended for solving a more general problem of minimizing weighted total flow time.

Keywords: Combinatorial optimization · Job scheduling · Precedence constraints · MDD

1 Introduction

A single machine scheduling problem to minimize total flow time under precedence constraints $(1|prec|\sum c_j)$ is fundamental in the scheduling literature. The problem is, given processing times of n jobs and precedence constraints for pairs of jobs, to find an order of n jobs (permutation) which minimizes the sum of wait times and process times of all the jobs (called flow time) among those satisfying the precedence constraints [2]. The problem is known to be NP-hard [12,13] and various 2-approximate polynomial time algorithms are proposed [4,5,10,14,20].

On the other hand, non-trivial exact algorithms for solving the problems are not known until recently, except standard methods on integer programming

© Springer Nature Switzerland AG 2019
C. J. Colbourn et al. (Eds.): IWOCA 2019, LNCS 11638, pp. 365–377, 2019.
https://doi.org/10.1007/978-3-030-25005-8_30

formulations. Matsumoto et al. [15] proposed an exact algorithm using a variant of the Multi-valued Decision Diagrams (MDD) [16]. It can efficiently construct a MDD expressing linear extensions from given precedence constraints, and solve the problem in linear time in terms of the size of the diagram, by reducing the scheduling problem to the shortest path problem on the diagram. The algorithm of Matsumoto et al. outperformed standard IP-based methods for synthetic problems with dense precedence constraints. However, Their algorithm performs worse when constraints are sparse, where the resulting MDDs have exponentially large size.

In this paper, we propose a more efficient exact algorithm which overcomes the weakness of Matsumoto et al.'s algorithm. Our key idea is to introduce some equivalence relations between linear extensions and exploit the equivalence properties to obtain a more succinct MDD which represents the equivalent feasible solutions.

In the experiments on synthetic data sets, our method performs more than 10 times faster than standard IP-based methods and improves Matsumoto et al.'s method for sparse constraints at the same time.

Moreover, we show that Matsumoto et al.'s algorithm can be extended for solving a more general problem of minimizing weighted total flow time under precedence constraints $(1|prec| \sum w_j c_j)$.

1.1 Comparison to Previous Work

Succinct data structures such as BDDs (Binary Decision Diagrams) [1,3], ZDDs (Zero Suppressed BDDs) [11,17,18] and MDDs are used for counting and solving NP-hard problems. To the best of our knowledge, there are few diagram-based approaches to scheduling problems except Matsumoto et al. [15], and Ciré and van Hoeve [6,7]. The work of Ciré and van Hoeve [6,7] deals with a different scheduling problem which has release times and deadlines and not applicable to ours.

A notable advantage of our method and other diagram-based methods over standard IP-based ones is that once the feasible set is represented by a diagram, we can reuse the diagram for different objectives, which save much computation time. This is also advantageous for the online setting, e.g. [8,19].

2   Preliminaries

Let $[n] = \{1, 2, \ldots, n\}$ denote the set of jobs and S_n denote the set of all permutations over $[n]$, where each permutation is represented by a vector $\pi = (\pi_1, \pi_2, \ldots, \pi_n)$. For example, $S_3 = \{(1,2,3), (1,3,2), (2,1,3), (2,3,1), (3,1,2), (3,2,1)\}$. For a permutation $\pi \in S_n$, we define the inverse permutation of π as $\pi^{-1} = (\pi_1^{-1}, \ldots, \pi_n^{-1})$, where $\pi_j^{-1} = i$ if and only if $\pi_i = j$. A permutation $\pi \in S_n$ specifies a job scheduling, i.e., an order of jobs to be processed, in the way that jobs i should be processed in the decreasing order of π_i. Here, π_i can be interpreted as the priority of job i. In other words, the order of jobs specified by a permutation π is $\pi_n^{-1} \to \pi_{n-1}^{-1} \to \cdots \to \pi_1^{-1}$.

For each job $i \in [n]$, let $p_i \in \mathbb{R}$ be the processing time, and define the completion time $c_i = \sum_{j:\pi_j \geq \pi_i} p_j$ which is sum of process times and wait times. Then the flow time of a job scheduling $\pi \in S_n$ is defined as $\sum_{i=1}^n c_i$. The flow time of π is the sum of completion times over all jobs under the order specified by π. For example, a permutation $\pi = (4, 2, 1, 3)$ specifies the order of jobs $1 \to 4 \to 2 \to 3$, and thus the sum of process times and wait times of all the jobs is $p_1 + (p_1 + p_4) + (p_1 + p_4 + p_2) + (p_1 + p_4 + p_2 + p_3)$, which amounts to $4p_1 + 3p_4 + 2p_2 + p_3 = \pi \cdot p$.

Precedence constraints over the jobs are represented as a directed acyclic graph (DAG) $G = ([n], E)$, where a directed edge $(i, j) \in E$ represents the constraint that job i has to be done before job j. We call G a constraint graph. We denote by S_G the set of permutations satisfying the precedence constraints specified by G (i.e., linear extensions of G). More specifically,

$$S_G = \{\pi \in S_n \mid \forall (i, j) \in E, \pi_i > \pi_j\}.$$

Similarly, the set S_G^{-1} of inverse permutations in S_G is defined as

$$S_G^{-1} = \{\pi^{-1} \mid \pi \in S_G\}.$$

An example is given in Fig. 1.

Now, we are ready to define a single machine scheduling problem to minimize total flow time under precedence constraintsas follows:

Input: DAG $G = ([n], E)$, process time vector $p \in \mathbb{R}^n$

$$\text{Output: } \pi = \underset{\pi \in S_G}{\text{argmin}} \sum_{i=1}^n c_i = \underset{\pi \in S_G}{\text{argmin}} \ \pi \cdot p \qquad (1)$$

$$\text{where, } c_i = \sum_{j:\pi_j \geq \pi_i} p_j$$

3   Previous Work of Matsumoto et al.

We review the previous work of Matsumoto et al. [15], where they propose an algorithm of finding an optimal solution of single machine scheduling problem to minimize total flow time under precedence constraints by using a data structure called a π-MDD.

In what follows, we identify a permutation $\pi = (\pi_1, \pi_2, \ldots, \pi_n)$ with the string $\pi_1 \pi_2 \cdots \pi_n$ of length n by concatenating all components of π. Then, S_G^{-1} can be regarded as a set of strings. A π-MDD for a constraint graph G is defined as the smallest DFA[1] that only accepts the strings in S_G^{-1}. Figure 2 shows an

---

[1] More precisely, we omit non-accepting states and transitions to non-accepting states in the automaton.

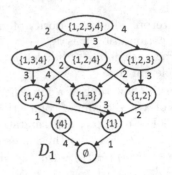

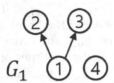

Fig. 1. DAG G_1 representing precedence constraints $S_{G_1}^{-1} = \{(2,3,1,4),$ $(2,3,4,1),(2,4,3,1),(3,2,1,4),(3,2,4,1),$ $(3,4,2,1),(4,2,3,1),(4,3,2,1)\}$.

Fig. 2. A π-MDD that represents linear extensions in S_{G_1} satisfying precedence constraints G_1 in Fig. 1

example of π-MDDs. We note that (i) a π-MDD is a DAG with the root (initial state) and a single leaf (unique accepting state) and every edge e is labeled with a job $l(e) \in [n]$. (ii) each path $(e_1, e_2, \ldots, e_n)$ from the root to the leaf corresponds to the permutation $(l(e_1), l(e_2), \ldots, l(e_n))$ that is accepted by the π-MDD. For a π-MDD D, we denote by $L(D)$ the set of all strings that D accepts. Clearly, a π-MDD D for G should satisfy $L(D) = S_G^{-1}$. We define the size of D, denoted by $|D|$, as the number of edges in D.

Matsumoto et al.'s method for solving problem (1) consists of the following two steps.

Step 1. Construct a π-MDD $D = (V_D, E_D)$ for a given DAG $G = ([n], E)$. See algorithm MakePiMDD in Algorithm 1. Note that $G|_{V'}$ is the subgraph of G induced by V'.

Step 2. (i) Assign weights $a_e = dp_{l(e)}$ to each edge $e \in E_D$ of π-MDD D, where d is the depth of edge e from the root of D. (ii) Find the shortest path $(e_1, e_2, \ldots, e_n)$ from the root to the leaf in the weighted graph obtained in (i). (iii) Output π^* where $(\pi^*)^{-1} = (l(e_1), l(e_2), \ldots, l(e_n)) \in S_G^{-1}$.

Below we describe an outline of Algorithm MakePiMDD for step 1.

Given a constraint graph $G = (V, E)$ with $V \subseteq [n]$, MakePiMDD recursively constructs a π-MDD D_G for G. Every node of D_G is associated with a subset of V and thus is identified with the subset. The root and the leaf of D_G correspond to the whole set V and the empty set $\emptyset$, respectively. The root V has outgoing edges to a node $V' \subset V$ if and only if there exists a node $v \in V$ such that (i) $v = V \backslash V'$ and (ii) the out-degree of v is zero in G.

The following fact is shown for Step 1.

Lemma 1 (Matsumoto et al. [15]). *Given a constraint graph $G = ([n], E)$, MakePiMDD outputs a π-MDD D such that $L(D) = S_G^{-1}$.*

Algorithm 1. MakePiMDD

Require: DAG $G = (V, E)$, where $V \subseteq [n]$
1: **if** $V = \emptyset$ **then**
2:     **return** node $\emptyset$
3: **end if**
4: $D \leftarrow (V_D, E_D)$ with $V_D = \{V\}$ and $E_D = \emptyset$.
5: **for each** $v \in V$ whose outdegree is 0 in $G = (V, E)$ **do**
6:     $V' \leftarrow V \setminus \{v\}$
7:     **if** have never memorized the π-MDD D_G for G **then**
8:         $D' = (V_{D'}, E_{D'}) \leftarrow$ MakePiMDD$(G|_{V'})$
9:     **end if**
10:     $V_D \leftarrow V_D \cup V_{D'}$
11:     $E_D \leftarrow E_D \cup E_{D'} \cup \{(V, V')\}$
12: **end for**
13: memorize D as D_G
14: **return** D_G

The correctness of Step 2 can be easily verified from the following observation. Assignment of weights in Step 2 (ii) ensures that for each path $(e_1, e_2, \ldots, e_n)$ from the root to the leaf and the corresponding permutation $\pi^{-1} = (l(e_1), l(e_2), \ldots, l(e_n))$, the weighted length of the path is exactly the flow time of π: $\sum_{d=1}^{n} a_{e_d} = \sum_{d=1}^{n} dp_{l(e_d)} = \sum_{d=1}^{n} dp_{\pi_d^{-1}} = \sum_{i=1}^{n} \pi_i p_i = \boldsymbol{\pi} \cdot \boldsymbol{p}$.

Now we give a characterization of π-MDDs, which is not explicitly given in [15]. Let $<_G$ be the partial order naturally defined by a constraing graph $G = ([n], E)$. That is, for any $u, v \in [n]$, $v \leq_G u$ if and only if there exists a directed path from u to v in G. A set $L \subseteq [n]$ is called a lower set of G if for any $u, v \in [n]$, $u \in L$ and $v <_G u$ imply $v \in L$. The family of lower sets of G is denoted as $\mathcal{L}_G \subseteq 2^{[n]}$. The following relationship between the family of lower sets $\mathcal{L}_G$ and linear extensions S_G of G is well-known and easily verified.

Proposition 1. *The partially ordered set* $(\mathcal{L}_G, \supseteq)$ *is a distributive lattice with the minimum element* $\emptyset$ *and the maximum element* $[n]$ *and for each maximal chain* $L_1 = [n] \supseteq L_2 \supseteq \cdots \supseteq L_{n+1} = \emptyset$, *there exists a linear extension* $\boldsymbol{\pi} = (\pi_1, \pi_2, \ldots, \pi_n) \in S_G$ *such that* $L_i = L_{i+1} \cup \{\pi_i^{-1}\}$.

From the proposition above, we immediately get the characterization as described in the following corollary.

Corollary 1. *The* π-*MDD* $D = (V_D, E_D)$ *for a constraint graph* G *is isomorphic to the Hasse diagram of the partially ordered set* $(\mathcal{L}_G, \supseteq)$. *That is, there exsits a one-to-one correspondence between* V_D *and the lower sets in* $\mathcal{L}_G$ *(the root and the leaf correspond to* $[n]$ *and* $\emptyset$, *respectively) and for any* $L, L' \in \mathcal{L}$ *such that* $L = L' \cup \{u\}$, *there exists a directed edge* $e = (L, L')$ *with* $l(e) = u$ *in* D.

Using the new characterization, we give a performance bound of Matsumoto et al.'s method in terms of the size of $\mathcal{L}_G$.

Theorem 1 ([15]). *For any constraint graph $G = ([n], E)$, we can find an optimal solution of problem (1) in $O(n|\mathcal{L}_G|)$ time.*

4   Main Result

Intuitively, when given fewer constraints in G we would have an exponentially many lower sets in $\mathcal{L}_G$ and thus Matsumoto et al.'s method would take exponential time as suggested in Theorem 1. In particular, the worst case occurs when no constraints are given (i.e., no edges in the constraint graph G). In this case, any subset in $[n]$ is a lower set and thus we have $\mathcal{L}_G = 2^{[n]}$. On the other hand, if no constraints are given, then we can easily obtain an optimal solution: process jobs i in the increasing order of the process time p_i. In other words, we can solve the problem in $O(n \log n)$ time by just sorting $(p_1, p_2, \ldots, p_n)$.

Our method is based on the observation above. Suppose we are given a subset A of jobs that can be somehow regarded as a no constraint set with respect to the linear extensions S_G, then we can find an optimal (partial) solution over A by sorting. So when constructing π-MDD for G, we do not need to consider all $|A|!$ permutations over A but it suffices to fix a representative permutation over A, which reduces the number of paths by a factor of $1/|A|!$. In this way, we construct a succinct π-MDD that accepts only representative permutations.

4.1   Equivalence Relation

Definition 1 (Input and output sets). *For a DAG $G = ([n], E)$, and each node $v \in [n]$, input set I_v and output set O_v of v are defined respectively as follows:*

$$I_v = \{v' \in [n] \mid (v', v) \in E\}, O_v = \{v' \in [n] \mid (v, v') \in E\}.$$

Definition 2 (Equivalence relations between jobs). *Given a DAG $G = ([n], E)$, job i and j are equivalent if and only if $I_i = I_j$ and $O_i = O_j$ and denoted as $i \simeq_G j$. If it is clear from the context, we abbreviate $i \simeq j$.*

The equivalence relation $\simeq_G$ implies a partition over $[n]$

$$[n] = A_1 \cup A_2 \cup \cdots \cup A_N,$$

where each A_j $(j = 1, \ldots, N)$ is an equivalence set defined by $\simeq_G$. We call each A_j a job equivalence set. Figure 3 shows an illustration of job equivalence sets.

For a permutation $\pi \in S_n$, a set $A \subseteq [n]$, and a bijection $\delta : A \to A$, let $\pi \circ \delta$ is defined as

$$(\pi \circ \delta)_i = \begin{cases} \pi_{\delta(i)} & i \in A, \\ \pi_i & \text{otherwise.} \end{cases}$$

Then we define a equivalence relation between linear extensions in S_G.

Definition 3 (Equivalence relations between linear extensions). *Permutations* $\pi, \pi' \in S_n$ *are equivalent w.t.t. a DAG* $G = ([n], E)$ *if and only if for each job equivalence set* A_j, *there exists a bijection* $\delta_{A_j} : A_j \to A_j$ *such that*

$$\pi' = \pi \circ \delta_{A_1} \circ \delta_{A_2} \circ \cdots \circ \delta_{A_N}.$$

We denote $\pi \simeq_G \pi'$ *if* π *and* π' *are equivalent w.r.t.* G. *When clear from the context, we abbreviate as* $\pi \simeq \pi'$.

Figure 4 shows an illustration of equivalence relation between permutations. For clarity, we used expression of inverse permutation.

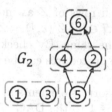

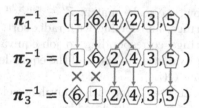

Fig. 3. Illustration of equivalence sets defined by a DAG G_2

Fig. 4. For a DAG G_2, π_1 and π_2 are equivalent, but π_2 and π_3 are not.

The following proposition holds for the permutations which are equivalence each other.

Proposition 2. *For any permutations* $\pi, \pi' \in S_G$, $\pi \simeq \pi' \iff \forall j \in [n], \pi_j^{-1} \simeq \pi'_j^{-1}$.

The following lemma holds for the equivalence relation on S_G.

Lemma 2. *For any permutations* $\pi, \pi' \in S_n$, $\pi \in S_G$ *and* $\pi' \simeq \pi$ *implies* $\pi' \in S_G$.

It can be proved easily by the symmetry between equivalence jobs on G.

For each equivalence class $[\pi]$, we define the representative $\tilde{\pi}$ as the permutation satisfying (i) $\tilde{\pi} \simeq \pi$, and (ii) for any $k, l \in A_j (k < l)$, $\tilde{\pi}_k > \tilde{\pi}_l$. In other words, $\tilde{\pi}$ is the particular representative of $[\pi]$ satisfying additional precedence constraints defined with job id numbers.

4.2 Our Algorithm

We propose an exact algorithm for solving problem (1). Our algorithm consists of the following steps.

1. Compute job equivalence sets $A_1, \ldots, A_N$.
2. Construct a DAG $\tilde{G}$ satisfying $S_{\tilde{G}} = \{\tilde{\pi} \mid \pi \in S_G\}$

3. Run the algorithm of Matsumoto et al. [15] with input $\tilde{G}$ and obtain a π-MDD $\tilde{D}$ representing $S_{\tilde{G}}$.
4. For some appropriate weights over edges in $\tilde{D}$, solve the shortest path problem and construct an optimal solution of (1) from it.

For step 1, we compute a adjacency matrix of G, $X \in \mathbb{R}^{n \times n}$, that is, $X(i,j) = 1$ when there is a directed edge from job i to job j, otherwise $X(i,j) = 0$. Then, we define matrix $Z \in \mathbb{R}^{n \times 2n}$ as follows

$$Z(i,j) = \begin{cases} X(i,j) & (j \le n) \\ X^{\top}(i,j-n) & (j > n). \end{cases}$$

Now, we make matrix $Z' \in \mathbb{R}^{n \times 2n}$ by regarding each rows of Z as individual strings, and sorting with radix-sort. Since $v \simeq v' \Leftrightarrow \forall i \in [n]\ Z'(v,i) = Z'(v',i)$, strings corresponding to job-equivalent nodes are the same. By checking the equivalence, we can find the equivalent classes. These procedures can be done in time $O(n^2)$.

In Step 2, for each equivalence class A_j ($j \in [N]$), sort its elements $v_1 < v_2 < \cdots < v_{|A_j|}$ and add edges (v_i, v_{i+1}) for $1 \le i \le |A_j| - 1$ to the DAG G, which we denote as $\tilde{G}$ (Fig. 5) shows an illustration of $\tilde{G}$. Computation time for Step 2 is $O(n)$.

Step 3 takes $O(n|\mathcal{L}_{\tilde{G}}|)$ time to obtain a π-MDD $\tilde{D}$ by using Matsumoto et al.'s algorithm.

In Step 4, we first sort the weight vector $\boldsymbol{p}$ for each job equivalent set. To do so, we prepare a n-dimensional array C and sort $\boldsymbol{p}$ in the following way.

1. For any $i \in [n]$, $i \simeq C[i]$.
2. For any $i,j \in [n]$ such that $i \simeq j$, $i < j$ implies $p_{C[i]} \le p_{C[j]}$.

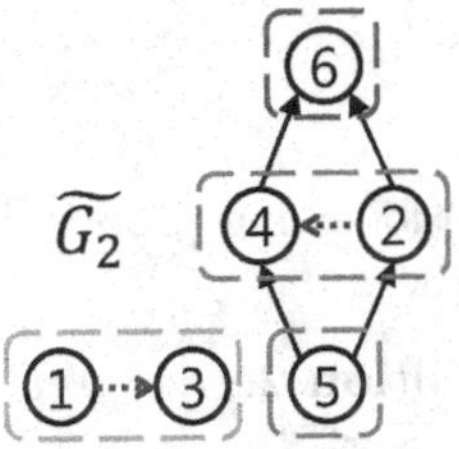

Then, for each edge $e \in E_{\tilde{D}}$, we assign a weight $dp_{C[l(e)]}$, where d is the depth of the edge e from the root node and $l(e)$ is label of e, respectively. For the assigned weights, we solve the shortest path problem over $\tilde{D}$ and obtain the shortest path $(e_1, \ldots, e_n)$ from the root node to the leaf node. Finally we output π^* such that $\pi^{*-1} = (C[l(e_1)], \ldots, C[l(e_n)]) \in S_G^{-1}$. Computation time for step 4 is $O(n \log n + n|\mathcal{L}_{\tilde{G}}|)$.

We now state our main result.

$\widetilde{G_2}$

Fig. 5. DAG $\tilde{G}_2$ obtained by adding edges to the DAG G_2 in Fig. 3

Theorem 2. *Our algorithm solves problem (1) in time* $O(n^2 + n|\mathcal{L}_{\tilde{G}}|) = O(n|\mathcal{L}_{\tilde{G}}|)$.

Proof. First of all, we show that the set $S_{\tilde{G}}$ of linear extensions of $\tilde{G}$ satisfies $S_{\tilde{G}} = \{\tilde{\pi} \mid \pi \in S_G\}$. For any $\pi \in S_{\tilde{G}}$, it is clear that $\pi \in S_G$ by the construction of $\tilde{G}$. Furthermore, since π satisfies additional precedence constraints regarding job equivalence classes A_js, for any jobs $k,l \in A_j$, $k < l$ implies $\pi_k > \pi_l$. This means that π is a representative for some $[\pi']$ in S_G. Therefore, $S_{\tilde{G}} \subseteq \{\tilde{\pi} \mid$

$\pi \in S_G$}. For any $\pi \in S_G$, by definition, the representative $\tilde{\pi} of [\pi]$ satisfies additional precedence constraints for job equivalence classes. Thus $\tilde{\pi} \in S_{\tilde{G}}$, implying $S_{\tilde{G}} \supseteq \{\tilde{\pi} \mid \pi \in S_G\}$.

Then, we prove the correctness of step 4.

From Lemma 2, for any permutation $\pi \in S_{\tilde{G}}$, it satisfies constraints represented with $\tilde{G}$ even if the elements of any equivalence classes are arbitrarily rearranged within the classes. Now, the order of elements of any equivalence class $A = \{i_1, i_2, \ldots, i_{|A|}\} \subseteq [n]$ with a smallest flow time is obviously the order such that for $i, j \in A (i \neq j)$ if $\pi_i > \pi j$, then $p_i < p_j$. Therefore, in the first half of step 4, for any equivalence class $A \subseteq [n]$, we make rules for converting the series of job number that job $i, j \in A (s.t. \ \pi_i > \pi_j)$ satisfies $i < j$ to the series of job number that job $i, j \in A (s.t. \ \pi_i > \pi_j)$ satisfies $p_i \leq p_j$ as array C. Accordingly, the conversion by array C can be regarded as playing the role of following function $F : S_{\tilde{G}} \to S_G$

$$F(\tilde{\pi}) = \operatorname*{argmin}_{\pi \in [\tilde{\pi}]} \pi \cdot p.$$

Thus, in π-MDD $\tilde{D}_C$ obtained by replace each edge label $l(e)$ of π-MDD $\tilde{D}$ with $C[l(e)]$, any path corresponds to the permutation with a smallest flow time in equivalence class which the permutation belongs. In the latter half of step 4, it can be regarded as solving the shortest path problem on π-MDD $\tilde{D}_C$, so the solution of the problem is also the inverse permutation of optimal solution of the original problem. Hence, by evaluating the inverse permutation of that permutation, we can get optimal solution.   □

At last, we show that the size of π-MDD constructed by our method doesn't become large than that of π-MDD constructed by Matsumoto et al.'s method.

Theorem 3. *Let D be the π-MDD representing linear extensions S_G of G, and let $\tilde{D}$ be the π-MDD representing linear extensions $S_{\tilde{G}}$ of $\tilde{G}$, $|\tilde{D}| \leq |D|$.*

Proof. The DAG $\tilde{G}$ can be obtained by adding constraints to G, so any lower set of $\tilde{G}$ is also a lower set of G. Accordingly, it holds that $\mathcal{L}_{\tilde{G}} \subseteq \mathcal{L}_G$. By Corollary 1, $\tilde{D}$ is isomorphic to the Hasse diagram of the partially ordered set $(\mathcal{L}_{\tilde{G}}, \supseteq)$. Clearly, the partially ordered set $(\mathcal{L}_{\tilde{G}}, \supseteq)$ is obtained by restricting $(\mathcal{L}_G, \supseteq)$ on the domain $\mathcal{L}_{\tilde{G}}$. Therefore, $\tilde{D}$ is the subgraph of D induced by the subset $\mathcal{L}_{\tilde{G}}$ of states.   □

5 Extension to $1|prec| \sum w_j c_j$

In this section, we show that Matsumoto et al.'s algorithm can be extended for solving a more general problem of minimizing weighted total flow time.

We consider about extensive setting which each job $i \in [n]$ has weight $w_i \in \mathbb{R}_+$. Now, we define a single machine scheduling problem to minimize weighted total flow time under precedence constraints($1|prec| \sum w_j c_j$) as follows [2].

Input: DAG $G = ([n], E)$ process time vector $\boldsymbol{p} \in \mathbb{R}^n$

weight vector $\boldsymbol{w} \in \mathbb{R}^n_+$

Output: $\boldsymbol{\pi} = \underset{\pi \in S_G}{\operatorname{argmin}} \ \boldsymbol{w} \cdot \mathbf{c}$ (2)

where, $c_i = \sum_{j : \pi_j \geq \pi_i} p_j$

Think about for every $i \in [n]$, $w_i = 1$, the flow time of (2) equals the flow time of (1). Thus, the problem (2) is clearly generalized setting of the problem (1).

For this problem, we propose simple extension of Matsumoto et al.'s method which there is only a difference point. The **Step 2** of Matsumoto et al.'s method in Sect. 3, assigns weights $v_e = dp_{l(e)}$ to each edge $e \in E_D$ of π-MDD $D = (V_D, E_D)$. For problem (2), instead, introduce cumulative weight $\hat{w}_e = \sum_{i:[n] \backslash V_e} w_i$ and assign $a_e = \hat{w}_e p_{l(e)}$. Then, for each path $(e_1, e_2, \ldots, e_n)$ from the root to the leaf and the corresponding permutation $\boldsymbol{\pi}^{-1} = (l(e_1), l(e_2), \ldots, l(e_n))$, the weighted length of the path is exactly the flow time of $\boldsymbol{\pi}$. Certainly, computation time is equal to the time of Matsumoto et al.'s method. Now, we show a theorem.

Theorem 4. *Our algorithm solves problem (2) in time $O(n|\mathcal{L}_G|)$.*

The proof is omitted and shown in Appendix.

6   Experiments

In this section, we compare the efficiency of proposed method and previous methods for the scheduling problem with precedence constraints on artificial data sets.

6.1   Settings of Artificial Data Sets and Methods

As artificial data sets, we generate Erdös-Rényi random graph $G_{n,q}$ as constraints $G = ([n], E)$. That is, over $G = ([n], E)$, for each job $i, j \in [n](i < j)$, there exists $(i, j) \in E$ with probability $0 \leq q \leq 1$[2]. Also, we chose processing time vectors $\boldsymbol{p}$ according to the uniform distribution over $[0, 1]^n$.

We compare the proposed method, Matsumoto et al.'s method, and integer programming (IP) with permutation matrices and comparison matrices[3], respectively. Let $n = 30$, and for each $q \in \{0.01, 0.02, \ldots, 1\}$, we generate 10 random graphs and processing time vectors, and observe the average of computation times of each method and sizes of π-MDD constructed by the proposed method and Matsumoto et al.'s method. These methods are implemented by C++ with Gurobi optimizer 8.1.0 [9] to solve integer programs. We run them in a machine with Intel(R) Xeon(R) Processor X5560 2.80 GHz and 198 GB memory.

---

[2] Note that, because of the constraint that $i < j$, the graph $G = ([n], E)$ is a DAG.

[3] Please refer to [15] for details.

6.2   Results and Discussion

Figure 6 shows the computation times(in the logarithmic scale) of each method for different choices of $q \in \{0.01, 0.02, \ldots, 1\}$. Figure 7 shows the size of π-MDD(logarithmic axis) generated by the proposed method and Matsumoto et al.'s method for different choices of $q \in \{0.01, 0.02, \ldots, 1\}$.

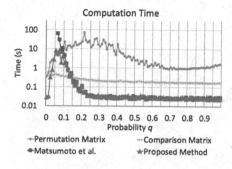

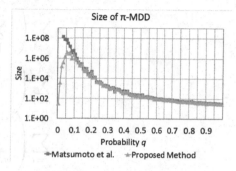

Fig. 6. Average computation time for $n = 24, q \in \{0.01, 0.02, \ldots, 1\}$

Fig. 7. Average size of π-MDD for $n = 24, q \in \{0.01, 0.02, \ldots, 1\}$

These results show that the proposed method is fastest and most space-efficient among others for any q. Also, the results of Matsumoto et al.'s method show that for sparse precedence constraints, its computational complexity become much worse. However, with proposed method, even for sparse constraints its computational time is moderately small and still smallest among others.

7   Conclusion and Future Work

We proposed an improved algorithm of Matsumoto et al. which exploits the symmetry in permutations satisfying precedence constraints by introducing a notion of equivalence class among jobs. Our future work includes improving our algorithm for the cases where conventional IP solvers are still advantageous. Also, extension of our algorithm for weighted flow time is still open. In addition, it may be possible to construct only necessary parts of π-MDD dynamically, which would further improve our algorithm.

Appendix

Proof of Theorem 4

We show the proof of Theorem 4.

Proof. We assign $a_e = \hat{w}_e p_{l(e)}$ to each edge $e \in E_D$ of π-MDD $D = (V_D, E_D)$. Then, for each path $(e_1, e_2, \ldots, e_n)$ from the root to the leaf and the corresponding permutation $\pi^{-1} = (l(e_1), l(e_2), \ldots, l(e_n))$, the weighted length of the path can be calculated as follows.

$$\sum_{d=1}^{n} a_{e_d} = \sum_{d=1}^{n} \hat{w}_{e_d} p_{l(e_d)}$$

$$= \sum_{d=1}^{n} \left(\sum_{i:[n]\backslash V_{e_d}} w_i \right) p_{l(e_d)}$$

$$= \sum_{d=1}^{n} \left(\sum_{i:\pi_i \leq \pi_{l(e_d)}} w_i \right) p_{l(e_d)}$$

$$= \sum_{d=1}^{n} \left(\sum_{i:\pi_i \leq \pi_d^{-1}} w_i \right) p_{\pi_d^{-1}}$$

$$= \sum_{j=1}^{n} \left(\sum_{i:\pi_i \leq \pi_j} w_i \right) p_j$$

$$= \sum_{i=1}^{n} w_i \left(\sum_{j:\pi_j \geq \pi_i} p_j \right) = \boldsymbol{w} \cdot \boldsymbol{c}$$

Now, we get the flow time of problem (2).   □

References

1. Akers, S.B.: Binary decision diagrams. IEEE Trans. Comput. **C–27**(6), 509–516 (1978)
2. Brucker, P.: Scheduling Algorithms, 4th edn. Springer, Heidelberg (2004). https://doi.org/10.1007/978-3-540-24804-0
3. Bryant, R.E.: Graph-based algorithms for Boolean function manipulation. IEEE Trans. Comput. **C–35**(8), 677–691 (1986)
4. Chekuri, C., Motwani, R.: Precedence constrained scheduling to minimize sum of weighted completion times on a single machine. Discrete Appl. Math. **98**(1–2), 29–38 (1999)
5. Chudak, F.A., Hochbaum, D.S.: A half-integral linear programming relaxation for scheduling precedence-constrained jobs on a single machine. Oper. Res. Lett. **25**(5), 199–204 (1999)
6. Ciré, A.A., van Hoeve, W.J.: MDD propagation for disjunctive scheduling. In: Proceedings of the 22nd International Conference on Automated Planning and Scheduling, ICAPS 2012 (2012)
7. Ciré, A.A., van Hoeve, W.J.: Multivalued decision diagrams for sequencing problems. Oper. Res. **61**(6), 1411–1428 (2013)

8. Fujita, T., Hatano, K., Kijima, S., Takimoto, E.: Online linear optimization for job scheduling under precedence constraints. In: Chaudhuri, K., Gentile, C., Zilles, S. (eds.) ALT 2015. LNCS (LNAI), vol. 9355, pp. 332–346. Springer, Cham (2015). https://doi.org/10.1007/978-3-319-24486-0_22
9. Gurobi Optimization Inc.: Gurobi optimizer reference manual (2018). http://www.gurobi.com
10. Hall, L.A., Schulz, A.S., Shmoys, D.B., Wein, J.: Scheduling to minimize average completion time: off-line and on-line approximation algorithms. Math. Oper. Res. **22**(3), 513–544 (1997)
11. Knuth, D.E.: The Art of Computer Programming. Volume 4A, Combinatorial Algorithms, Part 1. Addison-Wesley Professional, Boston (2011)
12. Lawler, E.L.: On sequencing jobs to minimize weighted completion time subject to precedence constraints (1978)
13. Lenstra, J.K., Kan, A.H.G.R.: Complexity of scheduling under precedence constraints. Oper. Res. **26**(1), 22–35 (1978)
14. Margot, F., Queyranne, M., Wang, Y.: Decompositions, network flows, and a precedence constrained single-machine scheduling problem. Oper. Res. **51**(6), 981–992 (2003)
15. Matsumoto, K., Hatano, K., Takimoto, E.: Decision diagrams for solving a job scheduling problem under precedence constraints. In: SEA 2018 (2018)
16. Miller, M.D., Drechsler, R.: Implementing a multiple-valued decision diagram package. In: Proceedings of the 28th IEEE International Symposium on Multiple-Valued Logic, ISMVL 1998, pp. 52 57 (1998)
17. Minato, S.: Zero-suppressed BDDs for set manipulation in combinatorial problems. In: Proceedings of the 30th International Design Automation Conference, DAC 1993, pp. 272–277 (1993)
18. Minato, S.: Zero-suppressed BDDs and their applications. Int. J. Softw. Tools Technol. Transf. **3**(2), 156–170 (2001)
19. Sakaue, S., Ishihata, M., Minato, S.: Efficient bandit combinatorial optimization algorithm with zero-suppressed binary decision diagrams. In: Proceedings of the Twenty-First International Conference on Artificial Intelligence and Statistics (AISTATS 2018). PMLR, vol. 84, pp. 585–594 (2018). http://proceedings.mlr.press/v84/sakaue18a.html
20. Schulz, A.S.: Scheduling to minimize total weighted completion time: performance guarantees of LP-based heuristics and lower bounds. In: Cunningham, W.H., McCormick, S.T., Queyranne, M. (eds.) IPCO 1996. LNCS, vol. 1084, pp. 301–315. Springer, Heidelberg (1996). https://doi.org/10.1007/3-540-61310-2_23

Maximum Clique Exhaustive Search in Circulant k-Hypergraphs

Lachlan Plant and Lucia Moura[✉]

School of Electrical Engineering and Computer Science, University of Ottawa,
Ottawa, Canada
{lplant053,lmoura}@uottawa.ca

Abstract. In this paper, we discuss two algorithms to solve the maximum clique problem in circulant k-hypergraphs, and do an experimental comparison between them. The first is the Necklace algorithm proposed by Tzanakis, Moura, Panario and Stevens [11] which uses an algorithm to generate binary necklaces by Ruskey, Savage and Wang [10]. The second algorithm is a new algorithm which we call Russian Necklace that is based on a Russian doll search for maximum cliques by Östergård [6] with extra pruning that takes advantages of properties specific to circulant k-hypergraphs. Our experiments indicate that the new algorithm is more effective for hypergraphs with higher edge density.

1 Introduction

Many problems in combinatorics can be modelled as finding a maximum clique in a certain graph or hypergraph [7]. Here, we consider hypergraphs with edges of the same cardinality $k \geq 2$, which we call k-hypergraphs. A *clique on a k-hypergraph* is a set of vertices such that every k-subset of vertices in the set forms an edge. A k-hypergraph is circulant if the hypergraph has an automorphism that is a cyclic permutation of its vertices. Cliques in circulant hypergraphs are relevant to combinatorial problems with some intrinsic cyclic structure, which appear quite often. For example, in [11] the authors search for covering arrays that correspond to a subset of columns of a matrix with a cyclic structure, which translates into searching for a maximum clique on a circulant k-hypergraph.

As with other NP-hard problems, a lot of effort has been done to push the boundary on the size of problems that can be solved via exhaustive search. Early versions of backtracking algorithms for maxclique have been given in [1, 2, 13].

The first author was supported by an OGS Scholarship and the second, by an NSERC discovery grant.

C. J. Colbourn et al. (Eds.): IWOCA 2019, LNCS 11638, pp. 378–392, 2019.
https://doi.org/10.1007/978-3-030-25005-8_31

A somewhat different and more recent algorithm based on the Russian Doll search introduced in [12] is given by Östergård [6]. This algorithm is the basis for the package Cliquer, which is an effective software for clique finding [5]. The generalization of clique algorithms from graphs to k-hypergraphs is straightforward, but since we have not found them in the literature, we present them in Sect. 2 for completeness.

In Sect. 3, we show how the algorithms for ordinary k-hypergraphs can be effectively modified to deal with circulant k-hypergraphs using the concept of *binary necklaces* to eliminate equivalent cliques. The algorithm in Sect. 3.2 is basically the same as the Necklace algorithm in [11], but presented in the context of cliques in circulant k-hypergraphs. In Sect. 3.3, we present the new algorithm, which we call the Russian Necklace algorithm.

In Sect. 4, we describe our implementation and do an experimental comparison of the four algorithms for two test sets of circulant k-hypergraphs. The experiments show that for the problems considered it is definitely much more efficient to take advantage of the cyclic structure using necklaces. When comparing the two necklace-based algorithms, our experiments with random circulant k-hypergraphs indicate that the Russian Necklace algorithm is more effective than the Necklace algorithm for hypergraphs with higher edge density, while the opposite is the case for lower edge density. The details of this threshold is analysed. In our second set of experiments, we attempt to make some progress on open problems from [11], specially to try to find a larger clique (and therefore a better covering array) for the cases where only a partial search was achieved. We were successful in completing one of their missing cases, but this complete search did not yield a clique larger than the ones they found. This harder problem was also an interesting test set to demonstrate the effectiveness of Russian Necklace over the Necklace algorithm for dense hypergraphs.

2   Max-Clique Algorithms for k-Hypergraphs

In this section, we show how to adapt two existing maximum clique algorithms for graphs to k-hypergraphs. These algorithms are Algorithm 1' and 2' given side-by-side as modifications of Algorithm 1 and 2, respectively, while these latter algorithms will only be fully described in Sect. 3.

Algorithm 1 Necklace Algorithm	▷ BT: **Algorithm 1'** Backtracking Algorithm

1: **procedure** NECKLACERECURSE(C, U)   ▷ BT: **procedure** BACKTRACKRECURSE(C, U)
2:     **if** $|U| = 0$ **then**
3:         **if** $|C| > |$maxClique$|$ **then**
4:             maxClique $\leftarrow C$
        return
5:     **while** $U \neq \emptyset$ **do**
6:         $i \leftarrow \min(U)$
7:         $U \leftarrow U \setminus \{i\}$
8:         **if** not(ISNECKLACE($C \cup \{i\}$)) **then return**   ▷ BT: remove this line
9:         $X \leftarrow$ HYPERGRAPHFILTER(C, i, U)
10:         $C \leftarrow C \cup \{i\}$
11:         **if** $|C| + |X| \leq |$maxClique$|$ **then return**
12:         NECKLACERECURSE(C, X)   ▷ BT: BACKTRACKRECURSE(C, X)
13:         $C \leftarrow C \setminus \{i\}$
    return

1: **procedure** NECKLACESEARCH($H = (V, E), k$)   ▷ BT: **procedure** BACKTRACKSEARCH(H, k)
2:     Global maxClique $\leftarrow \emptyset$
3:     Global edgeSize $\leftarrow k$
4:     $n \leftarrow |V|$
5:     **for** $e \in E$ **do**
6:         $C \leftarrow e$
7:         **if** not(ISNECKLACE(C)) **continue**   ▷ BT: remove this line
8:         $U \leftarrow \{\max(e) + 1, ..., n - 1\}$
9:         $X \leftarrow$INITIALFILTER(e, U)
10:         NECKLACERECURSE(C, X)   ▷ BT: BACKTRACKRECURSE(C, X)
    return maxClique

2.1   Standard Backtracking with Bounding

Backtracking for maximum cliques starts with an empty clique and larger cliques are recursively built by adding one possible vertex at a time. Backtracking Search is displayed as Algorithm 1' which differs from Algorithm 1 (described in Sect. 3.2) by eliminating line 8 of the recursive procedure and line 7 of the main procedure. In the main program BACKTRACKSEARCH, we initialize a clique with each edge at a time to avoid iterating through all k-subsets of the vertex set, some of which will not form an edge and thus cannot be together in a clique. The procedure INITIALFILTER(C, U) selects single vertices from U that form an edge together with each $(k - 1)$-subset of C. The recursive part BACKTRACK-RECURSE is similar to a standard clique algorithm (see [2,6]). Here we compute $X \leftarrow$HYPERGRAPHFILTER(C, i, U), which assigns to X the subset of U consisting of vertices such that each of those vertices together with i forms an edge with every $(k - 2)$-subset of C. We do not give this procedure explicitly, but this can be simply done by iterating through each element u of U and then through each $(k - 2)$-subset L of C and checking whether $\{i, u\} \cup L \in E$, immediately discarding u whenever this condition fails, and including u in X if it passes the test for each L. Thus, this is done in time $O(|U|^{k-1})$. Note that this procedure is much simpler in the case of graphs, since L is trivially empty, and we only need to check whether $\{i, u\}$ is an edge for each $u \in U$. In line 10, i is added to the current clique C. In Line 11, we use the size bound for abandoning the current search if the union of the current clique and set X of candidates that may be added to C cannot improve on the current best clique cardinality.

2.2 Russian Doll Backtracking for Max-Cliques

The Russian doll search applied to cliques is based on the idea of solving small (nested) maxclique subproblems which can then be used within the global max-clique problem to help with bounding the size of cliques on induced subgraphs. This algorithm was proposed by Östergård in [6] and here we adapt it to work for k-hypergraphs by initializing cliques with edges in the main program and by using algorithms HYPERGRAPHFILTER and INITIALFILTER in the same way as explained in Sect. 2.1. Russian Doll Algorithm is displayed as Algorithm 2' which differs from Algorithm 2 (Russian Necklace Algorithm discussed in Sect. 3.3) by deleting lines 6 and 7 from the recursive procedure and changing lines 11 and 12 in the main procedure.

An ordering of the vertices is set arbitrarily, let us assume vertices are already relabeled $\{0, \ldots, n-1\}$ to reflect this ordering. Let $H^i = H[\{i, \ldots, n-1\}]$ be the subhypergraph of H induced by $\{i, \ldots, n-1\}$. The maxclique problem is solved for each hypergraph in the following order $H^{n-1}, H^{n-2}, \ldots, H^0$, as each run of the loop in line 8 of the main procedure RUSSIANDOLLSEARCH; and the size of the maxclique in H^i is stored in the $dynamicValues[i]$ at the end of each such loop. Note that H^{i+1} is a subgraph of H^i. So, if C^i is a maxclique of H^i and C^{i+1} is a maxclique of H^{i+1} then $|C^{i+1}| \leq |C^i| \leq |C^{i+1}| + 1$ and moreover, if $|C^i| = |C^{i+1}| + 1$ then $i \in C^i$. For this reason, during the recursion to solve the problem for hypergraph H^i, it is only worth to consider cliques that contain vertex i, and if the current clique reaches size $dynamicValues[i+1] + 1$, then a maximum clique of C^i has been found; in this case, we can stop any further iterations for the loop in line 8 of RUSSIANDOLLSEARCH and any further recursive calls in procedure RUSSIANDOLLRECURSE (see the stopping criteria in line 18). Another bounding criteria is used in line 10 of RUSSIANDOLLRECURSE, before we add the vertex j as the next element to the current clique C. The clique being constructed is bound by the sum of $|C|$ and the size of the maximum clique in H^j, so we abandon the current search if this value does not improve on the best clique found so far.

Algorithm 2 Russian Necklace Algorithm   ▷ RD: **Algorithm 2'** Russian Doll Algorithm

1: **procedure** RUSSIANNECKLACERECURSE($C, U, step$)
2:   ▷ RD: **procedure** RUSSIANDOLLRECURSE($C, U, step$)
3:     **if** $|U| = 0$ **then**
4:         **if** $|C| > |$maxClique$|$ **then**
5:             maxClique $\leftarrow C$
        return $|C|$
6:     **if** LARGESTGAP(C, U) $> step$ **then return** $|C|$.   ▷ RD: remove line
7:     $x \leftarrow \max(C \setminus \{n-1\})$ ▷ RD: remove line
8:     **while** $U \neq \emptyset$ **do**   •
9:         $j \leftarrow \min(U)$
10:        **if** $dynamicValues[j] + |C| \leq |$maxClique$|$ **then return** $|C|$
11:        $U \leftarrow U \setminus \{j\}$
12:        $X \leftarrow$ HYPERGRAPHFILTER(C, j, U)
13:        $C \leftarrow C \cup \{j\}$
14:        **if** $|C| + |X| \leq |$maxClique$|$ **then return** $|C|$
15:        **if** $(j - x) > step$ **then return** $|C|$ ▷ RD: remove line
16:        $max \leftarrow$ RUSSIANNECKLACERECURSE($C, X, step$)
17:   ▷ RD: $max \leftarrow$ RUSSIANDOLLRECURSE($C, X, step$)
18:        **if** $max > dynamicValues[step + 1]$ **then return** max
19:        $C \leftarrow C \setminus \{j\}$
        return $|$maxClique$|$

1: **procedure** RUSSIANNECKLACESEARCH($H = (V, E), k$)
2:   ▷ RD: **procedure** RUSSIANDOLLSEARCH($H = (V, E), k$)
3:     Global maxClique $\leftarrow \emptyset$
4:     Global dynamicValues $\leftarrow int[|V|]$
5:     $n \leftarrow |V|$
6:     **for** $i = 0$ to $k - 1$ **do**
7:         $dynamicValues[n - 1 - i] \leftarrow i + 1$
8:     **for** $i = (n - 1 - k)$ downto 0 **do**
9:         $S \leftarrow \{e \in E : \min(e) = i\}$
10:        **for** $e \in S$ **do**
11:            $C \leftarrow e \cup \{n - 1\}$ ▷ RD: $C \leftarrow e$
12:            $U \leftarrow \{(\max(e) + 1), \ldots, n - 2\}$.   ▷ RD: $U \leftarrow \{(\max(e) + 1), \ldots, n - 1\}$
13:            $X \leftarrow$ INITIALFILTER(e, U)
14:            **if** RUSSIANNECKLACERECURSE(C, X, i)   ▷ RD: **if** RUSSIANDOLLRECURSE(C, X, i)
15:                $> dynamicValues[i + 1]$ **then break**
16:        $dynamicValues[i] \leftarrow |$maxClique$|$
        return maxClique

---

3   Max-Clique Algorithms for Circulant k-Hypegraphs

A few concepts related to necklaces are reviewed in Sect. 3.1, before we describe the algorithms for circulant k-hypergraphs, namely Algorithm 1 (Necklace Algorithm) in Sect. 3.2 and Algorithm 2 (Russian Necklace Algorithm) in Sect. 3.3.

3.1   Background on Circulant Graphs and Necklaces

Let $\alpha = \alpha_0 \alpha_1 \ldots \alpha_{n-1}$ be a string of length n. A string $\beta = \beta_0 \beta_1 \ldots \beta_{n-1}$ is a *rotation* of α if there exists a natural number s such that $\beta_i = \alpha_{(i+s) \bmod n}$ for all $0 \leq i \leq n - 1$. We denote by $R(\alpha, i)$ the string obtained from the rotation of string α, i spaces to the right. An n-bead binary *necklace* is an equivalence class of binary strings of length n under rotation. In other words strings α and β are equivalent if there exists i such that $R(\alpha, i) = \beta$ for some integer i. The *canonical necklace* of a necklace is the lexicographically smallest of the strings among the rotations of the necklace. In this paper, necklaces are used in the algorithms to represent equivalent classes of subsets of an n-set.

Let S be a subset of an n-set $\{p_1, p_2, ..., p_n\}$. The *characteristic string* of S is a string $x \in \{0,1\}^n$ such that $x_i = 1$ if and only if $p_i \in S$, $1 \leq i \leq n$. The *rotation of a subset S, i spaces to the right* is the subset S' which corresponds to the rotation i spaces to the right of the characteristic string of S. The *right justified characteristic string of a subset S* is the characteristic string of the subset $R(S, (n - \max(S)))$, where $\max(S) = \max\{i : p_i \in S\}$. The *canonical subset necklace* of a subset is either the empty set, or contains p_1 and its right justified characteristic string is a canonical necklace. From now on, we use the term *necklace* to refer to a canonical necklace instead of to its equivalence class. Similarly, we use *subset necklace* to refer to a canonical subset necklace. These concepts are illustrated in the following example.

Example 1. The following strings are necklaces: 010101, 001011, 001101, while their respective rotations 101010, 011001, 010011 are not necklaces since they are lexicographically larger. Consider subset $S_1 = \{0, 2, 3\} \subset \{0, 1, 2, 3, 4, 5\}$; its characteristic string is 101100 and its right justified characteristic string is 001011; since the latter is a necklace and $0 \in S_1$ then S_1 is a subset necklace. We note that the right justified characteristic string rotates the last 1 in the characteristic string to the right, making it possible to include element 0 in the subset while having a correspondence to a string that is a necklace (which would be impossible with trailing zeroes present in the characteristic string). Now, consider subset $S_2 = \{0, 1, 4\} \subset \{0, 1, 2, 3, 4, 5\}$; its characteristic string is 110010 and its right justified characteristic string is 011001; since the latter is not a necklace, then S_2 is not a subset necklace.

Here we present an alternative definition of circulant k-hypergraphs based on subset necklaces; the proof of this equivalence is based on using subset necklaces to represent all edges and can be found in [8]. A k-hypergraph $H = (V, E)$ is *circulant* if there exists a relabeling of the vertices $V = \{0, 1, ..., |V| - 1\}$ and a set $\mathcal{N}$ of subset necklaces, each being a subset of V and having cardinality k, such that, $E = \{R(N, i) : N \in \mathcal{N}, 0 \leq i \leq |V| - 1\}$.

We list a couple of simple results without proofs, which can be found in [8].

Proposition 1. *Let $H = (V, E)$ be a circulant k-hypergraph labeled $\{0, 1, ..., |V| - 1\}$ and let $0 \leq i < |V|$. Then, $C \subseteq V$ is a clique if and only if $R(C, i)$ is a clique.*

Corollary 1. *For a circulant k-hypergraph $H = (V, E)$ and $0 \leq i < |V|$, $C \subseteq V$ is a maximal clique if and only if $D = R(C, i)$ is a maximal clique.*

The consequence of this corollary is that it is sufficient to search for maximum cliques that are subset necklaces. This fact is the essential fact exploited in the algorithms to come.

3.2 The Necklace Algorithm for Circulant Hypergraphs

Tzanakis et al. [11] combined the necklace search algorithm by Ruskey et al. [10] with the standard backtracking for maxcliques to produce an algorithm to search for covering arrays. Their algorithm translated as a maxclique search in a k-hypergraph is presented in Algorithm 1, where changes with respect to the regular Backtracking Algorithm (Algorithm 1') are highlighted.

The direct application of Corollary 1 within procedure NECKLACERECURSE justifies skipping the processing of $C \cup \{i\}$ if it is not a necklace. The fact that we not only skip this clique but also any $C \cup \{j\}$ with $j \geq i$ in this loop comes from necklace properties observed by Ruskey et al. [10]. As shown in [11], by using the bijection between subset necklaces and their right-justified characteristic strings, the order in which the cliques are built in the clique algorithms is the same order as strings are considered in the necklace generation in [10]. Having a "return" in line 8 of procedure NECKLACERECURSE instead of simply continuing the loop with the next $C \cup \{j\}$ is justified by the correctness of the necklace generation by Ruskey et al. [10], this guarantees a number of search tree nodes bounded by twice the number of necklaces. In the main procedure NECKLACESEARCH, we also add a check in line 7 to ensure that each initial clique given to procedure NECKLACERECURSE is also a necklace. In Algorithm 1, we call procedure ISNECKLACE(C) that verifies if subset C of $\{0, \ldots, n-1\}$ is a subset necklace. This can be done in time $O(n)$ by checking if the corresponding right justified characteristic string is a necklace by using a well known modification of Duval's algorithm (see [8, Sect. 2.2] or [9, Sect. 7.4]).

3.3 The New Russian Necklace Algorithm for Circulant Hypegraphs

The objective of the new algorithm presented in this section (Algorithm 2) is to combine the advantages of the Russian doll algorithm with the fact that we only need to consider cliques that correspond to necklaces, which was also exploited in the Necklace algorithm (Algorithm 1). We state two key results for the optimizations used in the new algorithm. For a hypergraph $H = (V = \{0, 1, 2, \ldots, n-1\}, E)$, we continue to use H^i to denote $H[\{i, i+1, \ldots, n-1\}]$, the subgraph of H induced by $\{i, i+1, \ldots, n-1\}$.

Proposition 2. Let $H = (V = \{0, 1, 2, \ldots, n-1\}, E)$ be a circulant k-hypergraph. Let C_j be a maximum clique of H^j where $0 \leq j \leq n-1$, and let $0 \leq i < n-1$. If $|C_i| > |C_{i+1}|$ then $i \in C_i$ and $n-1 \in C_i$.

Proof. $C_i \subseteq \{i, i+1, \ldots, n-1\}$ and since $|C_i| > |C_{i+1}|$, we know $C_i \not\subseteq \{i+1, \ldots, n-1\}$ and it follows that $i \in C_i$. Assume that $n-1 \notin C_i$. Thus, $C_i \subseteq \{i, i+1, \ldots, n-2\}$ and $R(C_i, 1) \subseteq \{i+1, \ldots, n-1\}$. From Proposition 1, $R(C_i, 1)$

must be a clique in H, and since $R(C_i, 1) \subseteq \{i+1, ..., n-1\}$, $R(C_i, 1)$ must be a clique in $H[\{i+1, ..., n-1\}]$. This leads to a contradiction as $|R(C_i, 1)| > |C_{i+1}|$, where C_{i+1} is a maximum clique in $H[\{i+1, ..., n-1\}]$.   □

While for the Russian doll algorithm, we only need to consider cliques in H^i that include vertex i, the previous proposition allows us to force the inclusion of both i and $n-1$ in the case of circulant hypergraphs. This inclusion of $n-1$ in every clique gives modifications reflected in lines 11 and 12 of Procedure RUS-SIANNECKLACESEARCH in Algorithm 2, with respect to RUSSIANDOLLSEARCH.

Proposition 3. *Let* $H = (V = \{0, 1, 2, ..., n-1\}, E)$ *be a circulant k-hypergraph. Let C_j be a maximum clique of H^j where $0 \leq j \leq n-1$, and let $0 \leq i < n-1$. If $|C_i| > |C_{i+1}|$ then there does not exists a pair of vertices x, y where $i < x < y < n-1$ such that $y - x \geq i$ and $C_i \cap \{x, x+1, ..., y\} = \emptyset$.*

Proof. We assume by contradiction that there exists a pair of elements x, y where $i < x < y < n-1$ such that $y - x \geq i$ and $C_i \cap \{x, x+1, ..., y\} = \emptyset$. We let $j = y - x$. We take the rotation $A = R(C_i, -x)$. From Proposition 1, the set A must be a clique in H. Since $C_i \cap \{x, x+1, ..., y\} = \emptyset$, $A \cap \{0, 1, ..., j\} = \emptyset$ and A is a clique in H^{j+1}. Since $j \geq i$, H^{j+1} is a subgraph of H^{i+1}, and so A must be a clique in H^{i+1}. This contradicts the fact that $|C_i| = |A| > |C_{i+1}|$.   □

The previous proposition is a justification of what we call the *gap bound*: in H^i we only need to consider cliques that do not miss any interval of vertices $\{x, x+1, ..., y\}$ of cardinality $i+1$. A lower bound for the current largest gap is calculated in procedure LARGESTGAP(C, U) invoked in line 6 of procedure RUSSIANNECKLACERECURSE, taking into account the current clique C and set U containing the possible elements to be added to C. What this procedure returns is the largest gap in set $C \cup U$, since any clique that extends C must be contained in $C \cup U$ and its gap cannot be smaller than the gap in $C \cup U$. Note that in our implementation the largest gap is more efficiently updated within procedure HYPERGRAPHFILTER with no extra cost, since we can keep track of the gap as we filter the elements of U in order. The gap bound is invoked both in line 6 and in line 15 of procedure RUSSIANNECKLACERECURSE in Algorithm 2.

This concludes the description of the Russian Necklace Algorithm (Algorithm 2). We remark that inside individual calls to procedure RUSSIANNECK-LACERECURSE, some cliques considered may correspond to equivalent subset necklaces, unlike the Necklace Search algorithm, since in procedure RUSSIAN-NECKLACERECURSE no explicit check for the subset necklace property is done. However, we note that the gap bound for H^i is already eliminating a lot of non-necklaces that have already been considered in either of $H^{i+1}, ..., H^{n-1}$.

4   Experimental Comparison

We do an experimental evaluation of the Russian Necklace algorithm by comparing its performance to the other algorithms, namely Backtracking, Russian

Doll, and Necklace. These experiments are divided into two sets: randomly generated circulant k-hypergraphs, and circulant k-hypergraphs generated using linear feedback shift register sequences (LFSR). The first set of experiments involves many small problems varying the main parameters (edge size, density and number of vertices), as a way to compare the relative performance of the various algorithms. We verified that the two necklace-based algorithms indeed outperform the other algorithms. We are also able to verify a crossover in the value of the density that separates when the Russian Necklace outperforms the Necklace algorithm and vice-versa. The second set of experiments involves a difficult combinatorial problem of current interest for combinatorial design theory, chosen with the aim to compute new results and compare the Necklace and the Russian Necklace algorithms.

The second type of experiments was for a problem for which new results could yield improvements on new bounds for covering arrays. As mentioned in Sect. 3.2, Tzanakis et al. [11] designed the Necklace Algorithm to search for covering arrays as subsets of columns of a cyclic matrix defined from polynomials over finite fields. Several of their bounds still holds the current record in the covering array tables by Colbourn [3]. It is worth noticing that due to the large size of these problems, their search was only successfully completed for the smallest size of problems for finite fields $q \leq 4$. All the other values they reported for prime powers $5 \leq q \leq 23$ came from searches that they were not able to complete, but reported on best found cliques. Our hope was to use the new Russian Necklace algorithm to complete that search for a few more cases, while also using the data set to compare all algorithms on a challenging problem of combinatorial interest.

4.1 Implementation Details

All experiments were done on an Intel Core i7-4790 CPU @ 3.60 GHz CPU running Windows 7 OS and Java 7. In order to mitigate any differences in non-algorithm based segments of the implementation (logging, timing, graph storage, and edge lookups) the testing software was written in Java using abstract classes with only the search algorithm, and in the case of the Russian Necklace algorithm the filter, being different for each search. The hypergraphs were stored using the matrix adjacency map detailed in [8, Sect. 3.2.2]. In short, a $(k-2)$-dimensional matrix is used to index a list of vertices which together with the indexes of the matrix plus vertex 0 form an edge of the k-hypergraph; each of these lists is stored as a Hash map. This permits testing whether a k-subset of vertices is an edge in expected constant time. Memory usage was adequate for the range of k and n in our experiments, but some space should be saved with corresponding time tradeoff for larger sets of parameters.

4.2 Results for Random Circulant k-Hypergraphs, $k = 3, 4, 5$

There is no standard dataset for benchmarking the maxclique problem for circulant k-hypergraphs unlike clique problems for ordinary graphs [6]. Therefore, we use a general dataset to compare the algorithms based on randomly generated circulant k-hypergraphs. Three main parameters are used: number of vertices, edge size, and edge density. For each group of experiments, we first fix edge size $k = 3, 4, 5$; note that this is not too restrictive as $k = 2$ includes the class of all circulant graphs. In order to model the performance of each algorithm, for each parameter set of density and edge size, all algorithms were run with increasing number of vertices until the execution time of a single search surpassed 5 min. This process started with the smallest k-hypergraphs on 60 vertices and increased the number of vertices by 5, at each step. An increase of 5 vertices is equivalent to a potential search space 32 times larger than the previous step. Hypergraphs were generated at random and the same set of hypergraphs was used for all algorithms. Our generation of random circulant hypergraphs was implemented by selecting each of the possible subset necklaces to be an edge with a given probability d which we call "edge density". Because the necklace equivalence classes do not all have the same size, the expected value for the edge density of the k-hypergraph $(|E(H)|/\binom{n}{k})$ does not exactly match d, as it would be the case for k-hypergraphs that are not circulant. The word *density* should be understood as a density of necklace representatives rather an exact edge density.

The graphs in the appendix show a sample of our experiments for random k-hypergraphs with $k \in \{3, 4, 5\}$, where we compare the performance in terms of running time of the four algorithms presented in this paper. Note that Backtracking and Russian doll algorithms do not use the fact that the graphs are circulant, so here we refer to them as the naive algorithms; they both perform poorly with respect to the other algorithms and both perform similarly to each other. In most graphs the lines black and green overlap to the point they look like the same line.

For each $k = 3, 4, 5$, the Russian Necklace algorithm was much faster than the Necklace algorithm for higher density graphs. This can be seen in the first graphs for each k where the red line is much lower than the blue line. As an example for $k = 3$ and 95% density hypergraph, the Russian Necklace algorithm took 11 s to search the 100-vertex hypergraph while the Necklace algorithm took 239 s. Both naive algorithms surpassed the 5 min threshold at 90 vertices.

For all values of k considered, the relative advantage of the Russian Necklace algorithm with respect to the Necklace algorithm decreased consistently as the edge density decreased, until a point in which the performances inverted. This observation is based on more intermediate density data available in [8], from which a sample is included in the appendix. This crossover in performance happened at different densities for different k, namely around 50% for $k = 3$, 80% for $k = 4$ and between 85% and 90% for $k = 5$. This crossover point in the graphs can be observed in the second to last graph for each k.

The decrease of the density of graphs in successive experiments also leads to an increase in the overall size of the graphs which could be searched by all algorithms. As an example for $k = 3$, when the density was set to 95% the largest graphs searchable by the naive algorithms was 90, while the Necklace algorithm was able to search up to size 110 and Russian Necklace achieved a hypergraph of 130 vertices. At 50% density, the naive algorithms reached 565 vertices while the Necklace and Russian Necklace algorithms achieved 1160 and 1170 vertices respectively. The overall size (number of vertices) of the hypergraphs that were searchable within the alloted 5 min was greatly reduced when increasing the edge size of the hypergraphs. At 90% density the largest hypergraph searched dropped from 215 for $k = 3$ to 165 vertices for $k = 4$ and to under 120 vertices for $k = 5$. This is due to the increased complexity of the filtering operation. This decrease in hypergraph size was seen across all algorithms and densities. Finally, due to the system limitation on memory, a limit occurred in the ability of our implementation to test 4- and 5-hypergraphs beyond a certain size. For $k = 4$ this limitation started to manifest around 350 vertices and for $k = 5$ around 160 vertices.

The focus of this first set of experiment was not to invest a lot of time on solving large instances of the problem, but rather to investigate the general trend on how the various variables affect the performance of the algorithms. Our main conclusion is that the two algorithms based on necklace are much more efficient than the naive algorithms that ignore the circulant structure of the hypergraphs. For higher densities the algorithm of choice is our new Russian necklace algorithm, while for densities below a threshhold the Necklace algorithm is faster. Based on our experiment with random k-hypergraphs for $k = 3, 4, 5$, it seems this threshold gets higher as k increases.

4.3   Data Set 2: Circulant 4-Hypergraphs for Covering Arrays from LFSR

We refer the reader to [4] for an introduction to covering arrays, to [11] for more details about the types of covering array being constructed by this search and to [8] for details on how this relates to maxcliques on a 4-hypergraph; giving all the pertinent definitions is out of the scope of this extended abstract. In this section, we generally refer to this problem as "the covering array problem". The size of these problems is regulated by a parameter q which is a prime power, which yields a 4-hypergraph with $n = (q^4 - 1)/(q - 1)$ vertices with high edge density (for instance 90% density for $q = 5$). Each problem also depends on two primitive polynomials over finite fields that are used to generate two LFSR sequences that are used to specify which 4-subsets of the n vertices form an edge. The cyclic structure of the LFSR sequences makes the 4-hypergraphs circulant. The number of rows of the array is fixed as $2q^4 - 1$ and the objective is to maximize the number of columns while guaranteeing the "covering property". Roughly speaking, the "covering property" is equivalent to having a clique on the 4-hypergraph, while maximizing the number of columns corresponds to the clique being maximum. The problems are identified by columns q and α in Table 1, and correspond to the same problems searched by Tzanakis et al. [11] for the given values of q.

Table 1. Results for circulant 4-hypergraphs for the covering array problems.

q	α's	$|V|$	d (%)	MC	RN Nodes	RN time	NK Nodes	NK time
2	1,7	15	69.230	6	98	0:00:00:1174	171	0:00:00:1953
3	1,7	40	75.467	9	23,232	0:00:00:4687	36,792	0:00:00:7968
3	1,11	40	80.282	10	24,543	0:00:00:5248	39,063	0:00:00:8954
3	1,13	40	80.107	9	23,764	0:00:00:5421	45,538	0:00:00:8452
4	1,3	85	87.532	17	3,567,014	0:00:07:6440	9,898,739	0:00:27:8461
4	1,7	85	85.030	17	2,611,565	0:00:05:6160	7,009,781	0:00:19:5781
4	1,9	85	87.532	12	5,061,145	0:00:11:1228	15,069,882	0:00:41:1686
4	1,13	85	87.532	12	5,432,480	0:00:11:6064	15,404,529	0:00:42:1046
4	1,21	85	87.330	12	5,255,894	0:00:11:2927	14,187,758	0:00:38:5634
4	1,29	85	87.532	17	3,452,242	0:00:07:6440	10,005,604	0:00:28:9578
4	1,37	85	85.030	17	2,578,597	0:00:05:6940	6,985,333	0:00:19:4845
5	1,7	156	91.368	16	2,683,547,098	3:54:21:1033	6,134,969,456	8:39:42:7434
5	1,11	156	91.335	16	1,982,032,139	3:27:30:3462	6,208,479,757	8:22:55:8228
5	1,17	156	91.263	13	2,369,951,412	3:38:26:2212	8,899,220,743	10:57:54:4306
5	1,23	156	91.368	16	2,013,735,362	3:17:47:5576	5,937,207,993	7:59:51:4535
5	1,29	156	91.263	13	2,357,303,914	3:42:43:3005	8,825,998,885	10:34:30:4996
5	1,31	156	91.487	14	1,944,185,789	1:44:57:6043	7,444,111,830	10:01:27:4149
5	1,41	156	90.973	14	2,103,450,639	1:35:26:3755	7,391,899,093	5:25:18:3458
5	1,43	156	91.335	15	2,023,046,470	1:34:33:2259	6,017,721,671	8:08:33:7267
5	1,47	156	86.676	12	937,795,472	1:00:07:3437	3,129,362,174	3:33:59:8125
5	1,53	156	89.470	14	1,366,079,584	1:01:51:9345	4,810,185,753	6:05:18:5671
5	1,61	156	90.973	14	2,086,462,019	1:35:39:5419	7,453,156,714	9:20:50:8353

Tzanakis et al. [11] searched for max-cliques on $H_{q,f,g}$ for q a prime power in $[2, \ldots, 23]$ and every possible pair of inequivalent primitive polynomilal f, g with coefficients in a finite field of order q. For $q = 2, 3, 4$ they were able to complete the search for every relevant pair of polynomials, and determined the covering array with the largest number of columns for each of them. However for $q \geq 5$, each of their searches did not complete, and they reported the largest clique found at the point where the search was terminated. These results still yield current records of covering arrays in Colbourn's tables of covering arrays [3], but it was left as an open question whether these values could be improved upon. Here, we settle the case of $q = 5$ by performing a complete search of the $q = 5$ hypergraphs, and confirming that the best clique found in their partial search is optimal. We also reproduced their results of $q = 2, 3, 4$, using two completely different implementations. The next possible case, $q = 7$, appears to be out of reach for the current implementations. The remainder of this section describes how we used this difficult dataset to confirm the effectiveness of our new Russian Necklace algorithm for dense 4-hypergraphs and report on our results.

In this data set, we first ran the searches on $q = 2, 3, 4$ problems presented in [11] using both the Russian Necklace and Necklace algorithms. Table 1 shows the results of these searches. This table provides the time taken [RN Time, NK Time], and number of nodes [RN Nodes, NK Nodes] in the search tree for each search, as well as a reference to the LFSRs used [q, α's], density of the hypergraph [d], and maxclique sizes [MC]. Note that the values in column α are powers of the two primitive elements which give rise to the primitive polynomials, in accordance to the data in [11]. For the single hypergraph for $q = 2$, the search was trivial and took similar times for both algorithms. For the set of $q = 3$ hypergraphs, the Necklace search required approximately twice the number of nodes in the search tree, and took twice as long to run as the Russian Necklace algorithm. For $q = 4$ hypergraphs the Russian Necklace algorithm ran three times faster than the Necklace algorithm, and searches were completed in less than 12 s each. The full set of $q = 5$ hypergraphs were searched, taking between one and four hours for each search using the Russian Necklace search. The Necklace algorithm performed 4 times worse on average than the Russian Necklace algorithm. In summary, across all hypergraphs for the covering array problem, the Russian Necklace algorithm performs significantly better than the Necklace algorithm, which was expected from the results in the previous section, since we are dealing with dense hypergraphs.

The $q = 7$ hypergraph search using the Russian Necklace algorithm was aborted after 2 days. Based on the exponential trend in the runtime for each step of the Russian Necklace algorithm established from previous searches, and the established trend after the 2 days of running, the runtime for $q = 7$ is expected to be extremely large under the current system used for these experiments. One possible further direction is to use parallelization. The Necklace search can be more easily distributed among several processors since each branch of the search tree is somewhat independent of the others. For the Russian Necklace algorithm, there is more dependence between runs in the sense that the runs for later hypergraphs H^i depend on earlier runs for $H^{i+1}, \ldots, H^{n-k-1}$; however, paralelization within the same H^i is plausible. This is a topic of interesting further research on these methods.

Appendix

Runtimes of algorithms for $k = 3$ and decreasing edge density

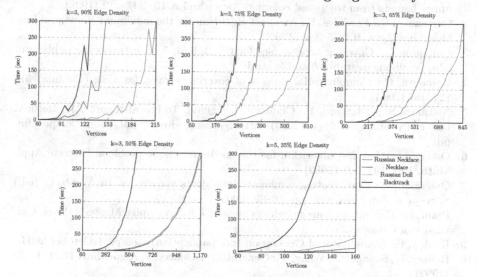

Runtimes of algorithms for $k = 4$ and decreasing edge density

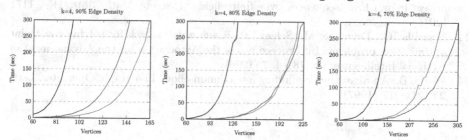

Runtimes of algorithms for $k = 5$ and decreasing edge density

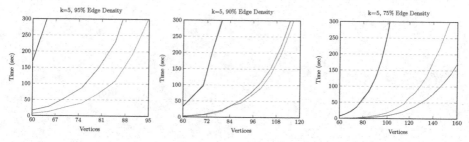

References

1. Balas, E., Xue, J.: Weighted and unweighted maximum clique algorithms with upper bounds from fractional coloring. Algorithmica **15**, 397–412 (1996)
2. Carraghan, R., Pardalos, P.: An exact algorithm for the maximum clique problem. Oper. Res. Lett. **9**, 375–382 (1990)
3. Colbourn, C.: Covering array tables for t = 2, 3, 4, 5, 6. http://www.public.asu.edu/~ccolbou/src/tabby/catable.html
4. Colbourn, C.: Combinatorial aspects of covering arrays. Le Mat. (Catania) **58**, 121–167 (2004)
5. Niskanen, S., Östergård, P.: Cliquer's user guide. Technical report T48, Helsinki University of Technology (2003). https://users.aalto.fi/~pat/cliquer/cliquer_fm.pdf
6. Östergård, P.: A fast algorithm for the maximum clique problem. Discrete Appl. Math. **120**, 197–207 (2002)
7. Östergård, P.: Constructing combinatorial objects via cliques. In: Webb, B. (ed.) Surveys in Combinatorics, pp. 56–82 (2005)
8. Plant, L.: Maximum clique search in circulant k-hypergraphs. Master's thesis, University of Ottawa (2018)
9. Ruskey, F.: Combinatorial Generation (pre-publication), 289 p., 1 October 2003
10. Ruskey, F., Savage, C., Wang, T.: Generating necklaces. J. Algorithms **13**, 414–430 (1992)
11. Tzanakis, G., Moura, L., Panario, D., Stevens, B.: Constructing new covering arrays from LFSR sequences over finite fields. Discrete Math. **339**, 1158–1171 (2016)
12. Verfaillie, G., Lemaitre, M., Schiex, T.: Russian doll search for solving constraint optimization problems. In: Proceedings of the Association for the Advancement of Artificial Intelligence, pp. 181–187 (1996)
13. Wood, D.: An algorithm for finding a maximum clique in a graph. Oper. Res. Lett. **21**, 211–217 (1997)

Burrows-Wheeler Transform of Words Defined by Morphisms

Srecko Brlek[1], Andrea Frosini[2], Ilaria Mancini[3(✉)], Elisa Pergola[2], and Simone Rinaldi[3]

[1] Laboratoire de combinatoire et d'informatique mathématique, UQAM, Montreal, Canada
brlek.srecko@uqam.ca
[2] Dipartimento di Matematica e Informatica "U. Dini", Università di Firenze, Florence, Italy
{andrea.frosini,elisa.pergola}@unifi.it
[3] Dipartimento di Ingegneria dell'Informazione e Scienze Matematiche, Università di Siena, Siena, Italy
ilaria.mancini@student.unisi.it, rinaldi@unisi.it

Abstract. The *Burrows-Wheeler transform* (BWT) is a popular method used for text compression. It was proved that BWT has optimal performance on standard words, i.e. the building blocks of Sturmian words. In this paper, we study the application of BWT on more general morphic words: the Thue-Morse word and to generalizations of the Fibonacci word to alphabets with more than two letters; then, we study morphisms obtained as composition of the Thue-Morse morphism with a Sturmian one. In all these cases, the BWT efficiently clusters the iterates of the morphisms generating prefixes of these infinite words, for which we determine the compression clustering ratio.

Keywords: Burrows-Wheeler transform · Morphisms · Thue-Morse word · Generalized Fibonacci words

1 Introduction

The *Burrows-Wheeler transform* (BWT) is a powerful technique used at the preprocessing stage in text compression algorithms [1, 4]. Actually, it produces a permutation of the characters of an input word and tends to group characters in runs, so that the output word is easier to compress because it has a lower number of runs (clusters). More precisely, the Burrows-Wheeler transform of a word w of length n is built by lexicographically sorting all its n conjugates and extracting the last character of each conjugate. The BWT is interesting in many aspects, and in particular because it is linked to a remarkable bijection due to Gessel and Reutenauer [8] on permutations, of which it is a special case [5].

In general it is difficult to determine a priori the compression ratio of the BWT on a given word, so recent studies use combinatorics on words tools in

© Springer Nature Switzerland AG 2019
C. J. Colbourn et al. (Eds.): IWOCA 2019, LNCS 11638, pp. 393–404, 2019.
https://doi.org/10.1007/978-3-030-25005-8_32

order to give a measure of efficiency of the BWT applied to some classes of words [7,10,15,19]. In order to investigate the clustering effect of the BWT from a combinatorial viewpoint, it is interesting to consider the structural properties of the words for which the BWT produces the maximal or the minimal compression ratio; a perfect clustering produced by the BWT corresponds to optimal performances of the run-length encoding.

The clustering effect of BWT on balanced words is studied in [18]. In particular, standard words - building blocks of infinite Sturmian words - have a BWT of the form $b^h a^j$ [14]. On a k-letter alphabet, words having a BWT with minimal number of clusters have been characterized in [17] in the case of balanced words, and in [6] in the general case.

The study on Sturmian words suggests that more general morphic words are expected to show a similar behaviour. In this paper, we support this thesis by studying the BWT performance on the building blocks of the Thue-Morse word and k-bonacci words, and, we determine the compression clustering ratio.

Finally, we study morphisms obtained by composition of the Thue-Morse morphism with a Sturmian one. In all these cases the BWT proves to efficiently cluster the building blocks of the considered infinite words.

2   Burrows-Wheeler Transform

We assume the reader familiar to the basic terminology on words defined on a finite alphabet (ref A [11]). The *Burrows-Wheeler transform* is defined as a map $\mathtt{bwt}(w) : A^* \to A^* \times \mathbb{N}$ such that $\mathtt{bwt}(w) = (L, I)$ where

- L is the last column of a matrix M whose lines are all the conjugates of w sorted lexicographically;
- I is the index of the line of M containing the original word w.

Example 1. The construction of BWT for $w = $ *filosofia* is

$$
\begin{array}{cccccccccc}
 & F & & & & & & & & L \\
 & \downarrow & & & & & & & & \downarrow \\
1 & a & f & i & l & o & s & o & f & i \\
2 & f & i & a & f & i & l & o & s & o \\
I \to 3 & f & i & l & o & s & o & f & i & a \\
4 & i & a & f & i & l & o & s & o & f \\
5 & i & l & o & s & o & f & i & a & f \\
6 & l & o & s & o & f & i & a & f & i \\
7 & o & f & i & a & f & i & l & o & s \\
8 & o & s & o & f & i & a & f & i & l \\
9 & s & o & f & i & a & f & i & l & o
\end{array}
$$

so $\mathtt{bwt}(w) = (ioaffislo, 3)$. In what follows we shall denote $\mathtt{bwt}_1(w)$, the first component, that is the column L.

Note that the first column F of the matrix M is the sequence of lexicographically sorted letters of w (see Example 1). The Burrows-Wheeler transform is reversible by using the properties described in the following proposition [4].

Proposition 1. *Let $w \in A^*$ be a word such that $\mathtt{bwt}(w) = (L, I)$ and let F be as above. The following properties hold:*

1. *$\forall i, 1 \le i \le n, i \neq I$, the letter $F[i]$ follows the letter $L[i]$ in w;*
2. *$\forall \alpha \in A$, the i-th occurrence of α in F matches the i-th occurrence of α in L;*
3. *the first letter of w is $F[I]$.*

According to property 2 of Proposition 1, there is a permutation τ_w giving the correspondence between the positions of letters in F and L. Hence, starting from the position I, the word w is obtained as follows:

$$w[i] = F[\tau_w^{i-1}(I)],$$

where $\tau_w^0(x) = x$ and $\tau_w^i(x) = \tau_w(\tau_w^{i-1}(x))$, with $1 \le i \le n$. For the word of Example 1, we have $\tau_w = (345168297)$.

Observation 1. For any two words u, v such that $|u| = |v|$ and any letter α, we have $\alpha u < \alpha v$ if and only if $u\alpha < v\alpha$ (if and only if $u < v$). Thus, given a word w, for all indices i, j if $i < j$ and $F[i] = F[j]$, then $\tau_w(i) < \tau_w(j)$.

The conjugation relation between words is denoted by $u \equiv v$. It is easy to see that in the BWT the column L is stable by conjugation.

Proposition 2. $u \equiv v$ *if and only if* $\mathtt{bwt}_1(u) = \mathtt{bwt}_1(v)$.

We denote by $\rho(u)$ the number of equal-letter runs or *clusters* of a word $u \in A^*$.

Definition 1. The *BWT-clustering ratio* of a word w is

$$\gamma(w) = \frac{\rho(\mathtt{bwt}_1(w))}{\rho(w)}.$$

For $w = filosofia$ we have $\rho(w) = 9$, $\rho(\mathtt{bwt}_1(w)) = \rho(ioaffislo) = 8$, $\gamma(w) = 8/9$. More details about bounds for the compression ratio can be found in [12, 13].

Burrows-Wheeler Transform of Sturmian Words. The family of Sturmian words has been extensively studied in several contexts, and recently in particular under the BWT transformation. For the unfamiliar reader we simply recall that Sturmian words are infinite words that approximate lines of irrational slopes on a square grid, so that they are conveniently encoded on the two-letter alphabet $A = \{a, b\}$. In [14], Mantaci et al. established that, given a word $w \in A^*$ then $\mathtt{bwt}_1(w) = b^k a^h$ if and only if u is the power of a conjugate of a standard word. In [20], Simpson and Puglisi provided an alternative proof of the previous statement and a characterization of words on the alphabet $\{a, b, c\}$ whose transforms have the form $c^i b^j a^k$.

Definition 2. A morphism f on A is *Sturmian* if $f(x)$ is a Sturmian word whenever x is a Sturmian word.

In particular, it is a well-known fact that each Sturmian morphism is obtained by composition from the following three morphisms

$$E: \begin{array}{l} a \mapsto b \\ b \mapsto a \end{array} \qquad \varphi: \begin{array}{l} a \mapsto ab \\ b \mapsto a \end{array} \qquad \widetilde{\varphi}: \begin{array}{l} a \mapsto ba \\ b \mapsto a. \end{array}$$

3   Burrows-Wheeler Transform of Non-Sturmian Words

The study on Sturmian words suggests that the compression ratio of words obtained by morphism iteration is likely to be high. Indeed, as these words are defined recursively, it is expected that their BWT can be expressed recursively as well, resulting in a lower number of clusters. In what follows we consider (right) infinite fixed points of morphisms on finite alphabets.

An alphabet A comes equipped with an order $\preceq$ which extends to A^* in the usual lexicographic way. Let $\Phi: A^* \to A^*$ be an order-preserving morphism on A, that is $\Phi(\alpha) \preceq \Phi(\beta)$ whenever $\alpha \preceq \beta$ for all $\alpha, \beta \in A$. Then we have

Lemma 1. *If Φ is an order-preseving uniform morphism then*

$$\forall x, y \in A^*, \; x \preceq y \implies \Phi(x) \preceq \Phi(y).$$

3.1   Burrows-Wheeler Transform of Standard Thue-Morse Words

The Thue-Morse word m is a recurrent cube-free infinite word having many combinatorial properties, and appearing in many contexts (see [3] for a primer). The word m is the limit of the sequence $(u_n)_{n \geq 0}$, with $u_n = \mu^n(a)$, obtained by iterating the morphism μ, defined by $\mu(a) = ab$, $\mu(b) = ba$, which is order preserving. So the first letters are the following

$$m = abbabaabbaababbabaab \cdots.$$

The word $u_n = \mu^n(a)$ is the n-standard Thue-Morse word. The BWT applied to the first n-standard Thue-Morse words shows some interesting regularities:

$n = 0 \; \mathsf{bwt}_1(a) = a$
$n = 1 \; \mathsf{bwt}_1(ab) = ba$
$n = 2 \; \mathsf{bwt}_1(abba) = baba$
$n = 3 \; \mathsf{bwt}_1(abbabaab) = b^2 ababa^2$
$n = 4 \; \mathsf{bwt}_1(abbabaabbaababba) = b^4 a^2 babab^2 a^4$
$n = 5 \; \mathsf{bwt}_1(abbabaabbaababbabaababbaabbabaab) = b^8 a^4 b^2 ababa^2 b^4 a^8$

To study the performance of the BWT on standard Thue-Morse words we need the following properties.

Proposition 3 (Thue, 1912). *For $n \geq 0$ the n-standard word u_n satisfies:*

(i) u_n does not contain the factors aaa and bbb for any $n \geq 0$.

(ii) $|u_n|_a = |u_n|_b = 2^{n-1}$.

Theorem 2. *For $n \geq 2$, it holds that:*

$$\mathtt{bwt}_1(u_n) = \begin{cases} b^{2^{n-2}} a^{2^{n-3}} \cdots b\,(ab)\,a \cdots b^{2^{n-3}} a^{2^{n-2}} & \text{if } n \text{ is even} \\ b^{2^{n-2}} a^{2^{n-3}} \cdots a\,(ba)\,b \cdots b^{2^{n-3}} a^{2^{n-2}} & \text{if } n \text{ is odd.} \end{cases}$$

Proof. The proof is obtained by induction on n. The statement holds for u_2 since $\mathtt{bwt}_1(u_2) = baba$ and $2^0 = 1$. Suppose the thesis holds for u_{n-1} and that n in even. So $\mathtt{bwt}_1(u_{n-1}) = b^{2^{n-3}} a^{2^{n-4}} \cdots a\,(ba)\,b \cdots b^{2^{n-4}} a^{2^{n-3}}$. The conjugates of u_{n-1} are words of the form

$$aaX_1b, \; abX_2b, \; abX_3a, \; baX_4a, \; baX_5b, \; bbX_6a \tag{1}$$

and, by inductive hypothesis, the final letters of the lexicographically ordered conjugates form the above $\mathtt{bwt}_1(u_{n-1})$. Applying μ to (1) we get some conjugates of u_n:

$$abab\mu(X_1)ba, \; abba\mu(X_2)ba, \; abba\mu(X_3)ab, \; baab\mu(X_4)ab, \; baab\mu(X_5)ba, \; baba\mu(X_6)ab.$$

Let $\mathcal{C}$ be the set of these conjugates obtained by μ. Lemma 1 ensures that the order of the conjugates in $\mathcal{C}$ is preserved by μ, and moreover, μ swaps the final letter of each conjugate from a to b and vice versa. The set of conjugates of u_n also includes

$$aabab\mu(X_1)b \prec aabba\mu(X_2)b \prec abaab\mu(X_5)b$$

and

$$babba\mu(X_3)a \prec bbaab\mu(X_4)a \prec bbaba\mu(X_6)a.$$

So all conjugates ending with b form a set $\mathcal{P}$ and precede those in $\mathcal{C}$, and all conjugates ending with a form a set $\mathcal{F}$ and follow those in $\mathcal{C}$ in lexicographic order. It follows that the central part of $\mathtt{bwt}_1(u_n)$ is $a^{2^{n-3}} b^{2^{n-4}} \cdots b\,(ab)\,a \cdots a^{2^{n-4}} b^{2^{n-3}}$. Finally, by Proposition 3(ii) $|\mathcal{P}| = 2^{n-2} = |\mathtt{bwt}_1(u_{n-1})|_a$, the number of occurrences of a in $\mathtt{bwt}_1(u_{n-1})$. Similarly, $|\mathcal{F}| = 2^{n-2} = |\mathtt{bwt}_1(u_{n-1})|_b$. The thesis follows:

$$\mathtt{bwt}_1(u_n) = b^{2^{n-2}} a^{2^{n-3}} b^{2^{n-4}} \cdots b\,(ab)\,a \cdots a^{2^{n-4}} b^{2^{n-3}} a^{2^{n-2}}.$$

The case n odd is similar.   □

Corollary 1. *For any $n > 0$ it holds that $I_{u_n} = 2^{n-1}$, i.e. the word u_n is the last word (in the lexicographical order) of all its conjugates starting with a.*

Proof. The proof is obtained by induction on n. For $n = 1$, $u_1 = ab$ so we have that $I_{u_1} = 1$. Suppose $I_{u_{n-1}} = 2^{n-2}$. By Lemma 1, $u_n = \mu(u_{n-1})$ keeps the same position with respect to the images of the conjugates of u_{n-1}. Moreover there are 2^{n-2} conjugates of u_n that precede u_n. So the index of u_n in the lexicographic order shifts to 2^{n-1}.   □

Denoting $\widetilde{w}$ the reversal of w, and $\overline{w} = E(w)$, observe that $\mathtt{bwt}_1(u_n)$ is also a pseudo-palindrome, that is

Corollary 2. *For any $n > 0$, we have $\widetilde{\mathtt{bwt}(u_n)} = \overline{\mathtt{bwt}(u_n)}$.*

Example 2. The conjugates of u_2 and u_3 are:

$$
\begin{aligned}
&\left.\begin{array}{l} 1\ a\ a\ b\ a\ b\ b\ a\ b \\ 2\ a\ b\ a\ a\ b\ a\ b\ b \end{array}\right\}\mathcal{P} \\[4pt]
\begin{array}{l} 1\ a\ a\ b\ b \\ I_{u_2} = 2\ a\ b\ b\ a \\ 3\ b\ a\ a\ b \\ 4\ b\ b\ a\ a \end{array}
\xrightarrow{\mu}
\ I_{u_3} =
&\left.\begin{array}{l} 3\ a\ b\ a\ b\ b\ a\ b\ a \\ 4\ a\ b\ b\ a\ b\ a\ a\ b \\ 5\ b\ a\ a\ b\ a\ b\ b\ a \\ 6\ b\ a\ b\ a\ a\ b\ a\ b \end{array}\right\}\mathcal{C} \\[4pt]
&\left.\begin{array}{l} 7\ b\ a\ b\ b\ a\ b\ a\ a \\ 8\ b\ b\ a\ b\ a\ a\ b\ a \end{array}\right\}\mathcal{F}
\end{aligned}
$$

An easy consequence of Theorem 2 shows that the number of clusters satisfies $\rho(\mathtt{bwt}_1(u_n)) = 2n$ and $\rho(u_n) = f_n$, where $f_0 = 1$, $f_1 = 2$, and:

$$
f_n = \begin{cases} 2\,f_{n-1} & \text{if } n \text{ is odd} \\ 2\,f_{n-1} - 1 & \text{if } n \text{ is even.} \end{cases}
$$

So $f_n \sim 2^n$, and therefore, for n large enough, the *clustering ratio* tends to zero:

$$
\gamma(u_n) = \frac{\rho(\mathtt{bwt}_1(u_n))}{\rho(u_n)} \ll 1,
$$

confirming the fact that the BWT is efficient on standard Thue-Morse words.

3.2   Generalizations of the Fibonacci Word

We already know that the BWT of the standard Fibonacci words (being Sturmian) have the minimum number of clusters. We show now that it does not hold for generalizations to a three-letters alphabet. Actually, in view of the characterization of words such that $\mathtt{bwt}(w) = c^i\,b^j\,a^k$ established in [20], the BWT of the n-standard Tribonacci word have more than three clusters, for n sufficiently large. Consider the morphism $\tau : A^* \to A^*$ defined by

$$
\tau(a) = ab,\ \tau(b) = ac,\ \tau(c) = a.
$$

The *Tribonacci word* is the limit of the sequence $(t_n)_{n>0}$, where $t_n = \tau^n(a)$ is the *n-standard Tribonacci word*. Some of its properties can be found in [2,21]. The sequence of Tribonacci numbers is defined as $(T_n)_{n \geq 0}$ with

$$
T_0 = 1,\ T_1 = 1,\ T_2 = 2,\ T_{n+3} = T_n + T_{n+1} + T_{n+2}.
$$

The first few terms are: $1, 1, 2, 4, 7, 13, 24, 44, 81, 149, \ldots$ (A000073 in [16]). The change of the initial conditions to $S_0 = 1$, $S_1 = 2$ and $S_2 = 3$ gives the integer sequence $S_n = T_n + T_{n-1}$.

Let us compute the BWT of the first standard Tribonacci words:

$n = 1$ $\mathsf{bwt}_1(a) = a$
$n = 2$ $\mathsf{bwt}_1(ab) = ba$
$n = 3$ $\mathsf{bwt}_1(abac) = cba^2$
$n = 4$ $\mathsf{bwt}_1(abacaba) = bcaba^3$
$n = 5$ $\mathsf{bwt}_1(abacabaabacab) = bc^2bab^2a^6$
$n = 6$ $\mathsf{bwt}_1(abacabaabacababacabaabac) = b^2c^4ba^2b^4a^{11}$
$n = 7$ $\mathsf{bwt}_1(abacabaabacababacabaabacabacabaabacababacaba) = b^4c^6bcaba^3b^7a^{20}$

The morphism τ is not order-preserving as $\tau(c) \prec \tau(a) \prec \tau(b)$ neither uniform, so that we cannot use Lemma 1 in order to compute the BWT for Tribonacci words. Instead, let us consider the morphism

$$\tau^3 : a \mapsto abacaba, b \mapsto abacab, c \mapsto abac,$$

which is order-reversing since $\tau^3(c) \prec \tau^3(b) \prec \tau^3(a)$. However, we can use the following result

Lemma 2. *Let $\alpha \prec \beta \in A$. Then $\tau^3(\alpha s) \preceq \tau^3(\beta t)$ for all $s, t \in A^+$.*

Proof. Since s and t are nonempty, $\tau^3(s)$ and $\tau^3(t)$ have $abac$ for prefix. It follows that $\tau^3(as) = abacaba \cdot abac \cdots \preceq ubacab \cdot abac \cdots = \tau^3(bt)$ and so $\tau^3(as) \preceq \tau^3(bt)$. The cases $(\alpha, \beta) = (a, c)$ and $(\alpha, \beta) = (b, c)$ are similar. $\square$

Theorem 3. *For every standard Tribonacci word t_n, with $n > 0$, we have*

$$\mathsf{bwt}_1(t_{n+3}) = b^{T_{n-1}} c^{S_{n-1}} \, \mathsf{bwt}_1(t_n) \, b^{T_n} \, a^{S_{n+1}}.$$

Proof. The proof follows a scheme similar to the one of Theorem 2 by applying Lemma 2 and the fact that the word t_n does not contain the factors bb, cc, bc, cb and so neither aaa, aac, caa, cac. $\square$

The number of clusters of $\mathsf{bwt}_1(t_n)$ satisfies the recurrence

$$\rho(\mathsf{bwt}_1(t_{n+3})) = \rho(\mathsf{bwt}_1(t_n)) + 4, \text{ for } n > 3.$$

The first terms are $(h_n)_{n>0} = 1, 2, 3, 5, 6, 6, 9, 10, 10, 13, \ldots$, and $h_n \sim n$.

Lemma 3. *The number of clusters in the n-standard Tribonacci word is $\rho(t_n) = f_n$, where $f_1 = 1$, and:*

$$f_n = \begin{cases} 2\,S_{n-2} + 1 & \text{if } n = 3\,m+1 \text{ for some } m \in \mathbb{N} \\ 2\,S_{n-2} & \text{otherwise.} \end{cases}$$

Proof. By definition of t_n and τ^3 we have that $\rho(t_n) = |t_n| - |t_n|_{aa}$ and every factor aa in t_n derives from an a in t_{n-3}, except the case where a is the final letter of t_{n-3}, which happens if $n = 3\,m+1$. So if $n \neq 3\,m+1$ we have that

$$\rho(t_n) = |t_n| - |t_{n-3}|_a = T_n - T_{n-4} = T_{n-1} + T_{n-2} + T_{n-3} - T_{n-4} = 2\,S_{n-2}.$$

Otherwise, if $n = 3\,m+1$ we have one additional cluster:

$$\rho(t_n) = |t_n| - |t_{n-3}|_a + 1 = 2\,S_{n-2} + 1.$$ $\square$

Since $f_n \sim 2^n$, we have that, for n large enough, the *clustering ratio* tends to 0:

$$\gamma(t_n) = \frac{\rho(\mathtt{bwt}_1(t_n))}{\rho(t_n)} \ll 1,$$

confirming that the BWT is remarkably efficient on standard Tribonacci words.

Further Generalizations. The results holding for Tribonacci words can be naturally extended to the generalization of Fibonacci words to an alphabet of cardinality $k \in \mathbb{N}$. So, with $A_k = \{a_1, \ldots, a_k\}$, let φ_k be defined as

$$\varphi_k(a_1) = a_1 a_2, \; \varphi_k(a_2) = a_1 a_3, \; \ldots \; \varphi_k(a_{k-1}) = a_1 a_k, \; \varphi_k(a_k) = a_1.$$

The *k-bonacci word* is the limit of the sequence $(g_{k,n})_{n>0}$, where $g_{k,n} = \varphi_k^n(a)$ is the *n-standard k-bonacci word* and the length of $g_{k,n}$ is the n-th *k-bonacci number* G_n^1:

$$G_0^1 = 1, \ldots, G_{k-1}^1 = \sum_{i=0}^{k-2} G_i^1 \text{ and } G_{n+k}^1 = \sum_{i=n}^{n+k-1} G_i^1.$$

From the sequence $(G_n^1)_{n \geq 0}$ we can define $k - 2$ other sequences:

$$G_n^2 = G_n^1 + G_{n-1}^1$$
$$G_n^3 = G_n^1 + G_{n-1}^1 + G_{n-2}^1$$
$$\vdots$$
$$G_n^{k-1} = G_n^1 + \ldots + G_{n-(k-2)}^1 = \sum_{i=n-k+2}^{n} G_i^1.$$

Theorem 4. *For every n-standard k-bonacci word $g_{k,n}$, it holds that*

$$\mathtt{bwt}_1(g_{k,n+k}) = a_2^{G_{n-1}^1} a_3^{G_{n-1}^2} \ldots a_k^{G_{n-1}^{k-1}} \mathtt{bwt}_1(g_{k,n}) a_{k-1}^{G_n^1} \ldots a_2^{G_{n+k-3}^{k-2}} a_1^{G_{n+k-2}^{k-1}}.$$

The proof is a simple generalization of the one of Theorem 3 to an alphabet with k letters. Finally, the number of clusters of $\mathtt{bwt}_1(g_{k,n})$ is $\rho(\mathtt{bwt}_1(g_{k,n+k})) = \rho(\mathtt{bwt}_1(g_{k,n})) + 2k - 2$ for each $n > k$. On the other hand the number of clusters in the standard *k*-bonacci word is $\rho(g_{k,n}) = f_n$, where $f_1 = 1$, and:

$$f_n = \begin{cases} 2 G_{n-2}^{k-1} + 1 & \text{if } n = km + 1 \text{ for some } m \in \mathbb{N} \\ 2 G_{n-2}^{k-1} & \text{otherwise.} \end{cases}$$

Since $f_n \sim 2^n$, again, for n large enough, the *clustering ratio* tends to 0:

$$\gamma(g_{k,n}) = \frac{\rho(\mathtt{bwt}_1(g_{k,n}))}{\rho(g_{k,n})} \ll 1,$$

and so, the BWT reduces the number of clusters in the general case as well.

Example 3. For $k = 4$, we have the *Tetranacci word* and the BWT of the standard Tetranacci words q_n, $n > 0$, has the following form:

$$\mathtt{bwt}_1(q_{n+4}) = b^{Q_{n-1}} c^{R_{n-1}} d^{V_{n-1}} \mathtt{bwt}_1(q_n) c^{Q_n} b^{R_{n+1}} a^{V_{n+2}}$$

where $(Q_n)_{n \geq 0}$ is the sequence of Tetranacci numbers (sequence A000078 in [16]) and $(R_n)_{n \geq 0}$, $(V_n)_{n \geq 0}$ are such that $R_n = Q_n + Q_{n-1}$, $V_n = Q_n + Q_{n-1} + Q_{n-2}$.

3.3 Composition of Thue-Morse with Sturmian Morphisms

A final result concerns the inspection of the BWT behaviour when applied to words obtained by composition of the Thue-Morse and Sturmian morphisms. We first consider the two different compositions of μ and φ.

The composition $\mu \circ \varphi$. It is defined on $A = \{a, b\}$ by

$$\mu \circ \varphi : a \mapsto abba; b \mapsto ab.$$

Let us consider the standard words $p_n = (\mu \circ \varphi)^n(a)$, whose first terms are: a, $abba$, $abbaababababba$, $\ldots$. In general, we have the recurrence formula:

Lemma 4. *For every $n > 1$, it holds that*

$$p_n = p_{n-1} \, p_{n-2} \cdots p_1 \, p_0 \, b \, p_{n-2} \, p_{n-3} \cdots p_1 \, p_0 \, b \, p_{n-1}.$$

Proof. By induction on n. If $n = 2$ we have $p_2 = abbaababababba = p_1 \, p_0 \, b \, p_0 \, b \, p_1$. Suppose that the statement holds for $n - 1$, i.e.:

$$p_{n-1} = p_{n-2} \, p_{n-3} \cdots p_1 \, p_0 \, b \, p_{n-3} \, p_{n-4} \cdots p_1 \, p_0 \, b \, p_{n-2}.$$

Then, by applying $\mu \circ \varphi$, the thesis follows, because $(\mu \circ \varphi)(b) = p_0 \, b$. □

Corollary 3. *For every word p_n with $n > 1$, we have that $|p_n| = 3 \, |p_{n-1}|$.*

The first terms of the sequence $(L_n)_{n \geq 0}$ of the lengths of the words p_n are: $1, 4, 12, 36, 108, 324, \ldots$ (sequence A003946 in [16]). The BWT applied to the first few terms of $(p_n)_{n \geq 0}$ yields:

$$B_0 = a$$
$$B_1 = baba$$
$$B_2 = b^2 abab^3 a^4$$
$$B_3 = b^6 a^4 b^3 abab^8 a^{12}$$
$$B_4 = b^{18} a^{12} b^8 abab^3 a^4 b^{24} a^{36}$$
$$B_5 = b^{54} a^{36} b^{24} a^4 b^3 abab^8 a^{12} b^{72} a^{108}$$

The Thue-Morse morphism μ being order-preserving, while the Fibonacci morphism φ is order reversing we immediately have

Lemma 5. *Let $\alpha \prec \beta \in A$. Then $(\mu \circ \varphi)(\alpha s) \succeq (\mu \circ \varphi)(\beta t)$ for all $s, t \in A^+$.*

Theorem 5. *For every word p_n, and $n > 0$, we have*

$$B_{n+1} = b^{\frac{L_n}{2}} \, \widetilde{B}_n \, b^{\frac{L_n}{2}} \, a^{L_n}.$$

Proof. The proof is similar to the proof of Theorem 2, and relies on: the application of Lemma 5, the fact that $|p_n|_a = |p_n|_b = L_n/2$ for $n > 0$, and the fact that p_1 has the same number of a's and b's, a property preserved by the morphism $\mu \circ \varphi$. □

For $n > 0$ the number of clusters of the BWT of the n-standard word is $\rho(B_{n+1}) = \rho(B_n) + 2$, as every reversal of B_n ends with the letter b. So the first terms of the sequence $(h_n)_{n \geq 0}$ are $1, 4, 6, 8, 10, 12, 14, \ldots$ and $h_n \sim n$.

Lemma 6. *The number of clusters in p_n for each $n \geq 0$ is $\rho(p_n) = f_n$ where $f_0 = 1$ and $f_n = L_n - L_{n-1} + 1$.*

Proof. By definition of $\mu \circ \varphi$, we have that in each p_n there are at most two consecutive a or b. It follows that

$$\rho(p_n) = |p_n| - |p_n|_{aa} - |p_n|_{bb}.$$

Every factor aa in p_n derives from an a in p_{n-1} except if a is the last letter, and every factor bb derives from an a. We have seen that $|p_{n-1}|_a = L_{n-1}/2$, so

$$\rho(p_n) = L_n - \left(\frac{L_{n-1}}{2} - 1 \right) - \frac{L_{n-1}}{2}$$

and the thesis follows. □

The first terms of the sequence $(f_n)_{n \geq 0}$ are $1, 2, 9, 25, 73, 217, \ldots$ and $f_n \sim 3^n$. It follows that, for n large enough, the *clustering ratio* tends to zero:

$$\gamma(p_n) = \frac{\rho(B_n)}{\rho(p_n)} \ll 1$$

which states that the BWT is very effective for the n-standard words p_n defined by the morphism $\mu \circ \varphi$.

Observe that if we replace φ with $\widetilde{\varphi}$, and we define $p'_n = (\mu \circ \widetilde{\varphi})^n(a)$, we obtain words that are conjugates of p_n for each $n \geq 0$ and so, by Proposition 2, they have the same BWT.

The composition $\varphi \circ \mu$. Let us consider now the composition:

$$\varphi \circ \mu : \begin{array}{l} a \mapsto aba \\ b \mapsto aab \end{array}$$

Let q_n denote the n-standard words associated with $\varphi \circ \mu$, i.e. $q_n = (\varphi \circ \mu)^n(a)$, $n \geq 0$. The first terms are: a, aba, $abaaababa$, $\ldots$. In general we have

Lemma 7. *For each $n > 1$ it holds that*

$$q_n = q_{n-1} \, q_{n-2} \, q_{n-2} \, q_{n-3} \, q_{n-3} \cdots q_1 \, q_1 \, q_0 \, q_0 \, b \, q_{n-1}.$$

The proof is similar to the one of Lemma 4.

As a consequence of the previous lemma, we have that $|q_n| = 3^n$ and, by definition of $\varphi \circ \mu$, $|q_n|_a = 2 |q_{n-1}| = 2 \cdot 3^{n-1}$ and $|q_n|_b = |q_{n-1}| = 3^{n-1}$. By applying the BWT to q_n, we get

$$B_{n+1} = b^{2 \cdot 3^{n-1}} \widetilde{B}_n \, a^{4 \cdot 3^{n-1}}$$

where $4 \cdot 3^{n-1}$ is obtained from $3^n + 3^{n-1}$.

The number of clusters of the BWT for each $n > 0$ is $\rho(B_{n+1}) = \rho(B_n) + 2$. So the first terms of the sequence $(h_n)_{n \geq 0}$ are $1, 4, 6, 8, 10, 12, 14, \ldots$ and $h_n \sim n$.

Lemma 8. *The number of clusters in q_n for each $n \geq 0$ is $\rho(q_n) = f_n$ where $f_0 = 1$, $f_1 = 3$ and $f_n = 3 f_{n-1} - 2$.*

Proof. As q_n begins and ends with an a for each $n \in \mathbb{N}$, by Lemma 7, we have that for each $n > 1$

$$f_n = 2 \left(f_{n-1} + f_{n-2} + \dots + f_1 + f_0 \right) + 1 - 2 \left(n - 1 \right)$$

so $f_n + 2 = 2 f_{n-1} + 2 \left(f_{n-2} + \dots + f_1 + f_0 \right) - 2 \left(n - 1 \right) + 3 = 3 f_{n-1}$. □

The first terms of the sequence $(f_n)_{n \geq 0}$ are $1, 3, 7, 19, 55, 163, 487, \dots$ and $f_n \sim 3^n$. It follows that, for n large enough, the *clustering ratio* tends to zero:

$$\gamma(q_n) = \frac{\rho(B_n)}{\rho(q_n)} \ll 1$$

which states that the BWT is again efficient for this word defined by the morphism $\varphi \circ \mu$.

3.4 Conclusions and Further Developments

The study done in this paper supports our initial idea that the BWT is particularly efficient on standard words associated with morphisms. Experimental evidence leads us to consider that this property also holds more generally. Further studies could investigate a formal way to express standard words defined by a generic composition of non-Sturmian morphisms, similarly to the case of Sturmian ones. For instance, consider the morphism $\mu \circ \varphi \circ \widetilde{\varphi}$:

$$\mu \circ \varphi \circ \widetilde{\varphi} : \begin{array}{l} a \mapsto ababba \\ b \mapsto abba \end{array}$$

with $w_n = (\mu \circ \varphi \circ \widetilde{\varphi})^n(a)$. The first iterates are: a, $ababba$, $\dots$ and

$$B_0 = a$$
$$B_1 = bab^2 a^2$$
$$B_2 = b^6 a^4 b a^2 b^8 a^9$$
$$B_3 = b^{30} a^{21} b^4 a b^2 a^8 b^{39} a^{45}$$
$$B_4 = b^{150} a^{105} b^{21} a^4 b a^2 b^8 a^{39} b^{195} a^{225}$$

In this case, though experimentally BWT shows a positive behaviour on these words, a formal proof of this fact is complicated to achieve since a generic expression for the terms w_n and B_n is not easy to find. The same problem shows up when we study several other morphisms such as: $\varphi \circ \mu \circ \widetilde{\varphi}$ and $\varphi \circ \widetilde{\varphi} \circ \mu$.

References

1. Adjeroh, D., Bell, T., Mukherjee, A.: The Burrows-Wheeler Transform: Data Compression, Suffix Arrays, and Pattern Matching. Springer, New York (2008). https://doi.org/10.1007/978-0-387-78909-5

2. Barcucci, E., Bélanger, L., Brlek, S.: On Tribonacci sequences. Fibonacci Quart. **42**, 314–319 (2004)
3. Berstel, J.: Axel Thue's papers on repetitions in words: a translation, Publications du LaCIM 20, Montreal (1995)
4. Burrows, M., Wheeler, D.J.: A Block-Sorting Lossless Data Compression Algorithm. Digital Systems Research Center Research Reports (1995)
5. Crochemore, M., Désarménien, J., Perrin, D.: A note on the Burrows-Wheeler transformation. Theor. Comput. Sci. **332**(1–3), 567–572 (2005)
6. Ferenczi, S., Zamboni, L.Q.: Clustering words and interval exchanges. J. Integer Sequences **16**(2), Article 13.2.1 (2013)
7. Ferragina, P., Giancarlo, R., Manzini, G., Sciortino, M.: Boosting textual compression in optimal linear time. J. ACM **52**(4), 688–713 (2005)
8. Gessel, I., Reutenauer, C.: Counting permutations with given cycle structure and descent set. J. Comb. Theory Ser. A **64**(2), 189–215 (1993)
9. Hedlund, G.A., Morse, M.: Symbolic dynamics. Am. J. Math. **60**, 815–866 (1938)
10. Kaplan, H., Landau, S., Verbin, E.: A simpler analysis of Burrows-Wheeler based compression. Theor. Comput. Sci. **387**(3), 220–235 (2007)
11. Lothaire, M.: Algebraic Combinatorics on Words (Encyclopedia of Mathematics and its Applications). Cambridge University Press, Cambridge (2002)
12. Mantaci, S., Restivo, A., Rosone, G., Sciortino, M.: Burrows-Wheeler transform and run-length enconding. In: Brlek, S., Dolce, F., Reutenauer, C., Vandomme, É. (eds.) WORDS 2017. LNCS, vol. 10432, pp. 228–239. Springer, Cham (2017). https://doi.org/10.1007/978-3-319-66396-8_21
13. Mantaci, S., Restivo, A., Rosone, G., Sciortino, M., Versari, L.: Measuring the clustering effect of BWT via RLE. Theor. Comput. Sci. **698**, 79–87 (2017)
14. Mantaci, S., Restivo, A., Sciortino, M.: Burrows-Wheeler transform and Sturmian words. Inf. Process. Lett. **86**, 241–246 (2003)
15. Manzini, G.: An analysis of the Burrows-Wheeler transform. J. ACM (JACM) **48**, 669–677 (1999)
16. OEIS Foundation Inc.: The On-line Encyclopedia of Integer Sequences (2011). http://oeis.org
17. Restivo, A., Rosone, G.: Balanced words having simple Burrows-Wheeler transform. In: Diekert, V., Nowotka, D. (eds.) DLT 2009. LNCS, vol. 5583, pp. 431–442. Springer, Heidelberg (2009). https://doi.org/10.1007/978-3-642-02737-6_35
18. Restivo, A., Rosone, G.: Balancing and clustering of words in the Burrows-Wheeler transform. Theor. Comput. Sci. **412**(27), 3019–3032 (2011)
19. Restivo, A., Rosone, G.: Burrows-Wheeler transform and palindromic richness. Theor. Comput. Sci. **410**(30–32), 3018–3026 (2009)
20. Simpson, J., Puglisi, S.J.: Words with simple Burrows-Wheeler transforms. Electr. J. Comb. **15**, 83 (2008)
21. Tan, B., Wen, Z.-Y.: Some properties of the Tribonacci sequence. Eur. J. Comb. **28**(6), 1703–1719 (2007)

Stable Noncrossing Matchings

Suthee Ruangwises$^{(\boxtimes)}$ (iD) and Toshiya Itoh

Department of Mathematical and Computing Science,
Tokyo Institute of Technology, Tokyo, Japan
ruangwises.s.aa@m.titech.ac.jp, titoh@c.titech.ac.jp

Abstract. Given a set of n men represented by n points lying on a line, and n women represented by n points lying on another parallel line, with each person having a list that ranks some people of opposite gender as his/her acceptable partners in strict order of preference. In this problem, we want to match people of opposite genders to satisfy people's preferences as well as making the edges not crossing one another geometrically. A *noncrossing blocking pair* w.r.t. a matching M is a pair (m, w) of a man and a woman such that they are not matched with each other but prefer each other to their own partners in M, and the segment (m, w) does not cross any edge in M. A *weakly stable noncrossing matching* (WSNM) is a noncrossing matching that does not admit any noncrossing blocking pair. In this paper, we prove the existence of a WSNM in any instance by developing an $O(n^2)$ algorithm to find one in a given instance.

Keywords: Stable matching · Stable marriage problem ·
Noncrossing matching · Geometric matching

1 Introduction

The stable marriage problem is one of the most classic and well-studied problems in the area of matching under preferences, with many applications in other fields including economics [5,10]. We have a set of n men and a set of n women, with each person having a list that ranks some people of opposite gender as his/her acceptable partners in order of preference. A *matching* is a set of disjoint man-woman pairs. A *blocking pair* w.r.t. a matching M is a pair of a man and a women that are not matched with each other in M but prefer each other to their own partners. The goal is to find a *stable matching*, a matching that does not admit any blocking pair.

On the other hand, the noncrossing matching problem is a problem in the area of geometric matching. We have a set of $2n$ points lying on two parallel lines, each containing n points. We also have some edges joining vertices on the opposite lines. The goal is to select a set of edges that do not cross one another subject to different objectives, e.g. maximum size, maximum weight, etc.

In this paper, we study a problem in geometric matching under preferences. In particular, we investigate the problem of having n men and n women represented

© Springer Nature Switzerland AG 2019
C. J. Colbourn et al. (Eds.): IWOCA 2019, LNCS 11638, pp. 405–416, 2019.
https://doi.org/10.1007/978-3-030-25005-8_33

by points lying on two parallel lines, with each line containing n people of one gender. Each person has a list that ranks some people of opposite gender in strict order of preference. A *noncrossing blocking pair* w.r.t. a matching M is a blocking pair w.r.t. M that does not cross any edge in M. Our goal is to find a noncrossing matching that does not admit any noncrossing blocking pair, called a *weakly stable noncrossing matching* (WSNM).

Note that the real-world applications of this geometric problem are more likely to involve immovable objects, e.g. construction of noncrossing bridges between cities on the two sides of a river, with each city having different preferences. In this paper, however, we keep the terminologies of men and women used in the original stable marriage problem in order to understand and relate to the original problem more easily.

1.1   Related Work

The stable marriage problem was first introduced by Gale and Shapley [3]. They proved that a stable matching always exists in an instance with n men and n women, with each person's preference list containing all n people of opposite gender and not containing ties. They also developed an $O(n^2)$ algorithm to find a stable matching in a given instance. Gusfield and Irving [5] later showed that the algorithm can be adapted to the setting where each person's preference list may not contain all people of opposite gender. The algorithm runs in $O(m)$ time in this setting, where m is the total length of people's preference lists. Gale and Sotomayor [4] proved that in this modified setting, a stable matching may have size less than n, but every stable matching must have equal size. Irving [7] then generalized the notion of a stable matching to the case where ties are allowed in people's preference lists. He introduced three types of stable matchings in this setting: *weakly stable*, *super-stable*, and *strongly stable*, as well as developing algorithms to determine whether each type of matching exists in a given instance and find one if it does.

The Stable Roommates problem is a generalization of the stable marriage problem to a non-bipartite setting where people can be matched regardless of gender. Unlike in the original problem, a stable matching in this setting does not always exist. Irving [6] developed an $O(n^2)$ algorithm to find a stable matching or report that none exists in a given instance, where n is the number of people.

On the other hand, the noncrossing matching problem in a bipartite graph where the points lie on two parallel lines, each containing n points, was encountered in many real-world situations such as in VLSI layout design [8]. In the special case where each point is adjacent to exactly one point on the opposite line, Fredman [2] presented an $O(n \log n)$ algorithm to find a maximum size noncrossing matching by computing the length of the longest increasing subsequence (LIS). Widmayer and Wong [11] developed another algorithm that runs in $O(k + (n - k) \log(k + 1))$ time, where k is the size of the solution. This algorithm has the same worst-case runtime as Fredman's, but runs faster in most general cases.

In a more general case where each point may be adjacent to more than one point, Malucelli et al. [9] developed an algorithm to find a maximum size noncrossing matching. The algorithm runs in either $O(m \log \log n)$ or $O(m + \min(nk, m \log k))$ time depending on implementation, where m is the number of edges and k is the size of the solution. In the case where each edge has a weight, they also showed that the algorithm can be adapted to find a maximum weight noncrossing matching with $O(m \log n)$ runtime.

1.2 Our Contribution

In this paper, we constructively prove that a weakly stable noncrossing matching always exists in any instance by developing an $O(n^2)$ algorithm to find one in a given instance.

2 Preliminaries

In this setting, we have a set of n men $m_1, ..., m_n$ represented by points lying on a vertical line in this order from top to bottom, and a set of n women $w_1, ..., w_n$ represented by points lying on another parallel line in this order from top to bottom. Only people of opposite genders can be matched with each other, and each person can be matched with at most one other person. A *matching* is a set of disjoint man-woman pairs.

For a person a and a matching M, define $M(a)$ to be the person matched with a (for convenience, let $M(a) = null$ for an unmatched person a). For each person a, let P_a be the preference list of a containing people of opposite gender to a as his/her acceptable partners in decreasing order of preference. Note that a preference list does not have to contain all n people of opposite gender. Throughout this paper, we assume that the preference lists are *strict* (containing no tie involving two or more people). Also, let $r_a(b)$ be the rank of a person b in P_a, with the first entry having rank 1, the second entry having rank 2, and so on (for convenience, let $r_a(null) = \infty$ and treat $null$ as the last entry appended to the end of P_a, as being matched is always better than being unmatched). A person a is said to prefer a person b to a person c if $r_a(b) < r_a(c)$.

A pair of edges cross each other if they intersect in the interior of both segments. Formally, an edge (m_i, w_x) crosses an edge (m_j, w_y) if and only if $(i - j)(x - y) < 0$. A matching is called *noncrossing* if it does not contain any pair of crossing edges.

The following are the formal definitions of a *blocking pair* given in the original stable marriage problem, and a *noncrossing blocking pair* introduced here.

Definition 1. *A blocking pair w.r.t. a matching M is a pair (m, w) of a man and a woman that are not matched with each other, but m prefers w to $M(m)$ and w prefers m to $M(w)$.*

Definition 2. *A noncrossing blocking pair w.r.t. a matching M is a blocking pair w.r.t. M that does not cross any edge in M.*

We also introduce two types of stable noncrossing matchings, distinguished as weakly and strongly stable.

Definition 3. *A matching M is called a weakly stable noncrossing matching (WSNM) if M is noncrossing and does not admit any noncrossing blocking pair.*

Definition 4. *A matching M is called a strongly stable noncrossing matching (SSNM) if M is noncrossing and does not admit any blocking pair.*

Note that an SSNM is a matching that is both noncrossing and stable, while a WSNM is "stable" in a weaker sense as it may admit a blocking pair, just not a noncrossing one.

An SSNM may not exist in some instances. For example, in an instance of two men and two women, with $P_{m_1} = (w_2, w_1)$, $P_{m_2} = (w_1, w_2)$, $P_{w_1} = (m_2, m_1)$, and $P_{w_2} = (m_1, m_2)$, the only stable matching is $\{(m_1, w_2), (m_2, w_1)\}$, and its two edges do cross each other. On the other hand, the above instance has two WSNMs $\{(m_1, w_2)\}$ and $\{(m_2, w_1)\}$. It also turns out that a WSNM always exists in every instance. Throughout this paper, we focus on the proof of existence of a WSNM by developing an algorithm to find one.

3   Our Algorithm

3.1   Outline

Without loss of generality, for each man m and each woman w, we assume that w is in m's preference list if and only if m is also in w's preference list (otherwise we can simply remove the entries that are not mutual from the lists). Initially, every person is unmatched.

Our algorithm uses proposals from men to women similarly to the Gale–Shapley algorithm in [3], but in a more constrained way. With M being the current noncrossing matching, when a woman w receives a proposal from a man m, if she prefers her current partner $M(w)$ to m, she rejects m; if she is currently unmatched or prefers m to $M(w)$, she dumps $M(w)$ and accepts m.

Consider a man m and a woman w not matched with each other. An entry w in P_m has the following possible states:

1. *accessible* (to m), if (m, w) does not cross any edge in M;
    1.1. *available* (to m), if w is accessible to m, and is currently unmatched or matched with a man she likes less than m, i.e. m is going to be accepted if he proposes to her (for convenience, if w is currently matched with m, we also call w accessible and available to m).
    1.2. *unavailable* (to m), if w is accessible to m, but is currently matched with a man she likes more than m, i.e. m is going to be rejected if he proposes to her;
2. *inaccessible* (to m), if w is not accessible to m;

For a man m, if every entry in P_m before $M(m)$ is either inaccessible or unavailable, then we say that m is *stable*; otherwise (there is at least one available entry before $M(m)$) we say that m is *unstable*.

The main idea of our algorithm is that, at any point, if there is at least one unstable man, we pick the topmost unstable man m_i (the unstable man m_i with least index i) and perform the following operations.

1. Let m_i *dump* his current partner $M(m_i)$ (if any), i.e. remove $(m_i, M(m_i))$ from M, and let him propose to the available woman w_j that he prefers most.
2. Let w_j *dump* her current partner $M(w_i)$ (if any), i.e. remove $(M(w_j), w_j)$ from M, and let her accept m_i's proposal.
3. Add the new pair (m_i, w_j) to M.

We repeatedly perform such operations until every man becomes stable. Note that throughout the algorithm, every proposal will result in acceptance and M will always be noncrossing since men propose only to women available to them.

3.2 Proof of Correctness

First, we will show that if our algorithm stops, then the matching M given by the algorithm must be a WSNM.

Assume, for the sake of contradiction, that M admits a noncrossing blocking pair (m_i, w_j). That means m_i prefers w_j to his current partner $M(m_i)$, w_j prefers m_i to her current partner $M(w_j)$, and (m_i, w_j) does not cross an edge in M, thus the entry w_j in P_{m_i} is available and is located before $M(m_i)$. However, by the description of our algorithm, the process stops when every man becomes stable, which means there cannot be an available entry before $M(m_i)$ in P_{m_i}, a contradiction. Therefore, we can conclude that our algorithm gives a WSNM as a result whenever it stops.

However, it is not trivial that our algorithm will eventually stop. In contrast to the Gale–Shapley algorithm in the original stable marriage problem, in this problem a woman is not guaranteed to get increasingly better partners throughout the process because a man can dump a woman too if he later finds a better available woman previously inaccessible to him (due to having an edge obstructing them). In fact, it is actually the case where the process may not stop if at each step we pick an arbitrary unstable man instead of the topmost one. For example, in an instance of two men and two women with $P_{m_1} = (w_2, w_1), P_{m_2} = (w_1, w_2), P_{w_1} = (m_1, m_2), P_{w_2} = (m_2, m_1)$, the order of picking $m_1, m_2, m_2, m_1, m_1, m_2, m_2, m_1, \dots$ results in the process continuing forever, with the matching M looping between $\{(m_1, w_2)\}, \{(m_2, w_2)\}, \{(m_2, w_1)\}$, and $\{(m_1, w_1)\}$ at each step.

We will prove that our algorithm must eventually stop and evaluate its worst-case runtime after we introduce the explicit implementation of the algorithm in the next subsection.

3.3 Implementation

To implement the above algorithm, we have to consider how to efficiently find the topmost unstable man at each step in order to perform the operations on him. Of course, a straightforward way to do this is to update the state of every entry in every man's preference list after each step, but that method will be very inefficient. Instead, we introduce the following scanning method.

Throughout the algorithm, we do not know exactly the set of all unstable men, but we instead keep a set S of men that are "possibly unstable." Initially, the set S contains all men, i.e. $S = \{m_1, m_2, ..., m_n\}$, and at each step we maintain the set S of the form $\{m_i, m_{i+1}, ...,, m_n\}$ for some $i \in [n]$ (that means $m_1, m_2, ..., m_{i-1}$ are guaranteed to be stable at that time). Note that in the actual implementation, we can store only the index of the topmost man in S instead of the whole set. At each step, we scan the topmost man m_i in S and check whether m_i is stable. If m_i is already stable, then we simply skip him by removing m_i from S and moving to scan the next man in S. If m_i is unstable, then m_i is indeed the topmost unstable man we want, so we perform the operations on m_i. Note that the operations may cause some men to become unstable, so after that we have to add all men that are possibly affected by the operations back to S. The details of the scanning and updating processes are as follows.

During the scan of m_i, let m_{prev} be the *matched* man closest to m_i that lies above him, and let $w_{first} = M(m_{prev})$ (we let $w_{first} = w_1$ if there is no m_{prev}). Also, let m_{next} be the *matched* man closest to m_i that lies below him, and let $w_{last} = M(m_{next})$ (we let $w_{last} = w_n$ if there is no m_{next}). Observe that matching m_i with anyone lying above w_{first} will cross the edge (m_{prev}, w_{first}), and matching m_i with anyone lying below w_{last} will cross the edge (m_{next}, w_{last}). Therefore, the range of all women accessible to m_i ranges exactly from w_{first} to w_{last}, hence the range of all women available to m_i ranges from either w_{first} or $w_{first+1}$ (depending on whether w_{first} prefers m_i to m_{prev}) to either w_{last} or w_{last-1} (depending on whether w_{last} prefers m_i to m_{next}). See Fig. 1.

Then in the available range, m_i selects the woman w_j that he prefers most.

Case 1: w_j does not exist or m_i is currently matched with w_j.

That means m_i is currently stable, so we can skip him. We remove m_i from S and proceed to scan m_{i+1} in the next step (called a *downward jump*).

Case 2: w_j exists and m_i is not currently matched with w_j

That means m_i is indeed the topmost unstable man we want, so we perform the operations on him by letting m_i propose to w_j and dump his current partner (if any).

Case 2.1: m_{prev} exists and $w_j = w_{first}$.

That means w_{first} dumps m_{prev} to get matched with m_i, which leaves m_{prev} unmatched and he may possibly become unstable. Furthermore, m_{prev+1}, $m_{prev+2}, ..., m_{i-1}$ as well as m_i himself may also possibly become unstable since they now gain access to women lying above w_{first} previously inaccessible to them (if $w_{first} \neq w_1$). On the other hand, $m_1, m_2, ..., m_{prev-1}$ clearly remain stable, hence we add $m_{prev}, m_{prev+1}, ..., m_{i-1}$ to S and proceed to scan m_{prev} in the next step (called an *upward jump*).

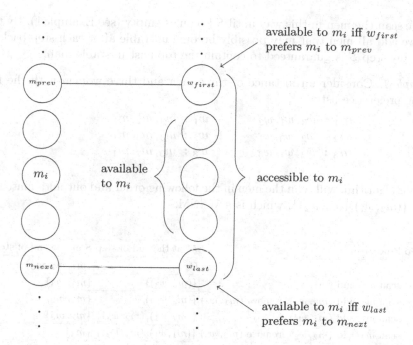

available to m_i iff w_{first}
prefers m_i to m_{prev}

available
to m_i

accessible to m_i

available to m_i iff w_{last}
prefers m_i to m_{next}

Fig. 1. Accessible and available women to m_i

Case 2.2: m_{prev} does not exist or $w_j \neq w_{first}$.

Case 2.2.1: m_i is currently matched and w_j lies geometrically below $M(m_i)$.

Then, $m_{prev}, m_{prev+1}, ..., m_{i-1}$ (or $m_1, m_2, ... m_{i-1}$ if m_{prev} does not exist) may possibly becomes unstable since they now gain access to women between $M(m_i)$ and w_j previously inaccessible to them. Therefore, we perform the upward jump to m_{prev} (or to m_1 if m_{prev} does not exist), adding $m_{prev}, m_{prev+1}, ..., m_{i-1}$ (or $m_1, m_2, ... m_{i-1}$) to S and proceed to scan m_{prev} (or m_1) in the next step, except when $m_i = m_1$ that we perform the downward jump to m_2.

It turns out that this case is impossible, which we will prove in the next subsection.

Case 2.2.2: m_i is currently unmatched or w_j lies geometrically above $M(m_i)$.

Then all men lying above m_i clearly remain stable (because the sets of available women to $m_1, ..., m_{i-1}$ either remain the same or become smaller). Also, m_i now becomes stable as well (because m_i selects a woman he prefers most in the available range), except in the case where $w_j = w_{last}$ (because the edge (m_{next}, w_{last}) is removed and m_i now has access to women lying below w_{last} previously inaccessible to him). Therefore, we perform the downward jump, removing m_i from S and moving to scan m_{i+1} in the next step, except when $w_j = w_{last}$ that we have to scan m_i again in the next step (this exception, however, turns out to be impossible, which we will prove in the next subsection).

We scan the men in this way until S becomes empty (see Example 1). By the way we add all men that may possibly become unstable after each step back to S, at any step S is guaranteed to contain the topmost unstable man.

Example 1. Consider an instance of three men and three women with the following preference lists.

$$m_1 : w_3, w_1, w_2 \qquad w_1 : m_3, m_2, m_1$$
$$m_2 : w_2, w_3, w_1 \qquad w_2 : m_3, m_2, m_1$$
$$m_3 : w_2, w_1, w_3 \qquad w_3 : m_3, m_2, m_1$$

Our algorithm will scan the men in the following order and output a matching $M = \{(m_2, w_1), (m_3, w_2)\}$, which is a WSNM.

Step	Process	M at the end of step	S at the end of step
0		$\varnothing$	$\{m_1, m_2, m_3\}$
1	scan m_1, add (m_1, w_3)	$\{(m_1, w_3)\}$	$\{m_2, m_3\}$
2	scan m_2, add (m_2, w_3), remove (m_1, w_3)	$\{(m_2, w_3)\}$	$\{m_1, m_2, m_3\}$
3	scan m_1, add (m_1, w_1)	$\{(m_1, w_1), (m_2, w_3)\}$	$\{m_2, m_3\}$
4	scan m_2, add (m_2, w_2), remove (m_2, w_3)	$\{(m_1, w_1), (m_2, w_2)\}$	$\{m_3\}$
5	scan m_3, add (m_3, w_2), remove (m_2, w_2)	$\{(m_1, w_1), (m_3, w_2)\}$	$\{m_2, m_3\}$
6	scan m_2, add (m_2, w_1), remove (m_1, w_1)	$\{(m_2, w_1), (m_3, w_2)\}$	$\{m_1, m_2, m_3\}$
7	scan m_1	$\{(m_2, w_1), (m_3, w_2)\}$	$\{m_2, m_3\}$
8	scan m_2	$\{(m_2, w_1), (m_3, w_2)\}$	$\{m_3\}$
9	scan m_3	$\{(m_2, w_1), (m_3, w_2)\}$	$\varnothing$

3.4   Observations

First, we will prove the following lemma about the algorithm described in the previous subsection.

Lemma 1. *During the scan of a man m_i, if m_i is currently matched, then m_i does not propose to any woman lying geometrically below $M(m_i)$.*

Proof. We call a situation when a man m_i proposes to a woman lying geometrically below $M(m_i)$ a *downward switch*. Assume, for the sake of contradiction, that a downward switch occurs at least once during the whole algorithm. Suppose that the first downward switch occurs at step s, when a man m_i is matched to $w_k = M(m_i)$ and proposes to w_j with $j > k$. We have m_i prefers w_j to w_k.

Consider the step $t < s$ when m_i proposed to w_k (if m_i proposed to w_k multiple times, consider the most recent one). At step t, w_j must be inaccessible or unavailable to m_i (otherwise he would choose w_j instead of w_k), meaning that there must be an edge (m_p, w_q) with $p > i$ and $k < q < j$ obstructing them in the inaccessible case, or an edge (m_p, w_q) with $p > i$, $q = j$, and w_j preferring m_p to m_i in the unavailable case.

We define a *dynamic edge* e as follows. First, at step t we set $e = (m_p, w_q)$. Then, throughout the process we update e by the following method: whenever the endpoints of e cease to be partners of each other, we update e to be the edge joining the endpoint that dumps his/her partner with his/her new partner. Formally, suppose that e is currently (m_x, w_y). If m_x dumps w_y to get matched with $w_{y'}$, we update e to be $(m_x, w_{y'})$; if w_y dumps m_x to get matched with $m_{x'}$, we update e to be $(m_{x'}, w_y)$.

By this updating method, the edge e will always exist after step t, but may change over time. Observe that from step t to step s, we always have $x > i$ because of the existence of (m_i, w_k). Moreover, before step s, if m_x dumps w_y to get matched with $w_{y'}$, by the assumption that a downward switch did not occur before step s, we have $y' < y$, which means the index of the women's side of e's endpoints never increases. Consider the edge $e = (m_x, w_y)$ at step s, we must have $x > i$ and $y \leq q \leq j$. If $y < j$, then the edge e obstructs m_i and w_j, making w_j inaccessible to m_i. If $y = j$, that means w_j never got dumped since step t, so she got only increasingly better partners, thus w_j prefers m_x to m_i, making w_j unavailable to m_i. Therefore, in both cases m_i could not propose to w_j, a contradiction. Hence, a downward switch cannot occur in our algorithm.

□

Lemma 1 shows that a woman cannot get her partner stolen by any woman that lies below her, which is equivalent to the following corollary.

Corollary 1. *If a man m_i dumps a woman w_j to propose to a woman w_k, then $k < j$.*

It also implies that Case 2.2.1 in the previous subsection never occurs. Therefore, the only case where an upward jump occurs is Case 2.1 (m_{prev} exists and m_i proposes to w_{first}). We will now prove the following lemma.

Lemma 2. *During the scan of a man m_i, if m_{next} exists, then m_i does not propose to w_{last}.*

Proof. Assume, for the sake of contradiction, that m_i proposes to w_{last}. Since m_{next} exists, this proposal obviously cannot occur in the very first step of the algorithm. Consider a man m_k we scanned in the previous step right before scanning m_i.

Case 1: m_k lies below m_i, i.e. $k > i$.

In order for the upward jump from m_k to m_i to occur, m_i must have been matched with a woman but got her stolen by m_k in the previous step. However, $m_{i+1}, m_{i+2}, ..., m_{next-1}$ are all currently unmatched (by the definition of m_{next}), so the only possibility is that $m_k = m_{next}$, and thus his partner that got stolen was w_{last}. Therefore, we can conclude that w_{last} prefers m_{next} to m_i, which means w_{last} is currently unavailable to m_i, a contradiction.

Case 2: m_k lies above m_i, i.e. $k < i$.

The jump before the current step was a downward jump, but since m_{next} has been scanned before, an upward jump over m_i must have occurred at some point before the current step. Consider the most recent upward jump over m_i before the current step. Suppose than it occurred at the end of step t and was a jump from $m_{k'}$ to m_j, with $k' > i$ and $j < i$. In order for this jump to occur, m_j must have been matched with a woman but got her stolen by $m_{k'}$ at step t. However, $m_{i+1}, m_{i+2}, ..., m_{next-1}$ are all currently unmatched (by the definition of m_{next}), so the only possibility is that $m_{k'} = m_{next}$, and thus m_j's partner that got stolen was w_{last}. We also have $m_{j+1}, m_{j+2}, ..., m_{next-1}$ were all unmatched during step t (otherwise w_{last} would be inaccessible to $m_{k'}$), and w_{last} prefers m_{next} to m_j.

Now, consider the most recent step before step t in which we scanned m_i. Suppose it occurred at step s. During step s, m_j was matched with w_{last} and w_{last} was accessible to m_i. However, m_i was still left unmatched after step s (otherwise an upward jump over m_i at step t could not occur), meaning that w_{last} must be unavailable to him back then due to w_{last} preferring m_j to m_i. Therefore, we can conclude that w_{last} prefers m_{next} to m_i, thus w_{last} is currently unavailable to m_i, a contradiction.   □

Lemma 2 shows that a man cannot get his partner stolen by any man lying above him, or equivalent to the following corollary.

Corollary 2. *If a woman w_j dumps a man m_i to accept a man m_k, then $k > i$.*

3.5   Proof of Finiteness

Now, we will show that the position of each woman's partner can only move downward throughout the process, which guarantees the finiteness of the number of steps in the entire process.

Lemma 3. *After a woman w_j ceases to be a partner of a man m_i, she cannot be matched with any man $m_{i'}$ with $i' \leq i$ afterwards.*

Proof. Suppose that w_j's next partner (if any) is m_a. It is sufficient to prove that $a > i$. First, consider the situation when m_i and w_j cease to be partners.

Case 1: w_j dumps m_i.

This means w_j dumps m_i to get matched with m_a right away. By Corollary 2, we have $a > i$ as required.

Case 2: m_i dumps w_j.

Suppose that m_i dumps w_j to get matched with w_k. By Corollary 1, we have $k < j$.

Case 2.1: m_i never gets dumped afterwards.

That means m_i will only get increasingly better partner, and the position of his partner can only move upwards (by Corollary 1), which means w_j cannot be matched with m_i again, or any man lying above m_i afterwards due to having an edge $(m_i, M(m_i))$ obstructing. Therefore, m_a must lie below m_i, i.e. $a > i$.

Case 2.2: m_i gets dumped afterwards.

Suppose that m_i first gets dumped by w_y at step s. By Corollary 1, we have $y \leq k < j$ (because m_i only gets increasingly better partners before getting dumped). Also suppose that w_y dumps m_i in order to get matched with m_x. By Corollary 2, we have $x > i$. Similarly to the proof of Lemma 1, consider a dynamic e first set to be (m_x, w_y) at step s. We have the index of the men's side of e's endpoints never decreases, and that of the women's side never increases. Therefore, since step s, there always exists an edge (m_x, w_y) with $x > i$ and $y < j$, obstructing w_j's access to m_i and all men lying above him. Therefore, m_a must lie below m_i, i.e. $a > i$.   □

3.6   Runtime Analysis

Consider any upward jump from m_i to m_k with $i > k$ that occurs right after m_i stole w_j from m_k. We call such a jump *associated to* w_j, and it has *size* $i - k$.

For any woman w_j, let U_j be the sum of the sizes of all upward jumps associated to w_j. From Lemma 3, we know that the position of w_j's partner can only move upward throughout the process, so we have $U_j \leq n - 1$. Therefore, the sum of the sizes of all upwards jumps is $\sum_{j=1}^{n} U_j \leq n(n-1) = O(n^2)$. Since the scan starts at m_1 and ends at m_n, the total number of downward jumps equals to the sum of the sizes of all upward jumps plus $n - 1$, hence the total number of steps in the whole algorithm is $O(n^2)$.

For each m_i, we keep an array of size n, with the jth entry storing the rank of w_j in P_{m_i}. Each time we scan m_i, we query the minimum rank of available women, which is a consecutive range in the array. Using an appropriate range minimum query (RMQ) data structure such as the one introduced by Fischer [1], we can perform the scan with $O(n)$ preprocessing time per array and $O(1)$ query time. Therefore, the total runtime of our algorithm is $O(n^2)$.

In conclusion, we proved that our developed algorithm is correct and can be implemented in $O(n^2)$ time, which also implicitly proves the existence of a WSNM in any instance.

Theorem 1. *A weakly stable noncrossing matching exists in any instance with n men and n women with strict preference lists.*

Theorem 2. *There is an $O(n^2)$ algorithm to find a weakly stable noncrossing matching in an instance with n men and n women with strict preference lists.*

Remarks: Our algorithm does not require the numbers of men and women to be equal. In the case that there are n_1 men and n_2 women, the algorithm works similarly with $O(n_1 n_2)$ runtime. Also, in the case that people's preference lists are not strict, we can modify the instance by breaking ties in an arbitrary way. Clearly, a WSNM in the modified instance will also be a WSNM in the original one (because every noncrossing blocking pair in the original instance will also be a noncrossing blocking pair in the modified instance).

4   Discussion

In this paper, we constructively prove that a WSNM always exists in any instance by developing an $O(n^2)$ algorithm to find one. Note that the definition of a

WSNM allows multiple answers with different sizes for an instance. For example, in an instance of three men and three women, with $P_{m_1} = (w_3, w_1, w_2), P_{m_2} = (w_1, w_2, w_3), P_{m_3} = (w_2, w_3, w_1), P_{w_1} = (m_2, m_3, m_1), P_{w_2} = (m_3, m_1, m_2)$, and $P_{w_3} = (m_1, m_2, m_3)$, both $\{(m_1, w_3)\}$ and $\{(m_2, w_1), (m_3, w_2)\}$ are WSNMs, but our algorithm only outputs the first one with smaller size. A possible future work is to develop an algorithm to find a WSNM with maximum size in a given instance, which seems to be a naturally better answer then other WSNMs as it satisfies more people. Another possible future work is to develop an algorithm to determine whether an SSNM exists in a given instance, and to find one if it does.

Other interesting problems include investigate the noncrossing matching in the geometric version of the stable roommates problem where people can be matched regardless of gender. The most basic and natural setting of this problem is where people are represented by points arranged on a circle. A possible future work is to develop an algorithm to determine whether a WSNM or an SSNM exists in a given instance, and to find one if it does.

References

1. Fischer, J.: Optimal succinctness for range minimum queries. In: Proceedings of the 9th Latin American Symposium on Theoretical Informatics (LATIN), pp. 158–169 (2010)
2. Fredman, M.L.: On computing the length of longest increasing subsequences. Discrete Appl. Math. **11**(1), 29–35 (1975)
3. Gale, D., Shapley, L.S.: College admissions and the stability of marriage. Am. Math. Mon. **69**, 9–15 (1962)
4. Gale, D., Sotomayor, M.: Some remarks on the stable matching problem. Discrete Appl. Math. **11**(3), 223–232 (1985)
5. Gusfield, D., Irving, R.W.: The Stable Marriage Problem: Structure and Algorithms. MIT Press, Cambridge (1989)
6. Irving, R.W.: An efficient algorithm for the "stable roommates" problem. J. Algorithms **6**, 577–595 (1985)
7. Irving, R.W.: Stable marriage and indifference. Discrete Appl. Math. **48**(3), 261–272 (1994)
8. Kajitami, Y., Takahashi, T.: The noncross matching and applications to the 3-side switch box routing in VLSI layout design. In: Proceedings of the IEEE International Symposium on Circuits and Systems, pp. 776–779 (1986)
9. Malucelli, F., Ottmann, T., Pretolani, D.: Efficient labelling algorithms for the maximum noncrossing matching problem. Discrete Appl. Math. **47**(2), 175–179 (1993)
10. Roth, A.E., Sotomayor, M.A.O.: Two-Sided Matching: A Study in Game-Theoretic Modeling and Analysis. Econometric Society Monographs. Cambridge University Press, Cambridge (1990)
11. Widmayer, P., Wong, C.K.: An optimal algorithm for the maximum alignment of terminals. Inf. Process. Lett. **10**, 75–82 (1985)

On the Average Case of MergeInsertion

Florian Stober and Armin Weiß[⊠] [ID]

FMI, Universität Stuttgart, Stuttgart, Germany
florian.stober@t-online.de, armin.weiss@fmi.uni-stuttgart.de

Abstract. MergeInsertion, also known as the Ford-Johnson algorithm, is a sorting algorithm which, up to today, for many input sizes achieves the best known upper bound on the number of comparisons. Indeed, it gets extremely close to the information-theoretic lower bound. While the worst-case behavior is well understood, only little is known about the average case. This work takes a closer look at the average case behavior. In particular, we establish an upper bound of $n \log n - 1.4005n + o(n)$ comparisons. We also give an exact description of the probability distribution of the length of the chain a given element is inserted into and use it to approximate the average number of comparisons numerically. Moreover, we compute the exact average number of comparisons for n up to 148. Furthermore, we experimentally explore the impact of different decision trees for binary insertion. To conclude, we conduct experiments showing that a slightly different insertion order leads to a better average case and we compare the algorithm to the recent combination with (1,2)-Insertionsort by Iwama and Teruyama.

Keywords: MergeInsertion · Minimum-comparison sort · Average case analysis

1 Introduction

Sorting a set of elements is an important operation frequently performed by many computer programs. Consequently there exist a variety of algorithms for sorting, each of which comes with its own advantages and disadvantages.

Here we focus on comparison based sorting and study a specific sorting algorithm known as MergeInsertion. It was discovered by Ford and Johnson in 1959 [5]. Before Knuth coined the term MergeInsertion in his study of the algorithm in his book "The Art of Computer Programming, Volume 3: Sorting and Searching" [7], it was known only as Ford-Johnson Algorithm, named after its creators. The one outstanding property of MergeInsertion is that the number of comparisons it requires is close to the information-theoretic lower bound of $\log(n!) \approx n \log n - 1.4427n$ (for sorting n elements). This sets it apart from many other sorting algorithms. MergeInsertion can be described in three steps: first

---

The second author has been supported by the German Research Foundation (DFG) under grant DI 435/7-1.

C. J. Colbourn et al. (Eds.): IWOCA 2019, LNCS 11638, pp. 417–429, 2019.
https://doi.org/10.1007/978-3-030-25005-8_34

pairs of elements are compared; in the second step the larger elements are sorted recursively; as a last step the elements belonging to the smaller half are inserted into the already sorted larger half using binary insertion.

In the worst case the number of comparisons of MergeInsertion is quite well understood [7] – it is $n \log n + b(n) \cdot n + o(n)$ where $b(n)$ oscillates between -1.415 and -1.3289. Moreover, for many n MergeInsertion is proved to be the optimal algorithm in the worst case (in particular, for $n \leq 15$ [9,10]). However, there are also n where it is not optimal [2,8]. One reason for this is the oscillating linear term in the number of comparisons, which allowed Manacher [8] to show that for certain n it is more efficient to split the input into two parts, sort both parts with MergeInsertion, and then merge the two parts into one array.

Regarding the average case not much is known: in [7] Knuth calculated the number of comparisons required on average for $n \in \{1, \ldots, 8\}$; an upper bound of $n \log n - 1.3999n + o(n)$ has been established in [3]. Most recently, Iwama and Teruyama [6] showed that in the average case MergeInsertion can be improved by combining it with their (1,2)-Insertion algorithm resulting in an upper bound of $n \log n - 1.4106n + O(\log n)$. This reduces the gap to the lower bound by around 25%. It is a fundamental open problem how close one can get to the information-theoretic lower bound of $n \log n - 1.4427n$ (see e. g. [6,11]).

The goal of this work is to study the number of comparisons required in the average case. In particular, we analyze the insertion step of MergeInsertion in greater detail. In general, MergeInsertion achieves its good performance by inserting elements in a specific order that in the worst case causes each element to be inserted into a sorted list of $2^k - 1$ elements (thus, using exactly k comparisons). When looking at the average case elements are often inserted into less than $2^k - 1$ elements which is slightly cheaper. By calculating those small savings we seek to achieve our goal of a better upper bound on the average case. Our results can be summarized as follows:

- We derive an exact formula for the probability distribution into how many elements a given element is inserted (Theorem 2). This is the crucial first step in order to obtain better bounds for the average case of MergeInsertion.
- We experimentally examine different decision trees for binary insertion. We obtain the best result when assigning shorter decision paths to positions located further to the left.
- We use Theorem 2 in order to compute quite precise numerical estimates for the average number of comparisons for n up to roughly 15000.
- We compute the exact average number of comparisons for n up to 148 – thus, going much further than [7].
- We improve the bound of [3] to $n \log n - 1.4005n + o(n)$ (Theorem 3). This partially answers a conjecture from [11] which asks for an in-place algorithm with $n \log n - 1.4n$ comparisons on average and $n \log n - 1.3n$ comparisons in the worst case. Although MergeInsertion is not in-place, the techniques from [3] or [11] can be used to make it so.
- We evaluate a slightly different insertion order decreasing the gap between the lower bound and the average number of comparisons of MergeInsertion by roughly 30% for $n \approx 2^k/3$.

– We compare MergeInsertion to the recent combination by Iwama and Teruyama [6] showing that, in fact, their combined algorithm is still better than the analysis and with the different insertion order can be further improved.

Due to space constraint, most proofs as well as additional explanations and experimental results can be found in the full version [13]. The code used in this work and the generated data is available on [12].

2   Preliminaries

Throughout, we assume that the input consists of n distinct elements. The average case complexity is the mean number of comparisons over all input permutations of n elements.

Description of MergeInsertion. The MergeInsertion algorithm consists of three phases: pairwise comparison, recursion, and insertion. Accompanying the explanations we give an example where $n = 21$. We call such a set of relations between individual elements a configuration.

1. **Pairwise comparison.** The elements are grouped into $\lfloor \frac{n}{2} \rfloor$ pairs. Each pair is sorted using one comparison. After that, the elements are called a_1 to $a_{\lfloor \frac{n}{2} \rfloor}$ and b_1 to $b_{\lceil \frac{n}{2} \rceil}$ with $a_i > b_i$ for all $1 \leq i \leq \lfloor \frac{n}{2} \rfloor$.

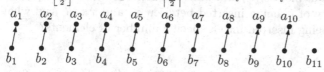

2. **Recursion.** The $\lfloor \frac{n}{2} \rfloor$ larger elements, i.e., a_1 to $a_{\lfloor \frac{n}{2} \rfloor}$ are sorted recursively. Then all elements (the $\lfloor \frac{n}{2} \rfloor$ larger ones as well as the corresponding smaller ones) are renamed accordingly such that $a_i < a_{i+1}$ and $a_i > b_i$ still holds.

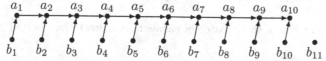

3. **Insertion.** The $\lceil \frac{n}{2} \rceil$ small elements, i.e., the b_i, are inserted into the main chain using binary insertion. The term "main chain" describes the set of elements containing $a_1, \ldots, a_{t_k}$ as well as the b_i that have already been inserted. The elements are inserted in batches starting with b_3, b_2. In the k-th batch the elements $b_{t_k}, b_{t_k-1}, \ldots, b_{t_{k-1}+1}$ where $t_k = \frac{2^{k+1}+(-1)^k}{3}$ are inserted in that order. Elements b_j where $j > \lceil \frac{n}{2} \rceil$ (which do not exist) are skipped. Note that technically b_1 is the first batch; but inserting b_1 does not need any comparison.

Because of the insertion order, every element b_i which is part of the k-th batch is inserted into at most $2^k - 1$ elements; thus, it can be inserted by binary insertion using at most k comparisons.

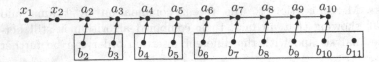

Regarding the average number of comparisons $F(n)$ we make the following observations: the first step always requires $\lfloor \frac{n}{2} \rfloor$ comparisons. The recursion step does not do any comparisons by itself but depends on the other steps. The average number of comparisons $G(n)$ required in the insertion step is not obvious. It will be studied closer in following chapters. Following [7], we obtain the recurrence (which is the same as for the worst-case number of comparisons)

$$F(n) = \left\lfloor \frac{n}{2} \right\rfloor + F\left(\left\lfloor \frac{n}{2} \right\rfloor\right) + G\left(\left\lceil \frac{n}{2} \right\rceil\right). \tag{1}$$

3   Average Case Analysis of the Insertion Step

In this section we have a look at different probabilities when inserting one batch of elements, i.e., the elements b_{t_k} to $b_{t_{k-1}+1}$. We assume that all elements of previous batches, i.e., b_1 to $b_{t_{k-1}}$, have already been inserted and together with the corresponding a_i they constitute the main chain and have been renamed to x_1 to $x_{2t_{k-1}}$ such that $x_i < x_{i+1}$. The situation is shown in Fig. 1.

We will look at the element b_{t_k+i} and want to answer the following questions: what is the probability of it being inserted between x_j and x_{j+1}? And what is the probability of it being inserted into a specific number of elements?

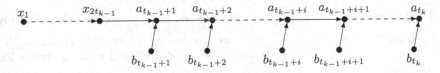

Fig. 1. Configuration where a single batch of elements remains to be inserted

We can ignore batches that are inserted after the batch we are looking at since those do not affect the probabilities we want to obtain.

First we define a probability space for the process of inserting one batch of elements: let Ω_k be the set of all possible outcomes (i.e., linear extensions) when sorting the partially ordered elements shown in Fig. 1 by inserting b_{t_k} to $b_{t_{k-1}+1}$. Each $\omega \in \Omega_k$ can be viewed as a function that maps an element e to its final position, i.e., $\omega(e) \in \{1, 2, \ldots, 2t_k\}$. While the algorithm mandates a specific order for inserting the elements $b_{t_{k-1}+1}$ to b_{t_k} during the insertion step, using a different order does not change the outcome, i.e., the elements are still sorted correctly. For this reason we can assume a different insertion in order to simplify calculating the likelihood of relations between individual elements.

Let us look at where an element will end up after it has been inserted. Not all positions are equally likely. For this purpose we define the random variable X_i as follows. To simplify notation we define $x_{t_{k-1}+j} := a_j$ for $t_{k-1} < j \leq t_k$ (hence, the main chain consists of $x_1, \ldots, x_{2^k}$).

$$X_i : \omega \mapsto \begin{cases} 0 & \text{if } \omega(b_{t_{k-1}+i}) < \omega(x_1) \\ j & \text{if } \omega(x_j) < \omega(b_{t_{k-1}+i}) < \omega(x_{j+1}) \text{ for } j \in \{1, \ldots, 2^k - 2\} \\ 2^k - 1 & \text{if } \omega(x_{2^k-1}) < \omega(b_{t_{k-1}+i}). \end{cases}$$

We are interested in the probabilities $P(X_i = j)$. These values follow a simple pattern (for $k = 4$ these are given in [13]).

Theorem 1. *The probability of $b_{t_{k-1}+i}$ being inserted between x_j and x_{j+1} is given by*

$$P(X_i = j) = \begin{cases} 2^{2i-2} \left(\frac{(t_{k-1}+i-1)!}{(t_{k-1})!} \right)^2 \frac{(2t_{k-1})!}{(2t_{k-1}+2i-1)!} & \text{if } 0 \leq j \leq 2t_{k-1} \\ 2^{4t_{k-1}-2j+2i-2} \left(\frac{(t_{k-1}+i-1)!}{(j-t_{k-1})!} \right)^2 \frac{(2j-2t_{k-1})!}{(2t_{k-1}+2i-1)!} & \text{if } 2t_{k-1} < j < 2t_{k-1}+i \\ 0 & \text{otherwise} \end{cases}$$

Next, our aim is to compute the probability that b_i is inserted into a particular number of elements. This is of particular interest because the difference between average and worst case comes from the fact that sometimes we insert into less than $2^k - 1$ elements. For that purpose we define the random variable Y_i.

$$Y_i : \omega \mapsto \left| \{ v \in \{ x_1, \ldots, x_{2^k} \} \cup \{ b_{t_{k-1}+i+1}, \ldots, b_{t_k} \} \mid \omega(v) < \omega(a_{t_{k-1}+i}) \} \right|$$

The elements in the main chain when inserting b_{t_k+i} are x_1 to $x_{2t_{k-1}+i-1}$ and those elements out of $b_{t_{k-1}+i+1}, \ldots, b_{t_k}$ which have been inserted before $a_{t_{k-1}+i}$ (which is $x_{2t_{k-1}+i}$). For computing the number of these, we introduce random variables $\tilde{Y}_{i,q}$ counting the elements in $\{ b_{t_{k-1}+i+1}, \ldots, b_{t_{k-1}+i+q} \}$ that are inserted before $a_{t_{k-1}+i}$:

$$\tilde{Y}_{i,q} : \omega \mapsto \left| \{ v \in \{ b_{t_{k-1}+i+1}, \ldots, b_{t_{k-1}+i+q} \} \mid \omega(v) < \omega(a_{t_{k-1}+i}) \} \right|.$$

By setting $q = t_k - t_{k-1} - i$, we obtain $Y_i = \tilde{Y}_{i,t_k-t_{k-1}-i} + 2t_{k-1} + i - 1$. Clearly we have $P(\tilde{Y}_{i,0} = j) = 1$ if $j = 0$ and $P(\tilde{Y}_{i,0} = j) = 0$ otherwise. For $q > 0$ there are two possibilities:

1. $\tilde{Y}_{i,q-1} = j-1$ and $X_{i+q} < 2t_{k-1}+i$: out of $\{ b_{t_{k-1}+i+1}, \ldots, b_{t_{k-1}+i+q-1} \}$ there have been $j - 1$ elements inserted before $a_{t_{k-1}+i}$ and $b_{t_{k-1}+i+q}$ is inserted before $a_{t_{k-1}+i}$.
2. $\tilde{Y}_{i,q-1} = j$ and $X_{i+q} \geq 2t_{k-1}+i$: out of $\{ b_{t_{k-1}+i+1}, \ldots, b_{t_{k-1}+i+q-1} \}$ there have been j elements inserted before $a_{t_{k-1}+i}$ and $b_{t_{k-1}+i+q}$ is inserted after $a_{t_{k-1}+i}$.

From these we obtain the following recurrence:

$$P(\tilde{Y}_{i,q} = j) = \quad P(X_{i+q} < 2t_{k-1} + i \mid \tilde{Y}_{i,q-1} = j - 1) \cdot P(\tilde{Y}_{i,q-1} = j - 1)$$
$$+ P(X_{i+q} \geq 2t_{k-1} + i \mid \tilde{Y}_{i,q-1} = j) \cdot P(\tilde{Y}_{i,q-1} = j)$$

The probability $P(X_{i+q} < 2t_{k-1} + i \mid \tilde{Y}_{i,q-1} = j - 1)$ can be obtained by looking at Fig. 1 and counting elements. When $b_{t_{k-1}+i+q}$ is inserted, the elements on the main chain which are smaller than $a_{t_{k-1}+i}$ are x_1 to $x_{2t_{k-1}}$, $a_{t_{k-1}+1}$ to $a_{t_{k-1}+i-1}$ and $j - 1$ elements out of $\{b_{t_{k-1}+i+1}, \ldots, b_{t_{k-1}+i+q-1}\}$ which is a total of $2t_{k-1} + 2i + j - 2$ elements. Combined with the fact that the main chain consists of $2t_{k-1} + 2i + 2q - 2$ elements smaller than $a_{t_{k-1}+i+q}$ we obtain the probability $\frac{2t_{k-1}+2i+j-1}{2t_{k-1}+2i+2q-1}$. We can calculate $P(X_{i+q} \geq 2t_{k-1} + i \mid \tilde{Y}_{i,q-1} = j)$ similarly leading to

$$P(\tilde{Y}_{i,q} = j) = \frac{2t_{k-1} + 2i + j - 1}{2t_{k-1} + 2i + 2q - 1} \cdot P(\tilde{Y}_{i,q-1} = j - 1) + \frac{2q - j - 1}{2t_{k-1} + 2i + 2q - 1} \cdot P(\tilde{Y}_{i,q-1} = j).$$

By solving the recurrence, we obtain a closed form for $P(\tilde{Y}_{i,q} = j)$ and, thus, for $P(Y_i = j)$. The complete proof is given in [13].

Theorem 2. For $1 \leq i \leq t_k - t_{k-1}$ and $2t_{k-1} + i - 1 \leq j \leq 2^k - 1$ the probability $P(Y_i = j)$, that $b_{t_{k-1}+i}$ is inserted into j elements is given by

$$P(Y_i = j) = 2^{j - 2t_{k-1} - i + 1} \frac{(2t_k - i - j - 1)!}{(j - 2t_{k-1} - i + 1)!(2^k - j - 1)!} \frac{(i + j)!}{(2t_k - 1)!} \frac{(t_k - 1)!}{(t_{k-1} + i - 1)!}.$$

Figure 2 shows the probability distribution for Y_1, Y_{21} and Y_{42} where $k = 7$. Y_{42} corresponds to the insertion of b_{t_k} (the first element of the batch). Y_1 corresponds to the insertion of $b_{t_{k-1}+1}$ (the last element of the batch). In addition to those three probability distributions Fig. 3 shows the mean of all Y_i for $k = 7$.

Binary Insertion and Different Decision Trees. The Binary Insertion step is an important part of MergeInsertion. In the average case many elements are inserted in less than $2^k - 1$ (which is the worst case). This leads to ambiguous decision trees where at some positions inserting an element requires only $k - 1$ instead of k comparisons. Since not all positions are equally likely (positions on the left have a slightly higher probability), this results in different average insertion costs. We compare four different strategies all satisfying that the corresponding decision trees have their leaves distributed across at most two layers. For an example with five elements see Fig. 4.

First there are the `center-left` and `center-right` strategies (the standard options for binary insertion): they compare the element to be inserted with the middle element, rounding down(up) in case of an odd number. The `left` strategy chooses the element to compare with in a way such that the positions where only $k - 1$ comparisons are required are at the very left. The `right` strategy is similar, here the positions where one can insert with just $k - 1$ comparisons are at the right. To summarize, the element to compare with is

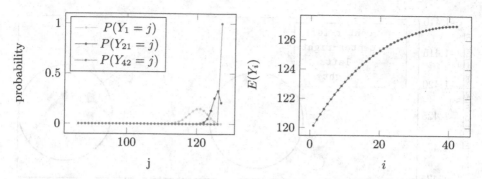

Fig. 2. Probability distribution of Y_i.   **Fig. 3.** Mean of Y_i for different i. $k = 7$.

$$\left\lfloor \frac{n+1}{2} \right\rfloor \qquad \text{strategy } \texttt{center-left}$$
$$\left\lceil \frac{n+1}{2} \right\rceil \qquad \text{strategy } \texttt{center-right}$$
$$\max\{n - 2^k + 1, 2^{k-1}\} \quad \text{strategy } \texttt{left}$$
$$\min\{2^k, n - 2^{k-1} + 1\} \quad \text{strategy } \texttt{right}$$

where $k = \lfloor \log n \rfloor$. Notice that the $\texttt{left}$ strategy is also used in [6], where it is called *right-hand-binary-search*. Figure 5 shows experimental results comparing the different strategies for binary insertion regarding their effect on the average-case of MergeInsertion. As we can see the $\texttt{left}$ strategy performs the best, closely followed by $\texttt{center-left}$ and $\texttt{center-right}$. $\texttt{right}$ performs the worst. The $\texttt{left}$ strategy performing best is no surprise since the probability that an element is inserted into one of the left positions is higher that it being inserted to the right. Therefore, in all further experiments we use the $\texttt{left}$ strategy.

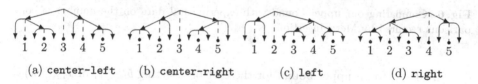

(a) $\texttt{center-left}$ (b) $\texttt{center-right}$ (c) $\texttt{left}$ (d) $\texttt{right}$

Fig. 4. Different strategies for binary insertion.

4   Improved Upper Bounds for MergeInsertion

Numeric Upper Bound. The goal of this section is to combine the probability given by Theorem 2 that an element $b_{t_{k-1}+i}$ is inserted into j elements with an upper bound for the number of comparisons required for binary insertion.

By [4], the number of comparisons required for binary insertion when inserting into $m - 1$ elements is $T_{\text{InsAvg}}(m) = \lceil \log m \rceil + 1 - \frac{2^{\lceil \log m \rceil}}{m}$. While only being exact in case of a uniform distribution, this formula acts as an upper bound in our case, where the probability is monotonically decreasing with the index.

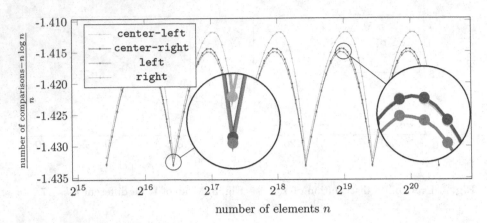

Fig. 5. Experimental results on the effect of different strategies for binary insertion on the number of comparisons.

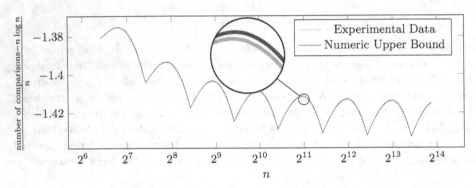

Fig. 6. Comparing our upper bound with experimental data on the number of comparisons required by MergeInsertion.

This leads to an upper bound for the cost of inserting $b_{t_{k-1}+i}$ of $T_{\mathrm{Ins}}(i,k) = \sum_j P(Y_i = j) \cdot T_{\mathrm{InsAvg}}(j+1)$. From there we calculated an upper bound for MergeInsertion. Figure 6 compares those results with experimental data on the number of comparisons required by MergeInsertion. We observe that the difference is rather small.

Computing the Exact Number of Comparisons. In this section we explore how to numerically calculate the exact number of comparisons required in the average case. The most straightforward way of doing this is to compute the external path length of the decision tree (sum of lengths of all paths from the root to leaves) and dividing by the number of leaves ($n!$ when sorting n elements), which unfortunately is only feasible for very small n. Instead we use Eq. (1), which describes the number of comparisons. The only unknown in that formula is $G(n)$ the number of comparisons required in the insertion step of the algorithm.

Since the insertion step of MergeInsertion works by inserting elements in batches, we write $G(n) = \left(\sum_{1 < k \leq k_n} \text{Cost}(t_{k-1}, t_k)\right) + \text{Cost}(t_{k_n}, n)$ for $t_{k_n} \leq n < t_{k_n+1}$. Here $\text{Cost}(s, e)$ is the cost of inserting one batch of elements starting from b_{s+1} up to b_e. The idea for computing $\text{Cost}(s, e)$ is to calculate the external path length of the decision tree corresponding to the insertion of that batch of elements and then dividing by the number of leaves. As this is still not feasible, we apply some optimizations which we describe in detail in [13].

For $n \in \{1, \ldots, 15\}$ the computed values are shown in Table 1, for larger n Fig. 7 shows the values we computed. The complete data set is provided in the file data/exact.csv in [12]. Our results match up with the values for $n \in \{1, \ldots, 8\}$ calculated in [7]. Note that for these values the chosen insertion strategy does not affect the average case (we use the left strategy).

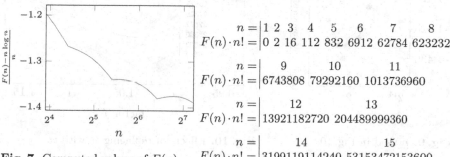

Fig. 7. Computed values of $F(n)$.

Table 1. Computed values of $F(n) \cdot n!$.

$n =$	1	2	3	4	5	6	7	8
$F(n) \cdot n! =$	0	2	16	112	832	6912	62784	623232

$n =$	9	10	11
$F(n) \cdot n! =$	6743808	79292160	1013736960

$n =$	12	13
$F(n) \cdot n! =$	13921182720	204489999360

$n =$	14	15
$F(n) \cdot n! =$	3199119114240	53153472153600

Improved Theoretical Upper Bounds. In this section we improve upon the upper bound from [3] leading to the following result:

Theorem 3. *The number of comparisons required in the average case of Merge-Insertion is at most $n \log n - c(x_n) \cdot n \pm \mathcal{O}(\log^2 n)$ where x_n is the fractional part of $\log(3n)$, i.e., the unique value in $[0, 1)$ such that $n = 2^{k - \log 3 + x_n}$ for some $k \in \mathbb{Z}$ and $c : [0, 1) \to \mathbb{R}$ is given by the following formula:*

$$c(x) = (3 - \log 3) - (2 - x - 2^{1-x}) + (1 - 2^{-x})\left(\frac{3}{2^x + 1} - 1\right) + \frac{2^{\log 3 - x}}{2292} \geq 1.4005$$

Hence we have obtained a new upper bound for the average case of MergeInsertion which is $n \log n - 1.4005n + \mathcal{O}(\log^2 n)$. A visual representation of $c(x)$ is provided in Fig. 8. The worst case is near $x = 0.6$ (i.e., n roughly a power of two) where $c(x)$ is just slightly larger than 1.4005.

The proof of Theorem 3 analyzes the insertion of one batch of elements more carefully than in [4]. The exact probability that $b_{t_{k-1}+i}$ is inserted into j elements is given by Theorem 2. We are especially interested in the case of $b_{t_{k-1}+u}$ where

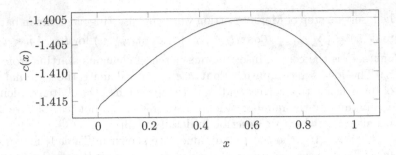

Fig. 8. Plot of $c(x)$.

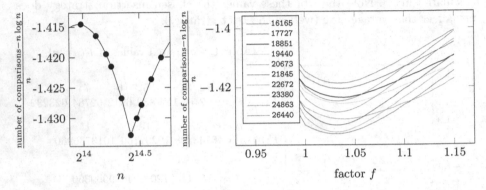

Fig. 9. n used in Fig. 10.

Fig. 10. Effects of replacing t_k with $\hat{t}_k$.

$u = \lfloor \frac{t_k - t_{k-1}}{2} \rfloor$, because, if we know $P(Y_u < m)$, then we can use that for all $q < u$ we have $P(Y_q < m) \geq P(Y_u < m)$.

However, the equation from Theorem 2 is hard to work with, so we approximate it with the binomial distribution $p(j) = \binom{\lceil \frac{u}{2} \rceil}{q}(\frac{\lfloor \frac{u}{2} \rfloor}{2t_k-1})^q(\frac{2t_k-1-\lfloor \frac{u}{2} \rfloor}{2t_k-1})^{\lceil \frac{u}{2} \rceil - q}$ with $q = 2^k - 1 - j$, that by construction fulfills $\sum_{j=0}^{j_0} p(j) \leq \sum_{j=0}^{j_0} P(Y_u = j) = P(Y_u \leq j_0)$ for all j_0. By using the approximation $P(Y_u = j) \approx p(j)$ we can calculate a lower bound for the median of $Y_{\frac{t_k - t_{k-1}}{2}}$ which is $2^k - 1 - \lfloor n_B \cdot p_B \rfloor \in 2^k - 1 - \frac{2^{k-6}}{3} + \mathcal{O}(1)$. Thus, with a probability of one half the elements $b_{t_{k-1}+i}$ for $1 \leq i \leq u$ are inserted in $\frac{2^{k-6}}{3}$ elements less compared to the worst case. Combining that with the bounds from [4] we obtain Theorem 3. The complete proof is given in [13].

5   Experiments

In this section we discuss our experiment, which consist of two parts: first, we evaluate how increasing t_k by some constant factor can reduce the number of comparisons, then we examine how the combination with the (1,2)-Insertion algorithm as proposed in [6] improves MergeInsertion.

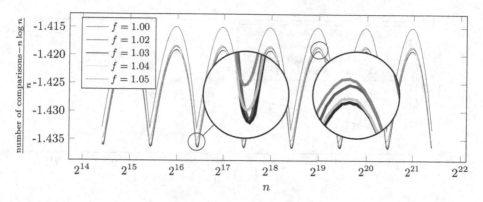

Fig. 11. Comparison of different factors f for $\hat{t}_k$.

We implemented MergeInsertion using a tree based data structure, similar to the Rope data structure [1] used in text processing, resulting in a comparably "fast" implementation. Implementation details can be found in [13]. All experiments use the `left` strategy for binary insertion (see Sect. 3). The number of comparisons has been averaged over 10 to 10000 runs, depending on the size of the input.

Increasing t_k by a Constant Factor. In this section we modify MergeInsertion by replacing t_k with $\hat{t}_k = \lfloor f \cdot t_k \rfloor$ – otherwise the algorithm is the same. Originally the numbers t_k have been chosen, such that each element b_i with $t_{k-1} < i \leq t_k$ is inserted into at most $2^k - 1$ elements (which is optimal for the worst case). As we have seen in previous sections many elements are inserted into slightly less than $2^k - 1$ elements. The idea behind increasing t_k by a constant factor f is to allow more elements to be inserted into close to $2^k - 1$ elements.

Figure 10 shows how different factors f affect the number of comparisons required by MergeInsertion. The different lines represent different input lengths. For instance, $n = 21845$ is an input size for which MergeInsertion works best. An overview of the different input lengths and how original MergeInsertion performs for these can be seen in Fig. 9. The chosen values are assumed to be representative for the entire algorithm. We observe that for all shown input lengths, multiplying t_k by a factor f between 1.02 and 1.05, leads to an improvement.

Figure 11 compares different factors from 1.02 to 1.05. The factor 1.0 (i. e., the original algorithm) is included as a reference. We observe that all the other factors lead to a considerable improvement compared to 1.0. The difference between the factors in the chosen range is rather small. However, 1.03 appears to be best out of the tested values. At $n \approx 2^k/3$ the difference to the information-theoretic lower bound is reduced to $0.007n$, improving upon the original algorithm, which has a difference of $0.01n$ to the optimum.

Another observation we make from Fig. 11 is that the plot periodically repeats itself with each power of two. Thus, we conclude that replacing t_k with $\hat{t}_k = \lfloor f \cdot t_k \rfloor$ with $f \in [1.02, 1.05]$ reduces the number of comparisons required per element by some constant.

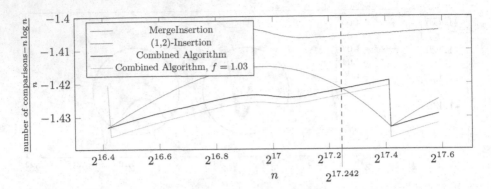

Fig. 12. Experimental results comparing MergeInsertion, (1,2)-Insertion and the combined algorithm.

Combination with (1,2)-Insertion. (1,2)-Insertion is a sorting algorithm presented in [6]. It works by inserting either a single element or two elements at once into an already sorted list. On its own (1,2)-Insertion is worse than MergeInsertion; however, it can be combined with MergeInsertion. The combined algorithm works by sorting $m = \max\{u_k \mid u_k \leq n\}$ elements with MergeInsertion. Then the remaining elements are inserted using (1,2)-Insertion. Let $u_k = \lfloor \left(\frac{4}{3}\right) 2^k \rfloor$ denote a point where MergeInsertion is optimal.

In Fig. 12 we can see that at the point u_k MergeInsertion and the combined algorithm perform the same. However, in the values following u_k the combined algorithm surpasses MergeInsertion until at one point close to the next optimum MergeInsertion is better once again. In their paper Iwama and Teruyama calculated that for $0.638 \leq \frac{n}{2^{\lceil \log n \rceil}} \leq \frac{2}{3}$ MergeInsertion is better than the combined algorithm. The fraction $\frac{2}{3}$ corresponds to the point where MergeInsertion is optimal. They derived the constant 0.638 from their theoretical analysis using the upper bound for MergeInsertion from [3]. Comparing this to our experimental results we observe that the range where MergeInsertion is better than the combined algorithm starts at $n \approx 2^{17.242}$. This yields $\frac{2^{17.242}}{2^{18}} = 2^{17.242-18} = 2^{-0.758} \approx 0.591$. Hence the range where MergeInsertion is better than the combined algorithm is $0.591 \leq \frac{n}{2^{\lceil \log n \rceil}} \leq \frac{2}{3}$, which is slightly larger than the theoretical analysis suggested. Also shown in Fig. 12 is the combined algorithm where we additionally apply our suggestion of replacing t_k by $\hat{t}_k = \lfloor f \cdot t_k \rfloor$ with $f = 1.03$. This leads to an additional improvement and comes even closer to the lower bound of $\log(n!)$.

Conclusion and Outlook. We improved the previous upper bound of $n \log n - 1.3999n + o(n)$ to $n \log n - 1.4005n + o(n)$ for the average number of comparisons of MergeInsertion. However, there still is a gap between the number of comparisons required by MergeInsertion and this upper bound.

In Sect. 4 we used a binomial distribution to approximate the probability of an element being inserted into a specific number of elements during the insertion step. However, the difference between our approximation and the actual probability distribution is rather large. Finding an approximation which reduces that gap while still being simple to analyze with respect to its mean would facilitate further improvements to the upper bound.

Our suggestion of increasing t_k by a constant factor f reduced the number of comparisons required per element by some constant. However, we do not have a proof for this. Thus, future research could try to determine the optimal value for the factor f as well as to study how this suggestion affects the worst-case.

References

1. Boehm, H.J., Atkinson, R., Plass, M.: Ropes: an alternative to strings. Softw. Pract. Exper. **25**(12), 1315–1330 (1995)
2. Bui, T., Thanh, M.: Significant improvements to the Ford-Johnson algorithm for sorting. BIT Numer. Math. **25**(1), 70–75 (1985)
3. Edelkamp, S., Weiß, A.: QuickXsort: efficient sorting with $n \log n - 1.399n + o(n)$ comparisons on average. In: Hirsch, E.A., Kuznetsov, S.O., Pin, J.É., Vereshchagin, N.K. (eds.) CSR 2014. LNCS, vol. 8476, pp. 139–152. Springer, Cham (2014). https://doi.org/10.1007/978-3-319-06686-8_11
4. Edelkamp, S., Weiß, A., Wild, S.: Quickxsort - A fast sorting scheme in theory and practice. CoRR abs/1811.01259 (2018)
5. Ford, L.R., Johnson, S.M.: A tournament problem. Am. Math. Monthly **66**(5), 387–389 (1959)
6. Iwama, K., Teruyama, J.: Improved average complexity for comparison-based sorting. Algorithms and Data Structures. LNCS, vol. 10389, pp. 485–496. Springer, Cham (2017). https://doi.org/10.1007/978-3-319-62127-2_41
7. Knuth, D.E.: The Art of Computer Programming. Sorting and Searching, 2nd Edn, vol. 3. Addison Wesley Longman, Redwood City (1998)
8. Manacher, G.K.: The Ford-Johnson sorting algorithm is not optimal. J. ACM **26**(3), 441–456 (1979)
9. Peczarski, M.: New results in minimum-comparison sorting. Algorithmica **40**(2), 133–145 (2004)
10. Peczarski, M.: The Ford-Johnson algorithm still unbeaten for less than 47 elements. Inf. Process. Lett. **101**(3), 126–128 (2007)
11. Reinhardt, K.: Sorting in-place with a worst case complexity of n log n-1.3n + O(logn) comparisons and epsilon n log n + O(1) transports. In: Proceedings of Algorithms and Computation, ISAAC 1992, pp. 489–498 (1992)
12. Stober, F.: Source code and generated data (2018). https://github.com/CodeCrafter47/merge-insertion
13. Stober, F., Weiß, A.: On the Average Case of MergeInsertion. arXiv e-prints abs/1905.09656 (2019)

Shortest Unique Palindromic Substring Queries on Run-Length Encoded Strings

Kiichi Watanabe[1(✉)], Yuto Nakashima[2], Shunsuke Inenaga[2], Hideo Bannai[2], and Masayuki Takeda[2]

[1] Department of Electrical Engineering and Computer Science,
Kyushu University, Fukuoka, Japan
`watanabe.kiichi.849@s.kyushu-u.ac.jp`
[2] Department of Informatics, Kyushu University, Fukuoka, Japan
`{yuto.nakashima,inenaga,bannai,takeda}@inf.kyushu-u.ac.jp`

Abstract. For a string S, a palindromic substring $S[i..j]$ is said to be a *shortest unique palindromic substring* (*SUPS*) for an interval $[s, t]$ in S, if $S[i..j]$ occurs exactly once in S, the interval $[i, j]$ contains $[s, t]$, and every palindromic substring containing $[s, t]$ which is shorter than $S[i..j]$ occurs at least twice in S. In this paper, we study the problem of answering *SUPS* queries on run-length encoded strings. We show how to preprocess a given run-length encoded string RLE_S of size m in $O(m)$ space and $O(m \log \sigma_{RLE_S} + m\sqrt{\log m / \log \log m})$ time so that all *SUPSs* for any subsequent query interval can be answered in $O(\sqrt{\log m / \log \log m} + \alpha)$ time, where α is the number of outputs, and σ_{RLE_S} is the number of distinct runs of RLE_S.

1 Introduction

The *shortest unique substring* (*SUS*) problem, which is formalized below, is a recent trend in the string processing community. Consider a string S of length n. A substring $X = S[i..j]$ of S is called a SUS for a position p ($1 \le p \le n$) iff the interval $[i..j]$ contains p, X occurs in S exactly once, and every substring containing p which is shorter than $S[i..j]$ occurs at least twice in S. The SUS problem is to preprocess a given string S so that SUSs for query positions p can be answered quickly. The study on the SUS problem was initiated by Pei et al., and is motivated by an application to bioinformatics e.g., designing polymerase chain reaction (PCR) primer [11]. Pei et al. [11] showed an $\Theta(n^2)$-time and space preprocessing scheme such that all k SUSs for a query position can be answered in $O(k)$ time. Later, two independent groups, Tsuruta et al. [13] and Ileri et al. [7], showed algorithms that use $\Theta(n)$ time and space[1] for preprocessing, and all SUSs can be answered in $O(k)$ time per query. To be able to handle huge text data where n can be massively large, there have been further efforts to reduce

---

[1] Throughout this paper, we measure the space complexity of an algorithm with the number of *words* that the algorithm occupies in the word RAM model, unless otherwise stated.

© Springer Nature Switzerland AG 2019
C. J. Colbourn et al. (Eds.): IWOCA 2019, LNCS 11638, pp. 430–441, 2019.
https://doi.org/10.1007/978-3-030-25005-8_35

the space usage. Hon et al. [5] proposed an "in-place" algorithm which works within space of the input string S and two output arrays A and B of length n each, namely, in $n \log_2 \sigma$ *bits* plus $2n$ words of space. After the execution of their algorithm that takes $O(n)$ time, the beginning and ending positions of a SUS for each text position i $(1 \le i \le n)$ are consequently stored in $A[i]$ and $B[i]$, respectively, and S remains unchanged. Hon et al.'s algorithm can be extended to handle SUSs with approximate matches, with a penalty of $O(n^2)$ preprocessing time. For a pre-determined parameter τ, Ganguly et al. [4] proposed a time-space trade-off algorithm for the SUS problem that uses $O(n/\tau)$ additional working space (apart from the input string S) and answers each query in $O(n\tau^2 \log \frac{n}{\tau})$ time. They also proposed a "succinct" data structure of $4n + o(n)$ *bits* of space that can be built in $O(n \log n)$ time and can answer a SUS for each given query position in $O(1)$ time. Another approach to reduce the space requirement for the SUS problem is to work on a "compressed" representation of the string S. Mieno et al. [10] developed a data structure of $\Theta(m)$ space (or $\Theta(m \log n)$ *bits* of space) that answers all k SUSs for a given position in $O(\sqrt{\log m / \log \log m} + k)$ time, where m is the size of the *run length encoding* (*RLE*) of the input string S. This data structure can be constructed in $O(m \log m)$ time with $O(m)$ words of working space if the input string S is already compressed by RLE, or in $O(n + m \log m)$ time with $O(m)$ working space if the input string S is given without being compressed.

A generalized version of the SUS problem, called the *interval* SUS problem, is to answer SUSs that contain a query interval $[s, t]$ with $1 \le s \le t \le n$. Hu et al. [6] proposed an optimal $\Theta(n)$ time and space algorithm to preprocess a given string S so that all k SUSs for a given query interval are reported in $O(k)$ time. Mieno et al.'s data structure [10] also can answer interval SUS queries with the same preprocessing time/space and query time as above.

Recently, a new variant of the SUS problem, called the *shortest unique palindromic substring* (*SUPS*) problem is considered [8]. A substring $P = S[i..j]$ is called a SUPS for an interval $[s, t]$ iff P occurs exactly once in S, $[s, t] \subseteq [i, j]$, and every palindromic substring of S which contains interval $[s, t]$ and is shorter than P occurs at least twice in S. The study on the SUPS problem is motivated by an application in molecular biology. Inoue et al. [8] showed how to preprocess a given string S of length n in $\Theta(n)$ time and space so that all α SUPSs (if any) for a given interval can be answered in $O(\alpha + 1)$ time[2]. While this solution is optimal in terms of the length n of the input string, no space-economical solutions for the SUPS problem were known.

In this paper, we present the *first* space-economical solution to the SUPS problem based on RLE. The proposed algorithm computes a data structure of $\Theta(m)$ space that answers each SUPS query in $O(\sqrt{\log m / \log \log m} + \alpha)$ time. The most interesting part of our algorithm is how to preprocess a given RLE string of length m in $O(m(\log \sigma_{RLE_S} + \sqrt{\log m / \log \log m}))$ time, where σ_{RLE_S} is the number of distinct runs in the RLE of S. Note that $\sigma_{RLE} \le m$ always holds. For this sake, we propose RLE versions of Manacher's maximal palindrome

---

[2] It is possible that $\alpha = 0$ for some intervals.

algorithm [9] and Rubinchik and Shur's eertree data structure [12], which may
be of independent interest. We remark that our preprocessing scheme is quite
different from Mieno et al.'s method [10] for the SUS problem on RLE strings
and Inoue et al.'s method [8] for the SUPS problem on plain strings.

2   Preliminaries

2.1   Strings

Let Σ be an ordered *alphabet* of size σ. An element of Σ^* is called a *string*.
The length of a string S is denoted by $|S|$. The empty string ε is a string of
length 0. For a string $S = XYZ$, X, Y and Z are called a *prefix*, *substring*, and
suffix of S, respectively. The i-th character of a string S is denoted by $S[i]$, for
$1 \leq i \leq |S|$. Let $S[i..j]$ denote the substring of S that begins at position i and
ends at position j, for $1 \leq i \leq j \leq |S|$. For convenience, let $S[i..j] = \varepsilon$ for $i > j$.

For any string S, let $S^R = S[|S|] \cdots S[1]$ denote the reversed string of S. A
string P is called a *palindrome* iff $P = P^R$. A substring $P = S[i..j]$ of a string S is
called a *palindromic substring* iff P is a palindrome. For a palindromic substring
$P = S[i..j]$, $\frac{i+j}{2}$ is called the *center* of P. A palindromic substring $P = S[i..j]$ is
said to be a *maximal palindrome* of S, iff $S[i-1] \neq S[j+1]$, $i = 1$ or $j = |S|$. A
suffix of string S that is a palindrome is called a *palindromic suffix* of S. Clearly
any palindromic suffix of S is a maximal palindrome of S.

We will use the following lemma in the analysis of our algorithm.

Lemma 1 ([3]). *Any string of length k can contain at most $k + 1$ distinct
palindromic substrings (including the empty string ε).*

2.2   MUPSs and SUPSs

For any strings X and S, let $occ_S(X)$ denote the number of occurrences of X
in S, i.e., $occ_S(X) = |\{i \mid S[i..i + |X| - 1] = X\}|$. A string X is called a *unique
substring* of a string S iff $occ_S(X) = 1$. A substring $P = S[i..j]$ of string S is
called a *minimal unique palindromic substring* (*MUPS*) of a string S iff (1) P is
a unique palindromic substring of S and (2) either $|P| \geq 3$ and the palindrome
$Q = S[i + 1..j - 1]$ satisfies $occ_S(Q) \geq 2$, or $1 \leq |P| \leq 2$.

Lemma 2 ([8]). *MUPSs do not nest, namely, for any pair of distinct MUPSs,
one cannot contain the other.*

Due to Lemma 2, both of the beginning positions and the ending positions
of MUPSs are monotonically increasing. Let $\mathcal{M}_S$ denote the list of MUPSs in S
sorted in increasing order of their beginning positions (or equivalently the ending
positions) in S.

Let $[s, t]$ be an integer interval over the positions in a string S, where $1 \leq s \leq
t \leq |S|$. A substring $P = S[i..j]$ of string S is called a *shortest unique palindromic
substring* (*SUPS*) for interval $[s, t]$ of S, iff (1) P is a unique palindromic substring
of S, (2) $[s, t] \subseteq [i, j]$, and (3) there is no unique palindromic substring $Q =
S[i'..j']$ such that $[s, t] \subseteq [i', j']$ and $|Q| < |P|$. We give an example in Fig. 1.

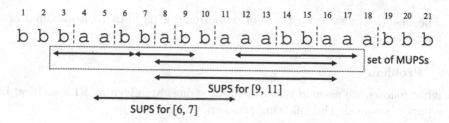

Fig. 1. This figure shows all *MUPS*s and some *SUPS*s of string $RLE_S = b^3a^2b^2a^1b^2a^3b^2a^3b^3$. There are 4 *MUPS*s illustrated in the box. The *SUPS* for interval $[6,7]$ is $S[5..11]$, and the *SUPS* for interval $[9,11]$ is $S[8..16]$.

2.3 Run Length Encoding (RLE)

The *run-length encoding* RLE_S of string S is a compact representation of S such that each maximal run of the same characters in S is represented by a pair of the character and the length of the run. More formally, let $\mathcal{N}$ denote the set of positive integers. For any non-empty string S, $RLE_S = (a_1, e_1), \ldots, (a_m, e_m)$, where $a_j \in \Sigma$ and $e_j \in \mathcal{N}$ for any $1 \le j \le m$, and $a_j \ne a_{j+1}$ for any $1 \le j < m$. E.g., if $S = \texttt{aaccccccccbbabbbb}$, then $RLE_S = (\texttt{a}, 2), (\texttt{c}, 7), (\texttt{b}, 2), (\texttt{a}, 1), (\texttt{b}, 4)$. Each (a, e) in RLE_S is called a (character) *run*, and e is called the exponent of this run. We also denote each run by a^e when it seems more convenient and intuitive. For example, we would write as (a, e) when it seems more convenient to treat it as a kind of character (called an RLE-character), and would write as a^e when it seems more convenient to treat it as a string consisting of e a's.

The *size* of RLE_S is the number m of runs in RLE_S. Let $Rb[j]$ (resp. $Re[j]$) denote the beginning (resp. ending) position of the jth run in the string S, i.e., $Rb[j] = 1 + \sum_{i=0}^{j-1} e_i$ with $e_0 = 0$ and $Re[j] = \sum_{i=1}^{j} e_i$. The *center* of the jth run is $\frac{Rb[j]+Re[j]}{2}$.

For any two ordered pairs $(a, e), (a', e') \in \Sigma \times \mathcal{N}$ of a character and positive integer, we define the equality such that $(a, e) = (a', e')$ iff $a = a'$ and $e = e'$ both hold. We also define a total order of these pairs such that $(a, e) < (a', e')$ iff $a < a'$, or $a = a'$ and $e < e'$.

An occurrence of a palindromic substring $P = S[i..i']$ of a string S with RLE_S of size m is said to be *RLE-bounded* if $i = Rb[j]$ and $i' = Re[j']$ for some $1 \le j \le j' \le m$, namely, if both ends of the occurrence touch the boundaries of runs. An RLE-bounded occurrence $P = S[i..i']$ is said to be *RLE-maximal* if $(a_{j-1}, e_{j-1}) \ne (a_{j'+1}, e_{j'+1})$, $j = 1$ or $j' = m$. Note that an RLE-maximal occurrence of a palindrome may not be maximal in the string S. E.g., consider string $S = \texttt{caabbcccbbaaaac}$ with $RLE_S = c^1a^2b^2c^3b^2a^4c^1$.

- The occurrence of palindrome c^3 is RLE-bounded but is neither RLE-maximal nor maximal.
- The occurrence of palindrome $b^2c^3b^2$ is RLE-maximal but is not maximal.
- The occurrence of palindrome $a^2b^2c^3b^2a^2$ is not RLE-maximal but is maximal.

- The first (leftmost) occurrence of palindrome a^2 is both RLE-maximal and maximal.

2.4 Problem

In what follows, we assume that our input strings are given as RLE strings. In this paper, we tackle the following problem.

Problem 1 (SUPS problem on run-length encoded strings).

Preprocess: $RLE_S = (a_1, e_1), \ldots, (a_m, e_m)$ *of size* m *representing a string* S *of length* n.
Query: *An integer interval* $[s, t]$ $(1 \le s \le t \le n)$.
Return: *All SUPSs for interval* $[s, t]$.

In case the string S is given as a plain string of length n, then the time complexity of our algorithm will be increased by an additive factor of n that is needed to compute RLE_S, while the space usage will stay the same since RLE_S can be computed in constant space.

3 Computing MUPSs from RLE Strings

The following known lemma suggests that it is helpful to compute the set $\mathcal{M}_S$ of MUPSs of S as a preprocessing for the SUPS problem.

Lemma 3 ([8]). *For any SUPS* $S[i..j]$ *for some interval, there exists exactly one MUPS that is contained in the interval* $[i, j]$. *Furthermore, the MUPS has the same center as the SUPS* $S[i..j]$.

3.1 Properties of MUPSs on RLE Strings

Now we present some useful properties of MUPSs on the run-length encoded string $RLE_S = (a_1, e_1), \ldots, (a_m, e_m)$.

Lemma 4. *For any MUPS* $S[i..j]$ *in* S, *there exists a unique integer* k $(1 \le k \le m)$ *such that* $\frac{i+j}{2} = \frac{Rb[k] + Re[k]}{2}$.

Proof. Suppose on the contrary that there is a MUPS $S[i..j]$ such that $\frac{i+j}{2} \ne \frac{Rb[k] + Re[k]}{2}$ for any $1 \le k \le m$. Let l be the integer that satisfies $Rb[l] \le \frac{i+j}{2} \le Re[l]$. By the assumption, the longest palindrome whose center is $\frac{i+j}{2}$ is $a_l^{\min\{i - Rb[l] + 1, Re[l] - j + 1\}}$. However, this palindrome $a_l^{\min\{i - Rb[l] + 1, Re[l] - j + 1\}}$ occurs at least twice in the lth run $a_l^{e_l}$. Hence MUPS $S[i..j]$ is not a unique palindromic substring, a contradiction.   □

The following corollary is immediate from Lemma 4.

Corollary 1. *For any string* S, $|\mathcal{M}_S| \le m$.

It is easy to see that the above bound is tight: for instance, any string where each run has a distinct character (i.e., $m = \sigma_{RLE}$) contains exactly m MUPSs. Our preprocessing and query algorithms which will follow are heavily dependent on this lemma and corollary.

3.2   RLE Version of Manacher's Algorithm

Due to Corollary 1, we can restrict ourselves to computing palindromic substrings whose center coincides with the center of each run. These palindromic substrings are called *run-centered* palindromes. Run-centered palindromes will be candidates of MUPSs of the string S.

To compute run-centered palindromes from RLE_S, we utilize Manacher's algorithm [9] that computes all maximal palindromes for a given (plain) string of length n in $O(n)$ time and space. Manacher's algorithm is based only on character equality comparisons, and hence it works with general alphabets.

Let us briefly recall how Manacher's algorithm works. It processes a given string S of length n from left to right. It computes an array **MaxPal** of length $2n + 1$ such that **MaxPal**$[c]$ stores the length of the maximal palindrome with center c for $c = 1, 1.5, 2, \ldots, n - 1, n - 0.5, n$. Namely, Manacher's algorithm processes a given string S in an online manner from left to right. This algorithm is also able to compute, for each position $i = 1, \ldots, n$, the longest palindromic suffix of $S[1..i]$ in an online manner.

Now we apply Manacher's algorithm to our run-length encoded input string $RLE_S = (a_1, e_1), \ldots, (a_m, e_m)$. Then, what we obtain after the execution of Manacher's algorithm over RLE_S is all *RLE-maximal* palindromes of S. Note that by definition all RLE-maximal palindromes are run-centered. Since RLE_S can be regarded as a string of length m over an alphabet $\Sigma \times \mathcal{N}$, this takes $O(m)$ time and space.

Remark 1. If wanted, we can compute all *maximal* palindromes of S in $O(m)$ time after the execution of Manacher's algorithm to RLE_S. First, we compute every run-centered maximal palindrome P_l that has its center in each lth run in RLE_S. For each already computed run-centered RLE-maximal palindrome $Q_l = S[Rb[i]..Re[j]]$ with $1 < i \leq j < m$, it is enough to first check whether $a_{i-1} = a_{j+1}$. If no, then $P_l = Q_l$, and if yes then we can further extend both ends of Q_l with $(a_{i-1}, \min\{e_{i-1}, e_{j+1}\})$ and obtain P_l. As a side remark, we note that any other maximal palindromes of S are not run-centered, which means that any of them consists only of the same characters and lie inside of one character run. Such maximal palindromes are trivial and need not be explicitly computed.

3.3   RLE Version of Eertree Data Structure

The *eertree* [12] of a string S, denoted $\mathsf{eertree}(S)$, is a pair of two rooted trees $\mathsf{T_{odd}}$ and $\mathsf{T_{even}}$ which represent all distinct palindromic substrings of S. The root of $\mathsf{T_{odd}}$ represents the empty string ε and each non-root node of $\mathsf{T_{odd}}$ represents a non-empty palindromic substring of S of odd length. Similarly, the root of $\mathsf{T_{even}}$ represents the empty string ε and each non-root node of $\mathsf{T_{even}}$ represents a non-empty palindromic substring of S of even length. From the root r of $\mathsf{T_{odd}}$, there is a labeled directed edge (r, a, v) if v represents a single character $a \in \Sigma$. For any non-root node u of $\mathsf{T_{odd}}$ or $\mathsf{T_{even}}$, there is a labeled directed edge (u, a, v) from u to node v with character label $a \in \Sigma$ if $aua = v$. For any node u, the labels of out-going edges of u must be mutually distinct.

By Lemma 1, any string S of length n can contain at most $n + 1$ distinct palindromic substrings (including the empty string ε). Thus, the size of eertree(S) is linear in the string length n. Rubinchik and Shur [12] showed how to construct eertree(S) in $O(n \log \sigma_S)$ time and $O(n)$ space, where σ_S is the number of distinct characters in S. They also showed how to compute the number of occurrences of each palindromic substring in $O(n \log \sigma_S)$ time and $O(n)$ space, using eertree(S).

Now we introduce a new data structure named *RLE-eertrees* based on eertrees. Let $RLE_S = (a_1, e_1), \ldots, (a_m, e_m)$, and let Σ_{RLE} be the set of maximal runs of S, namely, $\Sigma_{RLE} = \{(a, e) \mid (a, e) = (a_i, e_i) \text{ for some } 1 \le i \le m\}$. Let $\sigma_{RLE} = |\Sigma_{RLE}|$. Note that $\sigma_{RLE} \le m$ always holds. The RLE-eertree of string S, denoted by $\mathsf{e}^2\mathsf{rtre}^2(S)$, is a *single* eertree $\mathsf{T}_{\mathsf{odd}}$ *over the RLE alphabet* $\sigma_{RLE} \subset \Sigma \times \mathcal{N}$, which represents distinct run-centered palindromes of S which have an RLE-bounded occurrence $[i, i']$ such that $i = Rb[j]$ and $i' = Re[j']$ for some $1 \le j \le j' \le m$ (namely, the both ends of the occurrence touch the boundary of runs), or an occurrence as a maximal palindrome in S. We remark that the number of *runs* in each palindromes in $\mathsf{e}^2\mathsf{rtre}^2(S)$ is odd, but their decompressed string length may be odd or even. In $\mathsf{e}^2\mathsf{rtre}^2(S)$, there is a directed labeled edge (u, a^e, v) from node u to node v with label $a^e \in \Sigma_{RLE}$ if (1) $a^e u a^e = v$, or (2) $u = \varepsilon$ and $v = a^e \in \Sigma \times \mathcal{N}$. Note that if the in-coming edge of a node u is labeled with a^e, then any out-going edge of u cannot have a label a^f with the same character a. Since $\mathsf{e}^2\mathsf{rtre}^2 S$ is an eertree over the alphabet Σ_{RLE} of size $\sigma_{RLE} \le m$, it is clear that the number of out-going edges of each node is bounded by σ_{RLE}. We give an example of $\mathsf{e}^2\mathsf{rtre}^2(S)$ in Fig. 2.

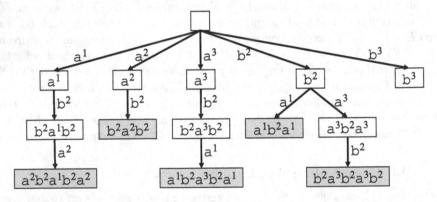

Fig. 2. The RLE-eertree $\mathsf{e}^2\mathsf{rtre}^2(S)$ of $RLE_S = \mathsf{b}^3\mathsf{a}^2\mathsf{b}^2\mathsf{a}^1\mathsf{b}^2\mathsf{a}^3\mathsf{b}^2\mathsf{a}^3\mathsf{b}^3$. Each white node represents a run-centered palindromic substring of S that has an RLE-bounded occurrence, while each gray node represents a run-centered palindromic substring of S that has a maximal occurrence in S.

Lemma 5. *Let S be any string of which the size of RLE_S is m. Then, the number of nodes in $\mathsf{e}^2\mathsf{rtre}^2(S)$ is at most $2m + 1$.*

Proof. First, we consider RLE_S as a string of length m over the alphabet Σ_{RLE}. It now follows from Lemma 1 that the number of non-empty distinct run-centered palindromic substrings of S that have an RLE-bounded occurrence is at most m. Each of these palindromic substrings are represented by a node of $\text{e}^2\text{rtre}^2(S)$, and let $\text{e}^2\text{rtre}^2(S)'$ denote the tree consisting only of these nodes (in the example of Fig. 2, $\text{e}^2\text{rtre}^2(S)'$ is the tree consisting only of the white nodes).

Now we count the number of nodes in $\text{e}^2\text{rtre}^2(S)$ that do not belong to $\text{e}^2\text{rtre}^2(S)'$ (the gray nodes in the running example of Fig. 2). Since each palindrome represented by this type of node has a run-centered maximal occurrence in S, the number of such palindromes is bounded by the number m of runs in RLE_S.

Hence, including the root that represent the empty string, there are at most $2m + 1$ nodes in $\text{e}^2\text{rtre}^2(S)$. $\qquad\Box$

Lemma 6. *Given RLE_S of size m, $\text{e}^2\text{rtre}^2(S)$ can be built in $O(m \log \sigma_{RLE})$ time and $O(m)$ space, where the out-going edges of each node are sorted according to the total order of their labels. Also, in the resulting $\text{e}^2\text{rtre}^2(S)$, each non-root node u stores the number of occurrences of u in S which are RLE-bounded or maximal.*

Proof. Our construction algorithm comprises three steps. We firstly build $\text{e}^2\text{rtre}^2(S)'$, secondly compute an auxiliary array $CPal$ that will be used for the next step, and thirdly we add some nodes that represent run-centered maximal palindromes which are not in $\text{e}^2\text{rtre}^2(S)'$ so that the resulting tree forms the final structure $\text{e}^2\text{rtre}^2(S)$.

Rubinchik and Shur [12] proposed an online algorithm to construct $\text{eertree}(T)$ of a string of length k in $O(k \log \sigma_T)$ time with $O(k)$ space, where σ_T denotes the number of distinct characters in T. They also showed how to store, in each node, the number of occurrences of the corresponding palindromic substring in T. Thus, the Rubinchik and Shur algorithm applied to RLE_S computes $\text{e}^2\text{rtre}^2(S)'$ in $O(m \log \sigma_{RLE})$ time with $O(m)$ space. Also, now each node u of $\text{e}^2\text{rtre}^2(S)'$ stores the number of *RLE-bounded* occurrence of u in S. This is the first step.

The second step is as follows: Let $CPal$ be an array of length m such that, for each $1 \leq i \leq m$, $CPal[i]$ stores a pointer to the node in $\text{e}^2\text{rtre}^2(S)'$ that represents the RLE-bounded palindrome centered at i. A simple application of the Rubinchik and Shur algorithm to RLE_S only gives us the leftmost occurrence of each RLE-bounded palindrome in $\text{e}^2\text{rtre}^2(S)'$. Hence, we only know the values of $CPal$ in the positions that are the centers of the leftmost occurrences of RLE-bounded palindromes. To compute the other values in $CPal$, we run Manacher's algorithm to RLE_S as in Sect. 3.2. Since Manacher's algorithm processes RLE_S in an online manner from left to right, and since we already know the leftmost occurrences of all RLE-bounded palindromes in RLE_S, we can copy the pointers from previous occurrences. In case where an RLE-bounded palindrome extends with a newly read RLE-character (a, e) after it is copied from a previous occurrence during the execution of Manacher's algorithm, then we traverse the edge labeled (a, e) from the current node of $\text{e}^2\text{rtre}^2(S)'$. By repeating this until the mismatch is found in extension of the current RLE-bounded palindrome, we can find the corresponding node for this RLE-bounded palindrome. This way we can compute $CPal[i]$ for all $1 \leq i \leq m$ in $O(m \log \sigma_{RLE})$ total time with $O(m)$ total space.

In the third step, we add new nodes that represent run-centered maximal (but not RLE-bounded) palindromic substrings. For this sake, we again apply Manacher's algorithm to RLE_S, but in this case it is done as in Remark 1 of Sect. 3.2. With the help of $CPal$ array, we can associate each run (a_l, e_l) with the RLE-bounded palindromic substring that has the center in (a_l, e_l). Let $Q_l = S[Rb[i]..Re[j]]$ denote this palindromic substring for (a_l, e_l), where $1 \leq i \leq l \leq j \leq m$, and u_l the node that represents Q_l in $\mathsf{e}^2\mathsf{rtre}^2(S)'$. We first check whether $a_{i-1} = a_{j+1}$. If no, then Q_l does not extend from this run (a_l, e_l), and if yes then we extend both ends of Q_l with $(a_{i-1}, \min\{e_{i-1}, e_{j+1}\})$. Assume w.l.o.g. that $e_{i-1} = \min\{e_{i-1}, e_{j+1}\}$. If there is no out-going edge of u_l with label (a_{i-1}, e_{i-1}), then we create a new child of u_l with an edge labeled (a_{i-1}, e_{i-1}). Otherwise, then let v be the existing child of u_l that represents $a_{i-1}^{e_{i-1}} u_l a_{i-1}^{e_{i-1}}$. We increase the number of occurrences of v by 1. This way, we can add all new nodes and we obtain $\mathsf{e}^2\mathsf{rtre}^2(S)$. Note that each node stores the number of RLE-bounded or maximal occurrence of the corresponding run-centered palindromic substring. It is easy to see that the second step takes a total of $O(m \log \sigma_{RLE})$ time and $O(m)$ space. □

It is clear that for any character $a \in \Sigma$, there can be only one MUPS of form a^e. Namely, a^e is a MUPS iff e is the largest exponent for all runs of a's in S and $occ_S(a^e) = 1$. Below, we consider other forms of MUPSs. Let P be a non-empty palindromic substring of string S that has a run-centered RLE-bounded occurrence. For any character $a \in \Sigma$, let emax and esec denote the largest and second largest positive integers such that $a^{\mathsf{emax}} P a^{\mathsf{emax}}$ and $a^{\mathsf{esec}} P a^{\mathsf{esec}}$ are palindromes that have run-centered RLE-bounded or maximal occurrences in S. If such integers do not exist, then let emax = nil and esec = nil.

Observation 1. *There is at most one MUPS of form $a^e P a^e$ in S. Namely,*

(1) *The palindrome $a^{\mathsf{esec}+1} P a^{\mathsf{esec}+1}$ is a MUPS of S iff emax $\neq$ nil, esec $\neq$ nil, and $occ_S(a^{\mathsf{emax}} P a^{\mathsf{emax}}) = 1$.*
(2) *The palindrome $a^1 P a^1$ is a MUPS of S iff emax $\neq$ nil, esec = nil, and $occ_S(a^{\mathsf{emax}} P a^{\mathsf{emax}}) = 1$.*
(3) *There is no MUPS of form $a^e P a^e$ with any $e \geq 1$ iff either emax = nil, or emax $\neq$ nil and $occ_S(a^{\mathsf{emax}} P a^{\mathsf{emax}}) > 1$.*

Lemma 7. $\mathcal{M}_S$ *can be computed in $O(m \log \sigma_{RLE})$ time and $O(m)$ space.*

Proof. For each node u of $\mathsf{e}^2\mathsf{rtre}^2(S)$, let Σ_u be the set of characters a such that there is an out-going edge of u labeled by (a, e) with some positive integer e. Due to Observation 1, for each character in Σ_u, it is enough to check the out-going edges which have the largest and second largest exponents with character a. Since the edges are sorted, we can find all children of u that represent MUPSs in time linear in the number of children of u. Hence, given $e^2rtre^2(S)$, it takes $O(m)$ total time to compute all MUPSs in S. $e^2rtre^2(S)$ can be computed in $O(m\log\sigma_{RLE})$ time and $O(m)$ space by Lemma 6.

What remains is how to sort the MUPSs in increasing order of their beginning positions. We associate each MUPS with the run where its center lies. Since each MUPS occurs in S exactly once and MUPSs do not nest (Lemma 2), each run cannot contain the centers of two or more MUPSs. We compute an array A of size m such that $A[j]$ contains the corresponding interval of the MUPS whose center lies in the jth run in RLE_S, if it exists. After computing A, we scan A from left to right. Since again MUPSs do not nest, this gives as the sorted list $\mathcal{M}_S$ of MUPSs. It is clear that this takes a total of $O(m)$ time and space.   □

4   SUPS Queries on RLE Strings

In this section, we present our algorithm for SUPS queries. Our algorithm is based on Inoue et al.'s algorithm [8] for SUPS queries on a plain string. The big difference is that the space that we are allowed for is limited to $O(m)$.

4.1   Data Structures

As was discussed in Sect. 3, we can compute the list $\mathcal{M}_S$ of all MUPSs of string S efficiently. We store $\mathcal{M}_S$ using the three following arrays:

- $M_{beg}[i]$: the beginning position of the ith MUPS in $\mathcal{M}_S$.
- $M_{end}[i]$: the ending position the ith MUPS in $\mathcal{M}_S$.
- $M_{len}[i]$: the length of the ith MUPS in $\mathcal{M}_S$.

Since the number of MUPSs in $\mathcal{M}_S$ is at most m (Corollary 1), the length of each array is at most m. In our algorithm, we use *range minimum queries* and *predecessor/successor queries* on integer arrays.

Let A be an integer array of length d. A range minimum query $\mathsf{RmQ}_A(i,j)$ returns one of $\arg\min_{i\le k\le j}\{A[k]\}$ for a given interval $[i,j]$ in A.

Lemma 8 (e.g. [2]). *We can construct an $O(d)$-space data structure in $O(d)$ time for an integer array A of length d which can answer $\mathsf{RmQ}_A(i,j)$ in constant time for any query $[i,j]$.*

Let B be an array of d positive integers in $[1,N]$ in increasing order. The predecessor and successor queries on B are defined for any $1\le k\le N$ as follows.

$$\mathsf{Pred}_B(k) = \begin{cases} \max\{i \mid B[i]\le k\} & \text{if it exists,} \\ 0 & \text{otherwise.} \end{cases}$$

$$\mathsf{Succ}_B(k) = \begin{cases} \min\{i \mid B[i]\ge k\} & \text{if it exists,} \\ N+1 & \text{otherwise.} \end{cases}$$

Lemma 9 ([1]). *We can construct, in $O(d\sqrt{\log d/\log\log d})$ time, an $O(d)$-space data structure for an array B of d positive integer in $[1,N]$ in increasing order which can answer $\mathsf{Pred}_B(k)$ and $\mathsf{Succ}_B(k)$ in $O(\sqrt{\log d/\log\log d})$ time for any query $k\in[1,N]$.*

4.2   Query Algorithm

Our algorithm simulates the query algorithm for a plain string [8] with $O(m)$-space data structures. We summarize our algorithm below.

Let $[s, t]$ be a query interval such that $1 \leq s \leq t \leq n$. Firstly, we compute the number of MUPSs contained in $[s, t]$. This operation can be done in $O(\sqrt{\log m / \log \log m})$ by using $\mathsf{Succ}_{M_{beg}}(s)$ and $\mathsf{Pred}_{M_{end}}(t)$.

Let num be the number of MUPSs contained in $[s, t]$. If $num \geq 2$, then there is no SUPS for this interval (Corollary 1 of [8]). Suppose that $num = 1$. Let $S[i..j]$ be the MUPS contained in $[s, t]$. If $S[i - z..j + z]$ is a palindromic substring, then $S[i - z..j + z]$ is the only $SUPS$ for $[s, t]$ where $z = \max\{i - s, t - j\}$. Otherwise, there is no $SUPS$ for $[s, t]$ (Lemma 6 of [8]). Since this candidate has a run as the center, we can check whether $S[i - z..j + z]$ is a palindromic substring or not in constant time after computing all run-centered maximal palindromes. Suppose that $num = 0$ (this case is based on Lemma 7 of [8]). Let $p = \mathsf{Pred}_{M_{end}}(t), q = \mathsf{Succ}_{M_{beg}}(s)$. We can check whether each of $S[M_{beg}[p] - t + M_{end}[p]..t]$ and $S[s..M_{end}[q] + M_{beg}[q] - s]$ is a palindrome or not. If so, the shorter one is a candidate of $SUPS$s. Let ℓ be the length of the candidates. Other candidates are the shortest $MUPS$s which contain the query interval $[s, t]$. If the length of these candidates is less than or equal to ℓ, we need to compute these candidates as $SUPS$s. We can compute these $MUPS$s by using range minimum queries on $M_{len}[p + 1, q - 1]$. Thus, we can compute all $SUPS$s in linear time w.r.t. the number of outputs (see [8] in more detail).

We conclude with the main theorem of this paper.

Theorem 1. *Given RLE_S of size m for a string S, we can compute a data structure of $O(m)$ space in $O(m(\log \sigma_{RLE} + \sqrt{\log m / \log \log m}))$ time so that subsequent $SUPS$ queries can be answered in $O(\alpha + \sqrt{\log m / \log \log m})$ time, where σ_{RLE} denotes the number of distinct RLE-characters in RLE_S and α the number of $SUPS$s to report.*

Acknowledgments. This work was supported by JSPS KAKENHI Grant Numbers JP18K18002 (YN), JP17H01697 (SI), JP16H02783 (HB), and JP18H04098 (MT).

References

1. Beame, P., Fich, F.E.: Optimal bounds for the predecessor problem and related problems. J. Comput. Syst. Sci. **65**(1), 38–72 (2002)
2. Bender, M.A., Farach-Colton, M.: The LCA problem revisited. In: Gonnet, G.H., Viola, A. (eds.) LATIN 2000. LNCS, vol. 1776, pp. 88–94. Springer, Heidelberg (2000). https://doi.org/10.1007/10719839_9
3. Droubay, X., Justin, J., Pirillo, G.: Episturmian words and some constructions of de Luca and Rauzy. Theor. Comput. Sci. **255**(1–2), 539–553 (2001)
4. Ganguly, A., Hon, W., Shah, R., Thankachan, S.V.: Space-time trade-offs for finding shortest unique substrings and maximal unique matches. Theor. Comput. Sci. **700**, 75–88 (2017)

5. Hon, W.-K., Thankachan, S.V., Xu, B.: An In-place framework for exact and approximate shortest unique substring queries. In: Elbassioni, K., Makino, K. (eds.) ISAAC 2015. LNCS, vol. 9472, pp. 755–767. Springer, Heidelberg (2015). https://doi.org/10.1007/978-3-662-48971-0_63

6. Hu, X., Pei, J., Tao, Y.: Shortest unique queries on strings. In: Moura, E., Crochemore, M. (eds.) SPIRE 2014. LNCS, vol. 8799, pp. 161–172. Springer, Cham (2014). https://doi.org/10.1007/978-3-319-11918-2_16

7. İleri, A.M., Külekci, M.O., Xu, B.: Shortest unique substring query revisited. In: Kulikov, A.S., Kuznetsov, S.O., Pevzner, P. (eds.) CPM 2014. LNCS, vol. 8486, pp. 172–181. Springer, Cham (2014). https://doi.org/10.1007/978-3-319-07566-2_18

8. Inoue, H., Nakashima, Y., Mieno, T., Inenaga, S., Bannai, H., Takeda, M.: Algorithms and combinatorial properties on shortest unique palindromic substrings. J. Discrete Algorithms 52–53, 122–132 (2018)

9. Manacher, G.: A new linear-time "on-line" algorithm for finding the smallest initial palindrome of a string. J. ACM 22, 346–351 (1975)

10. Mieno, T., Inenaga, S., Bannai, H., Takeda, M.: Shortest unique substring queries on run-length encoded strings. In: Proceedings of MFCS 2016, pp. 69:1–69:11 (2016)

11. Pei, J., Wu, W.C.H., Yeh, M.Y.: On shortest unique substring queries. Proc. ICDE 2013, 937–948 (2013)

12. Rubinchik, M., Shur, A.M.: EERTRE: an efficient data structure for processing palindromes in strings. Eur. J. Comb. 68, 249–265 (2018)

13. Tsuruta, K., Inenaga, S., Bannai, H., Takeda, M.: Shortest unique substrings queries in optimal time. In: Geffert, V., Preneel, B., Rovan, B., Štuller, J., Tjoa, A.M. (eds.) SOFSEM 2014. LNCS, vol. 8327, pp. 503–513. Springer, Cham (2014). https://doi.org/10.1007/978-3-319-04298-5_44

A Partition Approach to Lower Bounds for Zero-Visibility Cops and Robber

Yuan Xue[1], Boting Yang[1(⊠)], Farong Zhong[2], and Sandra Zilles[1]

[1] Department of Computer Science, University of Regina, Regina, Canada
{xue228,boting,zilles}@cs.uregina.ca
[2] College of Math, Physics and Information Technology, Zhejiang Normal University,
Jinhua, China
zfr@zjnu.edu.cn

Abstract. The zero-visibility cops and robber game is a variant of Cops and Robbers subject to the constraint that the cops have no information at any time about the location of the robber. We first study a partition problem in which for a given graph and an integer k, we want to find a partition of the vertex set such that the size of the boundary of the smaller subset in the partition is at most k while the size of this subset is as large as possible under some conditions. Then we apply such partitions to prove lower bounds on the zero-visibility cop numbers of graph products. We also investigate the monotonic zero-visibility cop numbers of graph products.

1 Introduction

Cops and Robbers is a pursuit and evasion game in graph theory, which was introduced independently by Nowakowski and Winkler [10] and Quilliot [11]. These two papers consider the game with one cop and one robber and characterize the cop-win graphs. More results can be found in [2,3].

The zero-visibility cops and robber game is a variant of Cops and Robbers, which was proposed by Tošić [12]. This game has the same setting as Cops and Robbers except that the cops have no information about the location of the robber. This feature makes the robber harder to capture, and the main question is to determine the minimum number of cops that can guarantee to capture the robber.

When the zero-visibility cops and robber game is played on a graph G with k cops and one robber, the robber has full information about the locations of all cops, but the cops have no information about the location of the robber at any time, i.e., the robber is invisible to the cops. The game is played in a sequence of rounds. Each *round* consists of a pair of turns, a cops' turn to move, followed by a robber's turn to move. At round 0, each of the k cops selects a vertex of G to occupy, and then the robber selects a vertex of G. At round i, $i \geq 1$, each cop either moves from the vertex currently occupied to one of its neighbors or stays still, then the robber does the same. The cops *capture* the robber if one of

© Springer Nature Switzerland AG 2019
C. J. Colbourn et al. (Eds.): IWOCA 2019, LNCS 11638, pp. 442–454, 2019.
https://doi.org/10.1007/978-3-030-25005-8_36

them occupies the same vertex as the robber. If this happens in a finite number of moves, then the *cops win*; otherwise, the *robber wins*. The *zero-visibility cop number* of G, denoted by $c_0(G)$, is the minimum number of cops required to capture the robber on G. A cop-win strategy for G is *optimal* if it uses $c_0(G)$ cops to capture the robber. We call a vertex *cleared* if it is certain that this vertex is not occupied by the robber, and *contaminated* otherwise.

Monotonicity (i.e., the property that each vertex or edge, once cleared, remains cleared forever) is an important issue in graph searching problems. Megiddo et al. [9] showed that the edge search problem is NP-hard. That this problem belongs to the NP class follows from the monotonicity result in [8]. Bienstock and Seymour [1] proposed a method that gives a succinct proof for the monotonicity of the mixed search problem. This method was extended to the monotonicity of digraph search problems [13,14]. The monotonic cop-win strategy in the zero-visibility cops and robber game is introduced in [5]. Let R_i be the set of vertices that are contaminated just after the cops' turn in the i-th round. We say that a cop-win strategy is *monotonic* if $R_{i+1} \subseteq R_i$ for any round $i \geq 0$. The *monotonic zero-visibility cop number* of G, denoted by $mc_0(G)$, is the minimum number of cops required by a monotonic cop-win strategy of G.

Although Cops and Robbers has been widely studied, there are not many results in the study of the zero-visibility cops and robber game. Tošić [12] gave characterizations of graphs for which one cop is sufficient, and computed the cop number of paths, cycles, complete graphs and complete bipartite graphs. Dereniowski et al. [5] proved that the zero-visibility cop number of a graph is bounded above by its pathwidth and the monotonic zero-visibility cop number can be bounded both above and below by multiples of the pathwidth. Dereniowski et al. [6] gave a linear-time algorithm for computing the zero-visibility cop number of trees. They also proved that the problem of determining the zero-visibility cop number of a graph is NP-complete.

Given two graphs G and H, the *cartesian product* of G and H, denoted $G \square H$, is the graph whose vertex set is the cartesian product $V(G) \times V(H)$, and in which two vertices (u, v) and (u', v') are adjacent if and only if $u = u'$ and v is adjacent to v' in H, or $v = v'$ and u is adjacent to u' in G. The *strong product* of G and H, denoted $G \boxtimes H$, has the same vertex set as $G \square H$. Two vertices (u, v) and (u', v') are adjacent in $G \boxtimes H$ if and only if $u = u'$ and v is adjacent to v' in H, or $v = v'$ and u is adjacent to u' in G, or u is adjacent to u' in G and v is adjacent to v' in H.

Note that the difference between the cop number in Cops and Robbers and the zero-visibility cop number can be arbitrarily large for product graphs. For example, in Cops and Robbers, we can use two cops to capture the robber on $P_m \square P_n$ or $P_m \boxtimes P_n$, where P_m is a path with m vertices and $n \geq m \geq 2$. But in the zero-visibility cops and robber game, we will show that $c_0(P_m \boxtimes P_n) \geq c_0(P_m \square P_n) = \lceil \frac{m+1}{2} \rceil$.

Dereniowski et al. [5] showed that the zero-visibility cops and robber game is highly non-monotonic. For any $k > 1$, they constructed a class of graphs G whose pathwidth is at least k but $c_0(G) = 2$. Because of these notorious graphs, it is difficult to apply the lower bound techniques for pathwidth or its related graph searching models to find nontrivial lower bounds for the zero-visibility cop number.

The main contribution of this paper is a new partition method that can be used to prove lower bounds for the zero-visibility cops and robber game. It is easy to see that at any moment of the zero-visibility cops and robber game, the set of cleared vertices and the set of contaminated vertices form a partition of the vertex set. This partition may change dynamically after each turn of either player. Since any subset of cops can move to their neighbors in a turn of the cops, we have many possible partitions to analyze after such a turn. To overcome this difficulty, we first establish general properties of partitions of vertex sets, independent of the game, in Sect. 2. We then apply those partition properties to show lower bounds on the zero-visibility cop number of cartesian products and strong products of graphs. The corresponding results are summarized in the following table.

Theorem #	G	m, n	Lower bound on $c_0(G)$	Upper bound on $mc_0(G)$
Theorem 6	$P_m \square P_n$	$n \geq m \geq 1$	$\lceil \frac{m+1}{2} \rceil$	$\lceil \frac{m+1}{2} \rceil$
Theorem 7	$C_m \square P_n$	$m \geq 3, n \geq 2$	$\min\{\lceil \frac{m+1}{2} \rceil, n+1\}$	$\min\{\lceil \frac{m+1}{2} \rceil, n+1\}$
Theorem 8	$C_m \square C_n$	$n \geq m \geq 3$	$\lceil \frac{m+1}{2} \rceil$	$m+1$
Theorem 9	$P_m \boxtimes P_n$	$n \geq m \geq 2$, m is even	$\frac{m}{2}+1$	$\frac{m}{2}+1$
Theorem 9	$P_m \boxtimes P_n$	$n \geq m \geq 3$, m is odd	$\frac{m+1}{2}$	$\frac{m+1}{2}+1$
Theorem 10	$C_m \boxtimes P_n$	$m \geq 3, n \geq 2$,	$\min\{\lceil \frac{m+1}{2} \rceil, n+1\}$	$\min\{\lceil \frac{m+1}{2} \rceil, 2\lfloor \frac{n+1}{2} \rfloor\}+1$
Theorem 11	$C_m \boxtimes C_n$	$n \geq m \geq 3$	$\lceil \frac{m+1}{2} \rceil$	$m+2$
Theorem 12	Q_n	$n \geq 2$	?	$\sum_{k=0}^{n-2} \binom{k}{\lfloor \frac{k}{2} \rfloor} + 1$

2   Partitions and Boundaries of Vertex Sets

Let $G = (V, E)$ be a graph with vertex set V and edge set E. A pair of subsets (V_1, V_2) is a *partition* of V if $V_1 \cup V_2 = V$ and $V_1 \cap V_2 = \emptyset$. For a partition (V_1, V_2) of V, the *boundary* of V_1, denoted ∂V_1, is the largest subset of V_1 such that each vertex in the subset is adjacent to a vertex in V_2 on G.

Theorem 1. *Let* $P_m \square P_n$, $n \geq m \geq 2$, *be a grid with m rows and n columns. Let (V_1, V_2) be a partition of the vertex set of $P_m \square P_n$. If $|\partial V_1| < m$ and there exists a row whose vertices all belong to V_2, then*

(i) *for any $\ell \in \{1, \ldots, m-1\}$, the number of rows that contain at least $m - \ell$ vertices in V_1 is at most ℓ;*

(ii) $|V_1| \leq \frac{m^2 - m}{2}$.

Proof. (i) For the sake of contradiction, assume that there exists ℓ', $1 \leq \ell' \leq m - 1$, such that the number of rows that contain at least $m - \ell'$ vertices in V_1 is at least $\ell' + 1$. Let i_1 be the index of a row in which all vertices are in V_2. Let i_2

be the index of a row that contains at least $m - \ell'$ vertices in V_1. Without loss of generality, suppose that there is no row between the i_1-th row and the i_2-th row which contains at least $m - \ell'$ vertices in V_1. Let $v_{i_2,j}$ be the vertex of V_1 on the i_2-th row and j-th column, and let $P' = v_{i_2,j} \ldots v_{i_1,j}$ be a subcolumn of the j-th column. Note that all the vertices in the i_1-th row are in V_2. Hence, there must exist a vertex v' of P' such that $v' \in \partial V_1$. Thus, there are at least $m - \ell'$ vertices in all the subcolumns from the i_1-th row to the i_2-th row, which belong to ∂V_1. Similar to the above, we can also show that there is no row containing at least m vertices in V_1; otherwise, $|\partial V_1|$ is at least m and this contradicts the condition that $|\partial V_1| < m$. So each row contains at most $m - 1$ vertices in V_1; and furthermore, if a row contains vertices in V_1, it also contains a vertex in ∂V_1. Note that on the i_2-th row, the rightmost vertex which belongs to ∂V_1 is counted twice, and this is the only doubly counted vertex in the above counting. Hence, $|\partial V_1| \geq (m - \ell') + (\ell' + 1) - 1 = m$, which contradicts the condition that $|\partial V_1| < m$. Therefore, for any $\ell \in \{1, \ldots, m - 1\}$, there are at most ℓ rows that contain $m - \ell$ or more vertices in V_1 each.

(ii) From (i), each row contains at most $m - 1$ vertices in V_1, and moreover, the number of rows that contain at least $m - \ell$ vertices in V_1 is at most ℓ, for each $\ell \in \{1, \ldots, m - 1\}$. Therefore, $|V_1| \leq \sum_{i=1}^{m-1} i = \frac{m^2 - m}{2}$.   □

Theorem 2. *Let $C_m \square P_n$ be a cylinder grid, where $m \geq 2n \geq 4$. Let (V_1, V_2) be a partition of the vertex set of $C_m \square P_n$. If $|\partial V_1| < 2n$ and there exists a copy of P_n whose vertices are all in V_2, then*

(i) *for any $\ell \in \{1, \ldots, n\}$, the number of copies of C_m that contain at least $2n - 2\ell + 1$ vertices in V_1 is at most ℓ;*

(ii) *for any $\ell \in \{2, \ldots, n\}$, the number of copies of C_m that contain at least $2n - 2\ell + 2$ vertices in V_1 is at most $\ell - 1$;*

(iii) *each copy of C_m contains at most $2n - 1$ vertices in V_1;*

(iv) $|V_1| \leq n^2$.

Theorem 3. *Let $C_m \square P_n$ be a cylinder grid, where $3 \leq m < 2n$. Let (V_1, V_2) be a partition of the vertex set of $C_m \square P_n$. If $|\partial V_1| < m$ and there exists a copy of P_n whose vertices all belong to V_2, then*

(i) *when m is even, for any $\ell \in \{0, \ldots, \frac{m-2}{2}\}$, the number of copies of P_n that contain at least $\frac{m}{2} - \ell$ vertices in V_1 is at most $2\ell + 1$;*

(ii) *when m is odd, for any $\ell \in \{1, \ldots, \frac{m-1}{2}\}$, the number of copies of P_n that contain at least $\frac{m+1}{2} - \ell$ vertices in V_1 is at most 2ℓ;*

(iii) $|V_1| \leq \frac{m^2}{4}$ *when m is even; and* $|V_1| \leq \frac{m^2-1}{4}$ *when m is odd.*

3   Lower Bounds

Consider the zero-visibility cops and robber game on a graph $G = (V, E)$. After any turn, the set of cleared vertices V_1 and the set of contaminated vertices V_2 form a partition of V. So the *boundary* of V_1 is the largest subset of V_1 such that each cleared vertex in the subset is adjacent to a contaminated vertex in V_2. Let V_B^i, $i \geq 1$, denote the boundary of V_1 after the cops' turn in round i, and let V_C^i be the set of cleared vertices after the robber's turn in round i. Note that V_B^i might not be a subset of V_C^i. After the cops' turn in round i, a vertex is called *non-boundary* if it is not in V_B^i.

Lemma 1. *Let G be a graph with $c_0(G) = k$. For any round i (≥ 1) of an optimal cop-win strategy of G, if $|V_B^i| \geq 2k$, then $|V_C^i| \leq |V_C^{i-1}|$.*

Proof. Since there are k cops, at most k contaminated vertices can be cleared by cops in each round. Note that at most k vertices in V_B^i are occupied by cops. Hence, after the cops' turn in round i, there are at least $|V_B^i| - k \geq k$ cleared vertices that are unoccupied and are adjacent to contaminated vertices, and so, these vertices get recontaminated after the robber's turn in round i. Therefore $|V_C^i| \leq |V_C^{i-1}|$. □

Lemma 2. *Let $P_m \square P_n$, $n \geq m \geq 1$, be a grid with m rows and n columns. For a cop-win strategy of $P_m \square P_n$, let i be a round in which $|V_B^i| < m$. Then either $|V_C^i| \leq \frac{m^2 - m}{2}$ or $|V_C^i| \geq mn - \frac{m^2 - m}{2}$.*

Proof. After the cops' turn in round i, if every row contains both cleared vertices and contaminated vertices, then $|V_B^i| \geq m$, which is a contradiction. So there is at least one row that contains only cleared vertices or only contaminated vertices. Similarly, there is at least one column that contains only cleared vertices or only contaminated vertices. Thus, we have two cases.

Case 1. There exist a row and a column that both contain only contaminated vertices. It follows from Theorem 1(ii) that $|V_C^i| \leq \frac{m^2 - m}{2}$.

Case 2. There exist a row and a column that both contain only cleared vertices. After the cops' turn in round i, let (U_1, U_2) be a partition of $V(P_m \square P_n)$, where U_1 is the set of cleared vertices, U_2 is the set of contaminated vertices, and V_B^i is the boundary of U_1. It is easy to see that every row and every column must contain a vertex in U_1. Note that $|V_B^i| < m$. Hence, there must exist a row that contains only non-boundary vertices. Since every row contains a vertex in U_1, there must exist a row that contains only cleared non-boundary vertices. Similarly, we can show that there must exist a column that contains only cleared non-boundary vertices. From Theorem 1(ii), we have $|U_2 \cup V_B^i| \leq \frac{m^2 - m}{2}$ (where $U_2 \cup V_B^i$ is considered as V_1 and $U_1 \setminus V_B^i$ is considered as V_2 in Theorem 1). Thus, $|U_1 \setminus V_B^i| \geq mn - \frac{m^2 - m}{2}$. Notice that the vertices in $U_1 \setminus V_B^i$ are all cleared non-boundary vertices. Hence, $|V_C^i| \geq |U_1 \setminus V_B^i| \geq mn - \frac{m^2 - m}{2}$. □

Theorem 4. *For $n \geq m \geq 1$, $c_0(P_m \square P_n) \geq \lceil \frac{m+1}{2} \rceil$.*

Proof. The claim is trivial when $m = 1$. Suppose that $c_0(P_m \square P_n) \leq \lceil \frac{m-1}{2} \rceil$, $m \geq 2$. Consider a cop-win strategy for $P_m \square P_n$ that uses at most $\lceil \frac{m-1}{2} \rceil$ cops. We will use mathematical induction to show that $|V_C^i| \leq \frac{m^2 - m}{2}$ for all $i \geq 0$.

When $i = 0$, it is easy to see that $|V_C^0| \leq \lceil \frac{m-1}{2} \rceil \leq \frac{m^2-m}{2}$. Assume that $|V_C^{i-1}| \leq \frac{m^2-m}{2}$ holds for round $i - 1$, where $i \geq 1$. There are two cases.

Case 1: $|V_B^i| < m$. Since $|V_C^{i-1}| \leq \frac{m^2-m}{2}$, there are at most $\frac{m^2-m}{2} + \lceil \frac{m-1}{2} \rceil \leq \frac{m^2}{2}$ cleared vertices after the cops' turn in round i. Since $\frac{m^2}{2} < mn - \frac{m^2-m}{2}$, it follows from Lemma 2 that $|V_C^i| \leq \frac{m^2-m}{2}$.

Case 2: $|V_B^i| \geq m$. It follows from Lemma 1 that $|V_C^i| \leq |V_C^{i-1}|$, and thus, $|V_C^i| \leq \frac{m^2-m}{2}$.

From the above, we have $|V_C^i| \leq \frac{m^2-m}{2}$ for all $i \geq 0$, which is a contradiction. Hence, $c_0(P_m \square P_n) \geq \lceil \frac{m+1}{2} \rceil$. □

Lemma 3. *For a cop-win strategy of $C_m \square P_n$ with $m \geq 2n \geq 4$, let i be a round in which $|V_B^i| < 2n$. Then either $|V_C^i| \leq n^2$ or $|V_C^i| \geq mn - n^2 + 2n - 1$.*

Proof. After the cops' turn in round i, if every copy of P_n contains both cleared and contaminated vertices, then $|V_B^i| \geq m \geq 2n$, which is a contradiction. So there must exist a path P_n^j, $1 \leq j \leq m$, which contains only cleared or only contaminated vertices. Hence, we have two cases.

Case 1. P_n^j contains only contaminated vertices. After the cops' turn in round i, let V_1 be the set of cleared vertices and V_2 be the set of contaminated vertices. It follows from Theorem 2(iv) that $|V_1| \leq n^2$. Since $|V_C^i| \leq |V_1|$, we have $|V_C^i| \leq n^2$.

Case 2. P_n^j contains only cleared vertices. Similar to the proof of Lemma 2, we know that the maximum number of contaminated vertices in Case 2 is equal to the maximum number of cleared non-boundary vertices in Case 1. So we need to find this number in Case 1.

Consider Case 1 again, that is, P_n^j contains only contaminated vertices. Let n_c be the maximum number of cleared non-boundary vertices over all possible cop-win strategies of $C_m \square P_n$ satisfying that $|V_B^i| \leq 2n - 1$. Let $k \leq 2n - 1$ be the largest size of the boundary such that the number of cleared non-boundary vertices is n_c. We will prove that $k = 2n - 1$. Assume that $k \leq 2n - 2$. If every copy of C_m contains at least two cleared vertices, then $|V_B^i| \geq 2n$, which is a contradiction. Thus, there exists a copy of C_m that contains at most one cleared vertex. Let u be a contaminated vertex on this cycle but not on P_n^j. We can obtain a new partition on $V(C_m \square P_n)$ by letting u become cleared. Notice that P_n^j still contains only contaminated vertices after changing u from contaminated to cleared. Further, the new partition must satisfy: (a) the size of the boundary is $k + 1$ and the number of cleared non-boundary vertices is n_c, or (b) the size of the boundary is at most k and the number of cleared non-boundary vertices is at least $n_c + 1$. Note that case (a) contradicts that k is the largest size of the boundary when the number of cleared non-boundary vertices is n_c; and case (b) contradicts that n_c is the maximum number of cleared non-boundary vertices when the boundary has size at most $2n - 1$. Thus, we know that k cannot be less than $2n - 1$. Hence, $k = 2n - 1$. Since there are at most n^2 cleared vertices when the size of the boundary is $2n - 1$, we have $n_c + (2n - 1) \leq n^2$, and thus $n_c \leq n^2 - 2n + 1$.

So in Case 2, the number of contaminated vertices is at most $n^2 - 2n + 1$, and therefore, the number of cleared vertices is at least $mn - n^2 + 2n - 1$. □

Lemma 4. *For $m \geq 2n \geq 4$, $c_0(C_m \square P_n) \geq n + 1$.*

Proof. Assume that $c_0(C_m \square P_n) \leq n$. Consider a cop-win strategy for $C_m \square P_n$ that uses n cops. We will show that $|V_C^i| \leq n^2$ for all $i \geq 0$. When $i = 0$, $|V_C^0| \leq n < n^2$. Assume that $|V_C^{i-1}| \leq n^2$, $i \geq 1$. There are two cases when we consider $|V_C^i|$.

Case 1: $|V_B^i| < 2n$. Since $|V_C^{i-1}| \leq n^2$, there are at most $n^2 + n$ cleared vertices after the cops' turn in round i. Since $n^2 + n < mn - n^2 + 2n - 1$, it follows from Lemma 3 that $|V_C^i| \leq n^2$.

Case 2: $|V_B^i| \geq 2n$. Similar to the proof of Lemma 1, we have $|V_C^i| \leq |V_C^{i-1}| \leq n^2$.

Thus, $|V_C^i| \leq n^2$ for all $i \geq 0$, which contradicts the assumption that n cops can clear $C_m \square P_n$. Hence, $c_0(C_m \square P_n) \geq n + 1$. □

Lemma 5. *For a cop-win strategy of $C_m \square P_n$ with $3 \leq m < 2n$, let i be a round in which $|V_B^i| < m$. Then either $|V_C^i| \leq \frac{m^2}{4}$ or $|V_C^i| \geq mn - \frac{m^2}{4} + m - 1$.*

Proof. After the cops' turn in round i, there is a path P_n^j, $1 \leq j \leq m$, which contains only cleared or only contaminated vertices; otherwise, $|V_B^i| \geq m \geq 2n$, which is a contradiction. So there are two cases for P_n^j.

Case 1. P_n^j contains only contaminated vertices. Let n_c be the maximum number of cleared non-boundary vertices over all partitions on $V(C_m \square P_n)$ when $|V_B^i| \leq m - 1$. Let $k \leq m - 1$ be the largest size of the boundary such that the number of all cleared non-boundary vertices is n_c. We will prove that $k = m - 1$. Assume that $k \leq m - 2$. Consider a partition (V_1, V_2) of $V(C_m \square P_n)$, where V_1 is the set of cleared vertices and V_2 is the set of contaminated vertices, such that $|\partial V_1| = k$ and $|V_1 \setminus \partial V_1| = n_c$. From the proof of Theorem 3, there must exist a copy of P_n containing both cleared and contaminated vertices. Let u be a contaminated vertex on this copy of P_n, and let $V_1' = V_1 \cup \{u\}$ and $V_2' = V_2 \setminus \{u\}$. Since P_n^j contains only contaminated vertices, on the new partition (V_1', V_2'), either the size of the boundary is $k + 1$ and the number of cleared non-boundary vertices is n_c, which is a contradiction, or the size of the boundary is at most k and the number of cleared non-boundary vertices is at least $n_c + 1$, which is also a contradiction. Hence, $k = m - 1$. It follows from Theorem 3(iii) that $|V_1| \leq \frac{m^2}{4}$ when m is even, and $|V_1| \leq \frac{m^2 - 1}{4}$ when m is odd. Since $|V_C^i| \leq |V_1|$, we have $|V_C^i| \leq \max\{\frac{m^2}{4}, \frac{m^2 - 1}{4}\} = \frac{m^2}{4}$.

Case 2. P_n^j contains only cleared vertices. In this case, the number of contaminated vertices can be considered as the number of cleared non-boundary vertices in Case 1. From Case 1, we know that $n_c + (m - 1) \leq \frac{m^2}{4}$, that is, $n_c \leq \frac{m^2}{4} - m + 1$. So the maximum number of contaminated vertices in Case 2 is at most $\frac{m^2}{4} - m + 1$. Hence, $|V_C^i| \geq mn - \frac{m^2}{4} + m - 1$. □

Lemma 6. *For $3 \leq m < 2n$, $c_0(C_m \square P_n) \geq \lceil \frac{m+1}{2} \rceil$.*

Proof. Assume that $c_0(C_m \square P_n) \leq \lceil \frac{m-1}{2} \rceil$. Consider a cop-win strategy that uses $\lceil \frac{m-1}{2} \rceil$ cops. We will show that $|V_C^i| \leq \frac{m^2}{4}$ for all $i \geq 0$. When $i = 0$, $|V_C^0| \leq \lceil \frac{m-1}{2} \rceil < \frac{m^2}{4}$. Assume that $|V_C^{i-1}| \leq \frac{m^2}{4}$, $i \geq 1$. There are two cases.

Case 1. $|V_B^i| \leq m-1$. Since $|V_C^{i-1}| \leq \frac{m^2}{4}$, there are at most $\lceil \frac{m-1}{2} \rceil + \frac{m^2}{4}$ cleared vertices after the cops' turn in round i. Since $\lceil \frac{m-1}{2} \rceil + \frac{m^2}{4} < mn - \frac{m^2}{4} + m - 1$, it follows from Lemma 3 that $|V_C^i| \leq \frac{m^2}{4}$.

Case 2. $|V_B^i| \geq m$. Note that in round i, at most $\lceil \frac{m-1}{2} \rceil$ vertices are cleared after the cops' turn, but at least $\lfloor \frac{m+1}{2} \rfloor$ cleared vertices get recontaminated after the robber's turn. Thus, $|V_C^i| \leq |V_C^{i-1}| \leq \frac{m^2}{4}$.

From the above cases, $|V_C^i| \leq \frac{m^2}{4}$ for all $i \geq 0$, which is a contradiction.   □

Theorem 5. *For $m \geq 3$ and $n \geq 2$, $c_0(C_m \square P_n) \geq \min\{\lceil \frac{m+1}{2} \rceil, n+1\}$.*

Lemma 7. *For $n \geq m \geq 3$, $c_0(C_m \square C_n) \geq \lceil \frac{m+1}{2} \rceil$.*

Let $\mathrm{pw}(G)$ denote the pathwidth of a graph G. The following result appears in [5], which gives a lower bound for $\mathrm{mc}_0(G)$.

Lemma 8. *([5]) For any connected graph G, $\mathrm{mc}_0(G) \geq \frac{1}{2}(\mathrm{pw}(G) + 1)$.*

Lemma 9. *For $n > m \geq 3$, both $\mathrm{mc}_0(C_m \square C_n)$ and $\mathrm{mc}_0(C_m \boxtimes C_n)$ are at least $m + 1$, and both $\mathrm{mc}_0(C_m \square C_m)$ and $\mathrm{mc}_0(C_m \boxtimes C_m)$ are at least m.*

Proof. From Theorem 7.1 in [7], we have

$$\mathrm{pw}(C_m \square C_n) = \begin{cases} 2m & \text{if } n > m \geq 3, \\ 2m - 1 & \text{if } n = m \geq 3. \end{cases}$$

From Lemma 8, we have

$$\mathrm{mc}_0(C_m \boxtimes C_n) \geq \frac{1}{2}(\mathrm{pw}(C_m \boxtimes C_n) + 1) \geq \frac{1}{2}(\mathrm{pw}(C_m \square C_n) + 1).$$

Hence, the claim holds for $\mathrm{mc}_0(C_m \boxtimes C_n)$. Similarly, the claim also holds for $\mathrm{mc}_0(C_m \square C_n)$.   □

4   Cartesian Products

Theorem 6. *For $n \geq m \geq 1$, $c_0(P_m \square P_n) = \mathrm{mc}_0(P_m \square P_n) = \lceil \frac{m+1}{2} \rceil$.*

Proof. The claim is trivial when $m = 1$. When m is odd and $m \geq 3$, we have the following monotonic strategy for $\frac{m+1}{2}$ cops to clear $P_m \square P_n$.

1. Place cop λ_i on $v_{2i-1,1}$ where $1 \leq i \leq \frac{m+1}{2}$. Hence, we use $\frac{m+1}{2}$ cops in total. Let $j = 1$ and $k = \frac{m+1}{2}$.
2. For $i = 1$ to $\frac{m+1}{2}$, take one of the following actions for the cop λ_i:
   (1) if $i = k$, then move λ_i to its right neighbor.
   (2) if $i \neq k$, then move λ_i to its lower neighbor.

3. If $k > 1$, then for $i = 1, \ldots, \frac{m+1}{2}$, move the cop λ_i to its upper neighbor. Set $k \leftarrow k - 1$. If $k \geq 1$, go to Step 2.
4. If $j < n - 1$, set $j \leftarrow j + 1$, $k \leftarrow \frac{m+1}{2}$, and go to Step 2. If $j = n - 1$, then all vertices of $P_m \square P_n$ are cleared.

When m is even and $m \geq 2$, we can easily modify the above strategy so that $\frac{m}{2} + 1$ cops can clear $P_m \square P_n$. Thus, from Theorem 4, the claim holds.   □

Theorem 7. *For $m \geq 3$ and $n \geq 2$,*

$$c_0(C_m \square P_n) = mc_0(C_m \square P_n) = \min\{\lceil \frac{m+1}{2} \rceil, n+1\}.$$

Proof. Applying a strategy similar to the one described in the proof of Theorem 6, we can use $\lceil \frac{m+1}{2} \rceil$ cops to clear $C_m \square P_n$. We now give a monotonic cop-win strategy that clears $C_m \square P_n$ with $n+1$ cops when n is odd (the strategy is similar if n is even).

1. For each $i \in \{1, \ldots, \frac{n+1}{2}\}$, place one cop on $v_{2i-1,1}$ and $v_{2i-1,2}$ respectively. Hence, we use $n + 1$ cops in total.
2. For each $i \in \{1, \ldots, \frac{n-1}{2}\}$, the cop on $v_{2i-1,1}$ vibrates between $v_{2i-1,1}$ and $v_{2i,1}$ throughout the strategy while the cop on $v_{n,1}$ stays still.
3. Using a strategy similar to the one in the proof of Theorem 6, the cops on $v_{2i-1,2}$, $1 \leq i \leq \frac{n+1}{2}$, can clear all paths from P_n^2 to P_n^m.

So the claim follows from Theorem 5.   □

Theorem 8. *(i) For $n \geq m \geq 3$, $\lceil \frac{m+1}{2} \rceil \leq c_0(C_m \square C_n) \leq m + 1$;*
(ii) for $n > m \geq 3$, $mc_0(C_m \square C_n) = m + 1$; and
(iii) for $m \geq 3$, $m \leq mc_0(C_m \square C_m) \leq m + 1$.

5   Strong Products

Lemma 10. *Consider two products $P_m \square P_n$ and $P_m \boxtimes P_n$. Let (V_1, V_2) be a partition of $V(P_m \square P_n)$ and let the same partition of $V(P_m \boxtimes P_n)$ be denoted by (V_1', V_2'). Then $|\partial V_1| \leq |\partial V_1'|$.*

Proof. Let $v \in V_1$ and let $v' \in V_1'$ be the corresponding vertex of v. If v has a neighbor in V_2, then v' also has a neighbor in V_2'. Thus $|\partial V_1| \leq |\partial V_1'|$.   □

Theorem 9. *For $n \geq m \geq 2$,*
(i) $c_0(P_m \boxtimes P_n) = mc_0(P_m \boxtimes P_n) = \frac{m}{2} + 1$ when m is even; and
(ii) $\frac{m+1}{2} \leq c_0(P_m \boxtimes P_n) \leq mc_0(P_m \boxtimes P_n) \leq \frac{m+1}{2} + 1$ when m is odd.

Proof. Suppose that m is even. Let $v_{i,j}$, $1 \leq i \leq m$, $1 \leq j \leq n$, be the vertex on the i-th row and j-th column of $P_m \boxtimes P_n$. It is easy to see that $mc_0(P_2 \boxtimes P_n) = 2$; when $m \geq 4$, the following monotonic strategy can clear $P_m \boxtimes P_n$ with $\frac{m}{2} + 1$ cops.

1. For each $i \in \{1, \ldots, \frac{m}{2}\}$, place a cop on $v_{2i-1,1}$. Place a cop on $v_{1,2}$. Hence, we use $\frac{m}{2} + 1$ cops in total. Let $j = 1$ and $k = 1$.
2. Slide all cops to their lower neighbors.
3. If $k < m - 1$, slide the cop on $v_{k+1,j}$ to $v_{k+2,j+1}$ and slide all other cops to their upper neighbors; otherwise, slide all cops to their upper neighbors. Set $k \leftarrow k + 2$. If $k \leq m - 1$, go to Step 2.
4. If $j = n - 1$, then stop and all vertices of $P_m \boxtimes P_n$ are cleared. If $j < n - 1$, then
   (a) slide the cop on $v_{m-1,j}$ to $v_{m,j+1}$;
   (b) slide the cop on $v_{1,j+1}$ to $v_{1,j+2}$; and
   (c) slide all other cops to their upper neighbors.
5. Slide all the cops to their upper neighbors, except the one on $v_{1,j+2}$. Set $j \leftarrow j + 1$, $k \leftarrow 1$, and go to Step 2.

When m is odd, we can clear $P_m \boxtimes P_n$ with $\frac{m+1}{2} + 1$ cops in a similar way. Hence, $\mathrm{mc}_0(P_m \boxtimes P_n) \leq \lfloor \frac{m+1}{2} \rfloor + 1$.

From Lemma 10, we can use a method similar to the one in the proof of Theorem 4 to prove that $c_0(P_m \boxtimes P_n) \geq \lceil \frac{m+1}{2} \rceil$. This completes the proof. $\square$

Theorem 10. *For $m \geq 3$ and $n \geq 2$,*

$$\min\{\lceil \frac{m+1}{2} \rceil, n+1\} \leq c_0(C_m \boxtimes P_n) \leq \mathrm{mc}_0(C_m \boxtimes P_n) \leq \min\{\lceil \frac{m+1}{2} \rceil, 2\lfloor \frac{n+1}{2} \rfloor\} + 1.$$

Proof. We first show that $C_m \boxtimes P_n$ can be cleared monotonically with $\lceil \frac{m+1}{2} \rceil + 1$ cops.

Suppose that m is odd. Let $v_{i,j}$ be the vertex of $C_m \boxtimes P_n$ that is on the i-th copy of C_m and the j-th copy of P_n. We place a cop on vertex $v_{1,m}$ and this cop will vibrate between the vertices $v_{1,m}$ and $v_{2,m}$ until all vertices of C_m^1 and C_m^2 are cleared. We use $\frac{m+1}{2}$ cops to clear vertices of C_m^1 and C_m^2 except $v_{1,m}$ and $v_{2,m}$ with the strategy described in the proof of Theorem 9. We then use $\frac{m+1}{2} + 1$ cops to clear vertices of C_m^2 and C_m^3. We continue like this until all vertices of C_m^n are cleared. If m is even, we can easily extend this strategy to clear $C_m \boxtimes P_n$ using $\frac{m}{2} + 2$ cops.

We now give a monotonic strategy to clear $C_m \boxtimes P_n$ with $2\lfloor \frac{n+1}{2} \rfloor + 1$ cops. Suppose that n is even (the strategy will be similar if n is odd). For each $i \in \{1, \ldots, \frac{n}{2}\}$, place a cop on $v_{2i-1,m}$ and let these $\frac{n}{2}$ cops vibrate between $v_{2i-1,m}$ and $v_{2i,m}$ until all vertices of the graph are cleared. We use $\frac{n}{2} + 1$ cops to clear $P_n^1, \ldots, P_n^{m-1}$ by the strategy in Theorem 9.

From Lemma 10 and the proofs of Lemmas 4 and 6, we have

$$c_0(C_m \boxtimes P_n) \geq \begin{cases} n+1 & \text{if } m \geq 2n, \\ \lceil \frac{m+1}{2} \rceil & \text{if } m \leq 2n - 1. \end{cases}$$

Thus the claim holds. $\square$

Theorem 11. (i) *For $n \geq m \geq 3$, $\lceil \frac{m+1}{2} \rceil \leq c_0(C_m \boxtimes C_n) \leq m + 2$;*
(ii) *for $n > m \geq 3$, $m + 1 \leq \mathrm{mc}_0(C_m \boxtimes C_n) \leq m + 2$; and*
(iii) *for $m \geq 3$, $m \leq \mathrm{mc}_0(C_m \boxtimes C_m) \leq m + 2$.*

Proof. (i) We have shown in Theorem 10 that $C_m \boxtimes P_n$ can be cleared with $\lceil \frac{m+1}{2} \rceil + 1$ cops. Suppose that m is odd. To clear $C_m \boxtimes C_n$, we extend the strategy in Theorem 10 by adding the following actions: in the beginning, for each $i \in \{1, \ldots, \frac{m-1}{2}\}$, place a cop on $v_{n,2i}$ that will only vibrate between $v_{n,2i-1}$ and $v_{n,2i}$, and place a cop on $v_{n,m}$ and this cop will stay still throughout the strategy. Clearly, C_m^n is cleared and protected by those $\frac{m-1}{2} + 1$ cops. We can use $\frac{m+1}{2} + 1$ cops to clear $C_m^1, \ldots, C_m^{n-1}$ sequentially. Hence, we can use $m + 2$ cops to clear $C_m \boxtimes P_n$ monotonically. The strategy is similar if m is even.

Similar to the proofs in Sect. 3, we can prove that $c_0(C_m \boxtimes C_n) \geq \lceil \frac{m+1}{2} \rceil$.

(ii) and (iii) follow from Lemma 9 and the monotonic strategy in the proof of (i). □

6   Hypercubes

Theorem 12. *For a hypercube Q_n, $n \geq 2$,*

$$\frac{1}{2} \sum_{k=0}^{n-1} \binom{k}{\lfloor \frac{k}{2} \rfloor} + \frac{1}{2} \leq mc_0(Q_n) \leq \sum_{k=0}^{n-2} \binom{k}{\lfloor \frac{k}{2} \rfloor} + 1.$$

Proof. From Lemma 8 and [4], we have

$$mc_0(Q_n) \geq \frac{1}{2}(pw(Q_n) + 1) = \frac{1}{2} \sum_{k=0}^{n-1} \binom{k}{\lfloor \frac{k}{2} \rfloor} + \frac{1}{2}.$$

We now consider the upper bound. Note that $Q_n = Q_{n-1} \square P_2$. For convenience, let Q_{n-1}^1 and Q_{n-1}^2 be the two copies of Q_{n-1} in Q_n. Let $(\mathcal{B}_1, \ldots, \mathcal{B}_m)$ be an optimal path decomposition of Q_{n-1}^1, where $\mathcal{B}_i \subseteq V(Q_{n-1}^1)$, $1 \leq i \leq m$. We give a monotonic cop-win strategy for clearing Q_n with $pw(Q_{n-1}) + 1$ cops.

1. In round 1, place $pw(Q_{n-1}) + 1$ cops on vertices in $\mathcal{B}_1$, such that every vertex in $\mathcal{B}_1$ contains at least one cop. Let $i = 1$.
2. In round 2, slide all cops to their neighbors in Q_{n-1}^2, and in round 3, slide all cops back to Q_{n-1}^1.
3. If all vertices are cleared, then stop.
4. If there exists a vertex v in $\mathcal{B}_i \setminus \mathcal{B}_{i+1}$ such that v is occupied by a cop and it is cleared in the last two rounds, then
   (a) select a vertex v' in $\mathcal{B}_{i+1}$ that contains the smallest number of cops among all other vertices in $\mathcal{B}_{i+1} \setminus \mathcal{B}_i$,
   (b) find a path between v and v', and
   (c) slide a cop on v along the path to v', during which this cop does not move to Q_{n-1}^2, but all other cops move to their neighbors in Q_{n-1}^2 in the even rounds and move back to Q_{n-1}^1 in the odd rounds.
   If the cop arrives at v' in an even round, then this cop stays on v' until all other cops move back to Q_{n-1}^1 in the next round; after that all cops move to Q_{n-1}^2 and then move back to Q_{n-1}^1 in the next two rounds. If the cop arrives at v' in an odd round, then all cops move to Q_{n-1}^2 and then move back to Q_{n-1}^1 in the next two rounds.

5. If $\mathcal{B}_i \setminus \mathcal{B}_{i+1}$ contains no cops, then set $i \leftarrow i + 1$ and go to Step 3; otherwise, go to Step 4.

From [4], we have $\mathrm{mc}_0(Q_n) \leq \mathrm{pw}(Q_{n-1}) + 1 = \sum_{k=0}^{n-2} \binom{k}{\lfloor \frac{k}{2} \rfloor} + 1.$ □

Since the strategy described in the proof of Theorem 12 can be easily extended to clear $G \square P_2$, where G is connected, we have the following result.

Corollary 1. *Let G be a connected graph and P_2 be a path with two vertices. Then $\mathrm{mc}_0(G \square P_2) \leq \mathrm{pw}(G) + 1$.*

For $c_0(Q_n)$, we conjecture that $c_0(Q_n) = \mathrm{mc}_0(Q_n)$. From the results in Sects. 4 and 5, we may think that $c_0(G \square H) = \mathrm{mc}_0(G \square H)$ for any graphs G and H. But this is not always true. The following result implies that their difference can be arbitrarily large for product graphs.

Lemma 11. *For any positive integer k, there exist graphs G and H such that $c_0(G \square H) \leq 4$ and $\mathrm{mc}_0(G \square H) \geq k$.*

Proof. For any positive integer ℓ, we can construct a graph G such that $c_0(G) = 2$ and $\mathrm{pw}(G) \geq \ell$ (see Theorem 4.1 in [5]). It is easy to see that we can use four cops to clear the two copies of G in $G \sqcap P_2$ synchronously. So $c_0(G \square P_2) \leq 4$. On the other hand, from Lemma 8, we have

$$\mathrm{mc}_0(G \square P_2) \geq \frac{1}{2}(\mathrm{pw}(G \square P_2) + 1) \geq \frac{1}{2}(\ell + 1).$$

□

7 Conclusions

In this paper, we introduce the partition method which is used to prove lower bounds for the zero-visibility cops and robber game. We apply this method to show lower bounds of graph products. We believe that this idea can be used for other classes of graphs.

We conclude this paper with the following conjectures.

(1) $c_0(C_m \square C_n) = \mathrm{mc}_0(C_m \square C_n) = m + 1$, for $n \geq m \geq 3$.

(2) $c_0(P_m \boxtimes P_n) = \mathrm{mc}_0(P_m \boxtimes P_n) = \frac{m+1}{2} + 1$, where m is odd and $n \geq m \geq 3$.

(3) $c_0(C_m \boxtimes P_n) = \mathrm{mc}_0(C_m \boxtimes P_n) = \min\{\lceil \frac{m+1}{2} \rceil, 2\lfloor \frac{n+1}{2} \rfloor\} + 1$, for $m \geq 3$ and $n \geq 2$.

(4) $c_0(C_m \boxtimes C_n) = \mathrm{mc}_0(C_m \boxtimes C_n) = m + 2$, for $n \geq m \geq 3$.

(5) For $n \geq 2$,

$$c_0(Q_n) = \mathrm{mc}_0(Q_n) = \sum_{k=0}^{n-2} \binom{k}{\lfloor \frac{k}{2} \rfloor} + 1.$$

References

1. Bienstock, D., Seymour, P.: Monotonicity in graph searching. J. Algorithms **12**, 239–245 (1991)
2. Bonato, A., Nowakowski, R.: The Game of Cops and Robbers on Graphs. American Mathematical Society, Providence (2011)
3. Bonato, A., Yang, B.: Graph searching and related problems. In: Pardalos, P.M., Du, D.-Z., Graham, R.L. (eds.) Handbook of Combinatorial Optimization, pp. 1511–1558. Springer, New York (2013). https://doi.org/10.1007/978-1-4419-7997-1_76
4. Chandran, L.S., Kavitha, T.: The treewidth and pathwidth of hypercubes. Discrete Math. **306**, 359–365 (2006)
5. Dereniowski, D., Dyer, D., Tifenbach, R., Yang, B.: Zero-visibility cops & robber and the pathwidth of a graph. J. Comb. Optim. **29**, 541–564 (2015)
6. Dereniowski, D., Dyer, D., Tifenbach, R., Yang, B.: The complexity of zero-visibility cops and robber. Theoret. Comput. Sci. **607**, 135–148 (2015)
7. Ellis, J., Warren, R.: Lower bounds on the pathwidth of some grid-like graphs. Discrete Appl. Math. **156**, 545–555 (2008)
8. LaPaugh, A.: Recontamination does not help to search a graph. J. ACM **40**, 224–245 (1993)
9. Megiddo, N., Hakimi, S., Garey, M., Johnson, D., Papadimitriou, C.: The complexity of searching a graph. J. ACM **35**, 18–44 (1998)
10. Nowakowski, R., Winkler, P.: Vertex to vertex pursuit in a graph. Discrete Math. **43**, 235–239 (1983)
11. Quilliot, A.: Jeux et pointes fixes sur les graphes. Thèse de 3ème cycle, Université de Paris VI, pp. 131–145 (1978)
12. Tošić, R.: Vertex-to-vertex search in a graph. In: Proceedings of the Sixth Yugoslav Seminar on Graph Theory, pp. 233–237. University of Novi Sad (1985)
13. Yang, B., Cao, Y.: Monotonicity of strong searching on digraphs. J. Comb. Optim. **14**, 411–425 (2007)
14. Yang, B., Cao, Y.: Monotonicity in digraph search problems. Theoret. Comput. Sci. **407**, 532–544 (2008)

Author Index

Amir, Amihood   70
Arimura, Hiroki   339

Bahoo, Yeganeh   10
Baig, Mirza Galib Anwarul Husain   22
Balko, Martin   35
Banerjee, Sandip   48
Banik, Aritra   61
Bannai, Hideo   430
Bataa, Magsarjav   70
Bentert, Matthias   85
Bhore, Sujoy   35, 48
Biswas, Arindam   97
Böckenhauer, Hans-Joachim   108
Bose, Prosenjit   10
Brlek, Srecko   393

Cazals, Pierre   122
Chateau, Annie   122
Chlebíková, Janka   136
Conte, Alessio   148
Cordasco, Gennaro   160
Corvelo Benz, Nina   108

Da Lozzo, Giordano   175
Dallard, Clément   136
Darties, Benoit   122
Das, Sandip   188
de Montgolfier, Fabien   251
Donovan, Zola   201
Dourado, Mitre C.   214
Durocher, Stephane   10

Froncek, Dalibor   229
Frosini, Andrea   393

Gahlawat, Harmender   188
Gargano, Luisa   160
Giroudeau, Rodolphe   122
Gupta, Rahul Raj   237

Haag, Roman   85
Habib, Michel   251

Hatano, Kohei   365
Haverkort, Herman   265
Henning, Michael A.   278
Hofer, Christian   85

Inenaga, Shunsuke   430
Itoh, Toshiya   405

Jacob, Ashwin   61
Jansson, Jesper   290

Kamatchi, Nainarraj   1
Kanté, Mamadou Moustapha   148
Kare, Anjeneya Swami   304
Karmakar, Sushanta   237
Kesh, Deepanjan   22
Khosravian Ghadikoalei, Mehdi   315
Koana, Tomohiro   85
Kobayashi, Yasuaki   327
Kobayashi, Yusuke   327
Komm, Dennis   108
Kübel, David   265
Kurita, Kazuhiro   339

Landau, Gad M.   70
Langetepe, Elmar   265
Lauri, Juho   352

Mampentzidis, Konstantinos   290
Mancini, Ilaria   393
Marino, Andrea   148
Martínez Sandoval, Leonardo   35
Melissinos, Nikolaos   315
Mitillos, Christodoulos   352
Miyake, Fumito   365
Miyazaki, Shuichi   327
Mkrtchyan, Vahan   201
Monnot, Jérôme   315
Mouatadid, Lalla   251
Moura, Lucia   378

Nakashima, Yuto   430
Nichterlein, André   85

Pagourtzis, Aris   315
Paliwal, Vijay Kumar   61
Pandey, Arti   278
Paramasivam, Krishnan   1
Park, Kunsoo   70
Park, Sung Gwan   70
Penso, Lucia D.   214
Pergola, Elisa   393
Plant, Lachlan   378
Prajeesh, Appattu Vallapil   1

Qiu, Jiangyi   229

Rajaby, Ramesh   290
Raman, Venkatesh   61, 97
Rautenbach, Dieter   214
Rescigno, Adele Anna   160
Rinaldi, Simone   393
Ruangwises, Suthee   405
Rutter, Ignaz   175

Sahoo, Uma Kant   188
Saurabh, Saket   97
Sen, Sagnik   188
Shermer, Thomas   10
Sodani, Chirag   22
Stober, Florian   417

Subramani, K.   201
Sung, Wing-Kin   290

Takeda, Masayuki   430
Takimoto, Eiji   365
Tamaki, Suguru   327
Tripathi, Vikash   278

Uno, Takeaki   148, 339

Valtr, Pavel   35
Vinod Reddy, I.   304

Wasa, Kunihiro   339
Watanabe, Kiichi   430
Weiß, Armin   417
Weller, Mathias   122

Xue, Yuan   442

Yang, Boting   442

Zhong, Farong   442
Zilles, Sandra   442
Zou, Mengchuan   251

Printed in the United States
By Bookmasters